# Stability and Stabilization of Linear Systems with Saturating Actuators

Sophie Tarbouriech · Germain Garcia ·
João Manoel Gomes da Silva Jr. ·
Isabelle Queinnec

# Stability and Stabilization of Linear Systems with Saturating Actuators

Springer

Sophie Tarbouriech
Laboratoire Analyse et Architecture
des Systèmes (LAAS)
CNRS
av. du Colonel Roche 7
31077 Toulouse CX 4
France
tarbour@laas.fr

João Manoel Gomes da Silva Jr.
Departamento de Engenharia Elétrica
Universidade Federal do Rio Grande do Sul
av. Osvaldo Aranha 103
90035-190 Porto Alegre
Rio Grande do Sul
Brazil
jmgomes@ece.ufrgs.br

Germain Garcia
Laboratoire Analyse et Architecture
des Systèmes (LAAS)
CNRS
av. du Colonel Roche
31077 Toulouse CX 4
France
garcia@laas.fr

Isabelle Queinnec
Laboratoire Analyse et Architecture
des Systèmes (LAAS)
CNRS
av. du Colonel Roche 7
31077 Toulouse CX 4
France
queinnec@laas.fr

ISBN 978-0-85729-940-6          e-ISBN 978-0-85729-941-3
DOI 10.1007/978-0-85729-941-3
Springer London Dordrecht Heidelberg New York

British Library Cataloguing in Publication Data
A catalogue record for this book is available from the British Library

Library of Congress Control Number: 2011935854

© Springer-Verlag London Limited 2011
Apart from any fair dealing for the purposes of research or private study, or criticism or review, as permitted under the Copyright, Designs and Patents Act 1988, this publication may only be reproduced, stored or transmitted, in any form or by any means, with the prior permission in writing of the publishers, or in the case of reprographic reproduction in accordance with the terms of licenses issued by the Copyright Licensing Agency. Enquiries concerning reproduction outside those terms should be sent to the publishers.
The use of registered names, trademarks, etc., in this publication does not imply, even in the absence of a specific statement, that such names are exempt from the relevant laws and regulations and therefore free for general use.
The publisher makes no representation, express or implied, with regard to the accuracy of the information contained in this book and cannot accept any legal responsibility or liability for any errors or omissions that may be made.

*Cover design*: VTeX UAB, Lithuania

Printed on acid-free paper

Springer is part of Springer Science+Business Media (www.springer.com)

*To Isabelle, Louise, Julie, Christiane and Pierre with my tenderness*
*Sophie Tarbouriech*

*To Lionel, Pauline, Sophie, Maria-Belen, Maria-Consuelo and Jaime in gratitude and affection*
*Germain Garcia*

*To Silvia and Gabriela. For their love, support and patience during the preparation of this book*
*João Manoel Gomes da Silva Jr.*

*To Sophie, Philippe, Nicole, Yvon. To those who are important for me*
*Isabelle Queinnec*

# Foreword

Actuator saturation presents the control system designer with a major challenge. It may cause loss in performance and in some cases unstable behaviour. Control engineers have long been aware of the need to guard against actuators hitting magnitude or rate limits and over the years a number of ad hoc methods have been developed to alleviate unwanted effects. In more recent years, researchers have sought to develop general systematic methods based on rigorous theory to give guarantees on stability and levels of performance. Valuable contributions have been made and there is now a wealth of information, technical results and synthesis techniques available in learned journals and conference proceedings. This book brings together many of these results, particularly those developed by the authors, into a single text.

There are three main approaches to tackling saturation problems. In the first, optimal control is used to synthesise a controller which directly takes into account the saturation constraints. This approach is underpinned by a strong theoretical foundation and important results on stability. For example, global stability is only possible for systems whose poles all lie in the closed left-half complex plane; global asymptotic stability with finite-gain stability is not possible using linear control laws if the system contains repeated poles on the imaginary axis; and if a system contains poles in the open right-half complex plain, then the most that can be achieved is local stability, with local gain properties. Part II of this book is devoted to this direct approach and presents the relevant theory leading to optimisation techniques for synthesising controllers that meet performance specifications while respecting the saturation constraints. The optimisation techniques are presented in terms of parameterised Riccati equations and parameterised linear matrix inequalities (LMIs).

The second main approach to handling saturation uses *anti-windup* to compensate an existing controller which has been designed to work well in the absence of saturation. Windup is the term used by early practising control engineers to describe the undesirable behaviour that occurred in integral (or PID) controllers for tracking problems when the control signal reached a physical limit. In these situations, the integrator state would "wind up" to a large value and the associated energy would then have to be dissipated once the control signal came out of saturation. The result was excessive overshoot and slow settling times. To overcome these problems, an additional compensator, an anti-windup compensator, was included which would modify

vii

the control signal if saturation occurred but would otherwise do nothing. A number of ad hoc methodologies were developed including high gain anti-windup, reset anti-windup and Hanus anti-windup, all with varying degrees of success depending on the application. A great advantage of this approach, is that the anti-windup compensator works with the existing controller and only becomes active if saturation occurs. The early anti-windup compensators did, however, have a number of weaknesses: there were no stability or performance guarantees, they were mainly restricted to single-input single-output systems, and their design or tuning was imprecise. Over the past decade or so, these weaknesses have been removed, and there are now systematic approaches available for the design of anti-windup compensators, along with reliable algorithms for their synthesis using Riccati equations or LMIs. Part III of this book covers these important developments to the handling of saturation using anti-windup compensation.

The third approach to saturation is to use model predictive control strategies. These are not addressed in this book but have been covered well in a variety of other books and so should not be viewed as an omission.

The book is completed by two other parts: Part I sets the scene by defining the problems of interest and the Appendices provide useful information on technical topics in the areas of stability theory, robust control, Riccati equations and LMIs. This reference material in the Appendices helps to make the book self-contained.

Previous books on this subject matter consist mainly of edited volumes, which are good at bringing the latest research results from key players to a wide readership, but which fall short in connecting the results in a coherent manner suitable for teaching purposes and then, hopefully, for use in real applications. This book has taken on this challenge. To make it work, the authors have had to restrict themselves to certain lines of research and methods. Given the vast array of results available there is no way that they could be exhaustive in their treatment. The authors therefore have been pragmatic and made the sensible decision to focus on what they know best, which is the research they have done themselves during the past ten years. By doing this, the work has a distinctive LAAS flavor; LAAS being the highly regarded Laboratoire d'Analyse et d'Architecture des Systèmes of CNRS, the French National Centre of Scientific Research, in Toulouse. LAAS is of course home for three of the four authors and the fourth is a frequent visitor to the laboratory, as have been many researchers on saturation control over the years.

I congratulate the authors on taking on such a difficult challenge and bringing together so many interesting research results. The inclusion of anti-windup is pleasing to see as in my experience this represents a very practical approach to reducing the undesirable effects of saturation.

I wish this book every possible success.

Northumbria University                                          Ian Postlethwaite
Newcastle upon Tyne, UK

# Preface

This book addresses both the problems of stability analysis and stabilization of linear systems when subject to control saturation. It is well known that actuator saturation is present in practically all control systems. The magnitude of the signal that an actuator can deliver is usually limited by physical or safety constraints. This limitation can be easily identified in the most common devices used in the process industry, such as proportional valves, heating actuators, power amplifiers, and electromechanical actuators. Common examples of such limits are the deflection limits in aircraft actuators, the voltage limits in electrical actuators and the limits on flow volume or rate in hydraulic actuators. While such limits obviously restrict the performance achievable in the systems of which they are part, if these limits are not treated carefully and if the relevant controllers do not account for them appropriately, peculiar and pernicious behaviors may be observed. In particular, actuator saturation both in magnitude and rate have been implicated in various aircraft crashes [8] and the meltdown of the Chernobyl nuclear power station [330]. Hence, an in-depth understanding of the phenomena caused by saturation is key to solve and avoid problems in industrial control. Actually, the operation problems which can be induced by actuator saturations are well known in industry. This is the reason why, frequently, the actuators are oversized with respect to the energy necessary to control the system and the control gains are limited with respect to what could be accepted while preserving the stability. The engineering solutions intend to avoid, as much as possible, occurrence of saturation during operation. Then, the study of saturated systems is of a real interest for engineers in various domains due to its potential to provide, for example:

- Reduction of validation costs of control laws;
- Better use of actuator (and/or sensor) capacity;
- The possibility for control engineers to become involved in the design process at a much earlier stage in order to help choose actuators/sensors which have a reduced size, mass etc.

Motivated by these practical issues, the actuator saturation problem has deservedly received the attention of many researchers since the second part of the twentieth century. An excellent overview on saturated systems has been published in 1995 by Bernstein and Michel [25]. More recently, great advances were made on this subject. This is attested by the large number of publications on this topic that can be found in important conferences and journals of control systems and applied mathematics.

Although an extensive literature in the form of articles can be found, only a small number of books devoted to this subject is available. Examples are the books edited by two of ourselves [346, 361] and the one edited by Kapila and Grigoriadis [208]. The former three books are presented in the form of a collection of chapters, written by different authors, with the particular aim of presenting new trends and specific advanced topics on the analysis and synthesis of control systems subject to actuator saturation. We believe now it is timely to write a textbook on the problem of control saturation, but given the diversity of approaches, it is certainly a challenge. To write a book addressing all the approaches and aspects related to this problem is a nearly impossible task. The main challenge then is to choose a set of specific topics or approaches and present them in a coherent and self-contained way. We think that this has been successfully achieved, for instance, in the book of Saberi et al. [310], where the main focus concerns the structural conditions for the solvability of many problems involving systems with input constraints. Other examples are the books [127, 188].

Roughly speaking, there are two approaches which one could adopt to avoid saturation problems in systems which are known to have actuator limits (i.e. the vast majority of practical systems). One approach, which we shall refer to as the one step approach is, as its name implies, an approach to controller design where a (possibly nonlinear) controller is designed "from scratch". This controller then attempts to ensure that all nominal performance specifications are met while also handling the saturation constraints imposed by the actuators. Let us emphasize that Part II of this book is dedicated to present several solutions in the scope of this approach in both analysis and synthesis frameworks.

An alternative approach to the above is to perform some separation in the controller such that one part is devoted to achieving nominal performance and the other part is devoted to control constraint handling. This is the approach taken in anti-windup compensation, which is the subject of Part III of this book. In anti-windup compensation, a linear controller which does not explicitly take into account the saturation constraints is first designed, usually using standard linear design tools. Then, after this controller has been designed, a so-called anti-windup compensator is designed to handle the saturation constraints. The anti-windup compensator is designed to ensure that stability is maintained (at least in some region near the origin) and that less performance degradation occurs than when no anti-windup is used.

The primary goal of the present work is to provide basic concepts and tools that we consider fundamental for the analysis and synthesis of linear systems subject to actuator saturation. Additionally, we aim to present in a formal, but also didactic way, some recent developments on the subject. As pointed out above, we do not intend to cover exhaustively all the approaches and the results on the subject. Naturally, we mainly focus on topics directly related to our own research in the field. In this sense, we consider a state space approach and we focus on the problems of

# Preface

stability analysis and synthesis of stabilizing control laws both in local and global contexts. In particular, different ways of modeling the saturation term and the behavior of the nonlinear closed-loop system are explored. Also, different kinds of Lyapunov functions such as polyhedral, quadratic and Lure type, are considered in order to present different stability and stabilization conditions. Results considering uncertain systems and performance in the presence of saturation are also presented. Associated with the different theoretical results, we propose methods and algorithms for computing estimates of the basin of attraction and for designing control systems taking into account, a priori, the control bounds and the saturation possibility. These methods and algorithms are mainly based on the use of linear programming and convex optimization problems with LMI constraints. Thus, they can easily be implemented with widely known mathematical software packages. We believe we address some aspects that are not yet covered, or not discussed in-depth, in the books currently available on this subject.

Finally, we hope this book will be a valuable reference for engineers (mainly aeronautical, electrical, mechanical and chemical), working on control applications, as well as graduate students and researchers interested in the development of new tools and theoretical results concerning systems with saturation. Due to its structure, it can also be used as a textbook in a specific course on saturation or as a complementary text in classical courses on control systems. The background required for the reader consists of basic concepts of linear and nonlinear systems, including linear algebra and matrix theory, differential equations, state space methods, and robust control theory, which can be found, for example, in classics like the books of Chen [64], Kailath [205], Khalil [215], Luenberger [252], O'Reilly [277], Sontag [327], Vidyasagar [385], Colaneri et al. [74], Zhou et al. [414], Skogestad and Postlethwaite [323]. Some familiarity with linear programming, convex analysis and LMIs is also desirable. As additional references for all these topics we can recommend the books of Maciejowski [255], Boyd et al. [45] and Zaccarian and Teel [406].

| | |
|---|---|
| Toulouse, France | Sophie Tarbouriech |
| Toulouse, France | Germain Garcia |
| Porto Alegre, Brazil | João Manoel Gomes da Silva Jr. |
| Toulouse, France | Isabelle Queinnec |

# Contents

**Part I    Generalities**

**1    Linear Systems Subject to Control Saturation—Problems and Modeling** . . . . . . . . . . . . . . . . . . . . . . . . . . . . 3
    1.1    Introduction . . . . . . . . . . . . . . . . . . . . . . . . . . . . 3
    1.2    The Open-Loop System . . . . . . . . . . . . . . . . . . . . . . 4
    1.3    The Closed-Loop System . . . . . . . . . . . . . . . . . . . . . 5
    1.4    The Region of Linearity . . . . . . . . . . . . . . . . . . . . . . 12
    1.5    The Region of Attraction . . . . . . . . . . . . . . . . . . . . . 13
    1.6    Problems Considered . . . . . . . . . . . . . . . . . . . . . . . 14
        1.6.1    Asymptotic Stability Analysis . . . . . . . . . . . . . . 15
        1.6.2    Asymptotic Stabilization . . . . . . . . . . . . . . . . 18
        1.6.3    External Stability Analysis . . . . . . . . . . . . . . . 28
        1.6.4    External Stabilization . . . . . . . . . . . . . . . . . . 30
        1.6.5    The Anti-windup Problem . . . . . . . . . . . . . . . 31
    1.7    Models for the Saturation Nonlinearity . . . . . . . . . . . . . . 32
        1.7.1    Polytopic Models . . . . . . . . . . . . . . . . . . . . 33
        1.7.2    Sector Nonlinearity Models . . . . . . . . . . . . . . . 40
        1.7.3    Regions of Saturation Models . . . . . . . . . . . . . . 44
    1.8    Equilibrium Points . . . . . . . . . . . . . . . . . . . . . . . . 46
    1.9    Conclusion . . . . . . . . . . . . . . . . . . . . . . . . . . . . 47

**Part II    Stability Analysis and Stabilization**

**2    Stability Analysis and Stabilization—Polytopic Representation Approach** . . . . . . . . . . . . . . . . . . . . . . . . . . . . . 51
    2.1    Introduction . . . . . . . . . . . . . . . . . . . . . . . . . . . . 51
    2.2    Asymptotic Stability Analysis . . . . . . . . . . . . . . . . . . 52
        2.2.1    Ellipsoidal Sets of Stability . . . . . . . . . . . . . . . 52
        2.2.2    Polytopic Approach I . . . . . . . . . . . . . . . . . . 54
        2.2.3    Polytopic Approach II . . . . . . . . . . . . . . . . . . 56

xiii

|  |  | 2.2.4 | Polytopic Approach III | 58 |
|--|--|-------|-----------------------|----|

2.2.4 Polytopic Approach III ... 58
2.2.5 Optimization Problems ... 59
2.3 External Stability ... 65
    2.3.1 Amplitude Bounded Exogenous Signals ... 66
    2.3.2 Energy Bounded Exogenous Signals ... 71
2.4 Stabilization ... 76
    2.4.1 State Feedback Stabilization ... 76
    2.4.2 Observer-Based Feedback Stabilization ... 91
    2.4.3 Dynamic Output Feedback Stabilization ... 95
    2.4.4 Global Stabilization ... 101
2.5 Uncertain Systems ... 103
    2.5.1 Stability Analysis ... 103
    2.5.2 Extensions ... 108
2.6 Discrete-Time Case ... 109
    2.6.1 Ellipsoidal Regions of Asymptotic Stability ... 110
    2.6.2 Polyhedral Regions of Asymptotic Stability ... 112
    2.6.3 Extensions ... 119
2.7 Conclusion ... 120

**3 Stability Analysis and Stabilization—Sector Nonlinearity Model Approach** ... 123
3.1 Introduction ... 123
3.2 Asymptotic Stability Analysis ... 124
    3.2.1 Quadratic Lyapunov Function ... 125
    3.2.2 Lure Lyapunov Function ... 129
    3.2.3 Computational Burden ... 132
    3.2.4 Optimization Problems ... 133
3.3 External Stability ... 136
    3.3.1 Amplitude Bounded Exogenous Signals ... 136
    3.3.2 Energy Bounded Exogenous Signals ... 141
    3.3.3 Optimization Issues ... 145
3.4 Stabilization ... 146
    3.4.1 State Feedback Stabilization ... 146
    3.4.2 Observer-Based Feedback Stabilization ... 155
    3.4.3 Dynamic Output Feedback Stabilization ... 161
3.5 Uncertain Systems ... 171
3.6 Discrete-Time Case ... 178
3.7 Extensions ... 180
    3.7.1 Nested Saturations ... 180
    3.7.2 Nested Nonlinearities ... 181
    3.7.3 Nonlinear and/or Hybrid Systems ... 182
3.8 Conclusion ... 183

**4 Analysis via the Regions of Saturation Model** ... 185
4.1 Introduction ... 185
4.2 Polyhedral Regions of Stability ... 186

Contents                                                                                              xv

        4.2.1   Positive Invariance . . . . . . . . . . . . . . . . . . . . . . . 186
        4.2.2   Contractivity—Compact Case . . . . . . . . . . . . . . . . 189
        4.2.3   Determination of Stability Regions . . . . . . . . . . . . 193
  4.3   Ellipsoidal Regions . . . . . . . . . . . . . . . . . . . . . . . . . . . 196
        4.3.1   Test Condition 1 . . . . . . . . . . . . . . . . . . . . . . . . 197
        4.3.2   Test Condition 2 . . . . . . . . . . . . . . . . . . . . . . . . 200
  4.4   Discrete-Time Case . . . . . . . . . . . . . . . . . . . . . . . . . . 202
        4.4.1   Positive Invariance . . . . . . . . . . . . . . . . . . . . . . 202
        4.4.2   Contractivity—Compact Case . . . . . . . . . . . . . . . 203
  4.5   Unbounded Sets of Stability . . . . . . . . . . . . . . . . . . . . . 204
        4.5.1   Unbounded Polyhedra . . . . . . . . . . . . . . . . . . . . 204
        4.5.2   Unbounded Ellipsoidal Sets . . . . . . . . . . . . . . . . 206
  4.6   Conclusion . . . . . . . . . . . . . . . . . . . . . . . . . . . . . . . . 207

**5   Synthesis via a Parameterized ARE Approach or a Parameterized**
**   LMI Approach** . . . . . . . . . . . . . . . . . . . . . . . . . . . . . . . . 209
  5.1   Introduction . . . . . . . . . . . . . . . . . . . . . . . . . . . . . . . 209
  5.2   The Parameterized ARE Approach . . . . . . . . . . . . . . . . 210
        5.2.1   Preliminaries . . . . . . . . . . . . . . . . . . . . . . . . . 211
        5.2.2   The Single-Input Case . . . . . . . . . . . . . . . . . . . 213
        5.2.3   The Multi-variable Case . . . . . . . . . . . . . . . . . . 216
        5.2.4   Control Computation and Implementation . . . . . . . . 219
  5.3   A Parameterized LMI Approach . . . . . . . . . . . . . . . . . . 224
  5.4   Multi-objective Control: Eigenvalues Placement and Guaranteed
      Cost . . . . . . . . . . . . . . . . . . . . . . . . . . . . . . . . . . . . 230
        5.4.1   Preliminaries . . . . . . . . . . . . . . . . . . . . . . . . . 230
        5.4.2   The Single-Input Case . . . . . . . . . . . . . . . . . . . 233
        5.4.3   The Multi-variable Case . . . . . . . . . . . . . . . . . . 238
  5.5   Disturbance Tolerance . . . . . . . . . . . . . . . . . . . . . . . . 243
        5.5.1   Preliminaries . . . . . . . . . . . . . . . . . . . . . . . . . 245
        5.5.2   $\tau$-Parameterized Control . . . . . . . . . . . . . . . . . . 247
        5.5.3   Control Law Computation and Implementation . . . . . 251
  5.6   Nonlinear Bounded Control for Time-Delay Systems . . . . . . 255
        5.6.1   Problem Statement . . . . . . . . . . . . . . . . . . . . . 255
        5.6.2   Preliminaries . . . . . . . . . . . . . . . . . . . . . . . . . 256
        5.6.3   Riccati Equation Approach . . . . . . . . . . . . . . . . . 257
        5.6.4   LMI Approach . . . . . . . . . . . . . . . . . . . . . . . . 260
  5.7   Conclusion . . . . . . . . . . . . . . . . . . . . . . . . . . . . . . . . 264

**Part III  Anti-windup**

**6   An Overview of Anti-windup Techniques** . . . . . . . . . . . . . . . 267
  6.1   Introduction—Philosophy . . . . . . . . . . . . . . . . . . . . . . 267
  6.2   General Anti-windup Architecture . . . . . . . . . . . . . . . . . 268
  6.3   A Bit of History . . . . . . . . . . . . . . . . . . . . . . . . . . . . 271
  6.4   Formulation of Problems . . . . . . . . . . . . . . . . . . . . . . 274

| | | |
|---|---|---|
| 6.5 | Regional (Local) Versus Global Strategies | 278 |
| | 6.5.1 A Quick Overview in the Regional (Local) Context | 278 |
| | 6.5.2 A Quick Overview in the Global Context | 279 |
| 6.6 | Mismatch-Based Anti-windup Synthesis | 280 |
| 6.7 | Conclusion | 281 |

**7 Anti-windup Compensator Synthesis** ... 283
- 7.1 Introduction ... 283
- 7.2 Problems Setup ... 283
  - 7.2.1 Direct Linear Anti-windup ... 285
  - 7.2.2 Model Recovery Anti-windup ... 286
- 7.3 Direct Linear Anti-windup Design ... 287
  - 7.3.1 Preliminary Elements ... 288
  - 7.3.2 DLAW Schemes with Global Stability Guarantees ... 290
  - 7.3.3 DLAW Schemes with Regional Stability Guarantees ... 294
  - 7.3.4 Some Algorithms ... 297
- 7.4 Model Recovery Anti-windup Design ... 300
  - 7.4.1 Preliminary Elements on the Architecture ... 300
  - 7.4.2 Some Algorithms ... 302
- 7.5 Anti-windup Algorithms Summary ... 304
- 7.6 Some Extensions ... 305
  - 7.6.1 Rate Saturation ... 306
  - 7.6.2 Sensor Saturations ... 306
  - 7.6.3 Nested Saturations ... 307
  - 7.6.4 Time-Delay Systems ... 308
  - 7.6.5 Anti-windup for Nonlinear Systems ... 309
- 7.7 Conclusion ... 309

**8 Applications of Anti-windup Techniques** ... 311
- 8.1 Introduction ... 311
- 8.2 Static Anti-windup Examples ... 311
- 8.3 Dynamic Anti-windup Examples ... 315
- 8.4 Toward More Complex Nonlinear Actuators ... 324
  - 8.4.1 Actuator with Position and Rate Saturations ... 325
  - 8.4.2 Dynamics Restricted Actuator ... 330
  - 8.4.3 Electro-Hydraulic Actuator ... 335
- 8.5 Pseudo Anti-windup Strategies when the Dead-Zone Element is not Accessible ... 339
  - 8.5.1 Anti-windup Scheme Using a Fictitious Linear Element ... 339
  - 8.5.2 Observer-Based Anti-windup Scheme ... 343
- 8.6 Conclusion ... 351

**Appendix A Some Concepts Related to Stability Theory** ... 353
- A.1 Introduction ... 353
- A.2 Stability of Linear Autonomous Systems ... 354
- A.3 Stability for Nonlinear Systems ... 355
  - A.3.1 Stability Definition (Lyapunov Stability) ... 355

Contents                                                                                                                                 xvii

A.3.2   Lyapunov's Stability Theorem . . . . . . . . . . . . . . . . . . 357
A.3.3   The Invariance Principle . . . . . . . . . . . . . . . . . . . . 359
A.3.4   Region of Attraction . . . . . . . . . . . . . . . . . . . . . . 360
A.3.5   Set of Equilibrium Points . . . . . . . . . . . . . . . . . . . 361
A.4   Application of Lyapunov Stability . . . . . . . . . . . . . . . . . . 361
A.4.1   Absolute Stability . . . . . . . . . . . . . . . . . . . . . . . 362
A.4.2   Positive Real Function Concept . . . . . . . . . . . . . . . 363
A.4.3   Circle Criterion . . . . . . . . . . . . . . . . . . . . . . . . 364
A.4.4   Popov Criterion . . . . . . . . . . . . . . . . . . . . . . . . 365
A.4.5   Quadratic Stability . . . . . . . . . . . . . . . . . . . . . . . 366

**Appendix B   Quadratic Approach for Robust Control** . . . . . . . . . 367
B.1   Introduction . . . . . . . . . . . . . . . . . . . . . . . . . . . . . 367
B.2   Uncertain Models and Quadratic Stability . . . . . . . . . . . . . . 368
B.2.1   Uncertain Models . . . . . . . . . . . . . . . . . . . . . . . 368
B.2.2   Quadratic Stability . . . . . . . . . . . . . . . . . . . . . . . 370
B.3   Quadratic Stabilizability . . . . . . . . . . . . . . . . . . . . . . . 373
B.3.1   P-Uncertainty . . . . . . . . . . . . . . . . . . . . . . . . . 374
B.3.2   NB-Uncertainty . . . . . . . . . . . . . . . . . . . . . . . . 375
B.4   Guaranteed Cost Control . . . . . . . . . . . . . . . . . . . . . . 377
B.4.1   P-Uncertainty . . . . . . . . . . . . . . . . . . . . . . . . . 379
B.4.2   NB-Uncertainty . . . . . . . . . . . . . . . . . . . . . . . . 380
B.5   Pole Placement . . . . . . . . . . . . . . . . . . . . . . . . . . . 382
B.5.1   Pole Placement in a Disk . . . . . . . . . . . . . . . . . . . 382
B.5.2   Pole Placement in LMI Regions . . . . . . . . . . . . . . . 385

**Appendix C   Linear Matrix Inequalities (LMI) and Riccati
Equations** . . . . . . . . . . . . . . . . . . . . . . . . . . . . . . . . . . 389
C.1   Introduction . . . . . . . . . . . . . . . . . . . . . . . . . . . . . 389
C.2   Algebraic Lyapunov Equation (ALE) . . . . . . . . . . . . . . . . 389
C.2.1   Continuous Algebraic Lyapunov Equation
(CALE) [414] . . . . . . . . . . . . . . . . . . . . . . . . . 390
C.2.2   Discrete Algebraic Lyapunov Equation (DALE) [414] . . . 391
C.3   Algebraic Riccati Equation (ARE) . . . . . . . . . . . . . . . . . 392
C.3.1   Continuous Algebraic Riccati Equation (CARE) [414] . . . 392
C.3.2   Discrete Algebraic Riccati Equation (DARE) [414] . . . . 395
C.4   Linear Matrix Inequalities (LMI) . . . . . . . . . . . . . . . . . . 397
C.5   Schur's Complement . . . . . . . . . . . . . . . . . . . . . . . . 398
C.6   Bilinear Matrix Inequalities (BMI) . . . . . . . . . . . . . . . . . 400
C.6.1   Eigenvalues and Generalized Eigenvalues Problems . . . . 401
C.7   The Elimination Lemma and the S-Procedure . . . . . . . . . . . 403
C.7.1   The Elimination Lemma [45] . . . . . . . . . . . . . . . . . 403
C.7.2   The S-Procedure [45] . . . . . . . . . . . . . . . . . . . . . 404
C.8   Ellipsoids and Polyhedral Sets . . . . . . . . . . . . . . . . . . . 405
C.8.1   Ellipsoid and Invariant Ellipsoid [291, 346] . . . . . . . . 405
C.8.2   Convex Polyhedron and Invariant Polyhedron
[135, 136, 344] . . . . . . . . . . . . . . . . . . . . . . . . . 406

|  | C.8.3 | Maximum Volume Ellipsoid Contained in a Symmetric Polytope | 407 |
|  | C.8.4 | Smallest Volume Ellipsoid Containing a Symmetric Polytope | 408 |

**References** . . . . . . . . . . . . . . . . . . . . . . . . . . . . . . . . . . 411

**Index** . . . . . . . . . . . . . . . . . . . . . . . . . . . . . . . . . . . . 429

# Notation

*Main notation*[1]

| | |
|---|---|
| $\mathfrak{R}$ | The set of real numbers |
| $\mathcal{C}$ | The set of complex numbers |
| $\mathcal{N}$ | The set of natural numbers |
| $\mathfrak{R}_+$ | The set of nonnegative real numbers |
| $\mathfrak{R}^n$ | The $n$-dimension real space |
| $\mathfrak{R}^{n \times m}$ | The set of real matrices of dimensions $n \times m$ |
| $A'$ | The transpose of matrix $A$ |
| $A_{(i)}$ | The $i$th row of matrix $A$ |
| $A_{(i,j)}$ | The element of the $i$th row and $j$th column of matrix $A$ |
| $A - B > 0$ | Means that matrix $A - B$ is positive definite |
| $A - B \geq 0$ | Means that matrix $A - B$ is semi-positive definite |
| $\lambda_{\max}(A)$ | The maximal eigenvalue of matrix $A$ |
| $\lambda_{\min}(A)$ | The minimal eigenvalue of matrix $A$ |
| $\mathfrak{R}e(\lambda_i(A))$ | The real part of the eigenvalue $\lambda_i$ of matrix $A$ |
| $\sigma(A)$ | The spectrum of matrix $A$ |
| $\mathrm{Diag}(A_1, A_2, \dots, A_p)$ | Denotes the block-diagonal matrix for which the block-diagonal matrices are the $A_i$, $j = 1, \dots, p$ |
| The symbol $\star$ stands for symmetric blocks in the expression of a matrix | |
| $\det(A)$ | The determinant of matrix $A$ |
| $\mathrm{trace}(A)$ | The trace of matrix $A$ |
| $\mathrm{rank}(A)$ | The rank of matrix $A$ |
| $\mathrm{Ker}(A)$ | The kernel of matrix $A$, that is, the set $\mathrm{Ker}(A) = \{x \in \mathfrak{R}^n; \ Ax = 0\}$ |
| $A^{\#}$, $A \in \mathfrak{R}^{m \times n}$, $\mathrm{rank}(A) = m$ | The pseudo-inverse of matrix $A$ such that $AA^{\#} = I_m$ |
| $I_n$ | The identity matrix of dimensions $n \times n$. The subscript may be removed when there is no ambiguity |

---

[1] Throughout the book, the notation used is quite standard. The main elements are described below.

| | |
|---|---|
| $1_n$ | The vector in $\Re^n$ defined by $1_n = [\,1 \ldots 1\,]'$ |
| $0$ | Denotes the null scalar or the null matrix of appropriate dimensions |
| $\mathrm{diag}(x)$ | The $n$-order diagonal matrix generated by the components of $x \in \Re^n$. For example for $n = 2$: $\mathrm{diag}(x) = \begin{bmatrix} x_{(1)} & 0 \\ 0 & x_{(2)} \end{bmatrix}$ |
| $x \succeq y$ with $x, y \in \Re^n$ | Means that $x_{(i)} - y_{(i)} \geq 0$, $\forall i = 1, \ldots, n$ |
| $\mathrm{Co}\{x_j, \ j = 1, \ldots, r\}$ | The convex hull defined by the vertices $x_j$ |
| $|x|$ | The vector composed by the absolute values of the components of $x$. The same notation will be used for a matrix $A$, i.e., $|A|$ |
| $\|x\|$ | The Euclidean norm of vector $x$, i.e., $\|x\| = \sqrt{x'x}$ |
| $\|x\|_2$ | The $\mathcal{L}_2$-norm of vector $x$, i.e., $\|x\|_2 = \sqrt{\int_0^\infty x'x\,dt}$ |
| $\mathrm{sign}(x)$ | The sign function of $x \in \Re$ |
| $\mathrm{sat}(z), z \in \Re^m$ | The saturation function of the vector $z$ defined by $\mathrm{sat}(z_{(i)}) = \mathrm{sign}(z_{(i)}) \min(u_{0(i)}, |z_{(i)}|)$, $\forall i = 1, \ldots, m$ |
| $x^{(j)}$ | Denotes the $j$th time-derivative of $x$ |
| $n!$ | Denotes the factorial function, i.e. $n! = n(n-1)\ldots 1$, $\forall n \in \mathcal{N}$ |
| $\otimes$ | Denotes the Kronecker product |

*Particular Notation*

Vectors and matrices are defined with brackets as follows

$$\text{For } x \in \Re^n \text{ one gets } x = \begin{bmatrix} x_{(1)} \\ \vdots \\ x_{(n)} \end{bmatrix}$$

$$\text{For } A \in \Re^{n \times m} \text{ one gets}$$

$$A = \begin{bmatrix} A_{(1,1)} & \cdots & A_{(1,m)} \\ \vdots & \vdots & \vdots \\ A_{(n,1)} & \cdots & A_{(n,m)} \end{bmatrix}$$

$S(R, \rho_1, \rho_2)$, with $R \in \Re^{m \times n}$, $\rho_1, \rho_2 \in \Re^m$, is the polyhedral set defined by

$$S(R, \rho_1, \rho_2) = \left\{ x \in \Re^n; \ -\rho_2 \preceq Rx \preceq \rho_1 \right\}$$

$S(|R|, \rho)$, with $R \in \Re^{m \times n}$, $\rho \in \Re^m_+$, is the symmetrical polyhedral set defined by

$$S(|R|, \rho) = \left\{ x \in \Re^n; \ -\rho \preceq Rx \preceq \rho \right\}$$

$S(R, \rho)$, with $R \in \Re^{m \times n}$, $\rho \in \Re^m$, is the polyhedral set defined by

$$S(R, \rho) = \left\{ x \in \Re^n; \ Rx \preceq \rho \right\}$$

Notation        xxi

$\mathcal{E}(P, \eta)$, with $P = P' > 0$, $P \in \mathfrak{R}^{n \times n}$, and $\eta > 0$, is the ellipsoid defined by

$$\mathcal{E}(P, \eta) = \left\{ x \in \mathfrak{R}^n; \ x'Px \leq \eta^{-1} \right\}$$

For two sets $S_1$ and $S_2$, $S_1 \backslash S_2$ denotes the set $S_1$ deprived of $S_2$

For a set $S$, $\partial S$ and int $S$ denote the boundary and the interior of $S$, respectively

For a linear operator $H$, $H \mid \mathcal{X}$ is a restriction of $H$ on the subspace $\mathcal{X}$

*Acronyms*

| | |
|---|---|
| ALE | Algebraic Lyapunov Equation |
| ARE | Algebraic Riccati Equation |
| AW | Anti-windup |
| BIBO | Bounded Input Bounded Output |
| BMI | Bilinear Matrix Inequality |
| CALE | Continuous Algebraic Lyapunov Equation |
| CARE | Continuous Algebraic Riccati Equation |
| DALE | Discrete Algebraic Lyapunov Equation |
| DARE | Discrete Algebraic Riccati Equation |
| DLAW | Direct Linear Anti-windup |
| EVP | EigenValue Problem |
| GAS | Global Asymptotic Stability |
| GES | Global Exponential Stability |
| GEVP | Generalized EigenValue Problem |
| HPPD | Homogeneous Polynomial Parameter-Dependent |
| IMC | Internal Model Control |
| LMI | Linear Matrix Inequality |
| MIMO | Multiple Input Multiple Output |
| MPC | Model Predictive Control |
| MRAW | Model Recovery Anti-windup |
| SISO | Single-Input Single-Output |
| PID | Proportional Integral Derivative |
| RAS | Region of Asymptotic Stability |
| RES | Regional Exponential Stability |
| SDP | Semi-definite Programming |

# Part I
# Generalities

# Chapter 1
# Linear Systems Subject to Control Saturation—Problems and Modeling

## 1.1 Introduction

In this chapter we state the main problems we are concerned within the book, with respect to the stability analysis and the stabilization of linear systems with saturating inputs.

Linear systems with saturating inputs sit on the boundary of the linear and nonlinear worlds. Even if the open-loop system is considered linear, the presence of the saturation makes the closed-loop system nonlinear. In this case, we should be aware of the nonlinear effects produced by saturation on performance and stability of the closed-loop system.

Some definitions like the region of linearity, region of attraction and regions of asymptotic stability are fundamental in the statement of the problems and also for solving them. In stark contrast to linear closed-loop systems, when control constraints are present, it may not be possible to ensure that trajectories converge to the origin for all initial conditions or that the trajectories are bounded if the input signals are bounded. Hence, the characterization of sets of admissible initial states and admissible disturbances plays a central role in stability analysis as well as in the synthesis of stabilizing control laws when control constraints exist.

On the other hand, although apparently simple, the saturation function presents hard discontinuities, which make its direct treatment, using Lyapunov-like inequalities for instance, quite difficult. Hence, appropriate models are particularly important to take saturation effects into account when deriving stability or stabilization conditions.

This chapter discusses the aspects above and can be seen as the basis for the methods and approaches, which are developed in the later chapters. With this aim the chapter is organized as follows. First, the open-loop and the saturating closed-loop systems are presented. Two examples to illustrate the nonlinear behaviors produced by control saturation are provided to motivate the definition of the problems of interest as well as the importance of the region of linearity and the region of attraction of the nonlinear closed-loop system. Then, the problems of stability analysis and stabilization as well as the so-called anti-windup problem are formally stated. Finally, some convenient methods to model the saturation term are presented.

S. Tarbouriech et al., *Stability and Stabilization of Linear Systems with Saturating Actuators*, DOI 10.1007/978-0-85729-941-3_1, © Springer-Verlag London Limited 2011

## 1.2 The Open-Loop System

In this book, we are mainly concerned with the treatment of stability and performance problems based on a state space approach and the use of linear differential equations. Basically, we suppose that the system to be controlled (the open-loop system or the *plant*) is described by a linear model, given by the following continuous-time state representation:

$$\dot{x}(t) = Ax(t) + Bu(t) + B_w w(t) \tag{1.1}$$

$$y(t) = Cx(t) + Du(t) \tag{1.2}$$

$$z(t) = C_z x(t) + D_z u(t) \tag{1.3}$$

where $x \in \Re^n$ is the state vector, $u \in \Re^m$ is the control vector, $y \in \Re^p$ is the measured output vector, $z \in \Re^{p_z}$ is the regulated output vector and $w \in \Re^q$ is an input exogenous signal. $A$, $B$, $B_w$, $C$, $D$, $C_z$ and $D_z$ are real matrices of appropriate dimensions.

The output $z$ corresponds to the outputs we want to control, i.e. the outputs that should respect performance requirements. In most of the practical cases, $z$ will be equal to $y$ or it will be composed by some elements of $y$.

The vector $w$ corresponds to all exogenous signals acting on the system. Hence, $w$ can represent disturbance signals as well as reference signals. Although not expressed in (1.2) and (1.3), it has to be noted that $w$ may also directly affect the outputs $y$ and $z$.

If matrix $D$ is null, the system is said to be *strictly proper*, otherwise it will be said to be *proper*. Matrix $D_z$ is a way to take into account performance requirements regarding the control signals given by $u$.

Concerning system (1.1)–(1.2), the following assumptions are made.

**Assumption 1.1** Pairs $(A, B)$ and $(C, A)$ are controllable and observable, respectively.

**Assumption 1.2** Vector $u$ takes values in a compact set denoted by $S(I_m, u_{\min}, u_{\max})$ defined by

$$S(I_m, u_{\min}, u_{\max}) = \left\{ u \in \Re^m : -u_{\min} \preceq u \preceq u_{\max} \right\} \tag{1.4}$$

where the symbol $\preceq$ denotes component-wise inequalities (see Notation), and $u_{\min}$ and $u_{\max}$ are vectors of $\Re^m$ with positive components, i.e. $u_{\min(i)} > 0$, $u_{\max(i)} > 0$, $\forall i = 1, \ldots, m$.

Assumption 1.1 is important in the controller synthesis problems. It can in fact be relaxed to $(A, B)$-stabilizability and $(C, A)$-detectability. For the sake of simplicity in the presentation of the stabilization results along the book, we suppose that Assumption 1.1 is satisfied.

Once the control signal is constrained, a fundamental issue regards the characterization of the states that can be steered to the origin using bounded controls, i.e. with

## 1.3 The Closed-Loop System

$u(t) \in S(I_m, u_{\min}, u_{\max})$. This corresponds to a generalization of the classical controllability concept [64] leading to the definition of the so-called *null controllable region* (see for instance [194, 236, 316, 326] and references therein).

**Definition 1.1** A state $x_0 \in \Re^n$ is said null controllable with respect to system (1.1) if there exists a time $T$ and an admissible control signal $u(t) \in S(I_m, u_{\min}, u_{\max})$ such that for $x(0) = x_0$ and $w(t) = 0$, it follows that $x(T) = 0$. The set of all null controllable states is called the *region of null controllability* of the system (1.1). If the region of null controllability is the whole state space, system (1.1) is said to be *null controllable* with bounded controls.

Actually, Assumption 1.2 corresponds to the source of the problems we are interested to address in the book. It represents the control amplitude constraints. These constraints are in general related to physical or security limits imposed by the actuators which provide the input signals to the system within maximal and minimum values. This is the case, for instance, in proportional valves, which present a maximal (100%) and a minimal (0%) opening values, which limits the flow to be delivered to the controlled plant; heating or cooling actuators, where the maximal power to be delivered or extracted from the system is limited; power converters and amplifiers, whose current and output voltages are always limited in order to avoid components to be damaged. Hence, we can conclude that the control limitation is a practical constraint present in almost all physical systems.

At a first glance, these constraints can seem as simple and not worthy of too much attention. Unfortunately, this is not the case. The importance of them in the behavior of closed-loop systems should not be underestimated. As will be seen in the next sections and throughout this book, these constraints can be modeled by a saturation nonlinearity in the control loop, which induces a nonlinear behavior in the closed-loop system, even if the open-loop system is linear. This fact can lead to performance degradation, the occurrence of limit cycles, multiple equilibria and even lead the closed-loop system to instability. These phenomena will be illustrated in the end of the next section.

## 1.3 The Closed-Loop System

In order to improve the dynamical response of a system, to achieve performance requirements in the regulated outputs as well as to make the operation robust to the action of disturbances and model uncertainties, the input signals to be applied to the system should be systematically computed by a controller.

The controller has as input the measured outputs of the plant (given by $y$) and provide a signal $v$ as output. This signal $v$ is therefore sent to an actuator which, in general, converts the low power signal provided by the controller to a high power signal. It should be noticed that input and output of the actuator can be different physical signals. For instance, the controller provides an electrical signal, which is converted by the actuator in energy, flow, position, etc. Since, in general, the actuator

**Fig. 1.1** The saturation function

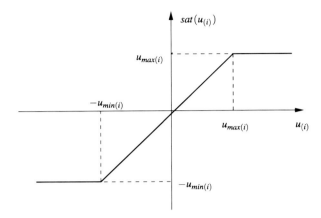

dynamics is much faster than that of the controlled system, it can be modeled by a simple static gain. For the sake of simplicity, in analysis or synthesis problems, this gain is in general normalized to 1. Different actuator gains or actuator linear dynamics can be directly encapsulated in the compensator or in the plant models.[1] Hence in the classical linear framework, i.e. in the absence of magnitude constraints, it is supposed that

$$u(t) = v(t)$$

However, due to the magnitude bounds, given by Assumption 1.2, the effective control signal provided by the actuator can be modeled by a saturation function, i.e.

$$u(t) = \text{sat}(v(t))$$

where sat($\cdot$) represents a vector function defined by

$$\text{sat}(v_{(i)}) = \begin{cases} u_{\max(i)} & \text{if } v_{(i)} > u_{\max(i)} \\ v_{(i)} & \text{if } -u_{\min(i)} \leq v_{(i)} \leq u_{\max(i)} \\ -u_{\min(i)} & \text{if } v_{(i)} < -u_{\min(i)} \end{cases} \quad (1.5)$$

for $i = 1, \ldots, m$.

The saturation function and the closed-loop system are, respectively, depicted in Figs. 1.1 and 1.2.

In the sequel, depending on the availability of the state, we consider three types of saturating control laws: state feedback, static output feedback and dynamic output feedback. These three control laws are succinctly discussed below.

*State Feedback Control Law* In this case we suppose that all the states of the system can be measured, i.e. $y = x$. The signal computed by the controller is therefore given by

$$v(t) = Kx(t)$$

---

[1]On the other hand, actuators presenting more complex nonlinear dynamics should be considered through dedicated models (see Chap. 8 for some examples in this sense).

## 1.3 The Closed-Loop System

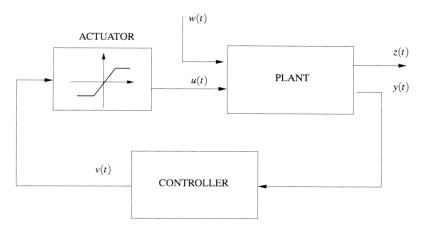

**Fig. 1.2** The closed-loop system

where $K \in \Re^{m \times n}$ is a constant matrix.
The effective control law applied to system (1.1) under Assumption 1.2 is

$$u(t) = \text{sat}(Kx(t)) \tag{1.6}$$

with each component, $i = 1, \ldots, m$, defined by

$$\text{sat}(K_{(i)}x(t)) = \begin{cases} u_{\max(i)} & \text{if } K_{(i)}x(t) > u_{\max(i)} \\ K_{(i)}x(t) & \text{if } -u_{\min(i)} \leq K_{(i)}x(t) \leq u_{\max(i)} \\ -u_{\min(i)} & \text{if } K_{(i)}x(t) < -u_{\min(i)} \end{cases} \tag{1.7}$$

The resulting closed-loop dynamics is

$$\dot{x}(t) = Ax(t) + B\,\text{sat}(Kx(t)) + B_w w(t) \tag{1.8}$$

*Static Output Feedback Control Law* By considering a change of variables, we define an auxiliary output as

$$\tilde{y}(t) = y(t) - Du(t) = Cx(t) \tag{1.9}$$

The static output feedback control applied to system (1.1) is given by

$$u(t) = \text{sat}(K\tilde{y}(t)) = \text{sat}(KCx(t)) \tag{1.10}$$

where $K \in \Re^{m \times p}$ is a constant matrix.
The closed-loop dynamics in this case is given by

$$\dot{x}(t) = Ax(t) + B\,\text{sat}(\tilde{K}x(t)) + B_w w(t) \tag{1.11}$$

with $\tilde{K} = KC$.

*Dynamic Output Feedback Control Law* In this case, the control signal is obtained from a dynamic linear system (or filter), whose input is the plant output $y$. Generically, the controller is represented by the following state equation:

8        1   Linear Systems Subject to Control Saturation—Problems and Modeling

$$\dot{x}_c(t) = A_c x_c(t) + B_c y(t) \tag{1.12}$$

$$v(t) = C_c x_c(t) + D_c y(t) \tag{1.13}$$

where $x_c(t) \in \Re^{n_c}$ is the controller state and $A_c$, $B_c$, $C_c$ and $D_c$ are matrices of appropriate dimensions.

It should be pointed out that an observer-based control law can be written in the form (1.12)–(1.13), i.e. it can be considered as a particular case of dynamic output feedback.

Define now an augmented state composed by the plant and the controller states:

$$\tilde{x} = \begin{bmatrix} x \\ x_c \end{bmatrix}$$

For the sake of simplicity, consider that the plant is strictly proper (i.e. $D = 0$). Hence, considering the dynamic controller (1.12)–(1.13), the closed-loop dynamics becomes

$$\dot{\tilde{x}}(t) = \mathcal{A}\tilde{x}(t) + \mathcal{B}\,\mathrm{sat}\big(\mathcal{K}\tilde{x}(t)\big) + \mathcal{B}_w w(t) \tag{1.14}$$

with

$$\mathcal{A} = \begin{bmatrix} A & 0 \\ B_c C & A_c \end{bmatrix}, \qquad \mathcal{B} = \begin{bmatrix} B \\ 0 \end{bmatrix}, \qquad \mathcal{B}_w = \begin{bmatrix} B_w \\ 0 \end{bmatrix} \quad \text{and} \quad \mathcal{K} = \begin{bmatrix} D_c C & C_c \end{bmatrix}$$

*Remark 1.1* The case of proper systems (i.e. $D \neq 0$) can be easily handled, as soon as the system is well-posed (see Part III).

It should be noticed that, without loss of generality, the three cases above lead to a closed-loop system in the form

$$\dot{x}(t) = A x(t) + B\,\mathrm{sat}\big(K x(t)\big) + B_w w(t) \tag{1.15}$$

Hence, in the sequel all the definitions, problem formulation and saturation modeling will be stated with reference to the closed-loop system (1.15).

In the two examples which follow, some nonlinear effects induced by the saturation on the closed-loop behavior are illustrated.

*Example 1.1* Consider the following simple system (balancing pointer) borrowed from [205]:

$$\dot{x}(t) = \begin{bmatrix} 0 & 1 \\ 1 & 0 \end{bmatrix} x(t) + \begin{bmatrix} 0 \\ -1 \end{bmatrix} u(t)$$

This system is open-loop unstable since the eigenvalues of $A$ are 1 and $-1$.

In order to stabilize this system consider the following state feedback control law $u(t) = K x(t)$ with:

$$K = \begin{bmatrix} 13 & 7 \end{bmatrix}$$

In the absence of the control bounds, the closed-loop system is linear and governed by the following dynamics:

$$\dot{x}(t) = (A + BK)x(t)$$

## 1.3 The Closed-Loop System

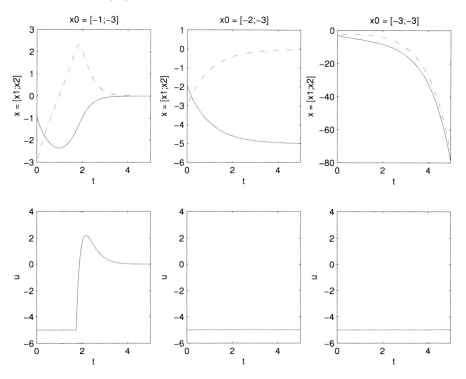

**Fig. 1.3** Example 1.1—time responses for three different initial conditions

Since the eigenvalues of $(A + BK)$ are $-3$ and $-4$, from the classical linear system theory, we can conclude that the origin of the closed-loop system is globally asymptotically stable, i.e. $\forall x(0) \in \Re^2$, $x(t, x(0)) \to 0$ as $t \to \infty$.

Suppose now that the control signal is bounded in magnitude:

$$-5 \leq u(t) \leq 5$$

The closed-loop system is now nonlinear:

$$\dot{x}(t) = Ax(t) + B\,\text{sat}\big(Kx(t)\big)$$

$$\text{with sat}\big(Kx(t)\big) = \begin{cases} 5 & \text{if } Kx(t) > 5 \\ Kx(t) & \text{if } |Kx(t)| \leq 5 \\ -5 & \text{if } Kx(t) < -5 \end{cases}$$

Figure 1.3 illustrates the time evolution of the closed-loop system considering three different initial conditions:

**Fig. 1.4** Example 1.1—presence of divergent and convergent trajectories as well as multiple equilibria. Equilibrium points (∘); initial conditions (∗)

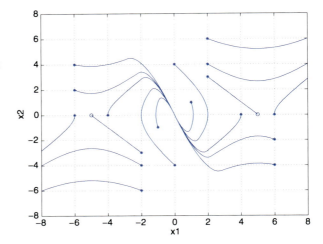

- $x_1(0) = [-1\ -3]'$—in spite of the fact that the control signal $u$ is saturated during the first instants, the trajectory converges asymptotically to the origin.
- $x_2(0) = [-2\ -3]'$—in spite of the fact that $(A + BK)$ is Hurwitz, the trajectory converges to an equilibrium point $x_e$ different from the origin ($x_e = [-5\ 0]'$). Note that in this case the control signal remains saturated all the time.
- $x_3(0) = [-3\ -3]'$—in spite of the fact that $(A + BK)$ is Hurwitz, the trajectory diverges. Note that in this case the control signal also remains saturated all the time.

Figure 1.4 shows a phase plane with many trajectories. Clearly, depending on the initial conditions, there are trajectories that converge asymptotically to the origin, converge to other equilibrium points or diverge. Hence, although the origin of the linear closed-loop system is supposed to be globally asymptotically stable in the absence of input constraints, this is not the case when these constraints are present.

*Example 1.2* Consider the same closed-loop system as in Example 1.1, with the same control constraints. Let us now evaluate the behavior of the system when it is excited by an external signal $w(t)$. With this aim, using the notation of (1.1)–(1.2), assume that $B_w = B$, $C = I_2$ and $x(0) = 0$. Consider that the signal $w(t)$ is a pulse, of amplitude $a$ and duration of $2s$:

$$w(t) = \begin{cases} a & \text{if } 0 < t \le 2s \\ 0 & \text{if } t > 2s \end{cases}$$

Note that since $(A + BK)$ is Hurwitz, we conclude that the linear system is BIBO-stable considering the input $w$ and the output $y$. This means that, in the absence of control constraints, for any bounded input $w(t)$ the outputs (and the states) are also bounded.

## 1.3 The Closed-Loop System

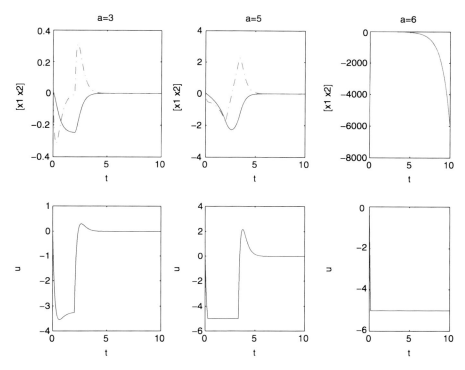

**Fig. 1.5** Example 1.2—time responses for different amplitudes of $w(t)$

Let us now show what happens if the control is constrained. Figure 1.5 illustrates the time response of the closed-loop system considering three different amplitudes for the pulse $w(t)$:

- $a = 3$—in this case, there is no control saturation and the system behaves linearly. Actually, since the amplitude of $w(t)$ is relatively small, there is enough control capacity to bring the state to the equilibrium once the signal $w(t)$ vanishes.
- $a = 5$—in this case, the control saturates during the first instants. Note, however, that, although there is enough control capacity to bring the state to the equilibrium, the settling time is degraded when compared with the linear case (when there are no control constraints).
- $a = 6$—in spite of the fact that the linear system is BIBO-stable, the trajectory diverges. It means that there is not enough control capacity to bring back the state to the origin when the disturbance vanishes.

We can conclude that the external stability, i.e. the input-to-state or the input-to-output stability, of a linear system subject to control constraints is highly dependent on the amplitude and/or the energy of the external signal. As will be seen in the sequel, it is worth to characterize the class of external signals for which bounded state behavior is ensured.

## 1.4 The Region of Linearity

Consider the definition of the saturation function (1.7) and the closed-loop system (1.15). For states near to the origin ($x = 0$) it follows that

$$u(t) = \text{sat}\big(Kx(t)\big) = Kx(t)$$

i.e. $u(t) = v(t)$ and we say that the *saturation is not active* or that the control signal *is not saturated*. In this case, the derivative of the state is given by the following linear dynamics:

$$\dot{x}(t) = (A + BK)x(t) + B_w w(t) \tag{1.16}$$

Hence, the following definition can be stated.

**Definition 1.2** The region of linearity of system (1.15), denoted $R_L$, is defined as the set of all states $x \in \Re^n$ such that $\text{sat}(Kx) = Kx$.

For simplicity, assume that $K$ is full row rank ($\text{rank}(K) = m$). From the saturation function definition, the linearity region can be defined as the intersection of the following $2m$ half-spaces

$$\mathcal{H}_i^{\max} = \big\{x \in \Re^n; \ K_{(i)}x \leq u_{\max(i)}\big\}, \quad i = 1, \dots, m$$

$$\mathcal{H}_i^{\min} = \big\{x \in \Re^n; \ K_{(i)}x \geq -u_{\min(i)}\big\}, \quad i = 1, \dots, m$$

or, in other words, $R_L$ is defined as the following polyhedral set in the state space:

$$R_L = S(K, u_{\min}, u_{\max}) = \big\{x \in \Re^n; \ -u_{\min} \preceq Kx(t) \preceq u_{\max}\big\} \tag{1.17}$$

The set $S(K, u_{\min}, u_{\max})$ presents the following geometric characteristics:

- It is symmetric if $u_{\min(i)} = u_{\max(i)}$, $\forall i = 1, \dots, m$ and asymmetric otherwise.
- It is bounded when $\text{rank}(K) = n$ and unbounded when $\text{rank}(K) < n$. In this last case, it is unbounded in the directions associated to $\text{Ker}(K)$.

Figure 1.6 illustrates the two cases mentioned above.

Hence, the state space is divided into two regions. The first one is defined by $S(K, u_{\min}, u_{\max})$ in which the closed-loop system can be described by the linear model (1.16). In the second subregion, $\Re^n \setminus S(K, u_{\min}, u_{\max})$, the saturation is active, i.e. at least one control entry is saturated, and the closed-loop system presents a nonlinear behavior.

Concerning the general behavior of the closed-loop system we can say that if $x(t) \in S(K, u_{\min}, u_{\max})$, then $\dot{x}(t)$ can be computed by (1.16). At this point, it is very important to remark that even if the system is initialized in $S(K, u_{\min}, u_{\max})$, without further assumptions, we cannot ensure that the corresponding trajectory $x(t, x(0))$ will be a trajectory of the linear system. Actually, a trajectory starting in $S(K, u_{\min}, u_{\max})$ can leave this domain, and so, it has to be characterized considering the nonlinear model (1.15). Furthermore, even if $(A + BK)$ is asymptotically stable (Hurwitz), we cannot ensure that a trajectory starting in $S(K, u_{\min}, u_{\max})$ converges asymptotically to the origin. Of course, this is a necessary, but not a sufficient condition to conclude on the stability of the system (1.15).

## 1.5 The Region of Attraction

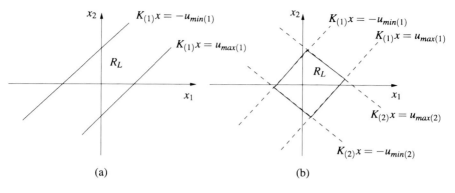

**Fig. 1.6** (a) Unbounded $R_L$ ($m = 1$, $n = 2$, rank($K$) $= 1$); (b) bounded $R_L$ ($m = n = 2$, rank($K$) $= 2$)

On the other hand, when $(A + BK)$ is asymptotically stable, it is always possible to compute invariant sets contained in the region of linearity. This is summarized in the following theorem [50].

**Theorem 1.1** *Given a matrix $K$ such that $A + BK$ is asymptotically stable, then there always exists a set, denoted $S_L$, for which the following properties hold*:

1. $S_L \subseteq S(K, u_{\min}, u_{\max})$;
2. $\forall x(0) \in S_L$, $x(t, x(0))$ *is a trajectory of linear system* (1.16);
3. $\forall x(0) \in S_L$, $x(t, x(0)) \in S_L$ *and* $x(t, x(0)) \to 0$ *as* $t \to \infty$.

Actually, as will be seen in the sequel, any positive invariant set contained in the region of linearity can be considered as a region $S_L$.

It should be noted that Theorem 1.1 does not provide any guarantee about the trajectories of the saturated system (1.15) initialized outside $S_L$.

## 1.5 The Region of Attraction

Due to the saturation, the closed-loop system (1.15) is nonlinear. Hence, assuming $w(t) = 0$, $\forall t$, the convergence of the closed-loop trajectories to the origin depends on the initial state of the system, even if $(A + BK)$ is Hurwitz. In other words, in general, there are initial conditions $x(0)$ such that the trajectories converge to the origin, i.e. $x(t) \to 0$ as $t \to \infty$, but also initial conditions leading to divergent trajectories, i.e. $x(t) \to \infty$ as $t \to \infty$. This fact has been illustrated in Example 1.1. The *region of attraction* or *the basin of attraction* of the closed-loop system is then defined as follows (see also Appendix A).

**Definition 1.3** The region of attraction $R_A$ of the origin for system (1.15) is defined as the set of all points $x \in \Re^n$ of the state space for which $x(0) = x$ leads to a trajectory $x(t, x(0))$ that converges asymptotically to the origin for $w = 0$. In other words, if $x(0) \in R_A$ then $x(t) \to 0$ as $t \to \infty$.

Under some particular assumptions on open-loop stability and gain matrix $K$, it is possible to ensure the global asymptotic stability of the saturated closed-loop system (1.15). In this case, the basin of attraction of the origin is the whole state space, i.e. $R_A = \Re^n$. However, in the general case, asymptotic convergence to the origin is not ensured for any initial condition of the state space.

The exact geometric characterization of the region of attraction is not an easy task [215, 300, 309, 334]. Actually, $R_A$ can be non-convex, open and unlimited. In general, it can be approximately determined only by intensive simulations, but its visualization is practically impossible for systems with more than 3 states. Hence, it is important to characterize subsets of the region of attraction with a well-defined analytical representation, such as ellipsoidal and polyhedral sets. These subsets are called *regions of asymptotic stability* (RAS), or simply *regions of stability*.

**Definition 1.4** A region $R_s$ is said to be a region of asymptotic stability (RAS) for system (1.15) if $R_s \subset R_A$ and $0 \in R_s$.

As a consequence of Definition 1.4, if the initial condition is taken in a region of stability, the convergence of the trajectory to the origin is ensured. Hence, a region of asymptotic stability (RAS) can be seen as a region where the initialization of the system leads for sure to a "safe" behavior.

As will be seen in the sequel, regions of stability can in general be associated with Lyapunov functions of the system (1.15), i.e. considered as Lyapunov level sets. In particular, these kinds of set also exhibit the properties of *positive invariance* and *contractivity* with respect to system (1.15). See Appendix A for more details about these concepts.

It should be pointed out here that sometimes in the literature the term region of attraction is inappropriately used to refer, in fact, to a region of asymptotic stability.

*Remark 1.2* It is important to make clear the difference between the region of null controllability (see Definition 1.1) and the region of attraction. Basically the first region is related to the open-loop system (1.1) while the second one is related to the closed-loop system (1.15). Actually, the region of null controllability regards the states that can be steered to the origin by some bounded control law. It presupposes only the existence of an appropriated control law. On the other hand, the region of attraction is defined considering a specific control law.

## 1.6 Problems Considered

Considering the generic closed-loop system (1.15), in this section, we state the main problems we are concerned with throughout the book, namely: asymptotic stability

1.6 Problems Considered 15

analysis, asymptotic stabilization, external stability analysis, external stabilization and anti-windup synthesis.

In particular, regarding the asymptotic stability analysis and stabilization problems, the system is considered free from the influence of external signals, that is, $w(t) = 0$. Only the influence of initial conditions is considered in this case. On the other hand, the problems of external stability analysis and stabilization take into account the influence of the external signals as well as of the initial conditions.

## 1.6.1 Asymptotic Stability Analysis

Control systems are often linearly designed. The plant is represented by a local linear model and the control law is generally linear (static or dynamic). Modern linear control theory furnishes efficient techniques and methods to compute such control laws guaranteeing both robust stability and robust performance for the closed-loop system. In general, these designs do not explicitly consider bounds on the control inputs. In this case, as seen in Example 1.1, the control saturation can be a source of parasitic equilibrium points and limit cycles, or even, can lead to divergent trajectories. Thus, given a control law, a problem of major interest is the stability analysis of the resulting saturated closed-loop system. Actually, it is important to determine regions of initial states that can be driven asymptotically to the origin, considering the possible occurrence of control saturation.

As mentioned in Sect. 1.5, only under some particular conditions on open-loop stability and the control law, is possible to ensure the global stability of the saturated closed-loop system [338]. In this case, the region (basin) of attraction of the origin is the whole state space. In the general case, as illustrated in the Example 1.1, asymptotic convergence to the origin is not ensured for any initial condition of the state space. On the other hand, the exact determination of the region of attraction of the origin is a very challenging issue. The idea is therefore to obtain convex inner approximations of the region of attraction, i.e., to obtain *regions of asymptotic stability*. These regions of stability can be seen as *estimates* of the region of attraction.

From the above, the stability analysis problem can be formulated as follows.

**Problem 1.1** (AS-Analysis) Given a control law generically represented by $v(t) = Kx(t)$ such that the linear system $\dot{x}(t) = (A + BK)x(t)$ is asymptotically stable and satisfies some performance and robustness requirements, determine regions of the state space in which convergence of the trajectories of the saturated closed-loop system (1.15) to the origin is guaranteed.

A natural objective is then to obtain a "good" estimate of the region of attraction by means of regions of asymptotic stability. In this case, we are interested in finding a region (or a union of regions) of stability which "best" fits in $R_A$. This is in general not an easy task, since the form of the region of attraction is not known. In practice, we try to find a RAS that is analytically well characterized and that can be "maximized" considering some specific geometric criteria. As will be seen in the next chapters, this will be the case of ellipsoidal and polyhedral regions, which can be associated with quadratic and polyhedral (Minkowski functionals) Lyapunov functions.

Of course, when the open-loop system satisfies specific hypotheses, it is worth determining conditions that allow one to conclude that the origin of the closed-loop system (1.15) is effectively globally asymptotically stable (GAS), i.e. to prove that the whole state space is a RAS.

Roughly speaking, we can classify the literature dealing with the determination of regions of stability for systems presenting control constraints in two groups of approaches, as follows.

### 1.6.1.1 Approaches Considering the RAS Included in the Region of Linearity

**Polyhedral RAS**   The basic idea in this case consists of computing invariant sets contained in the region of linearity of the closed-loop system. In particular, in the late 1980s and 1990s, great attention has been devoted in the literature to the study of conditions to ensure the positive invariance of polyhedral sets (see for instance [30, 31, 34, 36, 59, 85, 344]). Once computed an invariant polyhedra for the linear system (1.16), any scaled version of the polyhedra included in $R_L$, characterizes a RAS. In this case, the trajectories initialized in the polyhedral RAS are linear in the sense that the saturation is never active. Hence, if the matrix $(A + BK)$ is Hurwitz, convergence of the trajectories to the origin is ensured.

In this context, a key problem is the determination of the maximal invariant set contained in $R_L$. A solution to this problem has been given by Gilbert and Tan [126]. The authors consider a linear discrete-time system and a set of constraints $\mathcal{Y}$ defined by

$$\mathcal{Y} = \left\{ x \in \mathfrak{R}^n; \ f_i(x) \leq 0, \ i = 1, \ldots, p \right\}$$

where the $f_i(x) : \mathfrak{R}^n \to \mathfrak{R}$ are continuous functions such that $0 \in \mathcal{Y}$. They provide conditions under which the maximal invariant admissible set included in $\mathcal{Y}$ can be finitely determined. Considering linear systems with model uncertainty and additive perturbations, Blanchini [34] proposes a method to determine the maximal $\mu$-contractive set contained in a given convex and compact set. The application of the first or second above method allows to conclude that the maximal invariant set with respect to the system (1.16) included in $R_L = S(K, u_{\min}, u_{\max})$ is a polyhedral set and can be finitely determined, provided the pair $(K, A)$ is observable. Furthermore, since in this case $\mathcal{Y} = S(K, u_{\min}, u_{\max})$ is a polyhedral set, the computation of the maximal output set can be carried out by an algorithm based on linear programming schemes [85, 214, 348]. Let us underline that these approaches concern only linear discrete-time systems. In the continuous-time systems case, the problem remains still open: it is only possible to obtain some approximations of the maximal set [38, 126]. Other interesting works in this field are [37, 221].

## 1.6 Problems Considered

**Ellipsoidal RAS**   The determination of ellipsoidal RAS contained in $R_L$ is rather a trivial problem. It suffices to determine a matrix $P$ satisfying the Lyapunov inequality:

$$P(A + BK) + (A + BK)'P < 0$$

In this case, all the ellipsoidal sets $\mathcal{E}(P, \eta) = \{x \in \Re^n \ ; \ x'Px \leq \eta^{-1}\}$ included in $S(K, u_{\min}, u_{\max})$ are invariant and contractive sets with respect to the linear system (1.16), with $w = 0$.

The maximal ellipsoid associated to $P$ and included in $S(K, u_{\min}, u_{\max})$ can be determined from the optimization problem:

$$\max_{x} \ \eta^{-1} = x'Px$$

$$\text{subject to} \quad -u_{\min} \leq Kx \leq u_{\max}$$

The optimal solution of this problem is given by (see also Appendix C):

$$(\eta^\star)^{-1} = \min_{i} \left( \frac{(\min\{u_{\min(i)}, u_{\max(i)}\})^2}{K_{(i)} P^{-1} K'_{(i)}} \right)$$

where $K_{(i)}$ denote the $i$th line of $K$, $i = 1, \ldots, m$.

### 1.6.1.2 Approaches Considering the RAS not Included in the Region of Linearity

Since the region of attraction of system (1.15) is much larger than the linearity region, less conservative estimates are obtained considering regions that are not included in $R_L$. In this case the nonlinear behavior of the closed-loop system has to be taken explicitly into account. Moreover, only the invariance property of a set is not enough to ensure that this set is a RAS. Indeed, an invariant set with respect to the saturated system (1.15) can present equilibrium points or limit cycles inside it. To avoid this and ensure the asymptotic convergence of all trajectories initialized in the region to the origin, a more appropriate concept is the notion of set contractivity.

Thus, the general idea to address Problem 1.1 is to use Lyapunov functions to build regions of stability because they constitute powerful tools for characterizing contractive and invariant sets. If there exists a strictly decreasing Lyapunov function for system (1.15) in a set $\mathcal{D}$, it follows that all the level sets associated to the function, contained in $\mathcal{D}$, are invariant and contractive sets (see Appendix A). In particular, two main kinds of Lyapunov functions have been used to provide estimates of $R_A$ for systems presenting input saturation:

- Quadratic Lyapunov functions leading to ellipsoidal regions of stability (see for instance [2, 134, 172, 193]);
- Polyhedral Lyapunov functions leading to polyhedral regions of stability (see for instance [39, 92, 131, 135, 136, 258, 300, 348]).

More complex Lyapunov functions can also be used as, for example, piecewise quadratic functions [78, 198, 201, 202], convex hull quadratic function and max quadratic function [189, 196], Lure functions [3, 291], polynomial functions [380, 382] and rational functions [75].

## 1.6.2 Asymptotic Stabilization

Complementarily to Problem 1.1, the stabilization problem consists of taking explicitly into account the actuator amplitude limitations and the possibility of saturations in the conception of the control law.

Roughly speaking, it was proven in [326] that a linear system can be globally exponentially stabilized by a bounded input only if it is exponentially stable itself (see also [236]). The other important results established in [236, 326] were that

1. If the plant is not exponentially unstable (namely it has poles only in the closed left-half plane), then it can be globally asymptotically (not necessarily exponentially) stabilized by a bounded input.
2. If the plant is exponentially unstable, then the null controllability region is bounded and global results cannot be achieved. In this case, the null controllability region is unbounded only along the eigenspaces corresponding to eigenvalues with nonpositive real part and is bounded along the remaining ones. Another relevant result was that one of [99] where it was shown that no saturated linear controller can achieve global asymptotic stability for the triple integrator: this reveals the possible need for nonlinear control solutions when dealing with saturated plants (especially if global properties are sought).

Then, the general results on intrinsic limitations for saturated systems are mainly due to the mathematical formalization of reasonable intuition arising when looking at bounded stabilization: saturation can be well thought of as $\text{sat}(v) = g_E(v)v$, where $g_E(\cdot) \in (0, 1]$ is an equivalent gain that becomes smaller and smaller as $v$ grows larger. Then the large signal behavior of a bounded control system can be associated with an equivalent gain which is arbitrarily close to zero, namely the best that one can obtain globally corresponds to the properties of the system with zero control input. This explains why global exponential stability cannot be achieved unless the plant is already exponentially stable. The key to successful design of saturated control laws then becomes the use of one or both of the following two approaches.

1. Consider non-global stabilizing results that will not need to comply with the intrinsic limitations listed above so that the stability and performance/robustness can be achieved for a given set of states. In order words, we will design a control law ensuring that a certain set is included in the associated region of attraction of the closed-loop system. Of course, when the system is exponentially unstable, this is the only possible approach to be considered.
2. Even if the global stabilization is possible, apply preferable nonlinear solutions where the control gains are larger for initial conditions relatively close to the origin and smaller for initial conditions far from it.

To go further with the discussions above, in the following subsections the stabilization problem regarding linear systems subject to control saturation is formalized considering three contexts of stability, namely: global, semi-global and local (regional).

1.6 Problems Considered

## 1.6.2.1 Global Stabilization

First, let us state a formal definition of the global asymptotic stabilization with respect to system (1.1).

**Definition 1.5** System (1.1), with $w = 0$ is said to be globally asymptotically stabilizable (GAS) if and only if, there exists a control signal $u(t)$ satisfying Assumption 1.2 such that

$$\forall x(0) = x_0 \in \Re^n, \quad x(t; x_0) \to 0 \text{ as } t \to \infty$$

where $x(t; x_0)$ denotes the trajectory of system (1.1) initialized in $x_0$.

From the discussions above, it appears that this definition is closely related to the notion of null controllability. Actually, to achieve global stabilization, the region of null controllability of system must be the whole state space $\Re^n$, i.e. system (1.1) must be null controllable with bounded controls (see Definition 1.1). In this case, given a global stabilizing control law $v(t)$, the region of attraction of the closed-loop system

$$\dot{x}(t) = Ax(t) + B \operatorname{sat}(v(t)) \tag{1.18}$$

is the whole state space $\Re^n$.

Solvability conditions for the global asymptotic stabilization problem for a linear system subject to saturating controls have been stated in the seminal works of Sontag and his co-authors: see, for example [329, 338, 403] in the case of continuous-time systems and [402] for the discrete-time systems. These works provide the following formal result.

**Theorem 1.2** *System* (1.1) *subject to input constraints defined in Assumption* 1.2 *is globally stabilizable if and only if*

1. *the pair* $(A, B)$ *is stabilizable*;
2. *the eigenvalues of matrix* $A$ *do not have positive real parts.*

A system satisfying the conditions of Theorem 1.2 is said to be *asymptotically null controllable* [316, 329, 338]. It is important to note that the conditions of Theorem 1.2 only depend on structural properties of the pair $(A, B)$. Provided that conditions 1 and 2 of Theorem 1.2 are satisfied, the system is globally stabilizable for any control magnitude bounds. However, it should be noticed that such conditions are not constructive in terms of suitable control laws.

Hence, provided that conditions of Theorem 1.2 are satisfied, the problem of global stabilization can be stated as follows.

**Problem 1.2** (Global Stabilization) Compute a control law $v(t)$ which ensures that the origin of the closed-loop system (1.18) is globally asymptotically stable.

Regarding Problem 1.2, a natural question arises: is it always possible to determine state feedback control laws defined as

$$u(t) = \mathrm{sat}\big(v(t)\big) = \mathrm{sat}\big(Kx(t)\big), \qquad K \in \Re^{m \times n} \tag{1.19}$$

that globally stabilize system (1.18)? In other words, under the conditions of Theorem 1.2, does there always exist a matrix $K$ such that the closed-loop system (1.18) is globally asymptotically stable?

A negative answer to this question was first given by Fuller [99] and next by Sussmann and Sontag [337] and Ma [254]. In [99] and [337], the authors show that in the continuous-time case no saturating linear state feedback can globally stabilize a chain of integrators of order $n$, with $n \geq 3$. A similar result in the discrete-time case has been proven in [254]. Only the cases of order $n = 1$ and $n = 2$ can be globally stabilized by a control law (1.19) [338].

On the other hand, in the case where the open-loop system is stable in the Lyapunov sense, i.e. (for the continuous-time case) if matrix $A$ presents eigenvalues in the closed left-half complex plane and those on the imaginary axis have the algebraic multiplicity equal to the geometric one, it was proven that a control law as defined in (1.19) can always be determined [49, 50, 292, 341]. In particular, from the Kalman–Yakubovich Lemma and the circle criterion [215], a globally asymptotically stabilizing controller can be found as stated in the next theorem.

**Theorem 1.3** *Given a symmetric positive definite matrix $P \in \Re^{n \times n}$, a matrix $L \in \Re^{m \times n}$ and a matrix $E \in \Re^{n \times n}$ satisfying the following conditions*

1. $A'P + PA - P \leq -L'L - EP$;
2. $EP = PE' \geq 0$;
3. *Pair $(B'P + \sqrt{2}L, A)$ is observable (detectable)*;

*the gain*

$$K = -\Gamma\big(B'P + \sqrt{2}L\big)$$

*with $\Gamma \in \Re^{m \times m}$ being a diagonal matrix such that $0 < \Gamma \leq I_m$, ensures the global stability of the origin of the closed-loop system* (1.18)–(1.19).

In particular, in [292], it has been proven that this class of controllers contains those obtained in [220] or [343]. A similar study has been done through the use of the Popov criterion which has proven that the resulting class contains the two previous ones [292]. On the other hand, in the discrete-time context, if matrix $A$ is asymptotically stable it can be proven that it is always possible to determine a feedback gain which improves the rate of convergence at the origin of the closed-loop system (1.18) with respect to the open-loop system [50, 341]. See also [70, 297].

When $A$ presents eigenvalues on the imaginary axis with the algebraic multiplicity different from the geometric one, i.e. the Jordan blocks associated to the imaginary eigenvalues are of order 2 or greater, only nonlinear control laws can be

1.6 Problems Considered

synthesized. In this case, it appears that global stability can be obtained from the use of nonlinear control laws of the type

$$u_{(i)}(t) = \text{sat}\big(\Psi_i(x(t))\big), \quad \forall i = 1, \dots, m \tag{1.20}$$

where $\Psi_{(i)}(x(t))$ is a nonlinear function of the state at the instant $t$, or of the type

$$u(t) = K\big(x(t)\big)x(t) \tag{1.21}$$

Thus, from some Lyapunov functions (functions of controllability), Gavrilyako et al. [122] obtained a nonlinear controller in an implicit way, in the sense that it satisfies a certain nonlinear equation. Moreover, in [403], a generalization of a nice result due to Teel [365] is proposed and therefore global stability can be obtained from linear combinations and compositions of nested saturations. A new improvement of this second method leads to the use of more general saturation functions [338]. See also other improvements and generalization in this direction in [413]. Finally, in [335], the authors build an optimal state-dependent controller, from the solution of a $\tau$-parameterized family of algebraic Riccati equations. This approach is particularly interesting to cope with performance issues by applying strong (or high) control signals as the trajectory approaches to the origin.

Planar systems, and therefore the particular case of the double integrator, continued to attract the interest of research community due to their practical interest and their particular structure allowing new recipes for designing new families of global stabilizing controllers: see for example [97, 415] and references therein. Beyond the results on global stabilization for linear systems with saturating inputs, some results regarding similar problem for nonlinear open-loop systems can be cited as for instance [293, 333, 336].

### 1.6.2.2 Semi-global Stabilization

The semi-global stabilizability of system (1.1) can be defined as follows.

**Definition 1.6** System (1.1), with $w = 0$, is said to be semi-globally stabilizable, if for any given set of admissible initial conditions $\mathcal{X}_0$, no matter how large it is, it is possible to compute a control law satisfying Assumption 1.2 that ensures the convergence of the closed-loop trajectories to the origin provided that $x(0) \in \mathcal{X}_0$.

Roughly speaking, the concept of semi-global stabilizability can be viewed as an intermediate step between the global and the local stabilization. Indeed, it states that it is possible to ensure the asymptotic stability for an arbitrarily large set of initial conditions, but once this set is defined, nothing can be ensured for trajectories initialized outside it.

The semi-global stabilization problem is summarized as follows.

**Problem 1.3** (Semi-global Stabilization) Given a set of admissible initial states $\mathcal{X}_0$, as large as we want, find a control law $v(t)$ such that all the trajectories of the closed-loop system (1.18), initialized in $\mathcal{X}_0$, asymptotically converge to the origin.

Note that if Problem 1.3 presents a solution, we conclude that

1. the closed-loop system is locally asymptotically stable since $\mathcal{X}_0$ is not the whole state space $\mathfrak{R}^n$;
2. $\mathcal{X}_0$ is included in the region of attraction of the origin of the closed-loop system.

Historically, as evoked in [310], interest in semi-global stabilization derives from the fact that it is not possible to build a globally asymptotically stabilizable linear control law for systems (with bounded controls) with $n \geq 3$ integrators as mentioned in Sect. 1.6.2.1. In particular, in [242], the authors have shown that one can semi-globally (exponentially) stabilize a linear system subject to input magnitude saturation using linear feedback laws if and only if the system is asymptotically null controllable with bounded controls. However, differently from the global case, it is always possible to compute a control law that depends linearly on the state, i.e. $v(t) = Kx(t)$ [5, 242, 309, 366]. It means that the region of attraction of a linear system subject to input magnitude saturation can be made arbitrarily large using adequate tuned linear feedback laws provided that the system is asymptotically null controllable with bounded controls.

In the context of semi-global stabilization, we can extract three main approaches developed for the design of a control law: pole placement, Riccati equation and the use of high and low gains. Let us give a few elements regarding each of these approaches.

*Pole Placement Approach* This approach was developed in [243] in the case of discrete-time systems and in [242] and [5] for continuous-time systems.

Consider the continuous-time case. The first step consists of determining a similarity transformation matrix $T$ which puts matrices $A$ and $B$ of system (1.1) into the following forms:

$$
T^{-1}AT = \begin{bmatrix} A_1 & A_{12} & \cdots & A_{1q} & A_{1s} \\ 0 & A_2 & \cdots & A_{2q} & A_{2s} \\ \vdots & \vdots & \ddots & \vdots & \vdots \\ 0 & 0 & \cdots & A_q & A_{qs} \\ 0 & 0 & \cdots & 0 & A_o \end{bmatrix}
$$

$$
T^{-1}B = \begin{bmatrix} B_1 & 0 & \cdots & 0 & * \\ 0 & B_2 & \cdots & 0 & * \\ \vdots & \vdots & \ddots & \vdots & \vdots \\ 0 & 0 & \cdots & B_q & * \\ 0 & 0 & \cdots & 0 & 0 \end{bmatrix}
$$

where each sub-matrix $A_i$ is in controllable canonical form [64], $B_i = [0\,0\,\dots\,0\,1]'$. The pair $(A_i, B_i)$ is controllable. All the eigenvalues of matrices $A_i$ are supposed to be on the closed left half complex plane while the ones of $A_o$ are uncontrollable and have strictly negative real parts.

For each pair $(A_i, B_i)$, $i = 1, \dots, q$, a "decoupled" gain $K_i(\varepsilon)$ can be determined such that

$$
\lambda\left(A_i + B_i K_i(\varepsilon)\right) = -\varepsilon + \lambda(A_i), \quad 0 < \varepsilon \leq 1
$$

## 1.6 Problems Considered

By applying this procedure to each pair $(A_i, B_i)$, one obtains a state feedback gain parameterized in $\varepsilon$:

$$u(t) = \begin{bmatrix} \mathrm{diag}(K_i(\varepsilon)) & 0 \\ 0 & 0 \end{bmatrix} T^{-1} x(t)$$

It can then be proven that, given any bounded set $\mathcal{X}_0$ containing the origin, one can obtain an upper bound for $\varepsilon$, denoted $\varepsilon^*$, such that $\forall \varepsilon \in (0, \varepsilon^*]$, the semi-global stabilization of the resulting system is obtained without saturation of the control.

*Riccati Equation Approach*  The approach is based on the solution of a Riccati equation parameterized in $\varepsilon$, $\varepsilon > 0$ [309]:

$$A'P(\varepsilon) + P(\varepsilon)A + \varepsilon I_n - P(\varepsilon)BB'P(\varepsilon) = 0 \tag{1.22}$$

It can then be proven that, given any bounded set $\mathcal{X}_0$ containing the origin, it is possible to obtain an upper bound for $\varepsilon$, denoted $\varepsilon^*$, such that $\forall \varepsilon \in (0, \varepsilon^*]$, (1.22) admits a positive definite symmetric solution $P(\varepsilon)$ such that the control law:

$$v(t) = -B'P(\varepsilon)x(t) \tag{1.23}$$

semi-globally stabilizes system (1.18) without control saturation.

Regarding the same context, Teel [366] presents a proof based on the Lyapunov theory of the result proposed by Lin and Saberi [242] with the pole-placement approach (method above described). He builds his control law from the solution $P(\varepsilon)$ of a family of $H_\infty$ type Riccati equations parameterized in $\varepsilon$, $0 < \varepsilon \leq 1$. It is also shown how to use these results for semi-globally stabilizing some classes of nonlinear systems (nonlinear in the state).

For the counterpart of the results for discrete-time systems the reader can refer to [247].

*Low and High Gains Approach*  The main objective was to allow to go beyond semi-global stabilization with some performance issues. Furthermore, this technique tries to fully use the available control capacity [240, 244, 332].

Basically, the control law is computed in two steps:

1. First, the low-gain solution in the form (1.23) is computed in order to ensure the local asymptotic stability with respect to an arbitrarily large set of admissible initial states $\mathcal{X}_0$.
2. The second step consists of tuning a parameter $\rho$ in order to achieve performance, robustness and disturbance rejection.

The final control law takes the form

$$v(t) = -(1 + \rho)B'P(\varepsilon)x(t)$$

In particular, the use of the available control capacity increases with increasing $\rho$. As $\rho \to \infty$, $\mathrm{sat}(v(t))$ tends to bang–bang control.

Some papers dealing with generalization of the previously mentioned results have emerged, focusing for example, on structural properties of the system [311], on sandwich systems [392], and on saturated systems submitted to delays [416].

### 1.6.2.3 Local (Regional) Stabilization

When the open-loop system is exponentially unstable (matrix $A$ has eigenvalues on the open right half complex plane) or when some performance or robustness specifications should be met by the closed-loop system, only local (or regional) stabilization is possible.

In this case, we assume that a set of admissible initial states $\mathcal{X}_0$ is given as design data. This set is one in which the closed-loop system can be initialized or in which the state is driven by the action of additive nonpersistent disturbances.

Hence, the following problem can be stated.

**Problem 1.4** (Local Stabilization) Given a set of admissible initial states $\mathcal{X}_0$, find a control law $v(t)$ such that all the trajectories of the closed-loop system (1.18), initialized in $\mathcal{X}_0$, asymptotically converge to the origin.

Throughout the book we are particularly interested in the synthesis of control laws that depend linearly on the state, i.e. $v(t) = K x(t)$.

One approach to solve Problem 1.4 consists in finding a control law that ensures the positive invariance and contractivity of a set $\mathcal{X}$ containing the origin and such that $\mathcal{X}_0 \subseteq \mathcal{X}$.

Another related synthesis problem regards the computation of a control law that leads to the maximization of the region of attraction of the closed-loop system while ensuring some performance and robustness requirements. However, since the characterization of the region of attraction is in general impossible, the objective becomes one of synthesizing a control law which maximizes an estimate of it.

In the sequel we present a quick tour of the approaches, methods and techniques historically developed in the literature concerning stabilization problems for linear systems with saturating controls in a local (regional) context. As for the analysis problem, we can classify the results into two main approaches: the first one consists of determining control laws in order to avoid control saturation and thus involves to study the linear behavior of the closed-loop system; the second one handles the occurrence of the control saturation and therefore the nonlinear behavior of the closed-loop system. The objective of this second approach is to use all the capacity of the actuator to achieve a good compromise between the associated estimate of the region of attraction and the performance/robustness of the closed-loop system.

**Saturation Avoidance** To ease the presentation, we consider that the control law is a state feedback (i.e. in Problem 1.4 $v(t) = K x(t)$). In this case, according to Theorem 1.1, the solution to Problem 1.4 consists in determining a gain $K$ and a set $\mathcal{X}$ such that:

1. matrix $(A + BK)$ is Hurwitz;
2. $\mathcal{X}_0 \subseteq \mathcal{X} \subseteq S(K, u_{\min}, u_{\max})$;
3. $\mathcal{X}$ is positively invariant with respect to the closed-loop system $\dot{x}(t) = (A + BK)x(t)$.

## 1.6 Problems Considered

**Fig. 1.7** Solution to Problem 1.4: saturation avoidance

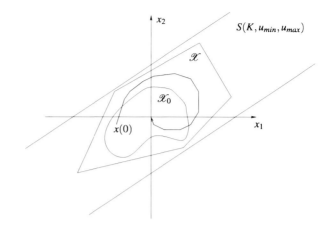

In this case, since $\mathcal{X}_0 \subset \mathcal{X}$ and $\mathcal{X}$ is positively invariant with respect to the linear system, all the trajectories initialized in $\mathcal{X}_0$ do not leave $\mathcal{X}$. Moreover, since $\mathcal{X}$ is included in the region of linearity, the control does not saturate and, provided $(A + BK)$ is Hurwitz, the trajectories converge to the origin. This behavior is depicted in Fig. 1.7.

In the literature, to exhibit such a solution, mainly polyhedral and ellipsoidal sets $\mathcal{X}$ have been considered. In the sequel, we provide a quick overview of the main approaches.

*Linear Programming* In this case, the set $\mathcal{X}$ is considered as a polyhedral set, defined generically as

$$\mathcal{X} = S(G, \omega) = \left\{ x \in \Re^n; \; Gx \leq \omega \right\}, \quad G \in \Re^{g \times n}, \; \omega \in \Re^g \tag{1.24}$$

A necessary and sufficient condition for the positive invariance of $S(G, \omega)$ regards the satisfaction of the relations [31, 58]

$$HG = G(A + BK) \tag{1.25}$$

$$H\omega \preceq 0 \tag{1.26}$$

by a matrix $H$ such that $H_{(i,l)} \geq 0$ if $i \neq l$. For $S(G, \omega)$ given, note that the relations (1.25) and (1.26) are linear in $K$ and $H$. Hence, they can be straightforwardly incorporated as constraints in linear programming schemes to exhibit a suitable matrix $K$ that renders $S(G, \omega)$ positively invariant with respect to the linear system $\dot{x}(t) = (A + BK)x(t)$. In this case, it suffices to add some constraints (also linear in $K$) to ensure the inclusion $S(G, \omega) \subset S(K, u_{\min}, u_{\max})$. Solutions using this approach have been proposed for instance in [166–168, 259, 384].

*Eigenstructure Assignment* The region of linearity $S(K, u_{\min}, u_{\max})$ being a polyhedral set, a problem well studied in the late 1980s consists of determining conditions and methods to compute a gain $K$ leading to the positive invariance of the region of linearity itself or of a polyhedral set having common faces with

$S(K, u_{min}, u_{max})$. The conditions in this case depend on the solution of the equation

$$HK = K(A + BK) \tag{1.27}$$

In particular, it can be shown that a gain $K$ satisfying (1.27) leads to a partial eigenstructure assignment problem. Hence, techniques of eigenstructure assignment can be used to exhibit a solution to the problem. In this context, note the following papers: [22, 23, 32, 58, 61, 169].

Regarding extensions of the approach to the case of output feedback control laws, we can also cite [60] and [345].

*Parameterized ARE and LMI* Regarding the use of a $\varepsilon$-parameterized Riccati equation, like (1.22), the main idea is to derive a nested family of ellipsoidal RAS parameterized by $\varepsilon$, associated to optimal gains $K(\varepsilon)$. Once initialized in the outermost ellipsoid of the previous family, system trajectories are steered to the origin by applying increasing linear gains without causing input saturation. In the innermost ellipsoid, a guaranteed cost can be applied to ensure performance and go beyond stabilization. Examples of works in this sense are [118, 174, 223, 395]. This approach will be further developed in Chap. 5.

On the other hand, a matrix $K$ leading to an ellipsoidal region of asymptotic stability (RAS), defined as

$$\mathcal{E}\left(W^{-1}, \eta\right) = \left\{ x \in \mathfrak{R}^n; \ x'W^{-1}x \leq \eta^{-1} \right\} \tag{1.28}$$

can be straightforwardly computed, thanks to the formulation of a convex problem with the following LMI constraints in variables $W \in \mathfrak{R}^{n \times n}$ and $Y \in \mathfrak{R}^{m \times n}$ [45]:

$$AW + WA' + BY + Y'B < 0 \tag{1.29}$$

$$\begin{bmatrix} W & Y'_{(i)} \\ Y_{(i)} & \eta u_{0(i)}^2 \end{bmatrix} \geq 0, \quad i = 1, \dots, m \tag{1.30}$$

with $u_{0(i)} = \min\{u_{min(i)}, u_{max(i)}\}$. The control gain is given by $K = YW^{-1}$. Note that (1.29) ensures that $V(x) = x'W^{-1}x$ is a Lyapunov function for the linear system $\dot{x}(t) = (A + BK)x(t)$. Moreover, (1.29) ensures that the ellipsoid $\mathcal{E}(W^{-1}, \eta)$ is contained in $S(K, u_{min}, u_{max})$. This approach has been considered for instance in [94], considering dynamic output feedback control laws and in [187], considering a piecewise linear control law.

*Polynomial Approach* In this case, the Youla–Kučera parameterization of all stabilizing controllers in the context of the polynomial approach to control systems is used. On the other hand, standard geometric properties of polyhedra and ellipsoids are considered to come up with a convex programming formulation of the constrained stabilization problem. Some references for this approach are [73, 176, 177].

**Saturation Allowance** In this context, the nonlinear behavior of the closed-loop system must be considered. Thus, the control design problem is solved by considering the effective occurrence of the control saturation.

## 1.6 Problems Considered

**Fig. 1.8** Solution to Problem 1.4: saturation allowance

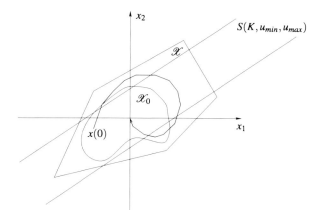

As previously, to ease presentation, we consider that the control law is a state feedback (i.e. in Problem 1.4 $v(t) = Kx(t)$). In this case, according to Definitions 1.3 and 1.4, the solution to Problem 1.4 consists in determining a gain $K$ and a set $\mathcal{X}$ such that:

1. matrix $(A + BK)$ is Hurwitz;
2. $\mathcal{X}_0 \subseteq \mathcal{X} \subseteq R_A$;
3. $\mathcal{X}$ is not included in $S(K, u_{\min}, u_{\max})$.

Since the set $\mathcal{X}$ contains $\mathcal{X}_0$ and is included in the region of attraction, it follows that any trajectory initialized in $\mathcal{X}$ converges asymptotically to the origin. In this case $\mathcal{X}$ is a region of asymptotic stability (RAS).

The conditions above can be fulfilled by computing a set $\mathcal{X}$ positively invariant and contractive with respect to the closed-loop system $\dot{x}(t) = Ax(t) + B\,\text{sat}(Kx(t))$. In this case, the trajectories initialized in $\mathcal{X}_0$ will never leave $\mathcal{X}$ and will converge asymptotically to the origin. Moreover, note that the trajectories can leave the region of linearity, i.e. the effective control saturation can occur. This situation is depicted in Fig. 1.8.

*Remark 1.3* It is important to stress that in this case, differently from the saturation avoidance one, only the guarantee of positive invariance is not sufficient to ensure that the set $\mathcal{X}$ is a RAS for the closed-loop saturated system. Actually, the possible presence of limit cycles or other equilibrium points inside $\mathcal{X}$ must be eliminated. An efficient way to do it is to ensure that $\mathcal{X}$ is indeed contractive. Moreover, it is also important to notice that a necessary condition to the asymptotic stability is that all the eigenvalues of $(A + BK)$ are in the open left-half complex plane, since the system is linear in a neighborhood of the origin.

For the control design problem, to take into account the control saturation occurrence allows, in general, to stabilize larger domains of admissible initial states without an important degradation of the required closed-loop performances. These aspects will be studied and illustrated in the next chapters of the book.

28    1   Linear Systems Subject to Control Saturation—Problems and Modeling

Let us now present a quick overview of the literature. We can first make reference to the approach in [158], where the authors develop a method decomposed into three steps and leading to the following control

$$u(t) = \text{sat}\big((K_1 + K_2(P, L))x(t)\big) \tag{1.31}$$

with $K_1$ and $K_2$ such that

- Matrix $(A + BK_1)$ is asymptotically stable and $\mathcal{X}_0 \subseteq S(K_1, u_{\min}, u_{\max})$.
- $V(x) = x'Px$ is a quadratic Lyapunov function for the system $\dot{x}(t) = (A + BK_1)x(t)$ such that the associated domain $\mathcal{E}(P, \eta) = \{x \in \mathfrak{R}^n;\ x'Px \leq \eta^{-1}\}$ satisfies $\mathcal{X}_0 \subseteq \mathcal{E}(P, \eta) \subseteq S(K_1, u_{\min}, u_{\max})$.
- Matrix gain $K_2(P, L)$ is a function of both $P$ and a matrix $L$ formed by some tuning parameters.

The authors show that this control law asymptotically stabilizes the closed-loop saturated system for any initial state belonging to $\mathcal{X}_0$. The saturation is used in order to improve the time response in terms of speed of convergence. This is done by modifying the parameters of matrix $L$. The main drawback of this method is the fact that it is not a systematic one. Moreover in the multivariable case, this method is not generic and the control law (1.31) is not always a stabilizing control law. Other more restrictive conditions have to be verified in this case.

From this preliminary technique, several new conditions have emerged thanks to more and more sophisticated modeling of the saturation nonlinearities: namely, polytopic models [1, 188, 346] and sector nonlinearity models [359, 361] (see Sect. 1.7 for a description of these models).

It is important to emphasize that most of the published results on regional (local) stability/stabilization use quadratic Lyapunov functions and ellipsoidal regions of stability (see, for example, [120, 137, 175, 183, 193, 389, 400]). The main advantages of using quadratic Lyapunov functions regard the fact that stabilization conditions can be easily expressed in form of LMIs or BMIs and then the synthesis of the control law can be carried out through the solution of convex optimization problems that can be solved by efficient commercial software packages. Some techniques and the key aspects regarding use of quadratic Lyapunov functions to synthesize stabilizing control laws will be deeply discussed in the later chapters.

*Remark 1.4* As will be discussed in the later chapters, in addition to the asymptotic stability objective, requirements on robustness and performance of the closed-loop system can also be considered in Problems 1.2, 1.3 and 1.4.

### 1.6.3  External Stability Analysis

Since the closed-loop system (1.15) is nonlinear, the action of the exogenous signal can produce trajectories that converge to other equilibrium points (different from the origin), to limit cycles, or that can even diverge, i.e. that leads to $\lim_{t \to \infty} x(t) = \infty$.

## 1.6 Problems Considered

**Fig. 1.9** External stability

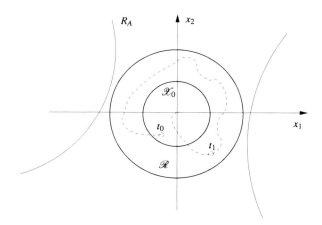

This fact has been illustrated in Example 1.2. In this case we are interested in determining sets of "admissible" exogenous signals. Hence, if $w(t)$ is admissible it means that the corresponding state trajectory is bounded, provided that the initial condition is also admissible. These admissible sets are in general characterized with respect to bounds on the amplitude ($\mathcal{L}_\infty$-norm) and/or the energy ($\mathcal{L}_2$-norm) of $w(t)$. This problem is also referred to as input-to-state stability analysis [215].

On the other hand, if the signal $w(t)$ is vanishing, i.e $w(t) = 0$ for some $t > t_1$ or $\lim_{t \to \infty} w(t) = 0$, the convergence of the trajectories to the equilibrium point $x = 0$ must be guaranteed. This happens if the origin of the saturated system is asymptotically stable and the trajectory generated by $w(t)$ does not leave the region of attraction of system (1.15).

Hence, considering a set $\mathcal{W}$ of admissible exogenous signals and a set $\mathcal{X}_0$ of admissible initial conditions we are generically interested in the following problem.

**Problem 1.5** Given sets $\mathcal{W}$ and $\mathcal{X}_0$, ensure that for all $w(t) \in \mathcal{W}$ and all $x(0) \in \mathcal{X}_0$:

1. the trajectories of the closed-loop system (1.15) are bounded, i.e. they are confined in some compact set $\mathcal{R}$;
2. if the disturbance is vanishing, then $\lim_{t \to \infty} x(t) = 0$.

The set $\mathcal{R}$ can be seen as an estimate of the reachable states considering $w(t) \in \mathcal{W}$ and $x(0) \in \mathcal{X}_0$. Figure 1.9 depicts the situations that can arise in Problem 1.5. It is supposed that at the instant $t_0$ the signal $w(t)$ starts to act. At this time the initial condition belongs to $\mathcal{X}_0$. The trajectory does not leave the region $\mathcal{R}$ which is contained in the region of attraction of the origin $R_A$. Hence, if $w(t) = 0$, for $t \geq t_1$ the state asymptotically converges to the origin.

Moreover, we may be interested in estimating the maximal set of admissible exogenous signals $\mathcal{W}$, defined for a class of disturbances $w(t)$. This problem can be viewed as a *disturbance tolerance* analysis problem.

Analogously, given a set $\mathcal{W}$ of admissible disturbances, we can be interested in determining an estimate of the maximal set of admissible states $\mathcal{X}_0$ for which the trajectories are bounded.

On the other hand, when the open-loop is null controllable, we may be interested in testing if the external stability holds in a global sense. In this case, for any bounded disturbance $w(t)$ and for any initial condition $x(0) \in \mathfrak{R}^n$ (i.e. for $\mathcal{X}_0 = \mathfrak{R}^n$), it should be ensured that the state trajectories are bounded.

## 1.6.4 External Stabilization

The external stabilization problem considers the synthesis of the control law taking explicitly into account the control saturation, in order to ensure the boundedness of the state trajectories when the system is subject to external bounded signals.

Hence, the following synthesis problem can be stated as follows.

**Problem 1.6** Given a set of admissible initial states $\mathcal{X}_0$ and a set $\mathcal{W}$ defined for a specified class of signals $w(t)$, determine a control law $v(t)$ such that

1. the trajectories of the closed-loop system (1.15) are bounded, i.e. they are confined in some compact set $\mathcal{R}$;
2. if the exogenous signal is vanishing, then $\lim_{t \to \infty} x(t) = 0$.

For the general Problem 1.6, some optimization sub-problems are of particular interest:

*Maximization of the Disturbance Tolerance* Given a set of admissible initial states $\mathcal{X}_0$, determine a control law $v(t)$ in order to maximize the set of admissible disturbances $\mathcal{W}$ for which the corresponding trajectories of the closed-loop system are guaranteed bounded.

*Maximization of the Set of Admissible Initial States* Given a set of admissible exogenous signals $\mathcal{W}$, determine a control law $v(t)$ in order to maximize the set of admissible initial conditions $\mathcal{X}_0$ for which the corresponding trajectories of the closed-loop system are guaranteed bounded.

*Maximization of the Disturbance Rejection* Given a set $\mathcal{W}$ of admissible exogenous signals and a regulated output $z(t)$, minimize the gain (measured with respect to some norm) between $w(t)$ and $z(t)$.

In particular, when the open-loop system is asymptotically stable (or null controllable), Problem 1.6 can be addressed in order to achieve *global external/internal asymptotic stabilization* of the origin. In this case, the control law $v(t)$ has to be computed in order to guarantee that for any bounded disturbance $w(t)$ and $\forall x(0) \in \mathfrak{R}^n$ the trajectories are bounded.

In the literature, most of the works dealing with Problem 1.6 follow a quadratic approach. Considering an ellipsoidal set of initial conditions, and a set of admissible

## 1.6 Problems Considered

disturbances, a control law is computed to ensure that the trajectories are confined in an ellipsoidal set (i.e. the reachable set is an ellipsoid). In this context, we can cite for instance [223, 271, 273], where magnitude bounded disturbances are considered and the problem of maximization of the disturbance rejection is treated in terms of the $\mathcal{L}_2$-gain; [193, 353], where the problems of disturbance tolerance and maximization of the set of admissible initial conditions are considered for a class of magnitude bounded disturbances; and [62, 89, 120, 146, 280] where the three optimization problems above described are addressed considering $\mathcal{L}_2$-bounded disturbances.

On the other hand, the particular case of Problem 1.6 when the exogenous signal is input-additive, i.e. the closed-loop system is given by

$$\dot{x}(t) = Ax(t) + B \operatorname{sat}\big(Kx(t) + w(t)\big)$$

has been studied in [246, 248]. In those papers, the so-called *almost disturbance decoupling problem* is addressed. Considering that the disturbance is $\mathcal{L}_p$-bounded, a low-and-high gain design method is proposed to compute a state feedback assuring:

- the (local, global or semi-global) asymptotic stability of the origin in the absence of the exogenous signal $w(t)$;
- the $\mathcal{L}_p$-gain from $w(t)$ to $x(t)$ is finite, i.e.

$$\|x\|_{\mathcal{L}_p} \le \gamma \|w\|_{\mathcal{L}_p}$$

### 1.6.5 The Anti-windup Problem

The "windup" term is historically related to the actuator saturation effects in control-loops containing integral action [127, 184]. In this case, when the saturation occurs the integral term remains charging unnecessarily. The output of the integrator is said to "wind-up". This unnecessary charge (above the saturation limits) produces undesirable effects on the transient response of the system, such as oscillations and slow convergence to the equilibrium point. The same effect can be noticed when the controller has slow modes or critically stable modes (poles on the imaginary axis). In order to avoid this performance degradation generated by the saturation, an additional control-loop, called *anti-windup* compensation can be added to the closed-loop system.

The basic principle of anti-windup strategies consists of determining the difference between the signal computed by the controller ($v(t)$) and the signal effectively applied to the controlled plant ($\operatorname{sat}(v(t))$). This difference is then fed back to the controller, through a static or a dynamic system called an *anti-windup compensator*.

Generically, the anti-windup compensator can be described by the following state space representation:

$$\dot{x}_{\mathrm{aw}}(t) = A_{\mathrm{aw}} x_{\mathrm{aw}}(t) + B_{\mathrm{aw}}\big(\operatorname{sat}(v(t)) - v(t)\big) \tag{1.32}$$

$$\begin{bmatrix} y_{\mathrm{aw}_1}(t) \\ y_{\mathrm{aw}_2}(t) \end{bmatrix} = C_{\mathrm{aw}} x_{\mathrm{aw}}(t) + D_{\mathrm{aw}}\big(\operatorname{sat}(v(t)) - v(t)\big) \tag{1.33}$$

32    1 Linear Systems Subject to Control Saturation—Problems and Modeling

If $A_{aw} = 0$, $B_{aw} = 0$ and $C_{aw} = 0$, the anti-windup scheme is said to be *static*. The signals $y_{aw_1}$ and $y_{aw_2}$ are therefore injected in the nominal controller:

$$\dot{x}_c(t) = A_c x_c(t) + B_c y(t) + y_{aw_1}(t) \qquad (1.34)$$

$$v(t) = C_c x_c(t) + D_c y(t) + y_{aw_2}(t) \qquad (1.35)$$

The anti-windup signal works in the sense of "correcting" the states of the nominal controller in order to provide a signal $v(t)$ that remains close to the saturation limits. It should be pointed out that in absence of saturation, since $\mathrm{sat}(v(t)) - v(t) = 0$, the anti-windup loop is not active and the closed-loop system behaves with the performance determined by the pre-computed nominal controller. The anti-windup compensator can then be seen as an "extra" control loop that is "appended" to a pre-computed controller.

Although anti-windup compensation has been mainly related to performance improvement, it can also be used to enlarge the region of attraction (or an estimate of it) of the saturated closed-loop system. Hence, the anti-windup problem can be generically defined as follows.

**Problem 1.7** (Anti-windup) Given a pre-computed nominal controller, which ensures that the closed-loop system is asymptotically stable in absence of control saturation, determine a compensator (1.32)–(1.33), in order to address one or more of the following objectives:

- minimize the difference between the behavior of the closed-loop system when the saturation occurs and the nominal (linear) behavior;
- improve the transient response of the states (or of a regulated output) due to the effect of disturbances;
- maximize the region of attraction of the closed-loop system.

Many different anti-windup methods, structures and design techniques can be found in the literature to address the problems above [108, 226, 352]. A general overview of the possible solutions for the anti-windup problem is given in Chaps. 6 and 7, with applications and extensions in Chap. 8.

## 1.7 Models for the Saturation Nonlinearity

Although at first glance the saturation looks like a simple nonlinearity, its mathematical treatment to obtain stability and stabilization conditions is rather complicated. In fact, the saturation can be classified as a "hard nonlinearity" as is the case of the dead-zone, the hysteresis, the backlash, etc.

Hence, in order to derive "tractable" conditions to provide solutions to the problems stated in Sect. 1.6, appropriate overboundings and representations of the saturation term are needed. For these purposes, we next present three useful representations for the closed-loop system in the presence of saturation:

1.7 Models for the Saturation Nonlinearity 33

(R1) The first one is based on the use of polytopic differential inclusions. It involves a local description of the saturated closed-loop system through a polytopic model. This allows stability and stabilization problems to be treated using robust control approaches. In particular, three types of polytopic modeling are presented.
(R2) The second representation involves re-writing the closed-loop system and re-placing the saturation term by a dead-zone nonlinearity. Hence, sector conditions, locally or globally valid, can be used to relax stability and stabilization conditions. In particular, two types of sector conditions are presented.
(R3) The third representation involves dividing the state space into what we call *regions of saturation*. Inside each one of these regions, the system dynamics is described by an affine linear system. This allows the problems to be treated using tools from hybrid systems.

For simplicity, the representations are presented considering the generic closed-loop system (1.15).

*Remark 1.5* We can also cite a fourth representation, which is, however, never used in this book. It consists in modeling the saturation as norm-bounded uncertainty (see for instance [141, 144, 302, 318]). In that case, the closed-loop system with saturation can be written in an LFT form.

## *1.7.1 Polytopic Models*

In this subsection, locally valid polytopic representations of the nonlinear system (1.15) are described. In particular, we present three kinds of polytopic differential inclusions. The first one can be considered "more intuitive". Historically, this approach was at the origin of the use of differential inclusions to study saturated systems, in the late 1990s. The second and the third ones can be seen as generalizations of the first one and are, in general, less conservative.

The use of these polytopic representations to derive stability and stabilization conditions for the saturated system (1.15) are detailed in Chap. 2.

### 1.7.1.1 Polytopic Model I

This type of representation based on the use of differential inclusions [13] was introduced in [262]. Its application to the study of stability of saturated systems (1.15) can be found in [130, 137, 141, 172, 272]. It uses the fact that the saturation term (1.6) can be written as

$$\text{sat}\big(Kx(t)\big) = \Gamma\big(\alpha(x)\big)Kx(t) \tag{1.36}$$

where $\Gamma(\alpha(x))$ is a diagonal matrix whose diagonal elements are defined for $i = 1, \ldots, m$ as

$$\alpha_{(i)}\big(x(t)\big) = \begin{cases} \frac{u_{\max(i)}}{K_{(i)}x(t)} & \text{if } K_{(i)}x(t) > u_{\max(i)} \\ 1 & \text{if } -u_{\min(i)} \leq K_{(i)}x(t) \leq u_{\max(i)} \\ \frac{-u_{\min(i)}}{K_{(i)}x(t)} & \text{if } K_{(i)}x(t) < -u_{\min(i)} \end{cases} \tag{1.37}$$

By definition, one gets $0 < \alpha_{(i)}(x(t)) \leq 1$, $\forall i = 1, \ldots, m$, $\forall x \in \Re^n$. Note that the scalar $\alpha_{(i)}(x(t))$ can be viewed as an indicator of the saturation degree of the $i$th entry of the control signal $u$. Actually, the smaller $\alpha_{(i)}(x(t))$ is, the farther $x(t)$ is from the region of linearity $S(K, u_{\min}, u_{\max})$ defined in (1.17). When $\alpha_{(i)}(x(t)) = 1$, $u_{(i)}$ is not saturated.

Thus from (1.36) and (1.37), system (1.15) becomes

$$\dot{x}(t) = \big(A + B\Gamma\big(\alpha\big(x(t)\big)\big)K\big)x(t) + B_w w(t) = \mathcal{A}_t x(t) + B_w w(t) \tag{1.38}$$

where the matrix $\mathcal{A}_t$ is function of $\alpha(x(t))$ and therefore of $x(t)$.

Consider now scalars $0 < \alpha_{l(i)} \leq 1$, $i = 1, \ldots, m$, and define the following polyhedral set:

$$S\big(K, u_{\min}^{\alpha}, u_{\max}^{\alpha}\big) = \big\{x \in \Re^n; \ -u_{\min}^{\alpha} \preceq Kx \preceq u_{\max}^{\alpha}\big\} \tag{1.39}$$

with

$$-u_{\min(i)}^{\alpha} = \frac{-u_{\min(i)}}{\alpha_{l(i)}} \quad \text{and} \quad u_{\max(i)}^{\alpha} = \frac{u_{\max(i)}}{\alpha_{l(i)}}, \quad i = 1, \ldots, m$$

From the definition of $\alpha_{(i)}(x(t))$ in (1.37), we can conclude that for all $x(t) \in S(K, u_{\min}^{\alpha}, u_{\max}^{\alpha})$, it follows that

$$0 < \alpha_{l(i)} \leq \alpha_{(i)}\big(x(t)\big) \leq 1, \quad \forall i = 1, \ldots, m$$

Associated with the set $S(K, u_{\min}^{\alpha}, u_{\max}^{\alpha})$, we can define $2^m$ diagonal matrices $\Gamma_j(\alpha_l)$, whose diagonal components take the value 1 or $\alpha_{l(i)}$. For instance, for $m = 2$, one has

$$\Gamma_1(\alpha_l) = \begin{bmatrix} 1 & 0 \\ 0 & 1 \end{bmatrix}; \qquad \Gamma_2(\alpha_l) = \begin{bmatrix} \alpha_{l(1)} & 0 \\ 0 & 1 \end{bmatrix}$$

$$\Gamma_3(\alpha_l) = \begin{bmatrix} \alpha_{l(1)} & 0 \\ 0 & \alpha_{l(2)} \end{bmatrix}; \qquad \Gamma_4(\alpha_l) = \begin{bmatrix} 1 & 0 \\ 0 & \alpha_{l(2)} \end{bmatrix}$$

Hence, for each $x \in S(K, u_{\min}^{\alpha}, u_{\max}^{\alpha})$ the value of $\text{sat}(Kx)$ can be exactly computed as a convex combination of the vectors $\Gamma_j(\alpha_l)Kx$, $j = 1, \ldots, 2^m$. The following lemma can thus be stated.

**Lemma 1.1** *If $x \in S(K, u_{\min}^{\alpha}, u_{\max}^{\alpha})$, then it follows that*

$$\text{sat}(Kx) \in \text{Co}\big\{\Gamma_j(\alpha_l)K, \ j = 1, \ldots, 2^m\big\} \tag{1.40}$$

1.7 Models for the Saturation Nonlinearity                                                35

It should be noticed that the equality $\text{sat}(Kx) = \sum_{j=1}^{2^m} \lambda_j(x)\Gamma_j(\alpha_l)Kx$ holds in fact for a particular convex combination that depends on the value of $x$, i.e. $\sum_{j=1}^{2^m} \lambda_j(x) = 1, 0 \le \lambda_j(x) \le 1, j = 1, \ldots, 2^m$.

From the matrices $\Gamma_j(\alpha_l)$, define now the following vertex matrices:

$$\mathbb{A}_j = A + B\Gamma_j(\alpha_l)K, \quad j = 1, \ldots, 2^m \tag{1.41}$$

Then, $\forall x(t) \in S(K, u_{\min}^\alpha, u_{\max}^\alpha)$, matrix $\mathcal{A}_t$ in (1.38) belongs to a convex hull of matrices whose vertices are $\mathbb{A}_j$. Consequently, $\forall x(t) \in S(K, u_{\min}^\alpha, u_{\max}^\alpha)$, $\dot{x}(t)$ can be determined from the following polytopic model

$$\dot{x}(t) = \sum_{j=1}^{2^m} \lambda_j\big(x(t)\big)\mathbb{A}_j x(t) + B_w w(t) \tag{1.42}$$

with $\sum_{j=1}^{2^m} \lambda_j(x(t)) = 1, 0 \le \lambda_j(x(t)) \le 1, j = 1, \ldots, 2^m$.

Note that $S(K, u_{\min}^\alpha, u_{\max}^\alpha)$ corresponds to an expansion of the linearity region. In fact $S(K, u_{\min}, u_{\max}) \subseteq S(K, u_{\min}^\alpha, u_{\max}^\alpha)$.

*Remark 1.6* The set $S(K, u_{\min}^\alpha, u_{\max}^\alpha)$ corresponds to the maximal set in which the polytopic model (1.42) can represent system (1.15). This means that all the trajectories of the system (1.15) that are confined in $S(K, u_{\min}^\alpha, u_{\max}^\alpha)$ can be generated by the polytopic model (1.42). This is why we refer to this representation as a "local" one. However, note that the converse is not true. Indeed, some trajectories of the polytopic model are not trajectories of the saturated system. Hence, this kind of modeling introduces some conservatism when it is used to derive stability and stabilization conditions for the saturated system (1.15).

It should be noticed that, without further conditions, if $x(0) \in S(K, u_{\min}^\alpha, u_{\max}^\alpha)$, we cannot ensure that the trajectory does not leave this set.

### 1.7.1.2 Polytopic Model II

This representation has been originally proposed by Hu et al. in [188, 192, 193]. The basic idea consists in using an auxiliary vector variable $h$, and to compose the output of the saturation function as a convex combination of the actual control signals $v$ and $h$.

In order to intuitively present the approach, let us consider first the mono-input case, i.e. $m = 1$. In this case, let $v$ and $h$ be scalars. Suppose now that

$$-u_{\min} \le h \le u_{\max}$$

Hence, it follows that the scalar function $\text{sat}(v)$ can be computed as a convex combination of $h$ and $v$, i.e.:

$$\text{sat}(v) = \lambda v + (1 - \lambda)h$$

with $0 \le \lambda \le 1$, or, equivalently

$$\text{sat}(v) \in \text{Co}\{v, h\}$$

This fact is illustrated in Fig. 1.10.

36                    1  Linear Systems Subject to Control Saturation—Problems and Modeling

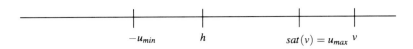

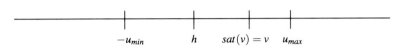

**Fig. 1.10** Polytopic model II—reasoning

Consider now $m = 2$ and define the vectors $v = \begin{bmatrix} v_{(1)} \\ v_{(2)} \end{bmatrix}$ and $h = \begin{bmatrix} h_{(1)} \\ h_{(2)} \end{bmatrix}$, such that

$$-u_{\min(1)} \leq h_{(1)} \leq u_{\max(1)}$$
$$-u_{\min(2)} \leq h_{(2)} \leq u_{\max(2)}$$

Hence, from convexity arguments, we can conclude that

$$sat(v) \in \text{Co}\left\{ \begin{bmatrix} v_{(1)} \\ v_{(2)} \end{bmatrix}, \begin{bmatrix} v_{(1)} \\ h_{(2)} \end{bmatrix}, \begin{bmatrix} h_{(1)} \\ v_{(2)} \end{bmatrix}, \begin{bmatrix} h_{(1)} \\ h_{(2)} \end{bmatrix} \right\}$$

Define now diagonal matrices $\Gamma_j^+$ and $\Gamma_j^-$ as follows:

- $\Gamma_j^+$ are diagonal matrices whose diagonal elements take the value 1 or 0, $j = 1, \ldots, 2^m$.
- $\Gamma_j^- = I_m - \Gamma_j^+$, $j = 1, \ldots, 2^m$.

For instance, for $m = 2$ the matrices $\Gamma_j^+$ and $\Gamma_j^-$ are defined by

$$\Gamma_1^+ = \begin{bmatrix} 1 & 0 \\ 0 & 1 \end{bmatrix}; \quad \Gamma_2^+ = \begin{bmatrix} 1 & 0 \\ 0 & 0 \end{bmatrix}; \quad \Gamma_3^+ = \begin{bmatrix} 0 & 0 \\ 0 & 1 \end{bmatrix}; \quad \Gamma_4^+ = \begin{bmatrix} 0 & 0 \\ 0 & 0 \end{bmatrix}$$

$$\Gamma_1^- = \begin{bmatrix} 0 & 0 \\ 0 & 0 \end{bmatrix}; \quad \Gamma_2^- = \begin{bmatrix} 0 & 0 \\ 0 & 1 \end{bmatrix}; \quad \Gamma_3^- = \begin{bmatrix} 1 & 0 \\ 0 & 0 \end{bmatrix}; \quad \Gamma_4^- = \begin{bmatrix} 1 & 0 \\ 0 & 1 \end{bmatrix}$$

The following Lemma generalizes the reasoning above for a vector-valued saturation function.

1.7 Models for the Saturation Nonlinearity 37

**Lemma 1.2** [193] *Consider two vectors $v \in \Re^m$ and $h \in \Re^m$. If $-u_{\min(i)} \leq h_{(i)} \leq u_{\max(i)}$, for $i = 1, \ldots, m$, then it follows that*

$$\text{sat}(v) \in \text{Co}\{\Gamma_j^+ v + \Gamma_j^- h, \; j = 1, \ldots, 2^m\}$$

Suppose now that $v(t) = Kx(t)$ and define the following polyhedral set:

$$S(H, u_{\min}, u_{\max}) = \{x \in \Re^n; \; -u_{\min} \preceq Hx \preceq u_{\max}\}$$

From Lemma 1.2, $\forall x(t) \in S(H, u_{\min}, u_{\max})$ it follows that

$$\text{sat}(Kx(t)) \in \text{Co}\{\Gamma_j^+ Kx(t) + \Gamma_j^- Hx(t), \; j = 1, \ldots, 2^m\}$$

and, as a consequence, $\dot{x}(t)$ can be computed from the following polytopic model:

$$\dot{x}(t) = \sum_{j=1}^{2^m} \lambda_j(x(t)) \mathbb{A}_j x(t) + B_w w(t) \tag{1.43}$$

with $\sum_{j=1}^{2^m} \lambda_j(x(t)) = 1, \, 0 \leq \lambda_j(x(t)) \leq 1, \, j = 1, \ldots, 2^m$, and

$$\mathbb{A}_j = A + B(\Gamma_j^+ K + \Gamma_j^- H)$$

The same comments stated at the end of Sect. 1.7.1.1 now apply to the polytopic model (1.43) and the region $S(H, u_{\min}, u_{\max})$.

Note, however, that now there is no direct relationship between the region of validity of the differential inclusion $S(H, u_{\min}, u_{\max})$ and the region of linearity. Furthermore, for the particular choice $H = \text{diag}(\alpha_l)K$, the polytopic model I is obtained. Hence, polytopic model II can be viewed as a generalization of polytopic model I. As will be seen in Chap. 2, the matrix $H$ introduces new degrees of freedom in the analysis and synthesis conditions leading to less conservative results.

### 1.7.1.3 Polytopic Model III

This model can be seen as a generalization of the polytopic approach II. It has been proposed originally by Alamo et al. in [1] and [2].

As seen in the previous polytopic models, supposing that we have $m$ control inputs, we can define $2^m$ possible combinations between saturated and non-saturated control entries. Associated to these combinations, define the sets $S_j, j = 1, \ldots, 2^m$, of indices $i \in \mathcal{M} = \{1, \ldots, m\}$, as follows:

- by definition, the combination associated to the case in which all the control inputs are not saturated corresponds to $S_1 = \emptyset$;
- for the $j$th combination, if the $i$th control input is saturated then $i \in S_j$.

Denote also $S_j^c$ as the complementary of $S_j$ in the set $\mathcal{M}$, i.e. if $i \notin S_j$ then $i \in S_j^c$. For instance, if $m = 2$ the following sets are defined:

$$S_1 = \emptyset; \qquad S_2 = \{1\}; \qquad S_3 = \{2\}; \qquad S_4 = \{1, 2\}$$
$$S_1^c = \{1, 2\}; \qquad S_2^c = \{2\}; \qquad S_3^c = \{1\}; \qquad S_4^c = \emptyset$$

Associated to the region $\mathcal{S}_j$, consider the following scalars $h_{j(i)}$, $i = 1, \ldots, m$:

$$
\begin{aligned}
-u_{\min(i)} &\leq h_{j(i)} \leq u_{\max(i)} \quad \text{if } i \in \mathcal{S}_j \\
h_{j(i)} &= 0 \quad \text{if } i \notin \mathcal{S}_j
\end{aligned}
\tag{1.44}
$$

From convexity arguments, similar to the ones used in Sect. 1.7.1.2, we can therefore write

$$
\text{sat}(v_{(i)}) = \lambda_1 v_{(i)} + \sum_{j=2}^{2^m} \lambda_{j(i)} h_{j(i)}
\tag{1.45}
$$

with $\sum_{j=1}^{2^m} \lambda_j = 1$, $0 \leq \lambda_j \leq 1$, $j = 1, \ldots, 2^m$.

To develop the reasoning, consider now $m = 2$. In this case, associated to each region $\mathcal{S}_j$, we can define the following vectors:

- $\mathcal{S}_1 : r_1 = B_1 v_{(1)} + B_2 v_{(2)} = \sum_{i \in \mathcal{S}_1^c} B_i v_{(i)} + \sum_{i \in \mathcal{S}_1} B_i h_{1(i)}$
- $\mathcal{S}_2 : r_2 = B_1 h_{2(1)} + B_2 v_{(2)} = \sum_{i \in \mathcal{S}_2^c} B_i v_{(i)} + \sum_{i \in \mathcal{S}_2} B_i h_{2(i)}$
- $\mathcal{S}_3 : r_3 = B_1 v_{(1)} + B_2 h_{3(2)} = \sum_{i \in \mathcal{S}_3^c} B_i v_{(i)} + \sum_{i \in \mathcal{S}_3} B_i h_{3(i)}$
- $\mathcal{S}_4 : r_4 = B_1 h_{4(1)} + B_2 h_{4(2)} = \sum_{i \in \mathcal{S}_4^c} B_i v_{(i)} + \sum_{i \in \mathcal{S}_4} B_i h_{4(i)}$

with $B_i$ standing for the $i$th column of $B$.

From (1.45), it follows that there exist $\alpha_j$ and $\beta_j$, $j = 1, \ldots, 4$, such that $\sum_{j=1}^4 \alpha_j = 1$, $0 \leq \alpha_j \leq 1$, $\sum_{j=1}^4 \beta_j = 1$, $0 \leq \beta_j \leq 1$, verifying the following equalities:

$$
\begin{aligned}
\text{sat}(v_{(1)}) &= \alpha_1 v_{(1)} + \alpha_2 h_{2(1)} + \alpha_3 v_{(1)} + \alpha_4 h_{4(1)} \\
\text{sat}(v_{(2)}) &= \beta_1 v_{(2)} + \beta_2 v_{(2)} + \beta_3 h_{3(2)} + \beta_4 h_{4(2)}
\end{aligned}
\tag{1.46}
$$

Noting that

$$
\sum_{j=1}^4 \sum_{k=1}^4 \alpha_j \beta_k = \sum_{j=1}^4 \alpha_j \sum_{k=1}^4 \beta_k = \sum_{k=1}^4 \beta_k \sum_{j=1}^4 \alpha_j = \sum_{j=1}^4 \lambda_j = 1
$$

with $\sum_{j=1}^4 \lambda_j = 1$, $0 \leq \lambda_j \leq 1$, we can write

$$
\begin{aligned}
\sum_{j=1}^4 \lambda_j r_j &= \sum_{k=1}^4 \beta_k B_1 (\alpha_1 v_{(1)} + \alpha_2 h_{2(1)} + \alpha_3 v_{(1)} + \alpha_4 h_{4(1)}) \\
&\quad + \sum_{j=1}^4 \alpha_j B_2 (\beta_1 v_{(2)} + \beta_2 v_{(2)} + \beta_3 h_{3(2)} + \beta_4 h_{4(2)}) \\
&= B_1 \text{sat}(v_{(1)}) + B_2 \text{sat}(v_{(2)}) \\
&= B \text{sat}(v)
\end{aligned}
$$

## 1.7 Models for the Saturation Nonlinearity

Hence, by induction, using convexity arguments, it follows that there exists a convex combination given by $\lambda_j$, $j = 1, \ldots, 2^m$ such that

$$B \, \text{sat}(v) = \sum_{j=1}^{2^m} \lambda_j \left( \sum_{i \in \mathcal{S}_j^c} B_i v_{(i)} + \sum_{i \in \mathcal{S}_j} B_i h_{j(i)} \right)$$

provided $h_{j(i)}$ satisfies (1.44), $\forall j = 1, \ldots, 2^m$.

Consider now $v = Kx$ and $h_{j(i)} = H_{j(i)}x$, with $H_{j(i)} \in \Re^{1 \times m}$ and define the following sets, for $j = 2, \ldots, 2^m$:

$$S(H_j, u_{\min}, u_{\max}) = \left\{ x \in \Re^n; \ -u_{\min(i)} \leq H_{j(i)}x \leq u_{\max(i)}, \ \forall i \in \mathcal{S}_j \right\}$$

The following lemma can be stated.

**Lemma 1.3** *If* $x \in \bigcap_{j=2}^{2^m} S(H_j, u_{\min}, u_{\max})$ *then*

$$B \, \text{sat}(Kx) \in \text{Co} \left\{ \sum_{i \in \mathcal{S}_j^c} B_i K_{(i)}x + \sum_{i \in \mathcal{S}_j} B_i H_{j(i)}x, \ j = 1, \ldots, 2^m \right\}$$

Actually, if $x \in \bigcap_{j=2}^{2^m} S(H_j, u_{\min}, u_{\max})$ we can conclude that there exists $\sum_{j=1}^{2^m} \lambda_j = 1, 0 \leq \lambda_j \leq 1, j = 1, \ldots, 2^m$ such that

$$B \, \text{sat}(Kx) = \sum_{j=1}^{2^m} \lambda_j(x) \left( \sum_{i \in \mathcal{S}_j^c} B_i K_{(i)}x + \sum_{i \in \mathcal{S}_j} B_i H_{j(i)}x \right)$$

or equivalently, the trajectories of the closed-loop system (1.15) can be generated from the following polytopic model:

$$\dot{x}(t) = \sum_{j=1}^{2^m} \lambda_j(x(t)) \mathbb{A}_j x(t) + B_w w(t) \tag{1.47}$$

with

$$\mathbb{A}_j = A + \sum_{i \in \mathcal{S}_j^c} B_i K_{(i)} + \sum_{i \in \mathcal{S}_j} B_i H_{j(i)}$$

The same comments stated at the end of Sect. 1.7.1.2 now apply to the polytopic model (1.47) and the regions $S(H_j, u_{\min}, u_{\max})$. In fact, polytopic model II appears as a particular case of polytopic model III. It suffices to consider $H_{j(i)} = H_{(i)}, \forall j$. In other words, the matrix $\mathbb{A}_j$ in the polytopic model II can be equivalently written as

$$\mathbb{A}_j = A + \sum_{i \in \mathcal{S}_j^c} B_i K_{(i)} + \sum_{i \in \mathcal{S}_j} B_i H_{(i)}$$

The fact to consider different $H_{j(i)}$ for each $j$ introduces new degrees of freedom in the analysis and synthesis conditions, leading in general to less conservative results.

## 1.7.2 Sector Nonlinearity Models

The problem of stability analysis of a saturated system (1.15) can be seen as a Lure problem [215, 290]. Basically, this problem considers the stability of a feedback loop containing a linear asymptotic stable system and a decentralized memoryless nonlinearity satisfying a sector condition (see Appendix A).

The saturation function satisfies these requirements, since it is memoryless and decentralized, and the closed-loop system can be re-written as

$$\dot{x}(t) = Ax(t) + Bu(t) + B_w w(t) \tag{1.48}$$

$$u(t) = \Psi\big(v(t)\big) \tag{1.49}$$

$$v(t) = Kx(t) \tag{1.50}$$

with the nonlinearity $\Psi(v(t)) = \mathrm{sat}(v(t))$. Thus, from (1.48)–(1.50), the connection between a linear system and the nonlinear element $\Psi$ is clear. Note, however, that the linear system (1.48) may not be asymptotically stable.

In order to overcome this problem and to fit the problem into the Lure's framework, let us define the following nonlinearity:

$$\phi\big(v(t)\big) = \mathrm{sat}\big(v(t)\big) - v(t)$$

or, equivalently,

$$\phi(v)_{(i)} = \phi(v_{(i)}) = \begin{cases} u_{\max(i)} - v_{(i)} & \text{if } v_{(i)} > u_{\max(i)} \\ 0 & \text{if } -u_{\min(i)} \leq v_{(i)} \leq u_{\max(i)} \\ -u_{\min(i)} - v_{(i)} & \text{if } v_{(i)} < -u_{\min(i)} \end{cases} \tag{1.51}$$

for $i = 1, \ldots, m$.

The nonlinearity $\phi(v(t))$ is in fact a *decentralized dead-zone* nonlinearity. Using $\phi(v(t))$, the closed-loop system (1.15) can now be represented by the following connection between a linear system and a nonlinearity $\phi(v(t))$:

$$\dot{x}(t) = (A + BK)x(t) + B\tilde{u}(t) + B_w w(t) \tag{1.52}$$

$$\tilde{u}(t) = \phi\big(v(t)\big) \tag{1.53}$$

$$v(t) = Kx(t) \tag{1.54}$$

Note that, if $v(t) = Kx(t)$ is a stabilizing control law, the linear system (1.52) is asymptotically stable (i.e. the matrix $(A + BK)$ is Hurwitz). Furthermore, considering a scalar $\lambda_{(i)} > 0$ and the nonlinearity $\phi(v_{(i)})$, two situations arise:

- If $\lambda_{(i)} = 1$, then it follows that

$$\begin{aligned} \text{for } v_{(i)} \geq 0 : \phi(v_{(i)}) \leq 0 \text{ and } \phi(v_{(i)}) \geq -\lambda_{(i)} v_{(i)} \\ \text{for } v_{(i)} \leq 0 : \phi(v_{(i)}) \geq 0 \text{ and } \phi(v_{(i)}) \leq -\lambda_{(i)} v_{(i)} \end{aligned} \tag{1.55}$$

or, equivalently, the nonlinearity $\phi(v_{(i)})$ is said to belong globally to the sector sec$[0, -1]$.

1.7 Models for the Saturation Nonlinearity

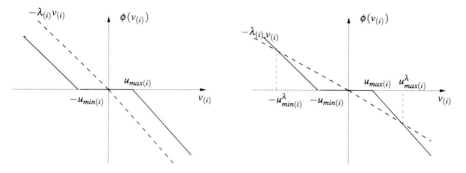

**Fig. 1.11** Dead-zone nonlinearity: (**a**) in the global sector; (**b**) in a local sector

- If $\lambda_{(i)} \leq 1$, then it follows that

$$\begin{aligned}&\text{for } 0 \leq v_{(i)} \leq \frac{u_{\max(i)}}{1-\lambda_{(i)}} : \phi(v_{(i)}) \leq 0 \text{ and } \phi(v_{(i)}) \geq -\lambda_{(i)} v_{(i)} \\ &\text{for } \frac{-u_{\min(i)}}{1-\lambda_{(i)}} \leq v_{(i)} \leq 0 : \phi(v_{(i)}) \geq 0 \text{ and } \phi(v_{(i)}) \leq -\lambda_{(i)} v_{(i)}\end{aligned} \quad (1.56)$$

or, equivalently, the nonlinearity $\phi(v_{(i)})$ is said to belong locally to the sector $\sec[0, -\lambda_{(i)}]$, i.e. it belongs to this sector provided that

$$-u^{\lambda}_{\min(i)} = \frac{-u_{\min(i)}}{1-\lambda_{(i)}} \leq v_{(i)} \leq \frac{u_{\max(i)}}{1-\lambda_{(i)}} = u^{\lambda}_{\max(i)}$$

These situations are illustrated in Fig. 1.11.

Based on the considerations above, we present in the sequel two sector conditions that will be used later in the book to derive stability and stabilization conditions for system (1.15).

#### 1.7.2.1 Classical Sector Condition

Since each component of $\phi(v)$ belongs to the sector $\sec[0, -1]$ globally, the following lemma can be stated.

**Lemma 1.4** *For all $v \in \Re^m$, the nonlinearity $\phi(v)$ satisfies the following inequality*:

$$\phi(v)' T \big( \phi(v) + v \big) \leq 0 \quad (1.57)$$

*for any diagonal positive definite matrix $T \in \Re^{m \times m}$.*

*Proof* From (1.55) it follows immediately that

$$\phi(v_{(i)})' T_{(i,i)} \big( \phi(v_{(i)}) + v_{(i)} \big) \leq 0, \quad i = 1, \ldots, m$$

provided that $T_{(i,i)} > 0$, which is equivalent to (1.57). □

42    1  Linear Systems Subject to Control Saturation—Problems and Modeling

Inequality (1.57) is a sector condition which is globally verified, i.e. it is verified for $\forall v \in \Re^m$.

On the other hand, considering the relations stated in (1.56), a local sector condition can be stated. With this aim, define the set

$$S\left(v, u^\lambda_{\min}, u^\lambda_{\max}\right) = \left\{v \in \Re^m; \; -u^\lambda_{\min} \preceq v \preceq u^\lambda_{\max}\right\} \tag{1.58}$$

with $u^\lambda_{\min(i)} = \frac{u_{\min(i)}}{1-\lambda_{(i)}}$ and $u^\lambda_{\max(i)} = \frac{u_{\max(i)}}{1-\lambda_{(i)}}$, $i = 1, \ldots, m$ and define a diagonal matrix $\Lambda \in \Re^{m \times m}$, whose $i$th diagonal element is $0 \leq \lambda_{(i)} \leq 1$. Thus the following Lemma can be stated.

**Lemma 1.5** *If $v \in S(v, u^\lambda_{\min}, u^\lambda_{\max})$, then the nonlinearity $\phi(v)$ satisfies the following inequality:*

$$\phi(v)'T\left(\phi(v) + \Lambda v\right) \leq 0 \tag{1.59}$$

*for any diagonal positive definite matrix $T \in \Re^{m \times m}$.*

*Proof* If $v \in S(v, u^\lambda_{\min}, u^\lambda_{\max})$ then $\frac{-u_{\min(i)}}{1-\lambda_{(i)}} \leq v_{(i)} \leq \frac{u_{\max(i)}}{1-\lambda_{(i)}}$, $i = 1, \ldots, m$. Hence, from (1.56), it immediately follows that

$$\phi(v_{(i)})'T_{(i,i)}\left(\phi(v_{(i)}) + \lambda_{(i)} v_{(i)}\right) \leq 0, \quad i = 1, \ldots, m$$

provided that $T_{(i,i)} > 0$, which is equivalent to (1.59). $\qquad\square$

Consider the control law $v(t) = Kx(t)$, then by applying Lemma 1.4, we can conclude that the relation

$$\phi(Kx)'T\left(\phi(Kx) + Kx\right) \leq 0 \tag{1.60}$$

is globally valid.

On the other hand, if we define the set

$$S\left(K, u^\lambda_{\min}, u^\lambda_{\max}\right) = \left\{x \in \Re^n; \; -u^\lambda_{\min} \preceq Kx \preceq u^\lambda_{\max}\right\} \tag{1.61}$$

it follows, from Lemma 1.5, that the sector condition

$$\phi(Kx)'T\left(\phi(Kx) + \Lambda Kx\right) \leq 0 \tag{1.62}$$

holds $\forall x \in S(K, u^\lambda_{\min}, u^\lambda_{\max})$.

Note that the set $S(K, u^\lambda_{\min}, u^\lambda_{\max})$, corresponds to an expansion of the region of linearity of the closed-loop system. In fact, when $\lambda_{(i)} = 0$, $i = 1, \ldots, m$, the set $S(K, u^\lambda_{\min}, u^\lambda_{\max})$ corresponds to $R_L$. On the other hand, when $\lambda_{(i)} \to 1$, $\forall i = 1, \ldots, m$, $S(K, u^\lambda_{\min}, u^\lambda_{\max})$ tends to cover all the state space and the global sector condition (1.60) is recovered.

It should be stressed that the sector condition (1.59) is in fact valid for any decentralized vector-valued nonlinearity, whose elements lie locally in the sectors $\sec[0, -\lambda_{(i)}]$, $i = 1, \ldots, m$. Hence, when this condition is used to prove stability of system (1.15), stability is in general implicitly ensured for a larger class of systems, which is potentially conservative. In the next section a different sector condition, which applies specifically to dead-zone nonlinearities, is presented.

1.7 Models for the Saturation Nonlinearity 43

### 1.7.2.2 Generalized Sector Condition

We state now a "generalized" or "modified" sector condition. This sector condition was proposed originally in [138, 139, 359]. The particular interest of this generalized sector condition is that it specifically applies to dead-zone nonlinearities, i.e. its use should result in less conservative conditions than the use of classical sector conditions.

In order to state the condition, let us first define the following set:

$$S(v - \omega, u_{\min}, u_{\max}) = \left\{ v \in \mathfrak{R}^m; \ \omega \in \mathfrak{R}^m; \ -u_{\min} \preceq v - \omega \preceq u_{\max} \right\} \qquad (1.63)$$

**Lemma 1.6** [359] *If $v$ and $\omega$ are elements of $S(v - \omega, u_{\min}, u_{\max})$, then the nonlinearity $\phi(v)$ satisfies the following inequality:*

$$\phi(v)' T \big( \phi(v) + \omega \big) \leq 0 \qquad (1.64)$$

*for any diagonal positive definite matrix $T \in \mathfrak{R}^{m \times m}$.*

*Proof* Assume that $v$ and $\omega$ are elements of $S(v - \omega, u_{\min}, u_{\max})$. In this case, it follows that $u_{\max(i)} - v_{(i)} + \omega_{(i)} \geq 0$ and $-u_{\min(i)} - v_{(i)} + \omega_{(i)} \leq 0$, $i = 1, \dots, m$.

Consider now the three cases below.

Case 1: $v_{(i)} > u_{\max(i)}$. It follows that $\phi(v_{(i)}) = u_{\max(i)} - v_{(i)} < 0$ and one gets $\phi(v_{(i)})T_{(i,i)}(\phi(v_{(i)}) + \omega_{(i)}) = \phi(v_{(i)})T_{(i,i)}(u_{\max(i)} - v_{(i)} + \omega_{(i)}) \leq 0$ provided that $T_{(i,i)} > 0$.

Case 2: $-u_{\min(i)} \leq v_{(i)} \leq u_{\max(i)}$. It follows that $\phi(v_{(i)}) = 0$ and $\phi(v_{(i)})T_{(i,i)} \times (\varphi(v_{(i)}) + \omega_{(i)}) = 0$, $\forall T_{(i,i)}$.

Case 3: $v_{(i)} < -u_{\min(i)}$. It follows that $\phi(v_{(i)}) = -u_{\min(i)} - v_{(i)} > 0$ and one gets $\phi(v_{(i)})T_{(i,i)}(\phi(v_{(i)}) + \omega_{(i)}) = \phi(v_{(i)})T_{(i,i)}(-u_{\min(i)} - v_{(i)} + \omega_{(i)}) \leq 0$ provided that $T_{(i,i)} > 0$.

Thus, once $v$ and $\omega$ are elements of $S(v - \omega, u_{\min}, u_{\max})$, we can conclude that $\phi(v_{(i)})T_{(i,i)}(\phi(v_{(i)}) + \omega_{(i)}) \leq 0$, $\forall T_{(i,i)} > 0$, $\forall i = 1, \dots, m$, from which follows (1.64). $\qquad \square$

Consider now the control law $v(t) = Kx(t)$, then, by applying Lemma 1.6. we can conclude that the relation

$$\phi(Kx)' T \big( \phi(Kx) + Gx \big) \leq 0 \qquad (1.65)$$

is verified provided that the state $x$ belongs to the following polyhedral set:

$$S(K - G, u_{\min}, u_{\max}) = \left\{ x \in \mathfrak{R}^n; \ -u_{\min} \preceq (K - G)x \preceq u_{\max} \right\} \qquad (1.66)$$

Note that the sector condition (1.62) corresponds to a particular case of the relation (1.65), obtained when $G = \Lambda K$. Hence, we can conclude that relation (1.65) is more general. In fact, as will be seen in Chap. 3, the matrix $G$ appears as an extra degree of freedom in the stability conditions. Furthermore, the bilinear terms generated when $\Lambda$ is a free variable are avoided when using (1.65). This is particularly important to obtain conditions in LMI form. Moreover, as will be seen in Chap. 7, the application of the generalized sector condition allows to convexify the problem of anti-windup synthesis when regional (local) stability is considered.

### 1.7.3 Regions of Saturation Models

The representation described in this section consists in dividing the state space in what we call *regions of saturation*. A region of saturation is defined as the intersection of half-spaces of type $K_{(i)}x \le d_{(i)}$ or $-K_{(i)}x \le d_{(i)}$, where $d_{(i)}$ can be $u_{\min(i)}$, $-u_{\min(i)}$, $u_{\max(i)}$ or $-u_{\max(i)}$. For a system with $m$ inputs, there exist $3^m$ regions of saturation. Considering $j = 1, \ldots, 3^m$, the $j$th region of saturation is a polyhedral set denoted generically as

$$S(R_j, d_j) = \left\{ x \in \Re^n : R_j x \preceq d_j \right\}$$

where $d_j \in \Re^{l_j}$ is defined from the entries of $u_{\max}$, $-u_{\max}$, $u_{\min}$ and $-u_{\min}$. $R_j \in \Re^{l_j \times n}$ is defined from the rows of $K$ and $-K$ (see Fig. 1.12 in Example 1.3).

Inside each region of saturation, system (1.15) can be modeled as an affine system, i.e. a linear system with an additive constant disturbance. This type of representation has been used, for instance, in [130, 136, 258, 299, 341].

Consider a vector $\xi \in \Re^m$ such that each entry $\xi_{(i)}$, $i = 1, \ldots, m$, takes the values 1, 0 or $-1$ as follows:

- if $u_{(i)}(t) = u_{\max(i)}$, then $\xi_{(i)} = 1$, that is, $x(t)$ is such that $K_{(i)}x(t) > u_{\max(i)}$;
- if $u_{(i)}(t) = K_{(i)}x(t)$, then $\xi_{(i)} = 0$, that is, $x(t)$ is such that $-u_{\min(i)} \le K_{(i)}x(t) \le u_{\max(i)}$;
- if $u_{(i)}(t) = -u_{\min(i)}$, then $\xi_{(i)} = -1$, that is, $x(t)$ is such that $K_{(i)}x(t) < -u_{\min(i)}$.

Each vector $\xi \in \Re^m$ represents therefore a possible combination between saturated and non-saturated control entries. There are $3^m$ different vectors $\xi : \xi_j \in \Re^m$, $j = 1, \ldots, 3^m$, and it is possible to associate each $\xi_j$ to a specific region of saturation $S(R_j, d_j)$. Note that the region corresponding to $\xi_j = 0$ is the region of linearity $S(K, u_{\min}, u_{\max})$. In the other regions, there is at least one control entry that is saturated.

From the definition of $\xi_j$ given above, the motion of system (1.15) inside the region $S(R_j, d_j)$ can be described by the following linear differential equation:

$$\dot{x}(t) = \left( A + B \operatorname{diag}\left( 1_m - |\xi_j| \right) K \right) x(t) + B u(\xi_j) + B_w w(t) \tag{1.67}$$

where, for $i = 1, \ldots, m$ and $j = 1, \ldots, 3^m$,

$$u_{(i)}(\xi_j) = \begin{cases} u_{\max(i)} & \text{if } \xi_{j(i)} = 1 \\ 0 & \text{if } \xi_{j(i)} = 0 \\ -u_{\min(i)} & \text{if } \xi_{j(i)} = -1 \end{cases} \tag{1.68}$$

Generically, if $x(t) \in S(R_j, d_j)$, it follows that [135, 136]

$$\dot{x}(t) = \bar{A}_j x(t) + p_j + B_w w(t) \tag{1.69}$$

with $\bar{A}_j = A + B \operatorname{diag}(1_m - |\xi_j|) K$, $p_j = B u(\xi_j)$.

Hence, considering any initial state $x(0)$, the trajectory of the system evolves in the state space and at each time instant may be locally described by a submodel $j$ valid in the region $j$. At the frontier between two regions, the local models are

1.7 Models for the Saturation Nonlinearity 45

identical and the trajectory of the system can be described by the local model of the new subregion. We conclude that this kind of representation describes the motion of the saturated system without any conservatism, i.e. it is an exact representation for the closed-loop system.

*Example 1.3* Consider system (1.15) described by the following data [135, 218]:

$$A = \begin{bmatrix} 0.1 & -0.1 \\ 0.1 & -3 \end{bmatrix}; \qquad B = \begin{bmatrix} 5 & 0 \\ 0 & 1 \end{bmatrix}; \qquad B_w = 0$$

$$K = \begin{bmatrix} -0.7283 & -0.0338 \\ -0.0135 & -1.3583 \end{bmatrix}; \qquad u_{\min} = u_{\max} = \begin{bmatrix} 5 \\ 2 \end{bmatrix}$$

Matrix $K$ and control constraints define nine regions of saturation. The polyhedron $S(K, u_{\min}, u_{\max})$ being symmetric, we need to analyze only five of these regions.

*Region 1.* $\xi_1 = 0 \Leftrightarrow$ *Region of linearity*

$$\bar{A}_1 = A + BK = \begin{bmatrix} -3.5415 & -0.2690 \\ 0.0865 & -4.3583 \end{bmatrix}; \qquad p_1 = \begin{bmatrix} 0 \\ 0 \end{bmatrix}$$

$$R_1 = \begin{bmatrix} K \\ -K \end{bmatrix}; \qquad d_1 = \begin{bmatrix} 5 \\ 2 \\ 5 \\ 2 \end{bmatrix}$$

*Region 2.* $\xi_2 = \begin{bmatrix} 0 \\ -1 \end{bmatrix}$

$$\bar{A}_2 = \begin{bmatrix} -3.5415 & -0.2690 \\ 0.1 & -3 \end{bmatrix}; \qquad p_2 = \begin{bmatrix} 0 \\ -2 \end{bmatrix}$$

$$R_2 = \begin{bmatrix} -0.7283 & -0.0338 \\ 0.7283 & 0.0338 \\ -0.0135 & -1.3583 \end{bmatrix}; \qquad d_2 = \begin{bmatrix} 5 \\ 5 \\ -2 \end{bmatrix}$$

*Region 3.* $\xi_3 = \begin{bmatrix} -1 \\ -1 \end{bmatrix}$

$$\bar{A}_3 = A; \qquad p_3 = \begin{bmatrix} -25 \\ -2 \end{bmatrix}$$

$$R_3 = \begin{bmatrix} -0.7283 & -0.0338 \\ -0.0135 & -1.3583 \end{bmatrix}; \qquad d_3 = \begin{bmatrix} -5 \\ -2 \end{bmatrix}$$

*Region 4.* $\xi_4 = \begin{bmatrix} -1 \\ 0 \end{bmatrix}$

$$\bar{A}_4 = \begin{bmatrix} 0.1 & -0.1 \\ 0.0865 & -4.3583 \end{bmatrix}; \qquad p_4 = \begin{bmatrix} -25 \\ 0 \end{bmatrix}$$

$$R_4 = \begin{bmatrix} -0.7283 & -0.0338 \\ -0.0135 & -1.3583 \\ 0.0135 & 1.3583 \end{bmatrix}; \qquad d_4 = \begin{bmatrix} -5 \\ 2 \\ 2 \end{bmatrix}$$

**Fig. 1.12** Example 1.3—regions of saturation

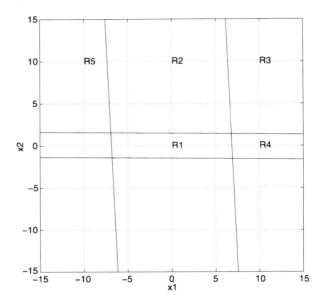

*Region 5.* $\xi_5 = \begin{bmatrix} 1 \\ -1 \end{bmatrix}$

$$\bar{A}_4 = A; \qquad p_5 = \begin{bmatrix} 25 \\ -2 \end{bmatrix}$$

$$R_5 = \begin{bmatrix} 0.7283 & 0.0338 \\ -0.0135 & -1.3583 \end{bmatrix}; \qquad d_5 = \begin{bmatrix} -5 \\ -2 \end{bmatrix}$$

These five regions are depicted in Fig. 1.12.

## 1.8 Equilibrium Points

As illustrated in Example 1.1, the saturated closed-loop system (1.15) may present multiple equilibrium points.

It is well known that the equilibrium points of a dynamic system are the points in the state space for which $\dot{x} = 0$ [215]. Hence, it follows straightforwardly that, for $w = 0$, the origin $x = 0$ is always an equilibrium point of the system (1.15). Furthermore, the origin is an asymptotically stable equilibrium point, at least locally, if and only if the matrix $(A + BK)$ has all its eigenvalues in the open left-half complex plane (i.e. if $(A + BK)$ is Hurwitz).

The other equilibrium points of the saturated system (1.15) can be obtained from the regions of saturation modeling presented in Sect. 1.7.3. This is formalized in the following definition, considering $w = 0$.

## 1.9 Conclusion

**Definition 1.7** The point $x_{\text{eq}}$ is an equilibrium point of system (1.15) with $w = 0$, if

$$\text{(a)} \quad x_{\text{eq}} = -\bar{A}_j^{-1} p_j \quad \text{for some } j = 1, \ldots, 3^m$$
$$\text{(b)} \quad x_{\text{eq}} \in S(R_j, d_j) \tag{1.70}$$

Note that just the satisfaction of (a) in (1.70) is not enough to conclude that $x_{\text{eq}}$ is an equilibrium point. In fact, the closed-loop system has $3^m$ "potential" equilibrium points. However, a point $x_{\text{eq}}$ is a "true" equilibrium point only if it belongs to the region of saturation associated with the local affine dynamic system (1.69) from which it has been computed.

*Example 1.4* Consider the same system as presented in Example 1.3. The "potential" equilibrium points are denoted by

$$x_{\text{eq}}^j = -\bar{A}_j^{-1} p_j$$

In this case, we obtain

$$x_{\text{eq}}^1 = \begin{bmatrix} 0 \\ 0 \end{bmatrix}; \quad x_{\text{eq}}^2 = -x_{\text{eq}}^6 = \begin{bmatrix} 0.0505 \\ -0.6650 \end{bmatrix}; \quad x_{\text{eq}}^3 = -x_{\text{eq}}^7 = \begin{bmatrix} 257.9310 \\ 7.9310 \end{bmatrix}$$

$$x_{\text{eq}}^4 = -x_{\text{eq}}^8 = \begin{bmatrix} 255.0623 \\ 5.0623 \end{bmatrix}; \quad x_{\text{eq}}^5 = -x_{\text{eq}}^9 = \begin{bmatrix} -259.3103 \\ -9.3103 \end{bmatrix}$$

Now testing if $R_j x_{\text{eq}}^j \leq d_j$ we can conclude that only the origin ($x_{\text{eq}}^1$) and $x_{\text{eq}}^4 = -x_{\text{eq}}^8$ are "true" equilibrium points of the closed-loop system. Actually,

$$R_4 x_{\text{eq}}^4 - d_4 = \begin{bmatrix} -183.1192 \\ -12.2548 \end{bmatrix} \preceq \begin{bmatrix} 0 \\ 0 \end{bmatrix}$$

which means that $x_{\text{eq}}^4$ belongs to the region $S(R_4, d_4)$.

## 1.9 Conclusion

In this chapter we focused on the presentation of some important definitions as well as on the formal statement of the main problems that we should be concerned with when controlling systems in presence of control saturation. It has been seen that, even if the open-loop system is linear, the closed-loop system presents a nonlinear behavior due to the saturation. In this case, two basic problems arise:

*Stability Analysis* Given a control law, it is important to characterize sets of initial conditions for which the internal asymptotic stability of the closed-loop system is ensured. These sets, called regions of asymptotic stability (RAS), can be seen as estimates of the region of attraction of the closed-loop system. On the other hand, in an external stability perspective, sets of admissible exogenous signals, for which the trajectories of the closed-loop system are guaranteed to remain bounded, should be properly characterized.

*Stabilization* This corresponds to a synthesis problem. In this case, the compensator should be designed taking into account the saturation effects with the following aims: ensure the internal asymptotic stability for a given set of initial states and ensure the external stability for a certain set of admissible exogenous signals. In parallel, performance as well as robustness requirements should be considered.

Another important and practical appealing issue is the so-called *anti-windup* problem. Considering a nominal controller designed to achieve some performance and robustness requirements, but disregarding the control bounds, the basic idea consists in adding a static or dynamic structure to this controller in order to mitigate the performance degradation due the control saturation or to increase the region of attraction of the closed-loop system. In particular, this structure acts only when effective control saturation occurs. In the absence of saturation the closed-loop dynamics is determined only by the nominal controller.

In order to derive tractable conditions and methods to address the problems above is of fundamental importance to model the saturation effects appropriately. With this aim, three kinds of modeling were presented, namely: the polytopic model, sector nonlinearity model and regions of saturation model. Based on these representations, in the next chapters of the book several methodologies are presented to address stability analysis, stabilization and anti-windup problems.

# Part II
# Stability Analysis and Stabilization

# Chapter 2
# Stability Analysis and Stabilization—Polytopic Representation Approach

## 2.1 Introduction

In this chapter, the problems of analysis and synthesis of linear control systems with input saturations are addressed by using the polytopic models for representing the saturated closed-loop system introduced in Sect. 1.7.1.

The basic idea behind using polytopic differential inclusions consists in guaranteeing some properties for a polytopic system that can be inherited by the actual saturated closed-loop system. The main advantage in this case is that stability and stabilization conditions can be obtained in a more tractable way, since the hard saturation nonlinearity disappears in the polytopic representation. However, it should be highlighted a crucial difference with respect to the case where polytopic differential inclusions are used to model uncertainties in a robust control framework. In that case, the polytopic model is in general valid in a global sense (i.e. it is valid in all state space) [45]. On the other hand, the key point when dealing with saturated systems consists in ensuring that the region of stability or the reachable region for the trajectories obtained using the polytopic model are included in the region of validity of the model, as discussed in Sect. 1.7.1. In other words, we should keep in mind that the polytopic differential inclusion represents the saturated system only locally.

First, we focus on the asymptotic stability analysis, i.e. the internal stability of the closed-loop system. Several conditions to ensure the local asymptotic stability of the closed-loop system are derived in the form of BMIs or LMIs. From these conditions, numerical algorithms (based on convex optimization problems) are proposed to determine regions of stability as large as possible, i.e. to obtain estimates of the region of attraction of the closed-loop system. We are mainly concerned with a quadratic approach and the consequent determination of ellipsoidal regions of stability in an LMI framework. In the end of the chapter, considering discrete-time systems, the determination of polyhedral regions of stability is also addressed. The results in this case are based on linear programming techniques.

From the proposed analysis conditions, results allowing the synthesis of control laws taking explicitly into account the actuator saturation possibility are derived.

S. Tarbouriech et al., *Stability and Stabilization of Linear Systems with Saturating Actuators*, DOI 10.1007/978-0-85729-941-3_2, © Springer-Verlag London Limited 2011

The synthesis of state feedback control laws aiming at enlarging an estimate of the region of attraction or at ensuring some time-domain performance with a guaranteed region of asymptotic stability (RAS) is addressed. Considering the action of exogenous input signals on the system, solutions to the problem of designing a control law aiming at maximizing the tolerance of the closed-loop system to disturbances or at maximizing the disturbance rejection are also presented. Although the results are mainly focused on the state feedback control laws, conditions regarding the design of observer-based and dynamic output feedback control laws are briefly presented.

The extensions of the approach to address the analysis and synthesis in a robust context are briefly discussed considering norm-bounded and polytopic uncertainties. Finally, the discrete-time counterpart of the results is presented. In this case, the determination of polyhedral regions of stability is also considered.

## 2.2 Asymptotic Stability Analysis

In this section, we focus on the internal asymptotic stability analysis problem formulated in Sect. 1.6.1. We consider that the saturated closed-loop system is given by

$$\dot{x}(t) = Ax(t) + B \operatorname{sat}(Kx(t)) \tag{2.1}$$

where matrices $A$ and $B$ are supposed to be real constant matrices of appropriate dimensions. The gain $K$ is also supposed to be a given real constant matrix of appropriate dimensions computed from an adequate design technique, such that all the eigenvalues of matrix $(A + BK)$ are placed in the left half complex plane. As seen in Sect. 1.3, (2.1) generically models closed-loop systems obtained from linear state feedback or output feedback control laws.

For simplicity, the saturation nonlinearity is assumed to be symmetric, that is $u_{\max} = u_{\min} = u_0$ with $u_0$ being a positive vector, i.e. $u_0 \succ 0$.

## 2.2.1 Ellipsoidal Sets of Stability

As seen in Chap. 1, due to actuator saturation, depending on the initial state the trajectory of the closed-loop system (2.1) may diverge, even if $(A + BK)$ is Hurwitz. Actually, the problem of finding the set of all initial conditions whose corresponding trajectories converge asymptotically to the origin, i.e. the exact determination of the region (or basin) of attraction of the closed-loop system is a very challenging issue which is still unsolved in the general case. The idea is therefore to estimate this region computing what we call "regions of stability" or "regions of guaranteed safe behavior". These regions correspond to sets of admissible initial conditions. Hence, if the initial condition belongs to a set of stability, the convergence of the associated trajectory to the origin is ensured.

## 2.2 Asymptotic Stability Analysis

A practical way of determining regions of stability comes from Lyapunov Theory [215]. It is well known that if $V(x)$ is a Lyapunov function, such that $\dot{V}(x) < 0$, $\forall x \in \mathcal{D} \subset \mathfrak{R}^n$, along the trajectories of the system, and $\mathcal{D}$ is a set containing the origin in its interior, then the level sets of $V(x)$, defined as

$$\mathcal{S}_V = \left\{ x \in \mathfrak{R}^n; \ V(x) \leq c \right\}$$

with $c \in \mathfrak{R}$, $c > 0$, are regions of asymptotic stability for the closed-loop system, provided that $\mathcal{S}_V \subset \mathcal{D}$ (see also Appendix A).

In particular, considering quadratic Lypaunov functions, i.e.

$$V(x) = x'Px \quad \text{with } P = P' > 0$$

the associated level sets are given by ellipsoidal domains defined as follows:

$$\mathcal{E}(P, \eta) = \left\{ x \in \mathfrak{R}^n; \ x'Px \leq \eta^{-1} \right\} \tag{2.2}$$

with $\eta > 0$.

Hence, the following lemma is instrumental for the derivation of results to determine estimates of the basin of attraction, using ellipsoidal domains.

**Lemma 2.1** *If there exist a positive definite symmetric matrix $P$ and a scalar $\eta$ such that*

$$\left( Ax(t) + B\,\mathrm{sat}\big(Kx(t)\big) \right)' Px(t) + x(t)' P\left( Ax(t) + B\,\mathrm{sat}\big(Kx(t)\big) \right) < 0, \quad \forall x(t) \in \mathcal{D} \tag{2.3}$$

*and*

$$\mathcal{E}(P, \eta) \subset \mathcal{D} \tag{2.4}$$

*then the set $\mathcal{E}(P, \eta)$ is a region of asymptotic stability (RAS) for system (2.1).*

The condition in Lemma 2.1 ensures in fact that the time derivative of the quadratic function $V(x(t)) = x(t)'Px(t)$ is strictly negative $\forall x(t) \in \mathcal{E}(P, \eta)$. Hence, from the Lyapunov theory, it follows that $\forall x(0) \in \mathcal{E}(P, \eta)$, $x(t) \in \mathcal{E}(P, \eta)$ and $\lim_{t \to \infty} x(t) = 0$. In other words, the ellipsoid $\mathcal{E}(P, \eta)$ is a positively invariant set and any trajectory initialized in this ellipsoid converges to the origin. We can say in this case that system (2.1) is locally (or regionally) asymptotically stable in $\mathcal{E}(P, \eta)$.

Moreover, since the inequality (2.3) is strictly verified, there exists a scalar $\sigma > 0$ such that

$$\left( Ax(t) + B\,\mathrm{sat}\big(Kx(t)\big) \right)' Px(t) + x(t)' P\left( Ax(t) + B\,\mathrm{sat}\big(Kx(t)\big) \right) < \sigma x(t)' Px(t) \tag{2.5}$$

which ensures that

$$V\big(x(t)\big) \leq e^{-\sigma t} V\big(x(0)\big)$$

i.e. the trajectories converge exponentially to the origin with a contraction rate $\sigma$. In this case, we say that the ellipsoid $\mathcal{E}(P, \eta)$ is also "contractive" or "$\sigma$-contractive" with respect to the trajectories of system (2.1).

Based on these considerations, the problem we want to solve can be stated as follows.

54      2 Stability Analysis and Stabilization—Polytopic Representation Approach

**Problem 2.1** Find a matrix $P$ and a positive scalar $\eta$ such that the set $\mathcal{E}(P, \eta)$ is a region of asymptotic stability for the closed-loop system (2.1).

Suppose that $\mathcal{E}(P, \eta)$ is a solution to the above analysis problem. Obviously, it follows that the sets $\mathcal{E}(P, \lambda\eta)$, with $\lambda > 1$ are also RAS. It may also exist sets $\mathcal{E}(P, \lambda\eta)$, with $0 < \lambda < 1$ that are RAS. Hence, given a matrix $P$, a relevant problem is to find the minimal value of $\lambda$ for which $\mathcal{E}(P, \lambda\eta)$ is effectively a RAS.

On the other hand, when we are concerned with the determination of an estimate of the region (basin) of attraction, the idea is to find $P$ and $\eta$ such that $\mathcal{E}(P, \eta)$ best fits in $R_A$ or, equivalently, that leads to a maximized region of stability considering some size criterion. In other words, we want to find the Lyapunov level set $\mathcal{E}(P, \eta)$ as large as possible. Our second problem originates from this remark.

**Problem 2.2** Find a matrix $P$ and a positive scalar $\eta$ such that $\mathcal{E}(P, \eta)$ is maximal considering some size criterion.

The way to measure the size of the set $\mathcal{E}(P, \eta)$ and the associated criteria for maximizing it will be discussed in the sequel.

### 2.2.2 Polytopic Approach I

According to Sect. 1.7.1.1, system (2.1) can be written as follows:

$$\dot{x}(t) = \left(A + B\Gamma\left(\alpha\left(x(t)\right)\right)K\right)x(t) \tag{2.6}$$

where $\Gamma(\alpha(x(t)))$ is a diagonal matrix whose diagonal elements are defined for $i = 1, \ldots, m$ as

$$0 < \alpha_{(i)}\left(x(t)\right) = \min\left(1, \frac{u_{0(i)}}{|K_{(i)}x(t)|}\right) \leq 1 \tag{2.7}$$

Note that when $\alpha_{(i)}(x)$ approaches $0$ there is almost no feedback injected in input $u_{(i)}$, whereas $\alpha_{(i)} = 1$ means that $u_{(i)}$ does not saturate.

It follows that for all $x(t)$ belonging to the region

$$S\left(|K|, u_0^\alpha\right) = \left\{x \in \Re^n; \ |Kx| \leq u_0^\alpha\right\} \tag{2.8}$$

with $u_{0(i)}^\alpha = \frac{u_{0(i)}}{\alpha_{\ell(i)}}, i = 1, \ldots, m$, one verifies

$$0 < \alpha_{l(i)} \leq \alpha_{(i)}\left(x(t)\right) \leq 1$$

Thus, $\dot{x}(t)$ can be determined from an appropriate convex linear combination of matrices $\mathbb{A}_j = A + B\Gamma_j(\alpha_l)K$ at time $t$, that is,

$$\dot{x}(t) = \sum_{j=1}^{2^m} \lambda_j\left(x(t)\right)\mathbb{A}_j x(t) \tag{2.9}$$

with $\sum_{j=1}^{2^m} \lambda_j(x(t)) = 1, \lambda_j(x(t)) \geq 0$.

## 2.2 Asymptotic Stability Analysis

It follows that all the trajectories of (2.1) that are confined in the region $S(|K|, u_0^\alpha)$, can be generated as trajectories of the system (2.9). As a consequence of this fact, the idea is to use the polytopic differential inclusion (2.9) to determine regions of stability for the saturated system (2.1).

Consider that such a region is an ellipsoid, i.e. it is defined by $\mathcal{E}(P, \eta)$. Hence, if

1. $V(x) = x'Px$ is strictly decreasing along the trajectories of the polytopic system (2.9) and
2. $\mathcal{E}(P, \eta) \subset S(|K|, u_0^\alpha)$,

then it follows that $\mathcal{E}(P, \eta)$ is a region of asymptotic stability (RAS) for system (2.1). The following formal result can therefore be stated [141, 172].

**Proposition 2.1** *If there exist a symmetric positive definite matrix $P \in \Re^{n \times n}$, a vector $\alpha_l \in \Re^m$ and a positive scalar $\eta$ satisfying:*

$$\left(A + B\Gamma_j(\alpha_l)K\right)'P + P\left(A + B\Gamma_j(\alpha_l)K\right) < 0, \quad j = 1, \ldots, 2^m \quad (2.10)$$

$$\begin{bmatrix} P & \alpha_{l(i)}K'_{(i)} \\ \alpha_{l(i)}K_{(i)} & \eta u_{0(i)}^2 \end{bmatrix} \geq 0, \quad i = 1, \ldots, m \quad (2.11)$$

$$0 < \alpha_{l(i)} \leq 1, \quad i = 1, \ldots, m \quad (2.12)$$

*then the ellipsoid $\mathcal{E}(P, \eta)$ is a region of asymptotic stability (RAS) for the saturated system (2.1).*

*Proof* Suppose that condition (2.10) is verified $\forall j = 1, \ldots, 2^m$. Then, for any scalars $0 \leq \lambda_j \leq 1$, such that $\sum_{j=1}^{2^m} \lambda_j = 1$, it follows that

$$\sum_{j=1}^{2^m} \lambda_j \left(\left(A + B\Gamma_j(\alpha_l)K\right)'P + P\left(A + B\Gamma_j(\alpha_l)K\right)\right) < 0$$

Hence, $\forall x \neq 0$, it follows that

$$x'\left[\left(\sum_{j=1}^{2^m} \lambda_j\left(A + B\Gamma_j(\alpha_l)K\right)\right)'P + P\left(\sum_{j=1}^{2^m} \lambda_j\left(A + B\Gamma_j(\alpha_l)K\right)\right)\right]x < 0$$

Since for any $x(t) \in S(|K|, u_0^\alpha)$,

$$\dot{x}(t) = Ax(t) + B\,\mathrm{sat}\left(Kx(t)\right) = \sum_{j=1}^{2^m} \lambda_j\left(A + B\Gamma_j(\alpha_l)K\right)$$

with appropriate $\lambda_j$, such that $0 \leq \lambda_j \leq 1$ and $\sum_{j=1}^{2^m} \lambda_j = 1$, if (2.10) is verified, we can therefore conclude that

$$\left(Ax(t) + B\,\mathrm{sat}\left(Kx(t)\right)\right)'Px(t) + x(t)'P\left(Ax(t) + B\,\mathrm{sat}\left(Kx(t)\right)\right) < 0$$
$$\forall x(t) \in S\left(|K|, u_0^\alpha\right)$$

On the other hand, the satisfaction of conditions (2.11) and (2.12) implies that $\mathcal{E}(P, \eta)$ is included in the set $S(|K|, u_0^\alpha)$. Thus, from Lemma 2.1 it follows that $\mathcal{E}(P, \eta)$ is a RAS. $\qquad\square$

In some cases, as we will see in the sequel of this book, we can be interested in conditions involving $P^{-1}$ instead of $P$. Hence setting $W = P^{-1}$ the following corollary to Proposition 2.1 can be stated.

**Corollary 2.1** *If there exist a symmetric positive definite matrix $W \in \Re^{n \times n}$, a vector $\alpha_l \in \Re^m$ and a positive scalar $\eta$ satisfying:*

$$W\big(A + B\Gamma_j(\alpha_l)K\big)' + \big(A + B\Gamma_j(\alpha_l)K\big)W < 0, \quad j = 1, \dots, 2^m \quad (2.13)$$

$$\begin{bmatrix} W & \alpha_{l(i)}WK'_{(i)} \\ \alpha_{l(i)}K_{(i)}W & \eta u^2_{0(i)} \end{bmatrix} \geq 0, \quad i = 1, \dots, m \quad (2.14)$$

$$0 < \alpha_{l(i)} \leq 1, \quad i = 1, \dots, m \quad (2.15)$$

*then the ellipsoid $\mathcal{E}(W^{-1}, \eta)$ is a region of asymptotic stability (RAS) for the saturated system (2.1).*

*Proof* The proof directly follows from the proof of Proposition 2.1. Relation (2.13) is obtained by right- and left-multiplying inequality (2.10) by $W$. Relation (2.14) is deduced by multiplying both sides of (2.11) by $\text{Diag}(W, 1)$. $\qquad\square$

*Remark 2.1* Proposition 2.1 and Corollary 2.1 give only sufficient conditions for the solution of Problem 2.1. In fact, the matrix inequality (2.10) ensures that the polytopic system is quadratically asymptotically stable in a global sense, which is not a necessary condition to ensure the asymptotic stability of system (2.1) in $\mathcal{E}(P, \eta)$. Note that, as mentioned in Remark 1.6, all the trajectories of system (2.1) that lie in $S(|K|, u_0^\alpha)$ are trajectories of (2.9), but the converse is not true.

*Remark 2.2* The problem of finding a positively invariant and contractive ellipsoidal set $\mathcal{E}(P, \eta)$ included in the region of linear behavior of the closed-loop system, $R_L$, can also be addressed with conditions of Proposition 2.1. For this it suffices to consider $\alpha_{l(i)} = 1, i = 1, \dots, m$. In this case, all the trajectories starting in $\mathcal{E}(P, \eta)$ will not generate control saturation, i.e., $\mathcal{E}(P, \eta)$ is a set of admissible initial conditions for which the behavior of the system is guaranteed to be linear. However, when we are interested in computing estimates of the region of attraction of the saturated system (2.1), it is important to have a set $\mathcal{E}(P, \eta)$ that spreads over the region of nonlinear behavior of the system, i.e., where control saturations are effectively active.

### 2.2.3 Polytopic Approach II

Define the set

$$S\big(|H|, u_0\big) = \big\{x \in \Re^n; \ |Hx| \preceq u_0\big\}$$

## 2.2 Asymptotic Stability Analysis

Recall the definition of matrices $\Gamma_j^+$ and $\Gamma_j^-$ given in Sect. 1.7.1.2:

- $\Gamma_j^+$ are diagonal matrices whose diagonal elements take the value 1 or 0, $j = 1, \ldots, 2^m$.
- $\Gamma_j^- = I_m - \Gamma_j^+$, $j = 1, \ldots, 2^m$.

As pointed in Sect. 1.7.1.2, if $x(t) \in S(|H|, u_0)$, there exists $\sum_{j=1}^{2^m} \lambda_j(x(t)) = 1$, $0 \leq \lambda_j(x(t)) \leq 1$, such that

$$\dot{x}(t) = \sum_{j=1}^{2^m} \lambda_j\big(x(t)\big)\big(A + B\Gamma_j^+ K + B\Gamma_j^- H\big)x(t) \tag{2.16}$$

Hence, following the same reasoning as performed in Sect. 2.2.2, if

1. $V(x) = x'Px$ is strictly decreasing along the trajectories of the polytopic system (2.16) and
2. $\mathcal{E}(P, \eta) \subset S(|H|, u_0)$,

then it follows that $\mathcal{E}(P, \eta)$ is a region of asymptotic stability (RAS) for system (2.6). The following formal result can therefore be stated [188, 193].

**Proposition 2.2** *If there exist a symmetric positive definite matrix $W \in \mathfrak{R}^{n \times n}$, a matrix $Q \in \mathfrak{R}^{m \times n}$ and a positive scalar $\eta$ satisfying:*

$$W\big(A + B\Gamma_j^+ K\big)' + Q'\Gamma_j^- B' + \big(A + B\Gamma_j^+ K\big)W$$
$$+ B\Gamma_j^- Q < 0, \quad j = 1, \ldots, 2^m \tag{2.17}$$

$$\begin{bmatrix} W & Q'_{(i)} \\ Q_{(i)} & \eta u_{0(i)}^2 \end{bmatrix} \geq 0, \quad i = 1, \ldots, m \tag{2.18}$$

*then the ellipsoid $\mathcal{E}(P, \eta) = \{x \in \mathfrak{R}^n; \ x'Px \leq \eta^{-1}\}$, with $P = W^{-1}$, is a region of asymptotic stability (RAS) for the saturated system (2.1).*

*Proof* By considering the change of variable $Q = HW$, the proof follows the same steps as the ones of Proposition 2.1 and Corollary 2.1. $\qquad\square$

Remark 2.1 also holds for the result of Proposition 2.2. In this case, it follows that all the trajectories of system (2.1) in $S(|H|, u_0)$ are trajectories of system (2.16), but the converse is not true. However, it is important to observe that the conditions of Proposition 2.2 encompass the ones of Proposition 2.1. Actually, if we choose $H = \text{diag}(\alpha_l)K$, the result of Corollary 2.1 is recovered. This fact means that Proposition 2.2 leads to less conservative results, although it is still only a sufficient condition to ensure that $\mathcal{E}(P, \eta)$ is a RAS.

### 2.2.4 Polytopic Approach III

Recall the definition of the sets $S_j$, $j = 1, \ldots, 2^m$, of indices $i \in \mathcal{M} = \{1, \ldots, m\}$ regarding the situations combining saturated and non-saturated entries, as given in Sect. 1.7.1.3.

- By definition, the combination associated to the case in which all the control inputs are not saturated corresponds to $S_1 = \emptyset$.
- For the $j$th combination, if the $i$th control input is saturated then $i \in S_j$.

Associated to each set $S_j$, $j = 2, \ldots, 2^m$, define the set

$$S(|H_j|, u_0) = \{x \in \mathfrak{R}^n; \; |H_{j(i)}x| \le u_{0(i)}, \; \forall i \in S_j\}$$

As pointed out in Sect. 1.7.1.3, if $x(t) \in \bigcap_{j=2}^{2^m} S(|H_j|, u_0)$, there exists $\sum_{j=1}^{2^m} \lambda_j(x(t)) = 1$, $0 \le \lambda_j(x(t)) \le 1$ such that

$$\dot{x}(t) = \sum_{j=1}^{2^m} \lambda_j(x(t)) \left( A + \sum_{i \in S_j^c} B_i K_{(i)} + \sum_{i \in S_j} B_i H_{j(i)} \right) x(t) \tag{2.19}$$

with $B_i$ standing for the $i$th column of $B$.

Hence, following the same reasoning as in Sect. 2.2.2, if

1. $V(x) = x'Px$ is strictly decreasing along the trajectories of the polytopic system (2.19) and
2. $\mathcal{E}(P, \eta) \subset \bigcap_{j=2}^{2^m} S(|H_j|, u_0)$,

then it follows that $\mathcal{E}(P, \eta)$ is a RAS for system (2.1).

The following formal result can therefore be stated [1].

**Proposition 2.3** *If there exist a symmetric positive definite matrix $W \in \mathfrak{R}^{n \times n}$, row vectors $Q_{j(i)} \in \mathfrak{R}^{1 \times n}$, $j = 2, \ldots, 2^m$, $i \in S_j$, and a positive scalar $\eta$ satisfying*

$$\left( AW + \sum_{i \in S_j^c} B_i K_{(i)} W + \sum_{i \in S_j} B_i Q_{j(i)} \right)$$

$$+ \left( AW + \sum_{i \in S_j^c} B_i K_{(i)} W + \sum_{i \in S_j} B_i Q_{j(i)} \right)' < 0, \quad \forall j = 1, \ldots, 2^m \tag{2.20}$$

$$\begin{bmatrix} W & Q'_{j(i)} \\ Q_{j(i)} & \eta u_{0(i)}^2 \end{bmatrix} \ge 0, \quad \forall j = 2, \ldots, 2^m, \; \forall i \in S_j \tag{2.21}$$

*then the ellipsoid $\mathcal{E}(P, \eta)$ with $P = W^{-1}$, is a region of asymptotic stability (RAS) for the saturated system (2.1).*

*Proof* By considering the change of variables $Q_{j(i)} = H_{j(i)} W$, the proof follows the same steps as the ones of Proposition 2.1 and Corollary 2.1. $\qquad\square$

## 2.2 Asymptotic Stability Analysis

Remark 2.1 also holds for the result of Proposition 2.3. In this case, it follows that all the trajectories of system (2.1) in $\bigcap_{j=2}^{2^m} S(|H_j|, u_0)$ are trajectories of the polytopic system (2.19), but the converse is not true. Moreover, the conditions of Proposition 2.3 encompass the ones of Propositions 2.1 and 2.2. Note that if we select a fixed $Q_{j(i)} = Q_{(i)}$, $\forall j$, conditions of Proposition 2.2 are recovered. This fact means that Proposition 2.3 leads in general to less conservative results, although it is still only a sufficient condition to ensure that $\mathcal{E}(P, \eta)$ is a RAS.

*Remark 2.3* In [1] and [417] different proofs for Proposition 2.3, not directly based on differential polytopic inclusion, are provided.

### 2.2.5 Optimization Problems

The results stated in Propositions 2.1, 2.2 and 2.3, allow to address the following optimization problems:

*P*1: Considering $P$ and $\eta$ given, test if $\mathcal{E}(P, \eta)$ is a RAS.
*P*2: Considering a given $P$, minimize $\eta$ for which $\mathcal{E}(P, \eta)$ is a RAS.
*P*3: Find $P$ and $\eta$ such that the RAS $\mathcal{E}(P, \eta)$ is maximized considering some size criterion.

The problem *P*1 is a feasibility problem, i.e., we have to test if the conditions in Propositions 2.1, 2.2 or 2.3 are actually satisfied.

Problem *P*2 corresponds to the following optimization problem:

$$\min \ \eta$$
$$\text{subject to} \quad \text{inequalities } (2.10)–(2.12) \text{ or } (2.17)–(2.18) \text{ or } (2.20)–(2.21) \tag{2.22}$$

Finally, problem *P*3 can be seen as a search of the best ellipsoidal estimate for the region of attraction, considering a given criterion. In order to address this problem, an optimization criterion, given by a function $f(\mathcal{E}(P, \eta))$, associated to the ellipsoidal region of stability that we want to determine must be conveniently defined. The minimization of this function implies the maximization of a geometric characteristic of the region, such as the volume, the length of the minor axis or even the maximization of the ellipsoid in certain directions. The maximization of the ellipsoid $\mathcal{E}(P, \eta)$ can therefore be tackled through the following generic optimization problem:

$$\min \ f\big(\mathcal{E}(P, \eta)\big)$$
$$\text{subject to} \quad \text{inequalities } (2.10)–(2.12) \text{ or } (2.17)–(2.18) \text{ or } (2.20)–(2.21) \tag{2.23}$$

At this point it should be noticed that inequalities (2.10)–(2.12) are BMIs if $P$ and $\alpha_l$ are decision variables and LMIs if $P$ or $\alpha_l$ are a priori fixed. On the other hand, the

60       2 Stability Analysis and Stabilization—Polytopic Representation Approach

inequalities (2.17)–(2.18) and (2.20)–(2.21) are LMIs in variables $W$, $Q$ (or $Q_{j(i)}$) and $\eta$. Hence, if $f(\mathcal{E}(P,\eta))$ is a convex function, it turns that the optimization problem can be solved in a convex framework. These issues are discussed in the following sections.

#### 2.2.5.1 Size Criteria

The objective function $f(\mathcal{E}(P,\eta))$, as above mentioned, should be associated to a geometric characteristic of the ellipsoidal domain. The most used size criteria are detailed in the sequel. Some of these criteria have to be associated to some auxiliary LMIs. In this case, the auxiliary LMIs must be added to the constraints of the optimization problem (2.23).

*Volume Maximization* The volume of the ellipsoid is proportional to

$$\sqrt{\det\left(P^{-1}\eta^{-1}\right)} = \sqrt{\det\left(W\eta^{-1}\right)}$$

Then it is possible to maximize the size of the ellipsoid by minimizing the function $\log(\det(\eta P))$ [45]. In this case, we can consider

$$f\left(\mathcal{E}(P,\eta)\right) = \log\left(\det(\eta P)\right) = \log\left(\eta^n \det(P)\right) = n\log(\eta) + \log\left(\det(P)\right)$$

Note that such a function is convex in the decision variables $P$ and $\eta$. When $W$ is a decision variable it suffices to consider the minimization of the following function:

$$f\left(\mathcal{E}(W^{-1},\eta)\right) = \log\left(\det(\eta W^{-1})\right) = n\log(\eta) - \log\left(\det(W)\right)$$

which is linear and convex in the decision variables $W$ and $\eta$.

It should be pointed out that the volume maximization can lead to ellipsoids that are "flat" in some directions. In this case, although the volume is maximized, the ellipsoidal region may be a bad estimate of the basin of attraction in those directions.

*Minor Axis Maximization* Noting that

$$\mathcal{E}(P,\eta) = \left\{x \in \Re^n;\ x'(\eta P)x \leq 1\right\}$$

the minimization of the greatest eigenvalue of $\eta P$ corresponds to maximize the minor axis of the ellipsoid $\mathcal{E}(P,\eta)$ [45].

Considering that $(\eta P)^{-1} = \eta^{-1}W$, it follows that $\lambda_{\max}(\eta P) = 1/\lambda_{\min}(\eta^{-1}W)$ Thus, depending on whether $W$ or $P$ are decision variables, the optimization criterion becomes

$$f\left(\mathcal{E}(W^{-1},\eta)\right) = -\lambda_{\min}\left(\eta^{-1}W\right) \quad \text{or} \quad f\left(\mathcal{E}(P,\eta)\right) = \lambda_{\max}(\eta P) \qquad (2.24)$$

Note that $\lambda_{\max}(\eta P) = \eta\lambda_{\max}(P)$. However, in general, $P$ and $\eta$ are simultaneously decision variables and it is desirable to have a linear criterion, in order to deal with convex optimization problems (this will made be clear in the next section).

## 2.2 Asymptotic Stability Analysis

In this case, we can consider the following linear criteria and additional constraints to be added in the optimization problem (2.23):

$$\min \ \beta_0 \eta + \beta_1 \lambda$$

$$\text{subject to} \quad \text{inequalities } (2.10)-(2.12)$$

$$P \leq \lambda I_n$$

or

$$\max \ -\beta_0 \eta + \beta_1 \lambda$$

$$\text{subject to} \quad \text{inequalities } (2.13)-(2.15) \text{ or } (2.17)-(2.18) \text{ or } (2.20)-(2.21)$$

$$W \geq \lambda I_n$$

where $\beta_0$ and $\beta_1$ are tuning parameters allowing to weight the effects of $\eta$ and $\lambda$ in the effective size of the minor axis.

*Trace Minimization* Each eigenvalue of $P$ is associated to the length of one ellipsoid axis. Since the trace of matrix $P$ is the sum of its eigenvalues, its minimization leads to ellipsoids that tend to be homogeneous in all directions. In fact, in this case, all the axis length have the same weight in the criterion.

Taking into account that $\text{trace}(\eta P) = \eta \, \text{trace}(P)$, the following optimization function can be chosen:

$$f\left(\mathcal{E}(P, \eta)\right) = \beta_0 \eta + \beta_1 \text{trace}(P)$$

where $\beta_0$ and $\beta_1$ are weighting parameters.

When $W$ is a decision variable one can consider the function

$$f\left(\mathcal{E}(W^{-1}, \eta)\right) = \beta_0 \eta + \beta_1 \text{trace}(M_W)$$

along with the following constraints in the optimization problems:

$$\begin{bmatrix} M_W & I_n \\ I_n & W \end{bmatrix} > 0; \qquad M_W = M'_W > 0 \tag{2.25}$$

Note that (2.25) ensures that $P < M_W$ and, in consequence, that $\text{trace}(P) < \text{trace}(M_W)$. Thus, the minimization of the $\text{trace}(M_W)$ implies the minimization of $\text{trace}(P)$.

*Maximization Along Certain Directions* It is possible that some preferential directions in which the system will be initialized or driven by the actions of disturbances are known. Thus, it is of interest to find a stability region as large as possible along these directions [130, 133, 137, 193].

With this aim, let us consider the set of vectors that define the directions in which the ellipsoid should be maximized:

$$\mathcal{D} = \{d_1, \ldots, d_s\}, \quad d_i \in \mathfrak{R}^n, \ i = 1, \ldots, s$$

For simplicity, consider $\eta = 1$. The idea in this case is to maximize scaling factors $\theta_i, i = 1, \ldots, s$, such that

$$\left(\theta_i d'_i\right) P (\theta_i d_i) \leq 1$$

or still

$$\begin{bmatrix} 1 & \theta_i d_i' \\ \theta_i d_i & W \end{bmatrix} \geq 0$$

Thus, the following optimization criterion and additional constraints can be considered:

$$\min \sum_{i=1}^{s} \beta_i \bar{\theta}_i$$

$$\text{subject to} \quad \text{inequalities (2.10)-(2.12)}$$

$$d_i' P d_i \leq \bar{\theta}_i, \quad i = 1, \ldots, s$$

with $\bar{\theta}_i = \theta_i^{-2}$, or

$$\max \sum_{i=1}^{s} \beta_i \theta_i$$

$$\text{subject to} \quad \text{inequalities (2.13)-(2.15) or (2.17)-(2.18) or (2.20)-(2.21)}$$

$$\begin{bmatrix} 1 & \theta_i d_i' \\ \theta_i d_i & W \end{bmatrix} \geq 0, \quad i = 1, \ldots, s$$

where $\beta_i$ are tuning parameters corresponding to a weight factor associated to the maximization in direction $d_i$. As a particular case it is possible to consider $\theta_i = \theta$, $i = 1, \ldots, s$, i.e. the same scaling factor or "amplification" in all directions.

### 2.2.5.2 BMI × LMI Problems

The inequalities (2.11) and (2.12) are LMIs in $P$, $\alpha_l$ and $\eta$. Note, however, that matrix inequality (2.10) is bilinear (i.e. it is a BMI) in decision variables $P$ and $\alpha_l$, due to the products involving these variables. This fact renders difficult the resolution of the optimization problem (2.23). The same type of remarks can be done with respect to inequalities (2.13) and (2.14), considering variables $W$ and $\alpha_l$.

In [129] it has been shown that many of the problems considered in the robust control literature can be formulated as BMIs. This is the case of the conditions in Proposition 2.1 and Corollary 2.1. A way to overcome this problem consists in relaxing the BMI constraints in an LMI form by fixing variables. In this case, a (probably) suboptimal solution can be searched by an iterative algorithm, where in each step some variables are fixed and an LMI problem is solved. An example of this procedure is given by the algorithm below.

### Algorithm 2.1

1. Initialize $\alpha_l$.
2. Solve the following optimization problem for $P$ and $\eta$:

$$\min \ f\big(\mathcal{E}(P, \eta)\big)$$

$$\text{subject to} \quad \text{inequalities (2.10)-(2.12)} \tag{2.26}$$

## 2.2 Asymptotic Stability Analysis

3. Keep the previous value of $P$, solve the following problem for $\eta$ and $\alpha_l$:

$$\min \ \eta$$
$$\text{subject to} \quad \text{inequalities (2.10)--(2.12)} \tag{2.27}$$

4. Go to step 2 until no significant change on the size of the ellipsoid $\mathcal{E}(P, \eta)$ is obtained. When no significant change arises then stop.

In particular, for the single-input case, it is also possible to seek the optimal solution of the considered optimization problem by performing a bisection (line) search on $\alpha_l$. For the general multi-input case, $P$ or $\alpha_l$ are fixed and a convex optimization problem with LMI constraints are solved as described in Algorithm 2.1. Two issues arise in this case: how to choose the initial vector $\alpha_l$ and how exactly to decrease the components of $\alpha_l$ if needed? One simple way of handling these issues is to apply trial and error procedures. On the other hand, since the matrix $A + BK$ is Hurwitz by assumption, there will always exist a solution to (2.26) for $\alpha_l = 1_m$. Hence, if we start the algorithm with $\alpha_l = 1_m$, the convergence of the algorithm is ensured. This follows from the fact that an optimal solution for one step is also a feasible solution for the next step. Although conservative (in the sense that, in general, the optimal solution is not achieved) this kind of approach solves, in part, the problem of the choice of vector $\alpha_l$ by using robust and available packages to solve LMIs [45, 231]. Of course, taking different initial vectors $\alpha_l$, the proposed algorithm can converge toward different values.

On the other hand, the inequalities of Propositions 2.2 and 2.3 are true LMIs in variables $W$, $Q$ (or $Q_{j(i)}$) and $\eta$. In this case, considering the criteria and the additional constraints presented in Sect. 2.2.5.1, it follows that problem (2.23) is a convex one and can be directly and efficiently solved by using available LMI solvers. In addition, without loss of generality, we can consider $\eta = 1$. Note that if $W$, $Q$ and $\eta$ satisfy LMIs (2.17) and (2.18), by multiplying both LMIs by $\eta^{-1}$ it follows that $\bar{W} = \eta^{-1} W$, $\bar{Q} = \eta^{-1} Q$ satisfy:

$$\bar{W} \left(A + B \Gamma_j^+ K\right)' + \bar{Q}' \Gamma_j^- B' + \left(A + B \Gamma_j^+ K\right) \bar{W} + B \Gamma_j^- \bar{Q} < 0, \quad j = 1, \ldots, 2^m$$

$$\begin{bmatrix} \bar{W} & \bar{Q}'_{(i)} \\ \bar{Q}_{(i)} & u_{0(i)}^2 \end{bmatrix} \geq 0, \quad i = 1, \ldots, m$$

The same reasoning can be done by considering LMIs (2.20) and (2.21).

*Remark 2.4* The conditions stated in Propositions 2.1, 2.2 and 2.3 regard only the guarantee of local (or regional) asymptotic stability of the closed-loop system. Indeed, the polytopic models are well defined only in $S(|K|, u_0^\alpha)$, $S(|H|, u_0)$ or $\bigcap_{j=2}^{2^m} S(|H_j|, u_0)$, respectively. In the case open-loop system is stable, the origin of the closed-loop system can be in fact globally asymptotically stable. In this case, the optimization problems can lead to numerical solutions with $\alpha_{l(i)}$ and the elements of $H$ with values tending to zero. This fact means that the obtained ellipsoid tends to cover all the state space and its size is in fact limited by the numerical precision of the solver.

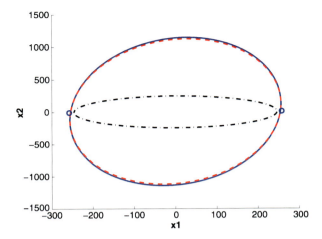

**Fig. 2.1** Example 2.1— regions of stability obtained with polytopic approaches I (*dot–dashed line*), II (*solid line*) and (III) (*dashed*)

*Example 2.1* Consider the system studied in [218], which is described by the following data:

$$A = \begin{bmatrix} 0.1 & -0.1 \\ 0.1 & -3 \end{bmatrix}; \qquad B = \begin{bmatrix} 5 & 0 \\ 0 & 1 \end{bmatrix}$$
$$K = \begin{bmatrix} -0.7283 & -0.0338 \\ -0.0135 & -1.3583 \end{bmatrix}; \qquad u_0 = \begin{bmatrix} 5 \\ 2 \end{bmatrix}$$

Consider first the polytopic approach I. In this case, we apply Algorithm 2.1 considering the maximization of the minor axis of the ellipsoidal region of stability. To apply the algorithm, we initialize $\alpha_l = 1_m$. Therefore, one obtains

$$\alpha_l = \begin{bmatrix} 0.0275 \\ 0.0034 \end{bmatrix}; \qquad P = 10^{-4} \times \begin{bmatrix} 0.1608 & 0.0001 \\ 0.0001 & 0.1592 \end{bmatrix}; \qquad \eta = 1$$

Considering now the same problem, but with the polytopic approach II we obtain

$$P = 10^{-4} \times \begin{bmatrix} 0.1542 & -0.0038 \\ -0.0038 & 0.0078 \end{bmatrix}; \qquad Q = 10^3 \times \begin{bmatrix} -1.2732 & -0.0135 \\ 0.2362 & -0.0155 \end{bmatrix}$$

Finally, with approach III the following is obtained:

$$P = 10^{-4} \times \begin{bmatrix} 0.1541 & -0.0038 \\ -0.0038 & 0.0081 \end{bmatrix}; \qquad Q_2 = 10^3 \times \begin{bmatrix} -1.2741 & -0.0400 \\ 0 & 0 \end{bmatrix}$$
$$Q_3 = 10^3 \times \begin{bmatrix} 0 & 0 \\ 0.0000 & -0.0007 \end{bmatrix}; \qquad Q_4 = 10^3 \times \begin{bmatrix} -1.2698 & 0.0759 \\ 0.2787 & -0.0167 \end{bmatrix}$$

Figure 2.1 depicts the obtained ellipsoidal regions of asymptotic stability (RAS). It is important to note that the saturated system has two additional unstable equilibrium points at

$$x_{\text{eq}} = \pm \begin{bmatrix} 257.931 \\ 7.931 \end{bmatrix}$$

2.3 External Stability

This fact allows to see that the provided estimates of the region of attraction are relatively good since the boundary of the three ellipsoids are very close to the equilibrium points above mentioned. The estimates obtained with the polytopic approaches II and III are clearly better. This enforces the fact that these approaches encompass the first one and provide, in general, less conservative estimates of the region of attraction. It should, however, be noticed that the directions in which the ellipsoid obtained with approach II and III are larger are not directly considered in the optimization criteria, i.e. considering the maximization of the minor axis of the ellipsoids, all the approaches practically lead to the same optimal value of the criterion (the size of the minor axis).

On the other hand, there is no practical difference between the obtained ellipsoid with approaches II and III. Nonetheless, the value of the optimal criterion for approach III ($-6.4800 \times 10^4$) is slightly smaller than that one for approach II ($-6.4763 \times 10^4$). This shows, as expected, a slight reduction in conservatism. Actually, this small difference can be justified by the fact that the considered example is of low dimension (second order, two inputs). Hence, the contribution of the extra degrees of freedom introduced by the different matrices $Q_j$ in polytopic approach III is less effective.

*Example 2.2* Recall the balancing pointer system studied in Example 1.1:

$$A = \begin{bmatrix} 0 & 1 \\ 1 & 0 \end{bmatrix}; \qquad B = \begin{bmatrix} 0 \\ -1 \end{bmatrix}; \qquad K = \begin{bmatrix} 13 & 7 \end{bmatrix}$$

Suppose that the control inputs are limited by $u_0 = 5$. The conditions of Proposition 2.2 are applied to find estimates of the region of attraction of the closed-loop saturated system. The results obtained with the following size criteria:

- maximization of $\lambda_{\min}(W)$
- minimization of trace($W^{-1}$)
- maximization in the direction $[-4.5\ 6]'$

are shown in Fig. 2.2. The region of linearity $R_L$ is also plotted in the figure (it is delimited by the dashed lines). Note that in all cases, the estimate of the region of attraction spreads over the region where the behavior of the system is actually nonlinear.

## 2.3 External Stability

In this section, we focus on the analysis of the external stability of the saturated closed-loop system when the exogenous signal $w(t)$ is present. This signal can be considered as a disturbance, a reference to track or a combination of both. Hence, we consider the closed-loop system is generically represented as

$$\dot{x}(t) = Ax(t) + B\,\text{sat}\big(Kx(t)\big) + B_w w(t) \tag{2.28}$$

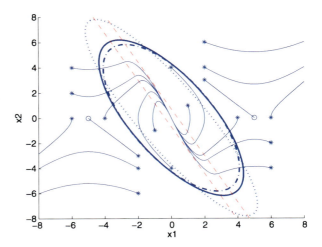

**Fig. 2.2** Example 2.2—region of asymptotic stability (RAS) considering: maximization of $\lambda_{\min}(W)$ (···), minimization of $\text{trace}(W^{-1})$ (—), maximization in the direction $[-4.5 \; 6]'$ (···); $R_L$ (- - -); equilibrium points (○); initial conditions (∗)

As illustrated in Example 1.2, due to the control signal saturation, in general it is not possible to ensure that the trajectories of system (2.28) will be bounded for any bounded signal $w(t)$. Hence, it is important to characterize admissible sets of exogenous signals as well as of initial conditions, for which the trajectories are guaranteed bounded. Furthermore, if $w(t)$ is vanishing, the convergence to the origin (equilibrium point of interest) must be ensured. In other words, for the admissible exogenous signals and initial conditions, the corresponding trajectories will never leave the region of attraction of system (2.1).

In the sequel we derive some results to address Problem 1.5. In particular, we consider two classes of disturbances: amplitude bounded and energy bounded ones. The results are based on a quadratic approach. The conditions are derived considering the polytopic model II. Formulations considering the polytopic approach I and III follow exactly the same reasoning and can be straightforwardly obtained.

## 2.3.1 Amplitude Bounded Exogenous Signals

Consider that the exogenous signal $w(t)$ belongs to the following set:

$$\mathcal{W} = \left\{ w \in \Re^q; \; w'Rw \leq \delta^{-1} \right\} \tag{2.29}$$

with $R = R' > 0$ and $\delta > 0$.

In this case, $w(t)$ is bounded by a quadratic norm which reflects amplitude bounds on $w(t)$ [193, 353]. Note that $\mathcal{L}_\infty$-norm constraints can be straightforwardly considered in this way. For instance, if $R$ is a diagonal matrix, it follows that $|w_i(t)| \leq \sqrt{1/(\delta r_i)}$, with $r_i$ denoting the $i$th diagonal element of $R$, $i = 1, \ldots, q$.

Considering a quadratic Lyapunov function $V(x) = x'Px$ and the application of the S-procedure (see Appendix C), a sufficient condition to obtain a solution

## 2.3 External Stability

to Problem 1.5 is achieved if the following relation is satisfied $\forall x \in \mathcal{E}(P, \eta)$ and $\forall w \in \mathcal{W}$ [45]:

$$\dot{V}(x) + \tau_1\left(x'Px - \eta^{-1}\right) + \tau_2\left(\delta^{-1} - w'Rw\right) < 0, \quad \tau_1 > 0, \ \tau_2 > 0 \qquad (2.30)$$

In fact, relation (2.30) ensures that $\dot{V}(x) < 0$, $\forall x(t) \notin \text{int}\,\mathcal{E}(P, \eta)$ and $\forall w(t) \in \mathcal{W}$. In this case, suppose that at time $t = t_1$, $x(t_1) \in \partial\mathcal{E}(P, \eta)$ and $w(t) \in \mathcal{W}$. It follows that $\dot{V}(x(t_1)) < 0$, which implies that $x(t_1 + \Delta t) \in \text{int}\,\mathcal{E}(P, \eta)$. Thus, we can conclude that (2.30) ensures that the trajectories initialized in $\mathcal{E}(P, \eta)$ do not escape from this domain $\forall w(t) \in \mathcal{W}$. In other words, the ellipsoid $\mathcal{E}(P, \eta)$ is said a $\mathcal{W}$-*positively invariant set* [33] or a *robustly positively invariant set* [36] with respect to the closed-loop system (2.28).

In particular, if

$$\dot{V}(x) + \tau_1 x'Px - \tau_2 w'Rw < 0 \qquad (2.31)$$

and

$$\delta^{-1}\tau_2 - \eta^{-1}\tau_1 < 0 \qquad (2.32)$$

are satisfied then condition (2.30) holds. On the other hand, when $w = 0$, it follows that (2.31) implies that

$$\dot{V}(x) < -\tau_1 x'Px < 0$$

which ensures that $\dot{V}(x) < 0$, $\forall x \in \mathcal{E}(P, \eta)$. Hence, we can conclude that $\mathcal{E}(P, \eta)$ is a region of asymptotic stability (RAS) for the saturated system. This means that if $w(t)$ vanishes, the trajectory will converge asymptotically to the origin. Moreover, this convergence is exponential with a rate given by $\tau_1$.

Considering a polytopic modeling for the saturated system (2.28), the idea therefore consists in ensuring (2.30) along the trajectories of a polytopic system. In addition, the positive invariant set must be included in the region of validity of the polytopic differential inclusion. From this reasoning, the following result can be stated considering the polytopic model II, and, without loss of generality, $\eta = 1$.

**Proposition 2.4** *If there exist a symmetric positive definite matrix $W \in \Re^{n \times n}$, a matrix $Q \in \Re^{m \times n}$ and positive scalars $\tau_1$ and $\tau_2$ satisfying:*

$$\begin{bmatrix} W(A + B\Gamma_j^+ K)' + Q'\Gamma_j^- B' + (A + B\Gamma_j^+ K)W + B\Gamma_j^- Q + \tau_1 W & B_w \\ B'_w & -\tau_2 R \end{bmatrix} < 0,$$
$$j = 1, \ldots, 2^m \qquad (2.33)$$

$$\begin{bmatrix} W & Q'_{(i)} \\ Q_{(i)} & u_{0(i)}^2 \end{bmatrix} \geq 0, \quad i = 1, \ldots, m \qquad (2.34)$$

$$\tau_2 - \delta\tau_1 < 0 \qquad (2.35)$$

*then*

1. *$\forall w \in \mathcal{W}$, the trajectories of the system (2.28) do not leave the set $\mathcal{E}(P, 1)$;*
2. *for $w = 0$, $\mathcal{E}(P, 1)$ is a region of asymptotic stability (RAS) for the system (2.28).*

*Proof* Left- and right-multiplying inequality (2.33) by the block diagonal matrix $\text{Diag}(P, I)$, with $P = W^{-1}$ it follows that (2.33) is equivalent to

$$\begin{bmatrix} (A + B\Gamma_j^+ K)'P + H'\Gamma_j^- B'P + P(A + B\Gamma_j^+ K) + PB\Gamma_j^- H + \tau_1 P & PB_w \\ B_w'P & -\tau_2 R \end{bmatrix}$$
$$< 0, \quad j = 1, \dots, 2^m \tag{2.36}$$

with $H = QP$. Now, left- and right-multiplying (2.36), $j = 1, \dots, 2^m$, respectively, by $[x(t)' \ w(t)']$ and $\begin{bmatrix} x(t) \\ w(t) \end{bmatrix}$, by convexity, it follows that

$$\dot{V}(x) + \tau_1 x'Px - \tau_2 w'Rw < 0$$

is verified along the trajectories of the polytopic system:

$$\dot{x}(t) = \sum_{j=1}^{2^m} \lambda_j \big( x(t) \big) \big( \mathbb{A}_j x(t) + B_w w(t) \big) \tag{2.37}$$

This fact along with the satisfaction of (2.35) implies that (2.30) is verified. Hence, if the set $\mathcal{E}(P, 1)$ is included in the region $S(|H|, u_0)$, it follows that (2.33) implies that (2.30) is verified along the trajectories of the saturated system (2.28).

Finally, left- and right-multiplying (2.34) by $\text{Diag}(P, I)$, it follows that this LMI ensures that $\mathcal{E}(P, 1) \subset S(|H|, u_0)$, which concludes the proof. $\qquad\square$

Given a set $\mathcal{W}$, the condition stated in Proposition 2.4 can be used to check if the trajectories of the system are bounded considering that $w(t) \in \mathcal{W}$. In this case $R$ and the scalar $\delta$ are given, and an optimization problem can be formulated to determine a region of admissible initial states, $\mathcal{E}(P, 1)$ as large as possible, considering some size criterion.

Considering now that a set of admissible initial conditions is given. In particular, let $\mathcal{X}_0$ be this set and assume that it is a polyhedral or an ellipsoid set (or a union of them) described, respectively, as follows.

(a) Polyhedral set:

$$\mathcal{X}_0 = \text{Co}\{v_1, \dots, v_{n_v}\}, \quad v_s \in \mathfrak{R}^n, \ \forall s = 1, \dots, n_v \tag{2.38}$$

where each vector $v_s$ denotes a vertex of the polyhedral set. In this case the constraint $\mathcal{X}_0 \subset \mathcal{E}(P, 1)$ can be expressed in LMI form as follows [45]:

$$\begin{bmatrix} 1 & v_s' \\ v_s & W \end{bmatrix} \geq 0, \quad \forall s = 1, \dots, n_v \tag{2.39}$$

(b) Ellipsoidal set:

$$\mathcal{X}_0 = \big\{ x \in \mathfrak{R}^n; \ x'P_0 x \leq 1 \big\} \tag{2.40}$$

where $P_0 = P_0' \in \mathfrak{R}^{n \times n}$. In this case the constraint $\mathcal{X}_0 \subset \mathcal{E}(P, 1)$ can be expressed in LMI form as follows [45]:

$$\begin{bmatrix} P_0 & I_n \\ I_n & W \end{bmatrix} \geq 0 \tag{2.41}$$

## 2.3  External Stability

**Table 2.1** Example 2.3—minimization of trace($R$)

| $\theta$ | trace($R$) | $(\tau_1; \tau_2)$ |
|---|---|---|
| 50 | 0.0736 | (0.4485; 0.3788) |
| 100 | 0.1828 | (0.1697; 0.1) |
| 130 | 0.6307 | (0.1697; 0.1) |

It should be pointed out that, even if $R$ is given, inequality (2.33) is not a "true" LMI in variables $W$ and $\tau_1$. However, a feasible solution to the problem can be directly searched by fixing the scalar $\tau_1$ and solving an LMI feasibility problem.

On the other hand, given $\mathcal{X}_0$, an interesting problem consists in maximizing the set $\mathcal{W}$, for which the trajectories are guaranteed bounded. In this case, without loss of generality, we can assume $\delta = 1$ and consider one of the size criteria for ellipsoidal sets, expressed in terms of matrix $R$, proposed in Sect. 2.2.5. For instance, the following optimization problem can be formulated:

$$
\begin{aligned}
&\min \ \text{trace}(R) \\
&\text{subject to} \quad \text{inequalities (2.33)–(2.35), (2.39) and/or (2.41)}
\end{aligned}
\tag{2.42}
$$

However, note that (2.42) is not a "true" LMI problem. Actually, there is a product between the variables $R$ and $\tau_2$ and also between $W$ and $\tau_1$. A way to overcome these bilinearities is to perform a search on a grid defined by parameters $\tau_1$ and $\tau_2$. In this case, for each fixed $(\tau_1, \tau_2)$, an LMI optimization problem can be solved.

*Example 2.3*  Consider the system treated in Example 2.1, with $B_w = B$, and the following set of admissible initial conditions:

$$
\mathcal{X}_0 = \text{Co}\left\{ \theta \begin{bmatrix} 1 \\ 1 \end{bmatrix}; \theta \begin{bmatrix} -1 \\ -1 \end{bmatrix} \right\}
$$

We aim at maximizing a set of admissible amplitude bounded disturbances, given by

$$
\mathcal{W} = \left\{ w \in \mathfrak{R}^q; \ w'Rw \leq 1 \right\}
$$

for which the trajectories of the system are guaranteed to be bounded in an ellipsoidal set $\mathcal{E}(P, 1) \supset \mathcal{X}_0$. For this, we consider the optimization problem (2.42). The optimal value of this problem is obtained by solving LMI problems in a grid on $\tau_1$ and $\tau_2$. Table 2.1 shows the obtained optimal values for trace($R$), considering different values for $\theta$. The values of $\tau_1$ and $\tau_2$ for which the optimal solution is achieved in each case are also shown.

The domains $\mathcal{W}$ and $\mathcal{E}(P, 1)$, obtained for the cases in Table 2.1 are depicted in Figs. 2.3 and 2.4 The trade-off between the admissible set of disturbances and the admissible set of initial conditions is clearly visible. The larger is the former, the smaller tends to be the latter. This means that, in general, the larger are the initial conditions to be considered, the smaller will be the admissible bounds on the tolerated disturbances, for which we can ensure that the closed-loop trajectories are bounded.

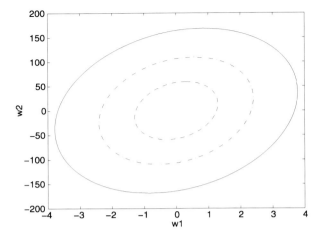

**Fig. 2.3** Example 2.3—obtained domains of admissible disturbances considering: $\theta = 50$ (—), $\theta = 100$ (- - -), $\theta = 130$ (- - -)

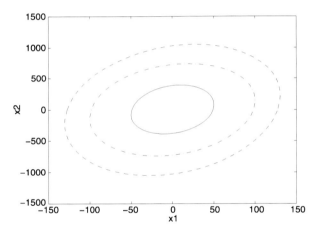

**Fig. 2.4** Example 2.3—obtained domains of admissible initial states considering: $\theta = 50$ (—), $\theta = 100$ (- - -), $\theta = 130$ (- - -)

Considering $\theta = 50$, the following matrices are obtained:

$$R = \begin{bmatrix} 0.0735 & -0.0004 \\ -0.0004 & 0.0000 \end{bmatrix}; \quad W = 10^5 \times \begin{bmatrix} 0.0251 & 0.0389 \\ 0.0389 & 1.5193 \end{bmatrix}$$

For this case, consider now an initial condition $x(0) = \begin{bmatrix} 50 \\ 50 \end{bmatrix}$ and the following disturbance constant amplitude signal:

$$w(t) = \begin{bmatrix} 0 \\ -164.3495 \end{bmatrix}, \quad \forall t \geq 0$$

In this case note that $x(0)'Px(0) = 1$ and $w(t)'Rw(t) = 1$. The response of the system is shown in Fig. 2.5. It appears that the state converges to an equilibrium point outside the linearity region. Note that control signal $u_{(2)}(t)$ remains saturated in steady state. Such a behavior is also depicted in Fig. 2.6. The dashed lines cor-

## 2.3 External Stability

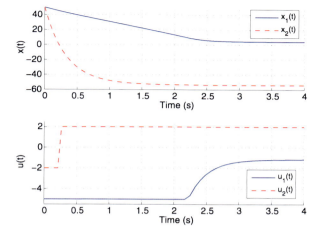

**Fig. 2.5** Example 2.3—time-domain simulation considering $x(0) = [50\ 50]'$ and $w(t) = [0\ -164.3495]'$

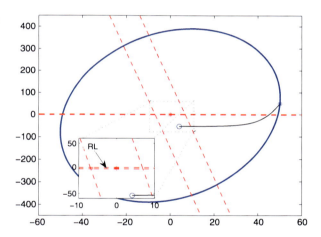

**Fig. 2.6** Example 2.3—state trajectory considering $x(0) = [50\ 50]'$ and $w(t) = [0\ -164.3495]'$, the set $\mathcal{E}(P, 1)$ (—) and the equilibrium point (o)

responds to the lines $K_{(1)}x = \pm u_0$ and $K_{(2)}x = \pm u_0$. As expected, note that the trajectory does not leave the set $\mathcal{E}(P, 1)$.

### 2.3.2 Energy Bounded Exogenous Signals

A generic energy measure of the signal $w(t)$ is given by

$$E_w = \int_0^\infty w(\tau)' R w(\tau)\, d\tau \tag{2.43}$$

with $R = R' > 0$.

Consider that the exogenous signal $w(t)$ is energy bounded, i.e. it belongs to the following set of functions:

$$\mathcal{W} = \left\{ w : [0, \infty) \to \Re^q; \ \int_0^\infty w(\tau)' R w(\tau) \, d\tau \leq \delta^{-1} \right\} \tag{2.44}$$

for some $\delta > 0$. In this case, the energy of $w(t)$ is bounded by $\delta^{-1}$. If $R = I_q$, the energy measure (2.43) corresponds to the square of the $\mathcal{L}_2$-norm of the signal $w(t)$, i.e

$$\|w\|_2 = \sqrt{\int_0^\infty w(\tau)' w(\tau) \, d\tau}$$

In this case, $\mathcal{W}$ represents a set of $\mathcal{L}_2$-bounded disturbances.

Consider a quadratic Lyapunov function $V(x) = x' P x$. If [45]

$$\dot{V}(x) - w' R w < 0 \tag{2.45}$$

is verified along the trajectories of system (2.28), then

$$V\big(x(T)\big) - V\big(x(0)\big) - \int_0^T w(t)' R w(t) \, dt < 0, \quad \forall T$$

Hence, $\forall x(0) \in \mathcal{E}(P, \beta)$ and $w(t) \in \mathcal{W}$, it follows that [89, 280]

- $V(x(T)) < \|w\|_2^2 + V(x(0)) \leq \delta^{-1} + \beta^{-1}$, $\forall T > 0$, i.e. the trajectories of the system do not leave the set $\mathcal{E}(P, (\delta^{-1} + \beta^{-1})^{-1})$;
- if $w(t) = 0$, $\forall t > t_1 \geq 0$, then $\dot{V}(x(t)) < 0$, which ensures that $x(t) \to 0$ as $t \to \infty$.

Based on the discussion above, the following formal result considering the polytopic model II can be stated.

**Proposition 2.5** *If there exist a symmetric positive definite matrix $W \in \Re^{n \times n}$, a matrix $Q \in \Re^{m \times n}$ and a positive scalar $\mu$, satisfying*

$$\begin{bmatrix} W(A + B\Gamma_j^+ K)' + Q'\Gamma_j^- B' + (A + B\Gamma_j^+ K)W + B\Gamma_j^- Q & B_w \\ B_w' & -R \end{bmatrix} < 0,$$
$$j = 1, \ldots, 2^m \tag{2.46}$$

$$\begin{bmatrix} W & Q'_{(i)} \\ Q_{(i)} & \mu u_{0(i)}^2 \end{bmatrix} \geq 0, \quad i = 1, \ldots, m \tag{2.47}$$

$$\delta - \mu \geq 0 \tag{2.48}$$

*then:*

1. *$\forall w(t) \in \mathcal{W}$ and $\forall x(0) \in \mathcal{E}(P, \beta)$, with $0 \leq \beta^{-1} \leq \mu^{-1} - \delta^{-1}$, the trajectories of the system (2.28) do not leave the set $\mathcal{E}(P, \mu)$, with $P = W^{-1}$;*
2. *$\mathcal{E}(P, \mu)$ is a region of asymptotic stability (RAS) for the system (2.28).*

## 2.3 External Stability

*Proof* Left- and right-multiplying the LMI (2.46) by the block diagonal matrix $\text{Diag}(P, I)$, with $P = W^{-1}$ and considering $H = QP$, it follows that (2.46) is equivalent to:

$$\begin{bmatrix} (A + B\Gamma_j^+ K)'P + H'\Gamma_j^- B'P + P(A + B\Gamma_j^+ K) + PB\Gamma_j^- H & PB_w \\ B_w'P & -R \end{bmatrix} < 0 \tag{2.49}$$

Moreover by left- and right-multiplying (2.49) by $[x(t)' \; w(t)']$ and $\begin{bmatrix} x(t) \\ w(t) \end{bmatrix}$, respectively, by convexity, it follows that (2.45) is verified along the trajectories of the polytopic system (2.37). Hence, if the set $\mathcal{E}(P, \mu)$ is included in the region $S(|H|, u_0)$, it follows that (2.46) implies that (2.45) is verified along the trajectories of the saturated system. Relation (2.48) ensures that $\mu^{-1} - \delta^{-1} \geq 0$, i.e. there exists $0 \leq \beta^{-1} \leq \mu^{-1} - \delta^{-1}$. Finally, the LMI (2.47) ensures that $\mathcal{E}(P, \mu) \subset S(|H|, u_0)$, which concludes the proof. $\qquad\square$

As commented in Sect. 1.6.3, for a given set of admissible initial states, we may be interested in estimating the maximal set of admissible exogenous signals $\mathcal{W}$, for which the trajectories are bounded and the internal asymptotic stability is preserved (disturbance tolerance analysis problem). This problem can be addressed by casting the conditions stated in Proposition 2.5 in some optimization problems as follows.

In particular, if $x(0) = 0$ (i.e. the system is in equilibrium), it follows that $\delta^{-1} = \mu^{-1}$ and the problem of maximization of the set $\mathcal{W}$ can be addressed by solving the following convex problem:

$$\min \mu \tag{2.50}$$
$$\text{subject to} \quad \text{inequalities (2.46)–(2.47)}$$

If $x(0) \neq 0$, a trade-off between the size of the admissible initial conditions and the maximal level of admissible disturbances, given by $\delta^{-1}$, appears [62, 280]. The larger is the admissible $\delta$ (i.e. the lower is the disturbance admissible energy), the larger is the "allowable" set of admissible initial conditions, given by $\mathcal{E}(P, \beta)$.

Differently from the amplitude bounded case discussed in Sect. 2.3.1, the set of admissible states does not coincide with the reachable set $\mathcal{E}(P, \beta)$, although both are defined from matrix $P$. Moreover, the set $\mathcal{E}(P, \mu)$ is not an invariant set. In fact, if $x(0) \in \mathcal{E}(P, \mu)$ but $x(0) \notin \mathcal{E}(P, \beta)$, there is no guarantee that the trajectories will lie in $\mathcal{E}(P, \mu)$, $\forall w \in \mathcal{W}$.

Hence, considering a generic given set of admissible initial states $\mathcal{X}_0$, described for instance as in (2.40) or (2.38), the solution of the problem of finding an estimate of the maximal bound on the admissible disturbances is not direct. In this case, two steps have to be performed:

1. solve (2.50);
2. from the obtained $\mu$ computed in the first step, compute $\beta^{-1} = \mu^{-1} - \delta^{-1}$, and test if $\mathcal{X}_0 \subset \mathcal{E}(P, \beta)$.

Of course, the scalar $\beta^{-1}$ in step 2 must be positive. Otherwise, we can conclude that conditions of Proposition 2.5 fail in providing a guarantee of external stability

**Fig. 2.7** Example 2.4—response to $x(0) = 0$ and $w(t)$ as defined in (2.51), with $\alpha = 40.8228$

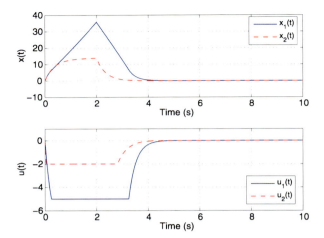

for the saturated system (2.28). Note that if $\mathcal{X}_0$ is described as (2.40) or as (2.38), it can be easily checked if $\mathcal{X}_0 \subset \mathcal{E}(P, \beta)$.

*Example 2.4* Consider the same system as in Example 2.3. We analyse now the case in which the disturbance is energy bounded, i.e.

$$\mathcal{W} = \left\{ w : [0, \infty) \to \Re; \int_0^\infty w(\tau)' w(\tau) d\tau \leq \delta^{-1} \right\}$$

In order to evaluate the disturbance tolerance of the closed-loop system, we consider the optimization problem (2.50), which gives, as optimal value

$$\mu = 3.0003 \times 10^{-4}$$

Hence, if the initial condition is zero, i.e. $x(0) = 0$, from Proposition 2.5, we can ensure that the closed-loop trajectories are bounded for $\|w\|_2 \leq 57.7322$, i.e. $\delta^{-1} = 3.3330 \times 10^3$. It should, however, be highlighted that this value is only an estimate of the actual allowable maximal admissible energy bounded disturbance.

Assuming $x(0) = 0$, Figs. 2.7 and 2.8, show the time response and the trajectory of the state for a disturbance given by

$$w(t) = \begin{cases} \alpha & \text{if } 0 \leq t \leq 2 \\ 0 & \text{if } t > 2 \end{cases} \quad (2.51)$$

with $\alpha = 40.8228$. In this case, it follows effectively that $\|w\|_2 = 57.7322$. It can be noticed that both control signals saturate, then once the disturbance action stops (from $t = 2$) the state converges to the origin. Moreover note that, as expected, the trajectory is bounded, but it is relatively "far" from the boundary of the computed reachable set $\mathcal{E}(P, \mu)$. Actually, optimization problem (2.50) may provide a very conservative estimate of the actual maximal disturbance for which the trajectories are bounded. This comes from the fact that Proposition 2.5 provides only a sufficient condition and that the polytopic model is also a conservative representation of the true saturated system. On the other hand, it is not possible to ensure

## 2.3 External Stability

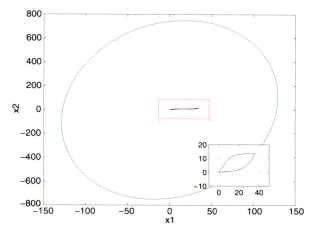

**Fig. 2.8** Example 2.4—state trajectory for $x(0) = 0$, $w(t)$ as defined in (2.51), with $\alpha = 40.8228$ and the set $\mathcal{E}(P, \mu)$

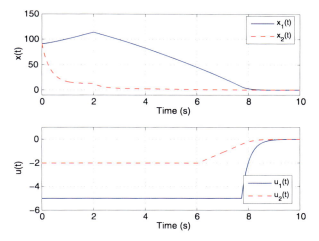

**Fig. 2.9** Example 2.4—response to $x(0) = [91.1859\ 91.1859]'$ and $w(t)$ as defined in (2.51), with $\alpha = 28.8661$

that there is no other particular signal with the same energy that takes the trajectories outside $\mathcal{E}(P, \mu)$ and, eventually, outside the region of attraction of the system.

Let us consider now a trade-off between the initial conditions and disturbances by setting:

$$\delta^{-1} = (2\mu)^{-1}; \qquad \beta^{-1} = (2\mu)^{-1}$$

The simulation results considering $x(0) = [91.1859\ 91.1859]'$ and $w(t)$ defined as in (2.51), with $\alpha = 28.8661$, are depicted in Figs. 2.9 and 2.10. Note that the initial condition is on the boundary of the set $\mathcal{E}(P, \beta)$ and $\|w\|_2^2 = \delta^{-1}$. As expected, the trajectory leaves the set $\mathcal{E}(P, \beta)$, but it does not leave the set $\mathcal{E}(P, \mu)$. Furthermore, when the disturbance goes to zero, the trajectory converges asymptotically to the origin.

**Fig. 2.10** Example 2.4—trajectory for $x(0) = [91.1859\ 91.1859]'$ and $w(t)$ as defined in (2.51), with $\alpha = 28.8661$, and sets $\mathcal{E}(P, \beta)$ and $\mathcal{E}(P, \mu)$

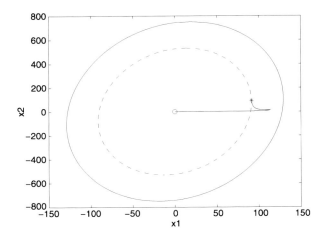

## 2.4 Stabilization

In this section the problems discussed in Sects. 1.6.2 and 1.6.4, regarding the synthesis of control laws taking into account the saturation effects, are addressed, by using the polytopic models. As previously discussed, this class of model is only locally valid. Hence, we are mainly concerned by stating local (regional) stabilization conditions.

In particular, we consider the synthesis of three types of control laws: state feedback, observed-based state feedback and dynamic output feedback. The stabilization conditions are derived in a quadratic framework, which allows to compute the controllers from BMI/LMI optimization problems.

### 2.4.1 State Feedback Stabilization

Regarding Problems 1.4 and 1.6, in this section we focus on the synthesis of a saturating state feedback control law:

$$u(t) = \text{sat}(Kx(t)) \quad (2.52)$$

In other words, the objective concerns the computation of the matrix $K$ in order to ensure the regional asymptotic stability and the external stability of the closed-loop systems (2.1) and (2.28), respectively.

#### 2.4.1.1 Regional Asymptotic Stabilization

Basically, considering a quadratic approach, given a set of admissible initial states $\mathcal{X}_0$, to solve Problem 1.4, the idea is to compute $K$ such that:

- $\dot{V}(x(t)) < 0$ along the trajectories of the polytopic model considered;

## 2.4 Stabilization

- $\mathcal{E}(P, \eta)$ is contained in the region of validity of the polytopic differential inclusion;
- $\mathcal{X}_0 \subseteq \mathcal{E}(P, \eta)$.

Hence, since $\mathcal{E}(P, \eta)$ is a region of asymptotic stability and $\mathcal{X}_0 \subseteq \mathcal{E}(P, \eta)$, it is ensured that $\forall x(0) \in \mathcal{X}_0$, $\lim_{t \to \infty} x(t) = 0$. Note, however, that $\mathcal{X}_0$ is not necessarily a positive invariant set.

On the other hand, we can be interested in computing $K$ in order to maximize the region of attraction of the origin. This problem can be indirectly addressed by computing the gain $K$ in order to maximize an ellipsoidal estimate $\mathcal{E}(P, \eta)$ of the region of attraction.

Regarding the stability conditions stated in Propositions 2.1, 2.2 and 2.3, as well as in Corollary 2.1, we can observe that there exists a product between the variable associated to the quadratic Lyapunov function ($P$ or $W$) and the gain matrix $K$. This bilinearity is unsuitable to solve the problem in a convex framework, considering LMI solvers. This apparent difficulty is easily overcome, by performing a classical linearizing change of variables [26, 45], which corresponds to the introduction of an auxiliary variable $Y$ defined as follows:

$$Y = KW$$

Hence, once we solve the matrix inequalities for $W$ and $Y$, the stabilizing gain is given by $K = YW^{-1}$.

A stabilization result, considering the polytopic approach I, can therefore be straightforwardly obtained from Corollary 2.1 as follows [141].

**Proposition 2.6** *If there exist a symmetric positive definite matrix $W \in \Re^{n \times n}$, a matrix $Y \in \Re^{m \times n}$, a vector $\alpha_l \in \Re^m$ and a positive scalar $\eta$ satisfying the following matrix inequalities:*

$$WA' + AW + B\Gamma_j(\alpha_l)Y + Y'\Gamma_j(\alpha_l)'B' < 0, \quad \forall j = 1, \ldots, 2^m \quad (2.53)$$

$$\begin{bmatrix} W & \alpha_{l(i)}Y'_{(i)} \\ \alpha_{l(i)}Y_{(i)} & \eta u_{0(i)}^2 \end{bmatrix} \geq 0, \quad \forall i = 1, \ldots, m \quad (2.54)$$

$$0 < \alpha_{l(i)} \leq 1, \quad i = 1, \ldots, m \quad (2.55)$$

*then, for $K = YW^{-1}$, the set $\mathcal{E}(P, \eta)$, with $P = W^{-1}$, is a region of asymptotic stability (RAS) for the closed-loop system (2.1).*

Considering polytopic approach II, the result of Proposition 2.2 can be extended as follows to cope with the synthesis of the stabilizing gain $K$ [193].

**Proposition 2.7** *If there exist a symmetric positive definite matrix $W \in \Re^{n \times n}$, matrices $Y \in \Re^{m \times n}$ and $Q \in \Re^{m \times n}$ and a positive scalar $\eta$ satisfying:*

$$WA' + (Y'\Gamma_j^+ + Q'\Gamma_j^-)B' + AW + B(\Gamma_j^+ Y + \Gamma_j^- Q) < 0$$

$$j = 1, \ldots, 2^m \quad (2.56)$$

$$\begin{bmatrix} W & Q'_{(i)} \\ Q_{(i)} & \eta u_{0(i)}^2 \end{bmatrix} \geq 0, \quad i = 1, \ldots, m \quad (2.57)$$

*then, for $K = YW^{-1}$, the set $\mathcal{E}(P, \eta)$, with $P = W^{-1}$, is a region of asymptotic stability (RAS) for the closed-loop system* (2.1).

Analogously, considering polytopic approach III, the following synthesis result can be stated [1].

**Proposition 2.8** *If there exist a symmetric positive definite matrix $W \in \mathfrak{R}^{n \times n}$, a matrix $Y \in \mathfrak{R}^{m \times n}$, row vectors $Q_{j(i)} \in \mathfrak{R}^{1 \times n}$, $j = 2, \ldots, 2^m$, $i \in \mathcal{S}_j$, and a positive scalar $\eta$ satisfying:*

$$\left( AW + \sum_{i \in \mathcal{S}_j^c} B_i Y_{(i)} + \sum_{i \in \mathcal{S}_j} B_i Q_{j(i)} \right)$$

$$+ \left( AW + \sum_{i \in \mathcal{S}_j^c} B_i Y_{(i)} + \sum_{i \in \mathcal{S}_j} B_i Q_{j(i)} \right)' < 0, \quad \forall j = 1, \ldots, 2^m \quad (2.58)$$

$$\begin{bmatrix} W & Q'_{j(i)} \\ Q_{j(i)} & \eta u_{0(i)}^2 \end{bmatrix} \geq 0, \quad \forall j = 2, \ldots, 2^m, \ \forall i \in \mathcal{S}_j \quad (2.59)$$

*then, for $K = YW^{-1}$, the ellipsoid $\mathcal{E}(P, \eta)$, with $P = W^{-1}$, is a region of asymptotic stability (RAS) for the saturated system* (2.1).

The results of Propositions 2.6, 2.7 and 2.8 can be straightforwardly cast in convex optimization problems aiming at the synthesis of $K$, as follows. When using approaches II and III, as pointed out in Sect. 2.2.5, the set of inequalities can be normalized with respect to $\eta$. We can therefore fix $\eta = 1$ without any additional conservatism. This is done in the sequel, i.e. we consider a normalized ellipsoid $\mathcal{E}(P, 1)$.

**A. Regional Guaranteed Stability** Suppose that the set $\mathcal{X}_0$ is given. In particular, this set can be described by the union of polyhedral and ellipsoidal sets described as in (2.38) and (2.40), respectively.

Hence, the problem of finding $K$ that ensures the asymptotic stability in $\mathcal{X}_0$ can be addressed by solving a feasibility problem of the set of inequalities of Propositions 2.6, 2.7 or 2.8 added to LMI constraints like (2.39) and (2.41).

Note that due to the product between $Y$ and $\alpha_l$, the set of matrix inequalities in Proposition 2.6 are bilinear. Thus, the feasibility problem can be addressed by iteratively solving LMI feasibility problems for $\alpha_l$ or $Y$ a priori fixed. On the other hand, the set of inequalities in Proposition 2.7 and 2.8 are linear on the variables $W$, $Y$ and $Q$ (or $Q_{j(i)}$) and a direct LMI feasibility problem can be solved.

**B. Maximization of an Estimate of the Region of Attraction** In this case the idea is to compute $K$ that leads to a region $\mathcal{E}(P, 1)$ as large as possible considering a size criterion, as discussed in Sect. 2.2.5.1.

2.4 Stabilization                                                                      79

For instance, the following optimization problems can be considered.

- Maximization of the minor axis of $\mathcal{E}(P, 1)$:

$$
\begin{aligned}
&\max \ \lambda \\
&\text{subject to} \quad W > \lambda I_n \\
&\qquad\qquad\quad \lambda > 0
\end{aligned}
$$
(2.60)

$$\text{inequalities (2.53)–(2.55) or (2.56)–(2.57) or (2.58)–(2.59)}$$

- Maximization of $\mathcal{E}(P, 1)$ in certain directions

  Sometimes, we are interested in maximizing the ellipsoidal set in some specific directions or following a given shape set $\mathcal{X}_0$. In particular, this shape set can be described as the union (or the convex hull) of polyhedral and ellipsoidal sets as defined in (2.40) or (2.38). Considering a scaling factor $\beta$, the idea is to maximize $\beta$ such that $\beta \mathcal{X}_0 \subset \mathcal{E}(P, 1)$ and subject to conditions of Propositions 2.6, 2.7 and 2.8 This can be accomplished by the following optimization problem [133, 137, 193]:

$$
\begin{aligned}
&\max \ \beta \\
&\text{subject to} \quad \begin{bmatrix} 1 & \beta v_i' \\ \beta v_i & W \end{bmatrix} \geq 0, \quad \forall i = 1, \ldots, n_v \quad \text{and/or} \\
&\qquad\qquad\quad \begin{bmatrix} P_0 & \beta I_n \\ \beta I_n & W \end{bmatrix} \geq 0
\end{aligned}
$$
(2.61)

$$\text{inequalities (2.53)–(2.55) or (2.56)–(2.57) or (2.58)–(2.59)}$$

**C. Optimization of the Actuator Size**    The control bounds are in general related to the size and the cost of the actuators. A simple (and frequently used in industry) strategy to tackle the problem of saturation is to oversize the actuators, i.e. specify actuators with large output limits. However, in general this leads to actuators that are large, which has serious implication in embedded systems, more expensive and less efficient (in terms of energy consumption), which increases project and production costs.

Hence the idea is to design the control law in order to achieve a guaranteed stability for a given set of admissible initial conditions $\mathcal{X}_0$, with less costly actuators. In other words, the control law and the control bounds are simultaneously computed to ensure the stability of the closed-loop system in a pre-specified set.

From the conditions stated in Propositions 2.6, 2.7 or 2.8, and supposing $\mathcal{X}_0$ given as the union (or the convex hull) of polyhedral and ellipsoidal sets as defined in (2.40) or (2.38), this problem can be addressed from the solution of the following optimization problem.

$$\min \sum_{i=1}^{m} c_i \bar{u}_{0(i)}$$

$$\text{subject to} \quad \begin{bmatrix} 1 & v'_i \\ v_i & W \end{bmatrix} \geq 0, \quad \forall i = 1, \ldots, n_v \quad \text{and/or}$$

$$\begin{bmatrix} P_0 & I_n \\ I_n & W \end{bmatrix} \geq 0$$

$$\text{inequalities} \begin{cases} (2.53), (2.55) \text{ and } \begin{bmatrix} W & \alpha_{l(i)} Y'_{(i)} \\ \alpha_{l(i)} Y_{(i)} & \bar{u}_{0(i)} \end{bmatrix} \geq 0, \\ \quad \forall i = 1, \ldots, m \quad \text{or} \\ (2.56) \text{ and } \begin{bmatrix} W & Q'_{(i)} \\ Q_{(i)} & \bar{u}_{0(i)} \end{bmatrix} \geq 0, \quad \forall i = 1, \ldots, m \quad \text{or} \\ (2.58) \text{ and } \begin{bmatrix} W & Q'_{j(i)} \\ Q_{j(i)} & \bar{u}_{0(i)} \end{bmatrix} \geq 0, \\ \quad \forall j = 2, \ldots, 2^m, \forall i \in \mathcal{S}_j \end{cases}$$

$$(2.62)$$

with the coefficients $c_i$ being the weights of the size/cost of each actuator in the composition of the cost function. In this case, it follows that $u_{0(i)} = \sqrt{\bar{u}_{0(i)}}$, $i = 1, \ldots, m$.

#### 2.4.1.2 Performance Issues

Suppose there exists a feasible solution for the inequalities of Propositions 2.6, 2.7 or 2.8 such that the inclusion $\mathcal{X}_0 \subset \mathcal{E}(P, 1)$ is satisfied.

A quite natural objective is the search, among all possible stabilizing gains (i.e. all the feasible solutions to the set of inequalities), of an optimal solution taking into account some performance requirement.

In the case where we are interested in the time-domain performance improvement, the idea is, in general, to use the available control to drive the initial state to the origin as fast as possible and with a good damping. Considering a Lyapunov function $V(x(t))$, the estimate of the convergence rate of a nonlinear system can be obtained by an upper bound of $\dot{V}(x(t))$. However, the exigence of the same performance level when the system presents control saturation and when it operates linearly is not realistic. Considering that a performance level is achieved with the unsaturated control law ($u(t) = Kx(t)$), in general it is not possible to keep the same performance level when saturation effectively occurs since the control availability is reduced. Hence, to impose the same performance level when the system operates inside and outside of the linearity region can lead to very conservative solutions. The idea is then to consider different performance criteria depending on the region of operation [141, 144].

The time-domain performance specification in the linearity region may be achieved by placing the poles of $(A + BK)$ in a suitable region of the left half

2.4 Stabilization 81

complex plane. This kind of region can be generically described as an LMI region as follows [68]:

$$\mathcal{D}_p = \left\{ s \in \mathcal{C}; \ \left( L + sZ + \bar{s}Z' \right) < 0 \right\} \tag{2.63}$$

where $L = L' \in \mathfrak{R}^{l \times l}$, $Z \in \mathfrak{R}^{l \times l}$ and $s$ is a complex number with its conjugate $\bar{s}$.

Henceforth, we assume that the time-domain requirements in the region of linear behavior are satisfied if the poles $(A + BK)$ are located in a region $\mathcal{D}_p$.

The following result can be used to incorporate this kind of performance constraint in the synthesis of the saturating control law.

**Theorem 2.1** [68] *Consider a region $\mathcal{D}_p$ described by (2.63). If there exists a positive definite matrix $W \in \mathfrak{R}^{n \times n}$ and a matrix $Y \in \mathfrak{R}^{m \times n}$ such that*

$$L \otimes W + Z \otimes (AW + BY) + Z' \otimes (AW + BY)' < 0 \tag{2.64}$$

*then $K = YW^{-1}$ places the poles of $(A + BK)$ in $\mathcal{D}_p$.*

The LMI constraint (2.64) can be straightforwardly incorporated in a feasibility problem or even in the optimization problems (2.60) and (2.61) [141]. On the other hand, if there is a feasible solution satisfying the conditions of Propositions 2.6, 2.7 or 2.8 and the inclusion constraint $\mathcal{X}_0 \subseteq \mathcal{E}(P, 1)$, it is possible to search for a gain $K$ in the sense of improving the performance in the region of linearity. Let us consider, for example, the region $\mathcal{D}_p$ be defined as follows:

$$\mathcal{D}_p = \left\{ s \in \mathcal{C}; \ \mathfrak{Re}\{s\} < -\sigma, \ \sigma > 0 \right\} \tag{2.65}$$

Notice that the larger is $\sigma$, the farther from the origin are the poles of $(A + BK)$ and the larger is the speed of convergence of the trajectories to the origin inside the region of linearity. Thus, the following optimization problem can be formulated in order to improve the convergence rate inside the linearity region [130, 137]:

max $\sigma$

subject to $\quad WA' + AW + BY + Y'B' + 2\sigma W < 0$

$$\begin{bmatrix} 1 & v_i' \\ v_i & W \end{bmatrix} \geq 0, \quad \forall i = 1, \dots, n_v \quad \text{and/or}$$

$$\begin{bmatrix} P_0 & I_n \\ I_n & W \end{bmatrix} \geq 0 \tag{2.66}$$

inequalities (2.53)–(2.55) or (2.56)–(2.57) or (2.58)–(2.59)

Considering (2.53)–(2.55) with a fixed $\alpha_l$, or directly (2.56)–(2.57) and (2.58)–(2.59), it follows that problem (2.66) is a *generalized eigenvalue problem* (GEVP) [45] and can be easily solved by any standard LMI solver package.

82      2 Stability Analysis and Stabilization—Polytopic Representation Approach

### 2.4.1.3 Trade-off Between Saturation, Size of the Stability Region and Time-Domain Performance

In some cases one can be interested in finding a gain $K$ such that for all $x(0) \in \mathcal{X}_0$, the trajectories never leave the linearity region. This solution, known as *saturation avoidance problem* (see Sect. 1.6.2.3), can be solved if the following constraints are satisfied:

(a) $WA' + AW + BY + Y'B' < 0$;
(b) $\mathcal{E}(P, 1) \subset S(|K|, u_0)$.

A fundamental issue is whether the use of effective saturating control laws can be advantageous or not. In [190, 219] it was shown that, at least in some cases, the use of the saturating control laws does not help in obtaining larger regions of stability. It is, however, very important to highlight that no constraints concerning neither the performance, nor the robustness, are taken into account in this analysis. In this case, although the optimal region of stability is obtained with a linear control law, the closed-loop poles associated to this solution can be very close to the imaginary axis, which implies a very slow behavior [130, 137, 144].

Hence, the idea is to allow saturation in order to ensure the stability for larger sets of admissible states, but in the presence of performance constraints. For instance, if the criterion is the maximization of $\mathcal{E}(P, 1)$ along a shape set $\mathcal{X}_0$ and a constraint $\sigma(A + BK) \subset \mathcal{D}_p$ is considered (with $\mathcal{D}_p$ defined in (2.65)), the following optimization problem can be solved to compute $K$:

$$\max \ \beta$$

$$\text{subject to} \quad WA' + AW + BY + Y'B' + 2\sigma W < 0$$

$$\begin{bmatrix} 1 & \beta v'_i \\ \beta v_i & W \end{bmatrix} \geq 0, \quad \forall i = 1, \dots, n_v \quad \text{and/or}$$

$$\begin{bmatrix} P_0 & \beta I_n \\ \beta I_n & W \end{bmatrix} \geq 0 \tag{2.67}$$

inequalities (2.53)–(2.55) or (2.56)–(2.57) or (2.58)–(2.59)

*Example 2.5*   Let the open-loop system be described by the following matrices:

$$A = \begin{bmatrix} 0 & 1 \\ 10 & -0.1 \end{bmatrix}; \qquad B = \begin{bmatrix} 0 \\ 1 \end{bmatrix}$$

Consider the control bounds given by $u_0 = 1$. Let the set of admissible initial conditions be given by an hypercube in $\Re^2$:

$$\mathcal{X}_0 = \left\{ x \in \Re^2; \ -1 \leq x_{(i)} \leq 1, \ \forall i = 1, \dots, 2 \right\}$$

We consider for the performance specification in the linearity region the placement of the poles of $(A + BK)$ in the region defined in (2.65).

Table 2.2 shows the optimal obtained values for the optimization Problem (2.67) considering the polytopic approach I. For different values of $\sigma$, the values of $\beta_{\text{lin}}$

## 2.4 Stabilization

**Table 2.2** Example 2.5—trade-off between the size of the region of stability and the performance constraint in terms of a pole placement

| $\sigma$ | $\beta_{\text{lin}}$ | $\alpha_l^\star$ | $\beta^\star$ | $K^\star$ | $\text{eig}_{\max}$ |
|---|---|---|---|---|---|
| 0 | 0.0653 | 1 | 0.0653 | $[-11.56\ -3.60\ -0.15]$ | $-0.0001$ |
| 1 | 0.0494 | 0.76 | 0.0653 | $[-15.29\ -4.76\ -0.206]$ | $-1.334$ |
| 5 | 0.0133 | 0.49 | 0.0266 | $[-50.42\ -12.28\ -0.50]$ | $-5.08 \pm j3.09$ |

(obtained considering $\alpha_l = 1$, i.e. $\mathcal{E}(P, 1) \subset S(|K|, u_0) = R_L$) and the optimal value of $\beta$ with the associated $\alpha_l$ are depicted. It is also shown the obtained gain $K$ and the maximal eigenvalue of $(A + BK)$. The maximum $\beta$ is obtained for the linear case ($\alpha_l = 1$) with $\sigma = 0$. This is in accordance with the results in [219] and [190]. However, note that in this case the eigenvalues of the linear system are very close to the imaginary axis. Considering $\sigma \neq 0$, the best $\beta$ is achieved considering saturation, i.e. $\alpha_l \neq 1$.

*Example 2.6* Consider the control of two inverted pendulums in cascade where the system matrices are

$$A = \begin{bmatrix} 0 & 1 & 0 & 0 \\ 9.8 & 0 & -9.8 & 0 \\ 0 & 0 & 0 & 1 \\ -9.8 & 0 & 2.94 & 0 \end{bmatrix}; \quad B = \begin{bmatrix} 0 & 0 \\ 1 & -2 \\ 0 & 0 \\ -2 & 5 \end{bmatrix}$$

The control bounds are given by $u_0 = [10\ 10]'$. Consider the following shape set:

$$\mathcal{X}_0 = \text{Co} \left\{ \begin{bmatrix} 0.5 \\ 0 \\ 0.5 \\ 0 \end{bmatrix}, \begin{bmatrix} 0.5 \\ 0 \\ -0.5 \\ 0 \end{bmatrix}, \begin{bmatrix} -0.5 \\ 0 \\ 0.5 \\ 0 \end{bmatrix}, \begin{bmatrix} -0.5 \\ 0 \\ -0.5 \\ 0 \end{bmatrix} \right\}$$

Notice that matrix $A$ is unstable (the eigenvalues of $A$ are 4.0930, $-4.0930$, $\pm 2.0032i$). As a performance requirement we consider the placement of the eigenvalues of $(A + BK)$ in the region $\mathcal{D}_p$ defined in (2.65). Hence the larger is $\sigma$, the larger tends to be the rate of convergence of the linear trajectories to the origin.

Considering the above data, and the polytopic approach I, Table 2.3 shows the final values of $\alpha_l$ and $\beta$ obtained from a stabilization version of the iterative algorithm proposed in Sect. 2.2.5.2 from different initial vectors $\alpha_l$ and scalars $\sigma$. $\beta_{\text{initial}}$ and $\beta_{\text{final}}$ denote the optimal value of $\beta$ obtained by the iterative algorithm from, respectively, $\alpha_{\text{initial}}$ and $\alpha_{\text{final}}$.

Regarding Table 2.3 we may notice the following: the smaller are the components of $\alpha_l$, the larger is the $\beta$ obtained from (2.67). This illustrates the fact that by allowing saturation we can stabilize the system for a larger set of initial conditions. Besides, the more stringent is the performance requirement (larger $\sigma$, in this

**Table 2.3** Example 2.6—algorithm performance and trade-off between performance requirement and size of the region of stability

| $\sigma$ | $\alpha_{initial}$ | $\beta_{initial}$ | $\alpha_{final}$ | $\beta_{final}$ |
|---|---|---|---|---|
| 1 | $[1.00\ 1.00]'$ | 1.9883 | $[0.7912\ 0.8950]'$ | 2.3112 |
| 1 | $[0.50\ 0.50]'$ | 2.3652 | $[0.4999\ 0.5000]'$ | 2.3653 |
| 1 | $[0.25\ 0.25]'$ | 2.5607 | $[0.2495\ 0.2500]'$ | 2.5613 |
| 2 | $[1.00\ 1.00]'$ | 0.9731 | $[0.7232\ 0.8430]'$ | 1.2154 |
| 2 | $[0.50\ 0.50]'$ | 1.2930 | $[0.4997\ 0.5000]'$ | 1.2931 |
| 2 | $[0.25\ 0.25]'$ | 1.5378 | $[0.2493\ 0.2500]'$ | 1.5388 |
| 3 | $[1.00\ 1.00]'$ | 0.5707 | $[0.6850\ 0.8039]'$ | 0.7519 |
| 3 | $[0.50\ 0.50]'$ | 0.8302 | $[0.5000\ 0.5000]'$ | 0.8316 |
| 3 | $[0.25\ 0.25]'$ | 1.0737 | $[0.2485\ 0.2500]'$ | 1.0747 |

case), the smaller is the region of admissible initial states for which we can find a solution.

Comparing the solution obtained by avoiding saturation (i.e. $\alpha_l = [1\ \ 1]'$) with the solution that takes into account the nonlinear behavior of the system, we can observe that:

- For $\sigma = 1$, it is possible to obtain a solution for a set of admissible initial states 28% larger.
- For $\sigma = 2$, it is possible to obtain a solution for a set of admissible initial states 58% larger.
- For $\sigma = 3$, it is possible to obtain a solution for a set of admissible initial states 88% larger.

In the examples above, the trade-off between the size of the stability region and the performance of the closed-loop system was illustrated. In general, the more demanding is the required time-domain performance, the smaller will be the region for which the stability can be ensured. Hence, an interesting idea is the application of a "low" gain when the state is far from the equilibrium point and a "high" gain when it approaches the equilibrium [309]. The switching between these gains should be done on a switching surface defined in the state space. This idea is developed in the next sub-section in the context of the present approach.

### 2.4.1.4 Piecewise Linear State Feedback

The philosophy of the piecewise linear state feedback control (or switched state feedback control) consists in applying higher feedback gains to the system as the state approaches the origin. This is an interesting way to deal with the problem of ensuring the stability for larger regions in the presence of saturation and, at the same time, to improve the rate of convergence of the closed-loop trajectories to the origin. This idea has been applied for instance in [187, 395], considering a saturation avoidance approach, and in [130, 133, 137], considering effective saturation.

2.4 Stabilization

The main problem of this kind of control law is to determine appropriate switching sets and the associated gains in order to avoid limit cycles or unstable behavior. In particular, it must be ensured that at a switching instant $t = t_s$, the state $x(t_s)$ belongs to the region of attraction associated to the state feedback gain to be applied. Following this principle, we show now how to compute a stabilizing piecewise *saturating* control law based on the conditions given in Propositions 2.6, 2.7 or 2.8.

Let $N$ be the number of desired switching sets and $\mathcal{X}_0$ the set of admissible states for which the stability should be ensured. A piecewise *saturating* control law can be computed as follows [137]:

*Step 1.* Define $N$ homothetical sets to $\mathcal{X}_0$ as follows:

$$\mathcal{X}_q = \beta_q \mathcal{X}_0, \quad 0 < \beta_q < 1, \ q = 1, \dots, N$$
$$\mathcal{X}_N \subset \mathcal{X}_{N-1} \subset \cdots \subset \mathcal{X}_1 \subset \mathcal{X}_0$$

*Step 2.* For each $q = 0, \dots, N$, solve a GEVP of type (2.66), by considering $\mathcal{X}_q$ as given and $W_q, Y_q, \sigma$ as the associated optimal solution.
*Step 3.* For each $q = 0, \dots, N$ define:

– the feedback matrix: $K_q = Y_q W_q^{-1}$;
– the switching set: $S_q = \{x \in \Re^n; \ x^T W_q^{-1} x \le 1\}$.

Hence, from Propositions 2.6, 2.7 or 2.8, it follows that the application of the control law defined as

$$u(t) = \begin{cases} \text{sat}(K_0 x(t)) & \text{if } x(t) \in S_0, \ x(t) \notin \{S_1, S_2, \dots, S_N\} \\ \text{sat}(K_1 x(t)) & \text{if } x(t) \in S_1, \ x(t) \notin \{S_2, S_3, \dots, S_N\} \\ \vdots & \vdots \\ \text{sat}(K_N x(t)) & \text{if } x(t) \in S_N \end{cases} \tag{2.68}$$

guarantees the asymptotic convergence to the origin of all the trajectories emanating from $\mathcal{X}_0$. By construction, if at instant $t = t_s$ the state feedback is switched from $K_{q-1}$ to $K_q$, the state $x(t_s)$ belongs to the region of attraction associated to $\dot{x}(t) = Ax(t) + B \, \text{sat}(K_q x(t))$. Note that we consider a different maximized $\sigma$ for each gain $K_q$. The idea is to accelerate even more the convergence by allowing the saturation each time the trajectory enters a new region $S_q$.

*Remark 2.5* Differently from [395], the ellipsoids do not need to be nested in the presented approach. Anyway, if for some reason one wants to ensure the nesting property for the switching surfaces, it suffices to consider inclusion constraints of type (2.41).

*Remark 2.6* Considering a parameterized Riccati equation approach and the avoidance saturation case, a similar philosophy of switching to "higher" gains as the state approaches the origin is developed in Chap. 5. In that case, each switching set is considered to be included in the region of linearity of the corresponding closed-loop system, i.e. $S_q \subset S(|K_q|, u_0)$.

## 2.4.1.5 Regional External Stabilization

Similarly to Sect. 2.4.1.1, considering the linearizing variable change $Y = KW$, stabilization conditions for the case where the system is subject to the action of additive exogenous signals can be straightforwardly derived. In this section we briefly present the resulting stabilization conditions stated considering the polytopic model II. Analogous results can be obtained considering the polytopic models I and III. In addition, some optimization problems are proposed in order to compute the state feedback gain from the proposed conditions.

**(A) Amplitude Bounded Exogenous Signals**  Considering that $w(t)$ is bounded in amplitude, i.e. $w(t) \in \mathcal{W}$ with $\mathcal{W}$ defined in (2.29), the following result follows from Proposition 2.4.

**Proposition 2.9**  *If there exist a symmetric positive definite matrix $W \in \Re^{n \times n}$, matrices $Q, Y \in \Re^{m \times n}$ and positive scalars $\tau_1$ and $\tau_2$ satisfying*

$$
\begin{bmatrix} WA' + AW + (Y'\Gamma_j^+ + Q'\Gamma_j^-)B' + B(\Gamma_j^+ Y + \Gamma_j^- Q) + \tau_1 W & B_w \\ B_w' & -\tau_2 R \end{bmatrix} < 0
$$

$$
j = 1, \dots, 2^m \tag{2.69}
$$

$$
\begin{bmatrix} W & Q'_{(i)} \\ Q_{(i)} & u_{0(i)}^2 \end{bmatrix} \geq 0, \quad i = 1, \dots, m \tag{2.70}
$$

$$
\tau_2 - \delta\tau_1 < 0 \tag{2.71}
$$

*then the gain $K = YW^{-1}$ is such that:*

1. *$\forall w \in \mathcal{W}$, the trajectories of system (2.28) do not leave the set $\mathcal{E}(P, 1)$;*
2. *for $w = 0$, $\mathcal{E}(P, 1)$ is a region of asymptotic stability (RAS) for system (2.28).*

From the conditions stated in Proposition 2.9, the following optimization problems can be used to compute a suitable stabilizing state feedback gain.

1. *Maximization of the Disturbance Tolerance with Guaranteed RAS*  Suppose that a set of admissible initial states $\mathcal{X}_0$ is given as a polyhedral or an ellipsoidal set, as defined, respectively, in (2.40) and (2.38), or a combination of both. The idea is to compute $K$ in order to maximize the disturbance tolerance [353], considering that $x(0) \in \mathcal{X}_0$. For the sake of simplicity consider $R = I$. In this case, the objective consists in maximizing the bound $\delta^{-1}$ on the exogenous signal, for which it is possible to find a stabilizing gain $K$. Hence, the following optimization problem can be considered:

$$
\min_{\delta, W, Y, \tau_1, \tau_2} \delta \tag{2.72}
$$

$$
\text{subject to} \quad \text{inequalities (2.39) and/or (2.41), (2.69)–(2.71)}
$$

Note that due to the products $\delta\tau_1$ and $\tau_1 W$, the inequalities (2.69) and (2.71) are not LMIs. In order to overcome this drawback, problem (2.72) can be solved by considering the solutions of LMI-based problems formulated on a grid in $\tau_1$, i.e. a line search considering subproblems with $\tau_1$ fixed.

## 2.4 Stabilization

2. *Maximization of the RAS for a Given Set of Admissible Disturbances* Suppose that matrix $R$ and the scalar $\delta$ defining the set of admissible exogenous signals are given. The idea is to find $K$ that maximizes the set $\mathcal{E}(P, 1)$, while ensuring that the trajectories are bounded in this set for every $w \in \mathcal{W}$ and $x(0) \in \mathcal{E}(P, 1)$. Thus, the following optimization problem can be formulated:

$$\max_{W, Y, \tau_1, \tau_2} f\left(\mathcal{E}(P, 1)\right)$$
$$\text{subject to} \quad \text{inequalities (2.69)–(2.71)} \tag{2.73}$$

with $f(\mathcal{E}(P, 1))$ being a size criterion as discussed in Sect. 2.2.5.1.

**(B) Energy Bounded Exogenous Signals** Considering now that $w(t)$ is bounded in energy, i.e. $w(t) \in \mathcal{W}$ with $\mathcal{W}$ defined in (2.44), the following result can be derived from Proposition 2.5.

**Proposition 2.10** *If there exist a symmetric positive definite matrix $W \in \Re^{n \times n}$, a matrices $Q, Y \in \Re^{m \times n}$ and a scalar $\mu$, satisfying*

$$\begin{bmatrix} WA' + AW + (Y'\Gamma_j^+ + Q'\Gamma_j^-)B' + B(\Gamma_j^+ Y + \Gamma_j^- Q) & B_w \\ B_w' & -R \end{bmatrix} < 0$$
$$j = 1, \ldots, 2^m \tag{2.74}$$

$$\begin{bmatrix} W & Q_{(i)}' \\ Q_{(i)} & \mu u_{0(i)}^2 \end{bmatrix} \geq 0, \quad i = 1, \ldots, m \tag{2.75}$$

$$\delta - \mu \geq 0 \tag{2.76}$$

*then the gain $K = YW^{-1}$ is such that:*

1. *$\forall w(t) \in \mathcal{W}$ and $\forall x(0) \in \mathcal{E}(P, \beta)$, with $0 \leq \beta^{-1} \leq \mu^{-1} - \delta^{-1}$ and $P = W^{-1}$, the trajectories of the system (2.28) do not leave the set $\mathcal{E}(P, \mu)$;*
2. *for $w = 0$, $\mathcal{E}(P, \mu)$ is a region of asymptotic stability (RAS) for the system (2.28).*

From Proposition 2.10, it is possible to compute $K$ in order to ensure that the trajectories of the system are bounded for any energy bounded admissible signal $w(t)$. Consider now that the regulated output of the system is given by

$$z(t) = C_z x(t) \tag{2.77}$$

An additional objective can be the determination of an upper bound for the $\mathcal{L}_2$-gain from the admissible disturbance $w$ to the regulated output $z$. In other words, when $w$ is viewed as a disturbance, we want to ensure some level (in the $\mathcal{L}_2$ sense) of disturbance rejection. In this case, we consider $R = I$. This objective can be achieved by defining a function [45, 215]:

$$\mathcal{J}(t) = \dot{V}\left(x(t)\right) - w(t)'w(t) + \frac{1}{\gamma} z(t)'z(t)$$

with $V(x(t)) = x(t)'Px(t)$, $P = P' > 0$. Hence, if $\mathcal{J}(t) < 0$, one obtains

$$\int_0^T \mathcal{J}(t)\,dt = V\big(x(T)\big) - V\big(x(0)\big) - \int_0^T w(t)'w(t)\,dt + \frac{1}{\gamma}\int_0^T z(t)'z(t)\,dt < 0$$
$$\forall\, T > 0$$

and it is possible to conclude that:

- $V(x(T)) < V(x(0)) + \|w\|_2^2$,
- for $T \to \infty$, $\|z\|_2^2 < \gamma(\|w\|_2^2 + V(x(0)))$,
- if $w(t) = 0$, $\forall t \geq t_1 \geq 0$, then $\dot{V}(x(t)) < -\frac{1}{\gamma}z(t)'z(t) < 0$.

The following result can be formulated in this case.

**Proposition 2.11** *If there exist a symmetric positive definite matrix $W \in \Re^{n\times n}$, a matrices $Q, Y \in \Re^{m\times n}$ and scalars $\mu$ and $\gamma$ satisfying*

$$\begin{bmatrix} WA' + AW + (Y'\Gamma_j^+ + Q'\Gamma_j^-)B' + B(\Gamma_j^+ Y + \Gamma_j^- Q) & B_w & WC_z' \\ B_w' & -I & 0 \\ C_z W & 0 & -\gamma I \end{bmatrix} < 0$$
$$j = 1, \ldots, 2^m \tag{2.78}$$

$$\begin{bmatrix} W & Q'_{(i)} \\ Q_{(i)} & \mu u_{0(i)}^2 \end{bmatrix} \geq 0, \quad i = 1, \ldots, m \tag{2.79}$$

$$\delta - \mu \geq 0 \tag{2.80}$$

*then the gain $K = YW^{-1}$ is such that:*

1. *$\forall w(t)$ such that $\|w(t)\|_2^2 \leq \delta^{-1}$ and $\forall x(0) \in \mathcal{E}(P, \beta)$, with $0 \leq \beta^{-1} \leq \mu^{-1} - \delta^{-1}$:*

   - *the trajectories of the system (2.28) do not leave the set $\mathcal{E}(P, \mu)$, with $P = W^{-1}$;*
   - *$\|z\|_2^2 < \gamma(\|w\|_2^2 + V(x(0)))$.*

2. *If $w(t) = 0$, $\forall t \geq t_1 \geq 0$, then $x(t) \to 0$, i.e. $\mathcal{E}(P, \mu)$ is a region of asymptotic stability (RAS) for the system (2.28).*

*Proof* Left- and right-multiplying (2.78) by $\mathrm{Diag}(P, I, I)$ and applying Schur's complement in the sequel, it follows that (2.78) is equivalent to

$$\begin{bmatrix} A'P + PA + (K'\Gamma_j^+ + H'\Gamma_j^-)B'P + PB(\Gamma_j^+ K + \Gamma_j^- H) + \gamma^{-1}C'C & PB_w \\ B_w'P & -I \end{bmatrix}$$
$$< 0, \quad j = 1, \ldots, 2^m$$

with $H = QP$, which implies that $\mathcal{J}(t) < 0$ along the trajectories of system (2.28), provided that $\mathcal{E}(P, \mu) \subset S(|H|, u_0)$. This inclusion is ensured by (2.79). The inequality (2.80) ensures that $\beta^{-1} \geq 0$. Hence, we can conclude that:

- For $w$ such that $\|w\|_2^2 \leq \delta^{-1}$ and $x(0) \in \mathcal{E}(P, \beta)$ it follows that $V(x(T)) < V(x(0)) + \|w\|_2^2 \leq \delta^{-1} + \beta^{-1} \leq \mu^{-1}$, $\forall T > 0$, i.e. the trajectories of the system are confined in the set $\mathcal{E}(P, \mu)$.

2.4 Stabilization     89

- For $T \to \infty$, $\|z\|_2^2 < \gamma(\|w\|_2^2 + V(x(0)))$.
- If $w(t) = 0$, $\forall t \geq t_1 \geq 0$, then $\dot{V}(x(t)) < -\frac{1}{\gamma} z(t)' z(t) < 0$, $\forall x(t) \in \mathcal{E}(P, \mu)$, which ensures that this set is a region of asymptotic stability.    □

Considering that $x(0) = 0$, the result of Proposition 2.11 ensures the $\mathcal{L}_2$ input-to-output stability of the closed-loop system. In this case $\mu^{-1} = \delta^{-1}$ and the $\mathcal{L}_2$-gain of the system is given by $1/\sqrt{\gamma}$, i.e. $\|z\|_2 \leq 1/\sqrt{\gamma}\|w\|_2$. Thus, the following convex optimization problems are of special interest.

1. *Maximization of the Disturbance Tolerance*
   The idea consists in maximizing the bound on the disturbance energy, for which we can ensure that the system trajectories remain bounded. This can be accomplished by the following optimization problem.

$$\begin{aligned} &\min \ \mu \\ &\text{subject to} \quad \text{inequalities (2.74)–(2.75)} \end{aligned} \qquad (2.81)$$

2. *Maximization of the Disturbance Rejection*
   For an *a priori* given bound on the $\mathcal{L}_2$-norm of the admissible disturbances (given by $\frac{1}{\delta}$), the idea consists in minimizing the upper bound for the $\mathcal{L}_2$-gain from $w(t)$ to $z(t)$. This can be obtained from the solution of the following optimization problem, considering $\mu = \delta$:

$$\begin{aligned} &\min \ \gamma \\ &\text{subject to} \quad \text{inequalities (2.78)–(2.79)} \end{aligned} \qquad (2.82)$$

*Remark 2.7* In the case of a non-null initial condition $x(0)$, since $\mu^{-1} = \delta^{-1} + \beta^{-1}$, there is a trade-off between the size of the set of admissible conditions (given basically by $\beta^{-1}$), and the size of the admissible norm of the exogenous signal (given by $\delta^{-1}$). In this case, the finite $\mathcal{L}_2$-gain from $w$ to $z$ presents a bias term [62]:

$$\|z\|_2^2 \leq \gamma \|w\|_2^2 + \gamma x(0)' P x(0) \leq \gamma(\|w\|_2^2 + \beta)$$

*Example 2.7* Recall the system of Example 2.2, with

$$B_w = \begin{bmatrix} 1 \\ 1 \end{bmatrix}$$

By solving the optimization problem (2.81) we obtain

$$\delta = \mu = 0.3200$$

which means that, provided $x(0) = 0$, the largest bound on the $\mathcal{L}_2$ disturbance, for which it is possible to determine a stabilizing gain using the result of Proposition 2.10, is given by $1/\sqrt{\delta} = 1.7678$.

In this case, the obtained state feedback matrix is

$$K = 10^4 \times [1.8799 \ 1.8801]$$

Clearly, this matrix is not suitable for implementation neither for simulation because its entries are too large, which can make the closed-loop system too much sensible

90    2  Stability Analysis and Stabilization—Polytopic Representation Approach

**Table 2.4** Example 2.7—minimization of $\gamma$

| $\delta = \mu$ | $\gamma$ | $K$ |
|---|---|---|
| 0.32100 | 318.02 | [67.874 67.879] |
| 0.35000 | 9.0317 | [67.772 67.863] |
| 0.40000 | 2.7262 | [68.043 68.046] |
| 0.50000 | 0.8809 | [67.782 67.783] |
| 0.60000 | 0.4468 | [68.327 68.356] |
| 0.70000 | 0.2723 | [68.498 68.498] |
| 0.80000 | 0.1839 | [68.669 68.669] |
| 1.0000 | 0.1004 | [69.017 69.017] |
| 1.2000 | 0.063276 | [69.376 69.376] |
| 1.4000 | 0.043544 | [69.713 69.714] |
| 10.000 | 0.00063389 | [97.131 97.131] |
| 20.000 | 0.00040000 | [101.00 101.00] |

to measurement noise, for instance. Note also that the eigenvalues of $(A + BK)$ in this case are given by $-0.9999$ and $-10^4 \times 1.8800$, i.e. there is a clear dominant mode. In order to overcome this problem, some constraints on the placement of the eigenvalues of $(A + BK)$ can be used. For instance, consider the pole placement in a strip in the complex plane between $-\alpha_1$ and $-\alpha_2$, $\alpha_1 < \alpha_2$, which can be achieved if the following LMI constraints are added to the problem:

$$
\begin{aligned}
WA' + AW + \left(Y'\Gamma_j^+ + Q'\Gamma_j^-\right)B' + B\left(\Gamma_j^+ Y + \Gamma_j^- Q\right) + 2\alpha_1 W < 0 \\
WA' + AW + \left(Y'\Gamma_j^+ + Q'\Gamma_j^-\right)B' + B\left(\Gamma_j^+ Y + \Gamma_j^- Q\right) + 2\alpha_2 W > 0
\end{aligned}
\tag{2.83}
$$

In this case, considering $\alpha_1 = 1$ and $\alpha_2 = 100$, the following result is obtained:

$$
\delta = \mu = 0.3200; \qquad K = [67.8306 \quad 67.8349]
$$

Note that the value of $\mu$ is practically the same while the gain coefficients are much smaller. The eigenvalues of $(A + BK)$ are now $-0.9999$ and $-66.8349$, i.e., the dominant dynamics is unchanged.

Let us now consider a regulated output given by (2.77) with

$$
C_z = [1 \quad 1]
$$

Considering now the optimization problem (2.82) with the pole placement constraints given in (2.83), Table 2.4 shows the optimal results for different values of $\delta = \mu > 0.3200$.

As pointed before, for $x(0) = 0$, the value of $\gamma$ corresponds to a bound on the $\mathcal{L}_2$-gain between the input $w$ and the output $z$. Considering in this case that the bound of the admissible disturbances is given by $\delta^{-1} = \mu^{-1}$, it appears a trade-off between the bound of the admissible disturbance and the achievable bound on the

2.4 Stabilization 91

$\mathcal{L}_2$-gain. Note that the greater is $\delta$ (the smaller is the admissible given disturbance bound), the smaller is $\gamma$ (i.e., the higher is the disturbance rejection).

It is also important to notice that the best achievable value of $\gamma$ is 0.0004, i.e. the value of $\gamma$ "saturates" as $\delta$ increases. This value corresponds in fact to the $\mathcal{L}_2$-gain bound obtained in the absence of saturation, and corresponds to the solution of the $H_\infty$ stabilization problem for the linear system [45].

## 2.4.2 Observer-Based Feedback Stabilization

In many practical cases, the state of the system is not fully available for measurement. In this case, an estimated state feedback can be considered. For linear systems, it is well known that the separation principle [64] holds: the state feedback and the observer gain can be computed separately in order to ensure the stability of the augmented closed-loop system.

This philosophy can also be applied to the case of systems presenting saturating actuators. From a state feedback solution to the stabilization problem, one can be interested in computing a state observer in order to ensure the stability of the closed-loop system under an estimated state feedback, while ensuring the same region of stability achieved with the actual state feedback. The main issue in this case is that the region of stability is now associated to the augmented state (composed by the states of the plant plus the states of the observer) and, in general, the dynamics of the plant state and the one of the observer are coupled.

In the sequel we discuss this issue and present solutions to the problem of computing a state-observer-based control law, considering explicitly the control saturation effects.

With this aim, we consider a full-order Luenberger state observer given by the following equation [64]:

$$\dot{\hat{x}}(t) = A\hat{x}(t) + Bu(t) - LC\big(x(t) - \hat{x}(t)\big) \tag{2.84}$$

where $\hat{x}(t) \in \Re^n$ is the estimate of the state and $L \in \Re^{n \times p}$ defines the estimation dynamics. The applied control is now given by

$$u(t) = \text{sat}\big(K\hat{x}(t)\big) \tag{2.85}$$

In order to derive conditions to compute the stabilizing observer, we give now the representation of the augmented closed-loop system (system + observer) in a different basis of the state space. For the sake of simplicity, we do not consider the time dependence explicitly. Let $e = x - \hat{x}$ be the estimate error and consider the following similarity transformation:

$$\begin{bmatrix} x \\ \hat{x} \end{bmatrix} = \begin{bmatrix} I_n & 0 \\ I_n & -I_n \end{bmatrix} \begin{bmatrix} x \\ e \end{bmatrix} \tag{2.86}$$

In the new basis the augmented closed-loop system is given by

$$\begin{bmatrix} \dot{x} \\ \dot{e} \end{bmatrix} = \begin{bmatrix} A & 0 \\ 0 & A + LC \end{bmatrix} \begin{bmatrix} x \\ e \end{bmatrix} + \begin{bmatrix} B \\ 0 \end{bmatrix} u \tag{2.87}$$

$$u = \mathrm{sat}\left( \begin{bmatrix} K & -K \end{bmatrix} \begin{bmatrix} x \\ e \end{bmatrix} \right) \tag{2.88}$$

The following result is based on the polytopic approach I.

**Proposition 2.12** *Suppose there exist matrices $P_1 = P_1' > 0$, $P_1 \in \Re^{n \times n}$, $P_2 = P_2' > 0$, $P_2 \in \Re^{n \times n}$, $U \in \Re^{n \times p}$, $K \in \Re^{m \times n}$ and a vector $\alpha_l \in \Re^m$, satisfying the following matrix inequalities*:

$$\begin{bmatrix} \mathbb{A}_j' P_1 + P_1 \mathbb{A}_j & -P_1 B_j K \\ -(P_1 B_j K)' & A' P_2 + C' U' + P_2 A + UC \end{bmatrix} < 0, \quad j = 1, \dots, 2^m \tag{2.89}$$

$$\begin{bmatrix} P_1 & 0 & \alpha_{l(i)} K_{(i)}' \\ 0 & P_2 & -\alpha_{l(i)} K_{(i)}' \\ \alpha_{l(i)} K_{(i)} & -\alpha_{l(i)} K_{(i)} & u_{0(i)}^2 \end{bmatrix} \geq 0, \quad i = 1, \dots, m \tag{2.90}$$

$$0 < \alpha_{l(i)} \leq 1, \quad i = 1, \dots, m \tag{2.91}$$

*where $\mathbb{A}_j = A + B\Gamma_j(\alpha_l) K$ and $B_j = B\Gamma_j(\alpha_l)$. Then the observer-based output feedback control law defined by (2.84)–(2.85), with $L = P_2^{-1} U$ and the gain $K$, ensures that the set*

$$\mathcal{E}(\mathcal{P}, 1) = \left\{ \tilde{x} \in \Re^{2n}; \ \tilde{x}' \mathcal{P} \tilde{x} \leq 1 \right\}$$

*with*

$$\tilde{x} = \begin{bmatrix} x \\ e \end{bmatrix} \quad and \quad \mathcal{P} = \begin{bmatrix} P_1 & 0 \\ 0 & P_2 \end{bmatrix}$$

*is a region of asymptotic stability (RAS) for the closed-loop system (2.87)–(2.88).*

*Proof* Define $\mathcal{K} = [ K \ -K ]$. The set of validity of the polytopic representation I for system (2.87)–(2.88) is then given in terms of the augmented state $\tilde{x}$:

$$S\left(|\mathcal{K}|, u_0^\alpha\right) = \left\{ \tilde{x} \in \Re^{2n}; \ |\mathcal{K}\tilde{x}| \preceq u_0^\alpha \right\}$$

Hence, if $\tilde{x} \in S(|\mathcal{K}|, u_0^\alpha)$, for an appropriated convex combination given by $\lambda_j$, $j = 1, \dots, 2^m$, it follows that

$$\begin{bmatrix} \dot{x} \\ \dot{e} \end{bmatrix} = \sum_{j=1}^{2^m} \lambda_j \begin{bmatrix} A + B\Gamma_j(\alpha_l) K & -B\Gamma_j(\alpha_l) K \\ 0 & A + LC \end{bmatrix} \begin{bmatrix} x \\ e \end{bmatrix} \tag{2.92}$$

Consider now the following definitions:

$$\mathcal{A}_j = \begin{bmatrix} A + B\Gamma_j(\alpha_l) K & -B\Gamma_j(\alpha_l) K \\ 0 & A + LC \end{bmatrix}$$

$$\mathcal{V}(\tilde{x}) = \tilde{x}' \mathcal{P} \tilde{x}$$

## 2.4 Stabilization

Then, provided $L = P_2^{-1}U$, if (2.89) and (2.91) are verified, we conclude that

$$\mathcal{P}\sum_{j=1}^{2^m}\lambda_j\mathcal{A}_j + \sum_{j=1}^{2^m}\lambda_j\mathcal{A}'_j\mathcal{P} < 0$$

The matrix inequality (2.90) ensures that $\mathcal{E}(\mathcal{P}, 1) \subset S(|\mathcal{K}|, u_0^\alpha)$. Suppose now that $\tilde{x} \in \mathcal{E}(\mathcal{P}, 1)$. Since $\mathcal{E}(\mathcal{P}, 1) \subset S(|\mathcal{K}|, u_0^\alpha)$ it follows that $\dot{\tilde{x}}$ can be computed by (2.92) and it follows that

$$\tilde{x}'\mathcal{P}\dot{\tilde{x}} + \dot{\tilde{x}}'\mathcal{P}\tilde{x} < 0$$

Since this reasoning is valid $\forall\tilde{x}(t) \in \mathcal{E}(\mathcal{P}, 1)$, $\tilde{x}(t) \neq 0$, we can conclude that $\tilde{\mathcal{V}}(\tilde{x}(t))$ is a locally strictly decreasing Lyapunov function for the system (2.87)–(2.88) in $\mathcal{E}(\mathcal{P}, 1)$ and thus this set is a RAS for the closed-loop system (2.87)–(2.88).    □

The conditions in Proposition 2.12 can therefore be used to compute the observer gain $L$. Note that, if $K$ is considered given, the inequalities (2.89)–(2.91) are LMIs for a fixed vector $\alpha_l$.

Consider that $K$ has been previously computed and that the control law $u(t) = \text{sat}(Kx(t))$ ensures that the system (2.1) is asymptotically stable for all $x(0) \in \mathcal{X}_0 \subset \mathfrak{R}^n$, with $\mathcal{X}_0$ given by (2.38) for instance. In particular, if the initial state of the observer is set to zero, i.e. $\hat{x}(0) = 0$, and the initial plant state $x(0) = x_0$, it follows that $e(0) = x_0$, i.e. $\tilde{x}(0) = [x'_0 \ x'_0]'$. The idea is therefore to compute $L$ in order to ensure that $\forall x(0) \in \mathcal{X}_0$ the trajectories of (2.87)–(2.88) converge asymptotically to the origin. However, since there is a coupling between the plant state and the observer state, this may be impossible. In such a case, we can try to find the best observer gain $L$, that ensures the closed-loop stability for a set as close as possible of the set $\mathcal{X}_0$. This solution can be accomplished using the following optimization problem:

$$\max_{\theta, \alpha_l, L, P_1, P_2} \theta$$

$$\text{subject to} \quad \begin{bmatrix} v'_i & v'_i \end{bmatrix}\begin{bmatrix} P_1 & 0 \\ 0 & P_2 \end{bmatrix}\begin{bmatrix} v_i \\ v_i \end{bmatrix} < \theta, \quad \forall i = 1, \dots, n_v \tag{2.93}$$

$$\text{inequalities (2.89)–(2.91)}$$

Considering that $\theta = \beta^{-2}$, note that the first inequality in (2.93) ensures that $\beta[v'_i \ v'_i]' \in \mathcal{E}(\mathcal{P}, 1)$. Hence, if $\mathcal{X}_0 = \text{Co}\{v_1, \dots, v_{n_v}\}$, $v_i \in \mathfrak{R}^n$, $\forall i = 1, \dots, n_v$, the solution of the optimization problem (2.93) ensures that the stability is guaranteed $\forall x(0) \in \beta\mathcal{X}_0$ and $\hat{x}(0) = 0$. In other words the region of asymptotic stability for the augmented system includes the set $\beta\tilde{\mathcal{X}}_0$, with $\tilde{\mathcal{X}}_0 = \text{Co}\{\begin{bmatrix} v_i \\ v_i \end{bmatrix}, i = 1, \dots, n_v\}$, i.e. $\beta\tilde{\mathcal{X}}_0 \subset \mathcal{E}(\mathcal{P}, 1)$.

Following the same steps, by using the polytopic approach II, the following result can be stated.

94     2 Stability Analysis and Stabilization—Polytopic Representation Approach

**Proposition 2.13** *Suppose there exist matrices* $W_1 = W_1' > 0$, $W_1 \in \Re^{n \times n}$, $P_2 = P_2' > 0$, $P_2 \in \Re^{n \times n}$, $U \in \Re^{n \times p}$, $K \in \Re^{m \times n}$ *and matrices* $Q_1$ *and* $Q_2 \in \Re^{m \times n}$, *satisfying the following matrix inequalities*:

$$\begin{bmatrix} W_1 \mathbb{A}_j' + \mathbb{A}_j W_1 + Q_1' \Gamma_j^- B' + B \Gamma_j^- Q_1 & -B \Gamma_j^+ K + B \Gamma_j^- Q_2 \\ \star & A' P_2 + C' U' + P_2 A + U C \end{bmatrix} < 0,$$

$$j = 1, \ldots, 2^m \tag{2.94}$$

$$\begin{bmatrix} W_1 & 0 & Q_{1(i)}' \\ 0 & P_2 & Q_{2(i)}' \\ \star & \star & u_{0(i)}^2 \end{bmatrix} \geq 0, \quad i = 1, \ldots, m \tag{2.95}$$

*where* $\mathbb{A}_j = A + B \Gamma_j^+ K$. *Then the observer-based output feedback control law defined by* (2.84)–(2.85), *with* $L = P_2^{-1} U$ *and the gain* $K$, *ensures that the set*

$$\mathcal{E}(\mathcal{P}, 1) = \left\{ \tilde{x} \in \Re^{2n}; \ \tilde{x}' \mathcal{P} \tilde{x} \leq 1 \right\}$$

*with*

$$\tilde{x} = \begin{bmatrix} x \\ e \end{bmatrix} \quad and \quad \mathcal{P} = \begin{bmatrix} P_1 & 0 \\ 0 & P_2 \end{bmatrix}$$

*where* $P_1 = W_1^{-1}$, *is a region of asymptotic stability (RAS) for the closed-loop system* (2.87)–(2.88).

*Proof* By left- and right-multiplying (2.94) by the matrix $\begin{bmatrix} P_1 & 0 \\ 0 & I \end{bmatrix}$ and considering $H_1 = P_1 Q_1$, $H_2 = Q_2$ and $P_2 L = U$ it follows that (2.94) is equivalent to:

$$\begin{bmatrix} \mathbb{A}_j' P_1 + P_1 \mathbb{A}_j + H_1' \Gamma_j^- B' + B \Gamma_j^- H_1 & -P_1 B \Gamma_j^+ K + P_1 B \Gamma_j^- H_2 \\ \star & (A + LC)' P_2 + P_2 (A + LC) \end{bmatrix} < 0,$$

$$j = 1, \ldots, 2^m \tag{2.96}$$

From (2.96), we can conclude that

$$\mathcal{P} \sum_{j=1}^{2^m} \lambda_j \mathcal{A}_j + \sum_{j=1}^{2^m} \lambda_j \mathcal{A}_j' \mathcal{P} < 0$$

with

$$\mathcal{A}_j = \begin{bmatrix} A + B \Gamma_j^+ K + B \Gamma_j^- H_1 & -B \Gamma_j^+ K + B \Gamma_j^- H_2 \\ 0 & A + LC \end{bmatrix}$$

On the other hand, by left- and right-multiplying (2.95) by the matrix

$$\begin{bmatrix} P_1 & 0 & 0 \\ 0 & I & 0 \\ 0 & 0 & 1 \end{bmatrix}$$

2.4 Stabilization 95

it follows that (2.95) is equivalent to

$$\begin{bmatrix} P_1 & 0 & H'_{1(i)} \\ 0 & P_2 & H'_{2(i)} \\ \star & \star & u^2_{0(i)} \end{bmatrix} \geq 0, \quad i = 1, \ldots, m$$

which ensures that $\mathcal{E}(\mathcal{P}, 1) \subset S(|\mathcal{H}|, u_0)$, with $\mathcal{H} = [H_1 \ H_2]$.

Hence, from the same arguments as used in the proof of Proposition 2.12, considering the polytopic modeling II for the augmented closed-loop system, it follows that $\mathcal{E}(\mathcal{P}, 1)$ is a RAS for the closed-loop system (2.87)–(2.88). $\square$

Using the same reasoning as done in the proof of Proposition 2.13, similar results can be straightforwardly derived considering the polytopic model III.

### 2.4.3 Dynamic Output Feedback Stabilization

In this section we show how to compute a parameter varying dynamic output feedback. Considering the polytopic approach I, the basic idea is to use the variable $\alpha(t)$ as a scheduling parameter for the controller [146, 273]. The approach is therefore similar to LPV control approaches [399].

Consider the strictly proper open-loop system

$$\dot{x}(t) = Ax(t) + Bu(t) \tag{2.97}$$

$$y(t) = Cx(t) \tag{2.98}$$

and a controller that furnishes an output denoted by $y_c(t)$.

Due to the saturation $u(t) = \Gamma(\alpha(t))y_c(t)$, with $\alpha(t)$ defined as

$$\alpha_{(i)}(t) = \min\left(1, \frac{u_{0(i)}}{|y_{c(i)}(t)|}\right), \quad i = 1, \ldots, m \tag{2.99}$$

Based now on the varying parameter $\alpha(t)$, define the following $n_c$-order nonlinear (or LPV) dynamic output controller:

$$\dot{x}_c(t) = A_c(\alpha(t))x_c(t)(t) + B_c(\alpha(t))y(t) \tag{2.100}$$

$$y_c(t) = C_c x_c(t) + D_c y(t) \tag{2.101}$$

where $A_c(\alpha(t))$ and $B_c(\alpha(t))$ denote real matrices of appropriate dimension whose elements depend on the value of $\alpha(t)$, i.e. they are "scheduled" by $\alpha(t)$.

The control input is therefore given by

$$u(t) = \text{sat}(y_c(t)) = \text{sat}(C_c x_c(t) + D_c y(t))$$
$$= \Gamma(\alpha(t))(C_c x_c(t) + D_c y(t))$$

and the closed-loop system can be written as

$$\dot{\tilde{x}}(t) = \mathcal{A}(\alpha(t))\tilde{x}(t) \tag{2.102}$$

96    2  Stability Analysis and Stabilization—Polytopic Representation Approach

with

$$\tilde{x} = \begin{bmatrix} x \\ x_c \end{bmatrix} \quad \text{and} \quad \mathcal{A}\big(\alpha(t)\big) = \begin{bmatrix} A + B\Gamma(\alpha(t))D_cC & B\Gamma(\alpha(t))C_c \\ B_c(\alpha(t))C & A_c(\alpha(t)) \end{bmatrix}$$

Define now the matrix $\mathcal{K} = \begin{bmatrix} D_cC & C_c \end{bmatrix}$ and the following set:

$$S\big(|\mathcal{K}|, u_0^\alpha\big) = \big\{ \tilde{x} \in \Re^{n+n_c}; \ |\mathcal{K}\tilde{x}| \preceq u_0^\alpha \big\}$$

From the polytopic approach I, if $\tilde{x} \in S(|\mathcal{K}|, u_0^\alpha)$ then it follows that

$$u(t) = \sum_{j=1}^{2^m} \lambda_j\big(\alpha(t)\big)\Gamma_j(\alpha_l)\mathcal{K}\tilde{x}(t) = \Gamma\big(\alpha(t)\big)\mathcal{K}\tilde{x}(t)$$

Based on the congruence transformations proposed in [314], the following proposition can be stated considering $n_c = n$.

**Proposition 2.14** *If there exist symmetric positive definite matrices $Y \in \Re^{n \times n}$, $X \in \Re^{n \times n}$, matrices $\hat{A} \in \Re^{n \times n}$, $\hat{B} \in \Re^{n \times p}$, $\hat{C} \in \Re^{m \times n}$ and $\hat{D} \in \Re^{m \times p}$ and a vector $\alpha_l \in \Re^m$, satisfying the following matrix inequalities:*

$$\begin{bmatrix} AX + XA' + B\Gamma_j(\alpha_l)\hat{C} + \hat{C}'\Gamma_j(\alpha_l)B' & \hat{A}' + (A + B\Gamma_j(\alpha_l)\hat{D}C) \\ \hat{A} + (A + B\Gamma_j(\alpha_l)\hat{D}C)' & Y'A + AY + \hat{B}C + C'\hat{B}' \end{bmatrix} < 0$$

$$j = 1, \dots, 2^m \tag{2.103}$$

$$\begin{bmatrix} X & I_n & \alpha_{l(i)}\hat{C}'_{(i)} \\ I_n & Y & \alpha_{l(i)}C'\hat{D}'_{(i)} \\ \alpha_{l(i)}\hat{C}_{(i)} & \alpha_{l(i)}\hat{D}_{(i)}C & u_{0(i)}^2 \end{bmatrix} \geq 0, \quad i = 1, \dots, m \tag{2.104}$$

$$0 < \alpha_{l(i)} \leq 1, \quad i = 1, \dots, m \tag{2.105}$$

*then the controller (2.100)–(2.101) with*

$$D_c = \hat{D}$$

$$C_c = (\hat{C} - D_cCX)\big(M'\big)^{-1}$$

$$B_c(\alpha(t)) = N^{-1}\big(\hat{B} - YB\Gamma\big(\alpha(t)\big)D_c\big) \tag{2.106}$$

$$A_c(\alpha(t)) = N^{-1}\big(\hat{A} - NB_c\big(\alpha(t)\big)CX - YB\Gamma\big(\alpha(t)\big)C_cM'$$

$$- Y\big(A + B\Gamma\big(\alpha(t)\big)D_cC\big)X\big)\big(M'\big)^{-1}$$

*where matrices $N$ and $M$ verify the relation*

$$MN' = I - XY \tag{2.107}$$

*ensures that the set*

$$\mathcal{E}(\mathcal{P}, 1) = \big\{ \tilde{x} \in \Re^{2n}; \ \tilde{x}'\mathcal{P}\tilde{x} \leq 1 \big\} \tag{2.108}$$

*with*

$$\mathcal{P} = \begin{bmatrix} Y & N \\ N' & F \end{bmatrix}$$

## 2.4 Stabilization

*and $F = -N'X(M')^{-1}$, is a region of asymptotic stability (RAS) for the closed-loop system* (2.102).

*Proof* Define the matrices [314]

$$\Pi_1 = \begin{bmatrix} X & I \\ M' & 0 \end{bmatrix} \quad \text{and} \quad \Pi_2 = \begin{bmatrix} I & Y \\ 0 & N' \end{bmatrix}$$

It follows that $\mathcal{P}\Pi_1 = \Pi_2$ and $\Pi_1'\mathcal{P}\Pi_1 = \Pi_1'\Pi_2 = \begin{bmatrix} X & I \\ I & Y \end{bmatrix}$.

Note that if (2.104) is verified, it follows that $X - Y^{-1} > 0$, which implies that $I - XY$ is nonsingular. Then it is always possible to compute square and nonsingular matrices $N$ and $M$ verifying the equation $NM' = I - XY$. This fact also ensures that $\Pi_1$ is nonsingular.

Since the inequality (2.103) is verified for $j = 1, \ldots, 2^m$, and considering that at the instant $t$ there exists $\lambda_j(\alpha(t))$ such that

$$\sum_{j=1}^{2^m} \lambda_j(\alpha(t)) \Gamma_j(\alpha_l) = \Gamma(\alpha(t)) \tag{2.109}$$

with $\sum_{j=1}^{2^m} \lambda_j(\alpha(t)) = 1, 0 \le \lambda_j(\alpha(t)) \le 1, \forall j = 1, \ldots, m$, it follows that

$$\begin{bmatrix} AX + XA' + B\Gamma(\alpha(t))\hat{C} + \hat{C}'\Gamma(\alpha(t))B' & \hat{A}' + (A + B\Gamma(\alpha(t))\hat{D}C) \\ \hat{A} + (A + B\Gamma(\alpha(t))\hat{D}C)' & Y'A + AY + \hat{B}C + C'\hat{B}' \end{bmatrix} < 0 \tag{2.110}$$

Considering the definitions in (2.106) it follows that

$$\hat{D} = D_c$$

$$\hat{C} = C_c M' + D_c CX$$

$$\hat{B} = NB_c(\alpha(t)) + YB\Gamma(\alpha(t))D_c \tag{2.111}$$

$$\hat{A} = NA_c(\alpha(t))M' + NB_c(\alpha(t))CX + YB\Gamma(\alpha(t))C_c M'$$
$$+ Y(A + B\Gamma(\alpha(t))D_c C)X$$

Substituting now (2.111) in (2.110) and after some algebraic manipulations it follows that (2.110) is equivalent to

$$\Pi_1'\mathcal{P}\mathcal{A}(\alpha(t))\Pi_1 + \Pi_1'\mathcal{A}(\alpha(t))'\mathcal{P}\Pi_1 < 0 \tag{2.112}$$

Thus, since $\Pi_1$ is nonsingular, (2.112) is equivalent to

$$\mathcal{P}\mathcal{A}(\alpha(t)) + \mathcal{A}(\alpha(t))'\mathcal{P} < 0 \tag{2.113}$$

By left- and right-multiplying (2.104) by $\begin{bmatrix} (\Pi_1^{-1})' & 0 \\ 0 & I \end{bmatrix}$ and $\begin{bmatrix} \Pi_1^{-1} & 0 \\ 0 & I \end{bmatrix}$, respectively, we can conclude that (2.104) is equivalent to

$$\begin{bmatrix} \mathcal{P} & \alpha_{l(i)}\mathcal{K}'_{(i)} \\ \alpha_{l(i)}\mathcal{K}_{(i)} & u_{0(i)}^2 \end{bmatrix} \ge 0, \quad i = 1, \ldots, m \tag{2.114}$$

Hence, inequality (2.104) ensures the inclusion $\mathcal{E}(\mathcal{P}, 1) \subset S(|\mathcal{K}|, u_0^\alpha)$.

98    2  Stability Analysis and Stabilization—Polytopic Representation Approach

Suppose now that $\tilde{x} \in \mathcal{E}(\mathcal{P}, 1)$. Since $\mathcal{E}(\mathcal{P}, 1) \subset S(|\mathcal{K}|, u_0^\alpha)$, it follows that $\Gamma(\alpha_l)$ can be computed by (2.109) and, we can conclude that (2.103) implies that

$$\tilde{x}'\mathcal{P}\dot{\tilde{x}} + \dot{\tilde{x}}'\mathcal{P}\tilde{x} < 0$$

Since this reasoning is valid $\forall \tilde{x}(t) \in \mathcal{E}(\mathcal{P}, 1)$, $\tilde{x}(t) \neq 0$, we can conclude that $\tilde{V}(\tilde{x}(t)) = \tilde{x}(t)'\mathcal{P}\tilde{x}(t)$ is a locally strictly decreasing Lyapunov function for the system (2.102) in $\mathcal{E}(\mathcal{P}, 1) \subset S(|\mathcal{K}|, u_0^\alpha)$ and thus this set is a RAS for the closed-loop system.    □

*Remark 2.8* As pointed out in the proof of Proposition 2.14, since the inequality (2.104) ensures that $X - Y^{-1} > 0$, it is always possible to compute nonsingular matrices $M$ and $N$, satisfying $NM' = I - YX$. However, given $X$ and $Y$ the choice of $N$ and $M$ is not unique. For example, for any given scalar $\kappa$, we can set $M = \kappa I$ and it follows that $N = (I - YX)\kappa^{-1}$ or we can even compute $M$ and $N$ from a LU or QR decomposition. This means, in fact, that the compensator realization is not unique [314].

*Remark 2.9* Note that for computing the controller matrices at time $t$ it is necessary to obtain $\Gamma(\alpha(t))$. Since the matrices $C_c$ and $D_c$ are time invariant and both the output $y(t)$ and the controller state $x_c(t)$ are available, it follows that $y_c(t)$ is available. The matrices $A_c(\alpha(t))$ and $B_c(\alpha(t))$ are in fact dependent on $y(t)$ and $x_c(t)$, since $\alpha(t)$ is computed from these variables. The controller can then be seen as a time-varying system in which the time-variation is "scheduled" by the values of $y(t)$ and $x_c(t)$.

Note that the conditions (2.103)–(2.105) are LMIs for a fixed vector $\alpha_l$. The result of Proposition 2.14 can then be cast in convex optimization problems to obtain a controller leading to a large region of asymptotic stability for the closed-loop system (2.102) or to ensure the regional stability with respect to a given set of admissible states while guaranteeing some time-domain performance in linear operation.

For instance, consider a shape set $\tilde{\mathcal{X}}_0 = \text{Co}\{v_1, v_2, \ldots, v_{n_r}\}$, $v_l \in \mathbb{R}^{2n}$, $l = 1, \ldots, n_r$. The idea is to compute the controller in order to satisfy $\beta \tilde{\mathcal{X}}_0 \subset \mathcal{E}(\mathcal{P}, 1)$ with $\beta$ as large as possible. As discussed in Sect. 2.2.5.1, the vectors $v_l$ can be viewed as directions in which we want to maximize the region of attraction. In particular, it is interesting to maximize the region of stability in directions associated to the states of the plant. In this case the vectors $v_l$ take the form $[v'_{l1}\ 0]'$.

Noticing that $\beta[v'_{l1}\ 0]' \in \mathcal{E}(\mathcal{P}, 1)$ is equivalent to $\beta[v'_{l1}\ 0]\mathcal{P}[v'_{l1}\ 0]'\beta \leq 1$ and considering $\theta = 1/\beta^2$, it follows that $\beta \tilde{\mathcal{X}}_0 \subset \mathcal{E}(\mathcal{P}, 1)$ is equivalent to:

$$v'_{l1} Y v_{l1} \leq \theta, \quad l = 1, \ldots, r \tag{2.115}$$

Hence, the maximization of the ellipsoid $\mathcal{E}(\mathcal{P}, 1)$ along the directions $[v'_{l1}\ 0]'$ is equivalent to the minimization of $\theta$.

Note that matrix $N$, which appears in matrix $\mathcal{P}$, is not optimized in the above problem. On the other hand, once one obtains matrices $X$ and $Y$, by maximizing $\mathcal{E}(\mathcal{P}, 1)$ in the space of the system state, it is possible to explore the degrees of freedom in the choice of matrix $N$. For instance, $\mathcal{E}(\mathcal{P}, 1)$ can be maximized along the directions given by generic vectors $v_l = [\, v'_{l1} \ v'_{l2} \,]'$ with $v_{l1} \in \mathfrak{R}^n$ and $v_{l2} \in \mathfrak{R}^n$, by solving the following problem:

$$\min_{N,\theta} \theta$$

$$\text{subject to} \quad \begin{bmatrix} \theta & v'_{l1} & v'_{l1}Y + v'_{l2}N' \\ v_{l1} & X & I_n \\ Yv_{l1} + Nv_{l2} & I_n & Y \end{bmatrix} > 0, \quad l = 1, \ldots, r \qquad (2.116)$$

where $X$ and $Y$ are given matrices verifying the conditions of Proposition 2.14.

Actually, the matrix inequality in (2.116) is equivalent to

$$\begin{bmatrix} \theta & [v'_{l1} \ v'_{l2}] \\ \begin{bmatrix} v_{l1} \\ v_{l2} \end{bmatrix} & \mathcal{P}^{-1} \end{bmatrix} > 0, \quad l = 1, \ldots, r \qquad (2.117)$$

In order to show that, note that

$$\mathcal{P}^{-1} = \begin{bmatrix} X & M \\ M' & U \end{bmatrix}$$

Then, it suffices to left- and right-multiply (2.117) by

$$\Pi_3 = \begin{bmatrix} 1 & 0 & 0 \\ 0 & I & 0 \\ 0 & Y & N \end{bmatrix}$$

and $\Pi'_3$, respectively, to obtain the matrix inequality in (2.116).

As seen in Sect. 2.4.1.2, in addition to the guarantee of stability for a region as large as possible, the controller can be designed in order to ensure some degree of time-domain performance in a neighborhood of the origin [137]. We consider this neighborhood as the region of linear behavior of the closed-loop system, i.e., the region where the control inputs do not saturate:

$$S(|\mathcal{K}|, u_0) = \left\{ \tilde{x} \in \mathfrak{R}^{2n}; \ |\mathcal{K}\tilde{x}| \preceq u_0 \right\}$$

When the system operates inside $S(|\mathcal{K}|, u_0)$ it follows that $\Gamma(\alpha(t)) = I$ and $A_c(\alpha(t))$ and $B_c(\alpha(t))$ are constant matrices which we denote as $A_c$ and $B_c$. In this case, the time-domain performance can be achieved if we consider the pole placement of the matrix $\mathcal{A} = \begin{bmatrix} A+BD_cC & BC_c \\ B_cC & A_c \end{bmatrix}$ in a suitable region in the left half complex plane. Considering an LMI framework, the results stated in [314] can be used to place the poles in a called LMI region in the complex plane. For example, if we verify the following LMI:

$$\begin{bmatrix} AX + XA' + B\hat{C} + \hat{C}'B' + 2\sigma X & \hat{A}' + (A + B\hat{D}C) + 2\sigma I \\ \hat{A} + (A + B\hat{D}C)' + 2\sigma I & Y'A + AY + \hat{B}C + C'\hat{B}' + 2\sigma Y \end{bmatrix} < 0 \qquad (2.118)$$

then it is easy to show that this inequality ensures that the real part of the poles of $\mathcal{A}$ are less than $-\sigma$, with $\sigma > 0$. Thus, the larger is $\sigma$, the faster will be the decay rate of the time response inside the linearity region.

100 2 Stability Analysis and Stabilization—Polytopic Representation Approach

*Remark 2.10* The controller (2.100)–(2.101) given by Proposition 2.14 is of order $n$. In fact, the choice $n_c = n$ leads to a simple and more direct procedure to obtain LMI conditions, since in this case the so-called *rank condition* is directly satisfied (see [314]). However, considering that $r$ states are available for direct measurement a reduced order controller can be obtained. The procedure follows the same lines as the ones proposed in [400].

*Remark 2.11* In Proposition 2.14, the same matrices $\hat{A}$ and $\hat{B}$ are supposed to verify all the matrix inequalities stated for each one of the $2^m$ vertices of the polytope defined by the matrices $\Gamma_j(\alpha_l)$. This can be seen as a source of conservatism. In order to reduce this conservatism, the result can be adapted in order to consider different matrices in each vertex, i.e. $\hat{A}_j$ and $\hat{B}_j$, $j = 1, \ldots, 2^m$. In this case, the matrices $A_c(\alpha(t))$ and $B_c(\alpha(t))$ can be obtained, at each instant, from an appropriate convex combination of matrices $\hat{A}_j$ and $\hat{B}_j$. The matrices of the dynamic controller can be obtained in this case as follows:

$$D_c = \hat{D},$$

$$C_c = (\hat{C} - D_c C X)(M')^{-1},$$

$$B_c(\alpha(t)) = N^{-1}\left(\sum_{j=1}^{2^m} \lambda_j(t)\hat{B}_j - Y B \Gamma(\alpha(t)) D_c\right),$$

$$A_c(\alpha(t)) = N^{-1}\left(\sum_{j=1}^{2^m} \lambda_j(t)\hat{A}_j - N B_c C X - Y B \Gamma(\alpha(t)) C_c M' \right.$$
$$\left. - Y(A + B \Gamma(\alpha(t)) D_c C) X\right)(M')^{-1} \qquad (2.119)$$

In this case it is necessary to determine explicitly the coefficients $\lambda_j(t)$ such that $\Gamma(\alpha(t)) = \sum_{j=1}^{2^m} \lambda_j(t)\Gamma_j(\alpha_l)$. This can be easily accomplished by obtaining a feasible solution for the following linear program:

$$\min \sum_{j=1}^{2^m} \lambda_j(t)$$

$$\text{subject to} \quad \sum_{j=1}^{2^m} \lambda_j(t)\Gamma_j(\alpha_l) = \Gamma(\alpha(t))$$

$$\sum_{j=1}^{2^m} \lambda_j(t) = 1$$

$$0 \le \lambda_j(t) \le 1, \quad j = 1, \ldots, 2^m \qquad (2.120)$$

*Remark 2.12* Similar results can be obtained using the polytopic approach II, as shown in [400]. For the discrete-time counterpart of the results the reader can refer to [146] and [410], considering the polytopic approaches I and II, respectively. In those papers, it is also established results concerning the external stabilization by

2.4 Stabilization 101

means of a dynamic compensator scheduled by $\alpha(t)$, considering energy bounded disturbances.

## 2.4.4 Global Stabilization

As pointed in Chap. 1, Sect. 1.6.2.1, the global stabilization of a linear system subject to bounded controls is possible if and only if the system is null controllable [338] (see in particular Theorem 1.2). In other words, the system must be $(A, B)$-controllable and no eigenvalue of $A$ can have strictly positive real part. Moreover, even if the null controllability assumption is verified, in general, the global stability cannot be obtained with a linear state feedback [99, 254, 337]. On the other hand, if the system is null controllable and if all the eigenvalues of $A$ that lie on the imaginary axis are associated to linear independent eigenvectors, or, equivalently if $A$ is asymptotically or critically stable, then it is always possible to find a global stabilizing control law in the form of a saturating static state feedback [52, 220]. This result and the way of computing this control law are presented in this section, using polytopic model I.

The following theorem shows that a quadratic Lyapunov function is a globally strictly decreasing Lyapunov function for system (2.1) only if this function is also a Lyapunov function for the open-loop system.

**Theorem 2.2** *If $V(x) = x'Px$, with $P = P' > 0$, is a Lyapunov function for the closed-loop system* (2.1) *such that*

- $\dot{V}(x) \leq 0$, $\forall x \in \mathfrak{R}^n$, *along the trajectories of system* (2.1),
- *the only invariant set belonging to the set $\mathcal{E} = \{x \in \mathfrak{R}^n; \dot{V}(x) = 0\}$ reduces to the origin $x = 0$,*

*then $V(x)$ is necessarily a Lyapunov function for the open-loop system $\dot{x} = Ax$.*

*Proof* Suppose that $\dot{V}(x) \leq 0$, $\forall x \in \mathfrak{R}^n$, along the trajectories of system (2.1). Consider a vector $x$ such that $x'x = \eta$, with $\eta > 0$. By hypothesis $\dot{V}(x)$, along the trajectories of the closed-loop system (2.1) for this vector $x$, satisfies

$$\dot{V}(x) = x'(A'P + PA)x + 2x'PB\,\text{sat}(Kx) \leq 0 \qquad (2.121)$$

Since (2.121) is valid $\forall x \in \mathfrak{R}^n$, it follows that it holds for $\tilde{x} = \rho x$, with $\rho > 0$. Hence, for any positive scalar $\rho$ it follows

$$\dot{V}(x) = \rho^2 x'(A'P + PA)x + 2\rho x'PB\,\text{sat}(\rho Kx) \leq 0$$

or equivalently

$$x'(A'P + PA)x + \frac{2}{\rho}x'PB\,\text{sat}(\rho Kx) \leq 0$$

By definition of the saturation term, one can write:

$$x'\big(A'P + PA\big)x - \frac{2}{\rho}\big|x'PB\big|u_0 \le x'\big(A'P + PA\big)x + \frac{2}{\rho}x'PB\,\mathrm{sat}(\rho Kx) \le 0$$

Then one gets:

$$x'\big(A'P + PA\big)x \le \frac{2}{\rho}\big|x'PB\big|u_0$$

and therefore, as $x'x = \eta$, when we consider that $\rho \to \infty$, one obtains $x'(A'P + PA)x \le 0$. As one can apply this reasoning for any $x$ such that $x'x = \eta$, for any $\eta > 0$, one can conclude that necessarily one verifies $x'(A'P + PA)x \le 0$, $\forall x \in \Re^n$. $\qquad\square$

Before stating the results that allow to determine a global stabilizing saturating state feedback gain, let us introduce the concept of $(A, B)_{P_{\mathrm{ol}}}$-stabilizability.

**Definition 2.1** Consider a matrix $P = P' > 0$ such that

$$A'P + PA = -Q_0 \le 0 \qquad (2.122)$$

The open-loop system

$$\dot{x} = Ax + Bu \qquad (2.123)$$

is said to be $(A, B)$-stabilizable with respect to the quadratic Lyapunov function $V(x) = x'Px$, with $P$ solution to (2.122), (shortly denoted as $(A, B)_{P_{\mathrm{ol}}}$), if there exists a state feedback gain $K$ such that one of the cases below is satisfied:

- $(A + BK)'P + P(A + BK) = -Q < 0$;
- $(A + BK)'P + P(A + BK) = -Q \le 0$ and $\{x = 0\}$ is the only invariant set contained in the set $\mathcal{E} = \{x \in \Re^n;\ x'(A + BK)'Px + x'P(A + BK)x = 0\}$.

*Remark 2.13* From Definition 2.1, by noting that $x'(A + BK)'Px + x'P(A + BK)x = -x'Q_0x + 2x'PBKx$, it appears that the property of $(A, B)_{P_{\mathrm{ol}}}$-stabilizability implies that $\mathrm{Ker}(Q_0) \cap (\mathrm{Ker}(K) \cup \mathrm{Ker}(B'P)) = \{0\}$.

We use this notion of $(A, B)_{P_{\mathrm{ol}}}$-stabilizability to build a class of global stabilizing controllers.

**Theorem 2.3** *If system* (2.123) *is* $(A, B)_{P_{\mathrm{ol}}}$*-stabilizable then system* (2.1) *with* $K = -\Gamma(\gamma)B'P$, *where* $\Gamma(\gamma)$ *is a positive diagonal matrix, is globally asymptotically stable.*

*Proof* Compute the time derivative of $V(x) = x'Px$, along the trajectories of the closed-loop system (2.1), that is along the trajectories of system (2.123) with $K = -\Gamma(\gamma)B'P$:

$$\dot{V}(x) = x'\big(A'P + PA\big)x + 2x'PB\,\mathrm{sat}\big(-\Gamma(\gamma)B'Px\big)$$

## 2.5 Uncertain Systems

Considering polytopic model I, from Sect. 1.7.1.1, and more precisely from relations (1.36) and (1.37), the closed-loop system can be written as (1.38) and therefore on gets

$$\dot{V}(x) = x'\left(A'P + PA\right)x - 2x'PB\Gamma\left(\alpha(x)\right)\Gamma(\gamma)B'Px$$
$$= -x'Q_0x - 2x'PB\Gamma\left(\alpha(x)\right)\Gamma(\gamma)B'Px$$

Hence, since $\Gamma(\alpha(x))$ and $\Gamma(\gamma)$ are positive diagonal matrices, it follows that $-2x'PB\Gamma(\alpha(x))\Gamma(\gamma)B'Px < 0$, for any $x \in \mathfrak{R}^n$, except for $x \in \text{Ker}(B'P)$. From the property of $(A, B)_{P_{\text{ol}}}$-stabilizability and Remark 2.13, it follows that either $\dot{V}(x) < 0$, for any $x \in \mathfrak{R}^n$, or $\dot{V}(x) \leq 0$ and the only invariant set belonging to the set $\mathcal{E} = \{x \in \mathfrak{R}^n; \dot{V}(x) = 0\}$ reduces to the origin $x = 0$, which proves the global asymptotic stability of the origin of the closed-loop system (2.1). $\qquad\square$

The diagonal elements of the matrix $\Gamma(\gamma)$ can be seen as design parameters. Indeed, they can be chosen to impose some performance or robustness requirement for the linear system $\dot{x} = (A - B\Gamma(\gamma)B'P)x(t)$. For instance, they can be selected to improve (with respect to the open-loop system) the speed of convergence of the trajectories to the origin [52, 220].

## 2.5 Uncertain Systems

### 2.5.1 Stability Analysis

Throughout this section, we discuss how to incorporate robustness requirements in the stability analysis and stabilization of systems with saturating inputs presented in the previous sections. This problem has been considered for instance in [112, 172, 175].

We consider that the matrices of system (2.1) are subject to uncertainties represented by a parameter $F \in \mathcal{F}$. Depending on the description of the set $\mathcal{F}$ different stability conditions can be derived. In particular, we focus here on two different types of uncertainties descriptions, namely (see also Appendix B):

- Norm-bounded uncertainty, where

$$A(F) = A_0 + DF(t)E_1, \qquad B(F) = B_0 + DF(t)E_2$$
$$F(t)'F(t) \leq I_r \tag{2.124}$$

with constant real matrices $D$, $E_1$ and $E_2$ of appropriate dimensions.

- Polytopic uncertainty, where matrices $A(F)$ and $B(F)$ belong to polytopes of matrices, i.e.

$$A(F) \in \mathcal{D}_A = \left\{ A(F) \in \Re^{n \times n}; \; A(F) = \sum_{i=1}^{N_A} \mu_i A_i, \; \sum_{i=1}^{N_A} \mu_i = 1, \; \mu_i \geq 0 \right\}$$

$$B(F) \in \mathcal{D}_B = \left\{ B(F) \in \Re^{n \times m}; \; B(F) = \sum_{k=1}^{N_B} \nu_k B_k, \; \sum_{k=1}^{N_B} \nu_k = 1, \; \nu_k \geq 0 \right\}$$

$$(2.125)$$

In presence of uncertainties the closed-loop system can therefore be written generically as

$$\dot{x}(t) = A(F)x(t) + B(F)\operatorname{sat}\big(Kx(t)\big) \tag{2.126}$$

Considering the uncertainty representations above, analogous results to the ones of Proposition 2.1 can be stated for the case of the polytopic model I. In particular, in order to provide straightforward extensions to the synthesis case in the conditions, the results are formulated in the dual framework, i.e. considering the matrix $W = P^{-1}$ as done in Corollary 2.1. The results considering explicitly the matrix $P$ can be found in [172].

**Proposition 2.15** *If there exist a symmetric positive definite matrix* $W \in \Re^{n \times n}$, *a vector* $\alpha_l \in \Re^m$ *and positive scalars* $\varepsilon$ *and* $\eta$ *satisfying*

$$\begin{bmatrix} W & \alpha_{l(i)} W K'_{(i)} \\ \alpha_{l(i)} K_{(i)} W & \eta u_{0(i)}^2 \end{bmatrix} \geq 0, \quad i = 1, \ldots, m \tag{2.127}$$

$$0 < \alpha_{l(i)} \leq 1, \quad i = 1, \ldots, m \tag{2.128}$$

*and*

- *for the norm-bounded uncertainty case*

$$\begin{bmatrix} W(A_0 + B_0\Gamma_j(\alpha_l)K)' + (A_0 + B_0\Gamma_j(\alpha_l)K)W + \varepsilon DD' & \star \\ E_1 W + E_2\Gamma_j(\alpha_l)KW & -\varepsilon I \end{bmatrix} < 0$$

$$j = 1, \ldots, 2^m \tag{2.129}$$

- *for the polytopic uncertainty case*

$$W\big(A_i + B_k\Gamma_j(\alpha_l)K\big)' + \big(A_i + B_k\Gamma_j(\alpha_l)K\big)W < 0,$$

$$i = 1, \ldots, N_A, \; k = 1, \ldots, N_B, \; j = 1, \ldots, 2^m \tag{2.130}$$

*then the ellipsoid* $\mathcal{E}(W^{-1}, \eta)$ *is a region of asymptotic stability (RAS) for the uncertain closed-loop saturated system* (2.126).

*Proof* The proof follows the same lines as the ones of Proposition 2.1 and Corollary 2.1.

## 2.5 Uncertain Systems

- Norm-bounded uncertainty case.
  By considering $V(x) = x'Px$, we have to prove that

$$\dot{V}(x) = 2 \sum_{j=1}^{2^m} \lambda_j x' \left[ P\left(A_0 + B_0 \Gamma_j(\alpha_l)K\right) + PDF(t)\left(E_1 + E_2 \Gamma_j(\alpha_l)K\right) \right] x$$

is strictly negative. Using the fact that $2u'v \le \varepsilon u'u + \varepsilon^{-1}v'v$, for all vectors $u$ and $v$ and a positive scalar $\varepsilon$ [288], and that $F(t)F(t)' \le I_r$, it follows that

$$2x'PDF(t)\left(E_1 + E_2 \Gamma_j(\alpha_l)K\right)x$$
$$\le x'\varepsilon PDF(t)F(t)'D'x + x'\varepsilon^{-1}\left(E_1 + E_2 \Gamma_j(\alpha_l)K\right)'\left(E_1 + E_2 \Gamma_j(\alpha_l)K\right)x$$
$$\le x'\varepsilon PDD'x + x'\varepsilon^{-1}\left(E_1 + E_2 \Gamma_j(\alpha_l)K\right)'\left(E_1 + E_2 \Gamma_j(\alpha_l)K\right)x$$

Hence, we can write that

$$\dot{V}(x) \le \sum_{j=1}^{2^m} \lambda_j x' \left[ P\left(A_0 + B_0 \Gamma_j(\alpha_l)K\right) + \left(A_0 + B_0 \Gamma_j(\alpha_l)K\right)'P + \varepsilon PDD'P \right.$$
$$\left. + \left(E_1 + E_2 \Gamma_j(\alpha_l)K\right)'\varepsilon^{-1}\left(E_1 + E_2 \Gamma_j(\alpha_l)K\right) \right] x$$

Considering now $W = P^{-1}$, left- and right-multiplying (2.129) by $\text{Diag}(P, I)$, it follows that

$$\left[ \begin{array}{cc} (A_0 + B_0 \Gamma_j(\alpha_l)K)'P + P(A_0 + B_0 \Gamma_j(\alpha_l)K) + \varepsilon PDD'P & \star \\ E_1 + E_2 \Gamma_j(\alpha_l)K & -\varepsilon I \end{array} \right] < 0$$
$$j = 1, \ldots, 2^m \tag{2.131}$$

By using now the Schur's complement, (2.131) is equivalent to

$$P\left(A_0 + B_0 \Gamma_j(\alpha_l)K\right) + \left(A_0 + B_0 \Gamma_j(\alpha_l)K\right)'P + \varepsilon PDD'P$$
$$+ \left(E_1 + E_2 \Gamma_j(\alpha_l)K\right)'\varepsilon^{-1}\left(E_1 + E_2 \Gamma_j(\alpha_l)K\right) < 0$$

which allows one to conclude that the satisfaction of relation (2.129) implies that $\dot{V}(x) < 0$, along the trajectories of the uncertain system (2.1), provided $x(t) \in S(|K|, u_0^\alpha)$.

- Polytopic uncertainty case.
  Considering $W = P^{-1}$, left and right-multiplying (2.130) by $P$, it follows by convexity that

$$\sum_{i=1}^{N_A} \mu_i \left( \sum_{k=1}^{N_B} \nu_k \left( \sum_{j=1}^{2^m} \lambda_j \left( \left(A_i + B_k \Gamma_j(\alpha_l)K\right)'P + P\left(A_i + B_k \Gamma_j(\alpha_l)K\right) \right) \right) \right) < 0 \tag{2.132}$$

for $\mu_i \ge 0$, $\nu_k \ge 0$ and $\lambda_j \ge 0$ such that $\sum_{i=1}^{N_A} \mu_i = 1$, $\sum_{k=1}^{N_B} \nu_k = 1$ and $\sum_{j=1}^{2^m} \lambda_j = 1$.

Hence, from the definition of $A(F)$ and $B(F)$, (2.132) ensures that

$$2 \sum_{j=1}^{2^m} \lambda_j x' P \big( A(F) + B(F) \Gamma_j(\alpha_l) K \big) x < 0$$

i.e. $\dot{V}(x) < 0$ along the trajectories of the uncertain system, provided that $x(t) \in S(|K|, u_0^\alpha)$.

Since the relations (2.127), (2.128) are verified, it follows that the set $\mathcal{E}(P, \eta) \subset S(|K|, u_0^\alpha)$ and we can conclude, in both cases, that $\mathcal{E}(P, \eta)$ is a RAS for the uncertain closed-loop saturated system (2.126). $\qquad\square$

Similarly, considering polytopic model II, the uncertain case, counterpart of Proposition 2.2 can be stated as follows:

**Proposition 2.16** *If there exist a symmetric positive definite matrix $W \in \Re^{n \times n}$, a matrix $Q \in \Re^{m \times n}$ and positive scalars $\varepsilon$ and $\eta$ satisfying*

$$\begin{bmatrix} W & Q'_{(i)} \\ Q_{(i)} & \eta u_{0(i)}^2 \end{bmatrix} \geq 0, \quad i = 1, \dots, m \tag{2.133}$$

*and*

- *for the norm-bounded uncertainty case*

$$\begin{bmatrix} W(A_0 + B_0 \Gamma_j^+ K)' + (A_0 + B_0 \Gamma_j^+ K)W + Q' \Gamma_j^- B_0' + B_0 \Gamma_j^- Q + \varepsilon DD' & \star \\ E_1 W + E_2 \Gamma_j^+ K W + E_2 \Gamma_j^- Q & -\varepsilon I \end{bmatrix}$$
$$< 0, \quad j = 1, \dots, 2^m \tag{2.134}$$

- *for the polytopic uncertainty case*

$$W \big( A_i + B_k \Gamma_j^+ K \big)' + \big( A_i + B_k \Gamma_j^+ K \big) W + Q' \Gamma_j^- B_k' + B_k \Gamma_j^- Q < 0,$$
$$i = 1, \dots, N_A, \ k = 1, \dots, N_B, \ j = 1, \dots, 2^m \tag{2.135}$$

*then the ellipsoid $\mathcal{E}(W^{-1}, \eta)$ is a region of asymptotic stability (RAS) for the uncertain saturated system (2.1).*

Quite similar conditions to the ones of Proposition 2.16 can be derived considering the polytopic approach III and the result given by Proposition 2.3.

*Remark 2.14* Considering the conditions given in Propositions 2.15 and 2.16, the same type of optimization problems and associated computational issues discussed in Sect. 2.2.5 apply.

*Remark 2.15* The results for the norm-bounded uncertainty case can also be derived considering a LFT representation of the closed-loop system:

## 2.5 Uncertain Systems

$$\begin{cases} \dot{x}(t) = A_0 x(t) + B_0 \operatorname{sat}\big(Kx(t)\big) + Dp(t) \\ z(t) = E_1 x(t) + E_2 \operatorname{sat}\big(Kx(t)\big) \\ p(t) = F(t)z(t), \quad F(t)'F(t) \le I_r \end{cases} \tag{2.136}$$

Hence, it follows that $p(t) = F(t)[E_1 \ E_2 \operatorname{sat}(Kx(t))]$.

Consider a Lyapunov candidate function $V(x(t)) = x(t)'Px(t)$, it follows that

$$\dot{V}(x(t)) = x(t)'P\left(A_0 + B_0 \sum_{j=1}^{2^m} \lambda_j(t)\Gamma_j(\alpha_l)K\right)x(t)$$

$$+ x(t)'\left(A_0 + B_0 \sum_{j=1}^{2^m} \lambda_j(t)\Gamma_j(\alpha_l)K\right)' Px(t) + x(t)'PDp(t)$$

$$+ p(t)'D'Px(t) \tag{2.137}$$

From (2.136), one has $p(t)'p(t) = z(t)'F(t)'F(t)z(t)$. Since $F(t)'F(t) \le I_r$, it follows that

$$\varepsilon^{-1}\left(p(t)'p(t) - x(t)'\left(E_1 + E_2 \sum_{j=1}^{2^m} \lambda_j(t)\Gamma_j(\alpha_l)K\right)'\right.$$

$$\left. \times \left(E_1 + E_2 \sum_{j=1}^{2^m} \lambda_j(t)\Gamma_j(\alpha_l)K\right)x(t)\right) \le 0 \tag{2.138}$$

for all $\varepsilon > 0$. Thus, from (2.137) and (2.138), the inequalities (2.129) can be obtained by applying the S-Procedure and Schur's complement [45]. Analogous procedure can be made considering polytopic models II and III.

*Example 2.8* The linearized equations of a satellite motion, given in [84], yield the following matrices for system (2.1):

$$A(F) = \begin{bmatrix} 0 & 1 & 0 & 0 \\ 3p(t)^2 & 0 & 0 & 2p(t) \\ 0 & 0 & 0 & 1 \\ 0 & -2p(t) & 0 & 0 \end{bmatrix}; \qquad B(F) = \begin{bmatrix} 0 & 0 \\ 1 & 0 \\ 0 & 0 \\ 0 & 1 \end{bmatrix}$$

Time-varying uncertain parameter $p(t)$ represents the period of rotation and lies within the interval $[0.5, 1.5]$. The saturation levels are $u_{0(1)} = u_{0(2)} = 15$. In order to represent the uncertainty on $p(t)$, we assume that matrix $A(F)$ belongs to the polytope[1]

---

[1] $p(t)$ and $p(t)^2$ are considered independently to satisfy the linearity condition of uncertain parameters in the polytopic modeling, i.e. two uncertain parameters are considered. This leads to a polytope of matrices with four vertices.

$$\mathcal{D}_A = \mathrm{Co}\left\{\begin{bmatrix} 0 & 1 & 0 & 0 \\ 0.75 & 0 & 0 & 1 \\ 0 & 0 & 0 & 1 \\ 0 & -1 & 0 & 0 \end{bmatrix}, \begin{bmatrix} 0 & 1 & 0 & 0 \\ 6.75 & 0 & 0 & 1 \\ 0 & 0 & 0 & 1 \\ 0 & -1 & 0 & 0 \end{bmatrix}, \right.$$
$$\left. \begin{bmatrix} 0 & 1 & 0 & 0 \\ 0.75 & 0 & 0 & 3 \\ 0 & 0 & 0 & 1 \\ 0 & -3 & 0 & 0 \end{bmatrix}, \begin{bmatrix} 0 & 1 & 0 & 0 \\ 6.75 & 0 & 0 & 3 \\ 0 & 0 & 0 & 1 \\ 0 & -3 & 0 & 0 \end{bmatrix}\right\}.$$

In [84], the authors show that the linear state feedback

$$u = \begin{bmatrix} -19 & -13 & -2 & 0 \\ 0 & 2 & -7 & -8 \end{bmatrix} x \tag{2.139}$$

stabilizes the saturated linear system for any initial condition in the unit sphere, i.e.

$$x(0) \in \left\{ x \in \mathfrak{R}^4; \; x'x \le 1 \right\},$$

when $p(t)$ is set to 1. However, state feedback (2.139) was not guaranteed to stabilize the uncertain closed-loop system for any initial condition in the unit sphere. Then, we apply the results of Proposition 2.15, using the same steps of Algorithm 2.1 with optimization criterion: min Trace$(P) + \eta$. Initializing the Algorithm 2.1 with $\alpha_l = \begin{bmatrix} 0.5 \\ 0.5 \end{bmatrix}$ in step 1, Algorithm 2.1 converges to the solution,

$$P = \begin{bmatrix} 0.4192 & 0.1872 & -0.0435 & -0.0037 \\ 0.1872 & 0.2036 & 0.0681 & -0.0219 \\ -0.0435 & 0.0681 & 0.2233 & 0.0293 \\ -0.0037 & -0.0219 & 0.0293 & 0.0761 \end{bmatrix}$$

$$\eta = 0.9149; \qquad \alpha_l = \begin{bmatrix} 0.4442 \\ 0.4610 \end{bmatrix}$$

The cuts of ellipsoid $\mathcal{E}(P, \eta)$ of safe initial conditions is plotted in Fig. 2.11. Note that this set encloses the unit sphere. This shows that the state feedback proposed by [84] for the nominal certain case is a robust state feedback for the uncertain system for any initial condition belonging to a domain larger than the unit sphere.

### 2.5.2 Extensions

To address the external stability analysis problem, such as done in Sect. 2.3, inequalities (2.30), for the bounded amplitude case, and (2.45), for the energy bounded case, should be considered with respect to the trajectories of the uncertain system (2.126). In this case, considering the reasoning done in the proof of Propositions 2.15 and 2.16 to bound $\dot{V}(x(t))$ considering both norm-bound and polytopic uncertainties, extensions of the results stated in Propositions 2.4 and 2.5 can be straightforwardly derived to cope with the external stability analysis problem.

## 2.6 Discrete-Time Case

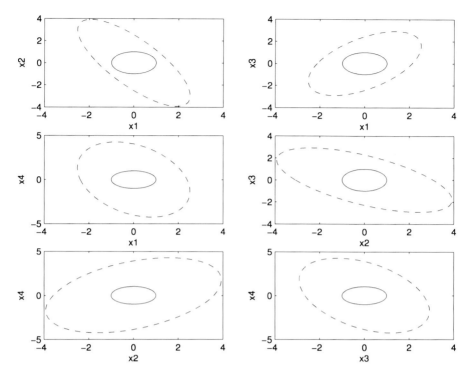

**Fig. 2.11** Example 2.8—the region of stability of the uncertain saturated system (- - -) and the unit ball considered in [84] (—)

The extension of the results to the synthesis problem is also immediate. It suffices to apply the variable change $Y = KW$ in the conditions of Propositions 2.15 and 2.16. In this case, the idea is to compute a stabilizing gain that will ensure the stability for any matrix belonging to the uncertainty set. The same optimization criteria presented in Sect. 2.4.1 can thus be used to compute a robust stabilizing gain to ensure internal or external stability of the uncertain closed-loop system.

## 2.6 Discrete-Time Case

Consider a linear discrete-time system subject to control saturation generically described by the following state equation:

$$x(k+1) = Ax(k) + B\,\mathrm{sat}\bigl(Kx(k)\bigr) \qquad (2.140)$$

As in the continuous-time case, a polytopic differential inclusion can be used to model the system dynamics under saturation effects. In fact, considering the polytopic representation I, II or III we have the following, respectively:

- if $x(k) \in S(|K|, u_0^\alpha)$ then

$$x(k+1) = \sum_{j=1}^{2^m} \lambda_j \big( A + B\Gamma_j(\alpha_l)K \big) x(k) \qquad (2.141)$$

- if $x(k) \in S(|H|, u_0)$ then

$$x(k+1) = \sum_{j=1}^{2^m} \lambda_j \big( A + B\Gamma_j^+ K + \Gamma_j^- H \big) x(k) \qquad (2.142)$$

- $x(k) \in \bigcap_{j=2}^{2^m} S(|H_j|, u_0)$ then

$$x(k+1) = \sum_{j=1}^{2^m} \lambda_j \left( A + \sum_{i \in \mathcal{S}_j^c} B_i K_{(i)} + \sum_{i \in \mathcal{S}_j} B_i H_{j(i)} \right) x(k) \qquad (2.143)$$

with appropriate scalars $0 \le \lambda_j \le 1$, $\sum_{j=1}^{2^m} \lambda_j = 1$.

### 2.6.1 Ellipsoidal Regions of Asymptotic Stability

Ellipsoidal sets can be associated to a discrete-time Lyapunov quadratic function defined as

$$V\big(x(k)\big) = x(k)' P x(k)$$

Regarding this function, we define

$$\Delta V\big(x(k)\big) = V\big(x(k+1)\big) - V\big(x(k)\big)$$

Following the same reasoning as developed in Sect. 2.2.2, an ellipsoidal set $\mathcal{E}(P, \eta)$ will be a region of asymptotic stability for system (2.140) if $\Delta V(x(k)) < 0$, $\forall x(k) \in \mathcal{E}(P, \eta)$. Hence, considering the polytopic representations (2.141), (2.142) and (2.143) for the system (2.140), we can conclude that $\mathcal{E}(P, \eta)$ is a region of asymptotic stability if:

1. $\Delta V(x(k)) < 0$ along the trajectories of the polytopic system (2.141), (2.142) or (2.143);
2. $\mathcal{E}(P, \eta) \subset S(|K|, u_0^\alpha)$, if polytopic approach I is considered, $\mathcal{E}(P, \eta) \subset S(|H|, u_0)$, if polytopic approach II is used, or $\mathcal{E}(P, \eta) \subset \bigcap_{j=2}^{2^m} S(|H_j|, u_0)$, if polytopic approach III is used.

Hence, the following results can be stated considering the polytopic model I [130, 134].

**Proposition 2.17** *If there exist a symmetric positive definite matrix $P \in \mathfrak{R}^{n \times n}$, a vector $\alpha_l \in \mathfrak{R}^m$ and a positive scalar $\eta$ satisfying*

## 2.6 Discrete-Time Case

$$\left(A + B\Gamma_j(\alpha_l)K\right)'P\left(A + B\Gamma_j(\alpha_l)K\right) - P < 0, \quad j = 1, \ldots, 2^m \quad (2.144)$$

$$\begin{bmatrix} P & \alpha_{l(i)}K'_{(i)} \\ \alpha_{l(i)}K_{(i)} & \eta u^2_{0(i)} \end{bmatrix} \geq 0, \quad i = 1, \ldots, m \quad (2.145)$$

$$0 < \alpha_{l(i)} \leq 1, \quad i = 1, \ldots, m \quad (2.146)$$

*then the ellipsoid $\mathcal{E}(P, \eta)$ is a region of asymptotic stability (RAS) for the discrete-time saturated system* (2.140).

*Proof* From the Schur's complement and convexity arguments, if (2.144) is verified for $j = 1, \ldots, 2^m$, it follows that

$$\begin{bmatrix} P & \sum_{j=1}^{2^m} \lambda_j (A + B\Gamma_j(\alpha_l)K)' \\ \sum_{j=1}^{2^m} \lambda_j (A + B\Gamma_j(\alpha_l)K) & P^{-1} \end{bmatrix} > 0$$

with $\sum_{j=1}^{2^m} \lambda_j = 1, 0 \leq \lambda_j \leq 1$.

Applying Schur's complement again, it follows that

$$\sum_{j=1}^{2^m} \lambda_j \left(A + B\Gamma_j(\alpha_l)K\right)' P \sum_{j=1}^{2^m} \lambda_j \left(A + B\Gamma_j(\alpha_l)K\right) - P < 0$$

This implies, $\forall x(k) \neq 0$, that

$$x(k)' \sum_{j=1}^{2^m} \lambda_j \left(A + B\Gamma_j(\alpha_l)K\right)' P \sum_{j=1}^{2^m} \lambda_j \left(A + B\Gamma_j(\alpha_l)K\right)x(k) - x(k)'Px(k) < 0$$

Then, provided that $x(k) \in S(|K|, u_0^\alpha)$ it follows that

$$x(k+1)'Px(k+1) - x(k)'Px(k) \leq 0, \quad \forall x(k) \neq 0$$

with respect to the dynamics given by the saturated system (2.140).

On the other hand (2.145) and (2.146) guarantee that $\mathcal{E}(P, \eta) \subset S(|K|, u_0^\alpha)$, Hence, if (2.144), (2.145) and (2.146) are verified we have indeed that $\Delta V(x(k)) < 0, \forall x(k) \in \mathcal{E}(P, \eta)$, which ensures that $\mathcal{E}(P, \eta)$ is a region of asymptotic stability (RAS). $\qquad\square$

Considering the polytopic model (2.142), the following result can be stated [192].

**Proposition 2.18** *If there exist a symmetric positive definite matrix $W \in \Re^{n \times n}$, a matrix $Q \in \Re^{m \times n}$ and a positive scalar $\eta$ satisfying*

$$\begin{bmatrix} W & W(A + B\Gamma_j^+ K)' + Q'\Gamma_j^- B' \\ (A + B\Gamma_j^+ K)W + B\Gamma_j^- Q & W \end{bmatrix} > 0$$

$$j = 1, \ldots, 2^m \quad (2.147)$$

$$\begin{bmatrix} W & Q'_{(i)} \\ Q_{(i)} & \eta u^2_{0(i)} \end{bmatrix} \geq 0, \quad i = 1, \ldots, m \quad (2.148)$$

*then the ellipsoid $\mathcal{E}(W^{-1}, \eta)$ is a region of stability (RAS) for the discrete-time saturated system* (2.140).

112        2 Stability Analysis and Stabilization—Polytopic Representation Approach

*Proof* Consider now the change of variables $W = P^{-1}$ and $Q = HW$. Pre- and post-multiplying (2.147) by matrix $\begin{bmatrix} P & 0 \\ 0 & I_n \end{bmatrix}$, it follows that

$$\begin{bmatrix} P & \sum_{j=1}^{2^m} \lambda_j (A + B\Gamma_j^+ K + B\Gamma_j^- H)' \\ \sum_{j=1}^{2^m} \lambda_j (A + B\Gamma_j^+ K + B\Gamma_j^- H) & W \end{bmatrix} > 0$$

From here the proof follows the same steps as in the proof of Proposition 2.17.   □

An analogous result considering the polytopic model (2.143) can be stated as follows [2].

**Proposition 2.19** *If there exist a symmetric positive definite matrix $W \in \mathfrak{R}^{n \times n}$, row vectors $Q_{j(i)} \in \mathfrak{R}^{1 \times n}$, $j = 2, \ldots, 2^m$, $i \in \mathcal{S}_j$ and $\eta$ satisfying*

$$\begin{bmatrix} W & (AW + \sum_{i \in \mathcal{S}_j^c} B_i K_{(i)} W + \sum_{i \in \mathcal{S}_j} B_i Q_{j(i)})' \\ (AW + \sum_{i \in \mathcal{S}_j^c} B_i K_{(i)} W + \sum_{i \in \mathcal{S}_j} B_i Q_{j(i)}) & W \end{bmatrix}$$
$$> 0, \quad j = 1, \ldots, 2^m \qquad (2.149)$$

$$\begin{bmatrix} W & Q'_{j(i)} \\ Q_{j(i)} & \eta u_{0(i)}^2 \end{bmatrix} \geq 0, \quad \forall j = 2, \ldots, 2^m, \ \forall i \in \mathcal{S}_j \qquad (2.150)$$

*then the ellipsoid $\mathcal{E}(W^{-1}, \eta)$ is a region of stability (RAS) for the discrete-time saturated system (2.140).*

## 2.6.2 Polyhedral Regions of Asymptotic Stability

In this section, we present a complementary technique to the one presented in the previous section. We will show that if a contractive ellipsoidal region of asymptotic stability has been obtained for system (2.140) using a polytopic differential inclusion, it is possible to obtain a polyhedral region of asymptotic stability that includes the ellipsoidal one. The methodology described below uses some tools based on the positive invariance and contractvity of polyhedra (see also Chap. 4) and the computation of maximal invariant sets (see for instance [34, 126, 214]). In particular, the results can be seen as an extension of the ones presented in [37] and [348].

For the sake of simplicity, we derive the results considering the polytopic approach I. In this case, the idea is to compute the maximal invariant set, with respect to the polytopic model I, included in the region $S(|K|, u_0^\alpha)$. Analogous results can be obtained by using the polytopic approaches II and III.

### 2.6.2.1 Preliminaries

Let us consider the following definitions:

$$\mathbb{A}_j = A + B\Gamma_j(\alpha_l)K, \quad j = 1, \ldots, 2^m$$

## 2.6 Discrete-Time Case

and

$$S(G_0, \psi_0) = S\big(|K|, u_0^\alpha\big)$$
$$G_0 = \begin{bmatrix} K \\ -K \end{bmatrix} : \psi_0 = \begin{bmatrix} u_0^\alpha \\ u_0^\alpha \end{bmatrix}$$
$$\mathcal{J} = \{1, \ldots, 2^m\}; \qquad \mathcal{I}_0 = \{1, \ldots, m_0\}$$

with $m_0 = 2m$.

A series of polytopes $S(G_k, \psi_k)$ can be defined. The first element is $S(G_0, \psi_0)$ and the other elements are generically denoted by $S(G_k, \psi_k)$ where

$$G_k = \begin{bmatrix} G_{k-1} \\ T_k \end{bmatrix}; \qquad \psi_k = \begin{bmatrix} \psi_{k-1} \\ r_k \end{bmatrix}$$
$$T_k \in \mathfrak{R}^{m_k \times n}; \ G_k \in \mathfrak{R}^{l_k \times n}; \ r_k \in \mathfrak{R}^{m_k}; \ \psi_k \in \mathfrak{R}^{l_k}$$
$$\mathcal{I}_k = \{1, \ldots, m_k\}$$

By definition $G_0 = T_0$ and $\psi_0 = r_0$.

The determination of $T_{k+1}$ and $r_{k+1}$, is based on the solution of the following linear programs:

$$LP_k(i, j) = \begin{cases} y_{k(i,j)} = \max_x T_{k(i)} \mathbb{A}_j x, \\ \text{subject to} \quad x \in S(G_k, \psi_k), \end{cases} \quad i \in \mathcal{I}_k, \ j \in \mathcal{J} \quad (2.151)$$

From the solution of (2.151), consider the definition of the following index sets:

$$\mathcal{J}_{k,i} = \{j \in \mathcal{J}; \ y_{k(i,j)} > r_{k(i)}\}; \qquad \tilde{\mathcal{I}}_k = \{i \in \mathcal{I}_k \ ; \ \mathcal{J}_{k,i} \neq \emptyset\}$$

Define now the set

$$S(T_{k+1}, r_{k+1})_{i,j} = \big\{x \in \mathfrak{R}^n; \ T_{k(i)} \mathbb{A}_j x \leq r_{k(i)}\big\}$$

If $\tilde{\mathcal{I}}_k \neq \emptyset$, consider the construction of the sets

$$S(T_{k+1}, r_{k+1})_i = \bigcap_{j \in \mathcal{J}_{k,i}} S(T_{k+1}, r_{k+1})(i, j)$$
$$S(T_{k+1}, r_{k+1}) = \bigcap_{i \in \tilde{\mathcal{I}}_k} S(T_{k+1}, r_{k+1})_i$$

The set $S(G_{k+1}, \psi_{k+1})$ is therefore defined as follows:

- if $\tilde{\mathcal{I}}_k \neq \emptyset$: $S(G_{k+1}, \psi_{k+1}) = S(G_k, \psi_k) \cap S(T_{k+1}, r_{k+1})$
- if $\tilde{\mathcal{I}}_k = \emptyset$: $S(G_{k+1}, \psi_{k+1}) = S(G_k, \psi_k)$

Denote $\prod^k \mathbb{A}_j$ as any product of any $k$ matrices $\mathbb{A}_j$.

Recalling that $S(G_0, \psi_0) = S(|K|, u_0^\alpha)$, by construction, the following properties with respect to sets $S(G_k, \psi_k)$ and $S(G_{k+1}, \psi_{k+1})$ hold.

*Property 2.1* Considering the definition of the set $S(G_k, \psi_k)$ the following properties hold:

1. $S(G_{k+1}, \psi_{k+1}) \subseteq S(G_k, \psi_k)$,
2. if $x \in S(G_k, \psi_k)$ then $\prod^k \mathbb{A}_j x \in S(|K|, u_0^\alpha)$, or equivalently, $|K \prod^k \mathbb{A}_j x| \preceq u_0^\alpha$,
3. $S(G_k, \psi_k) = \{x \in S(|K|, u_0^\alpha); \ \prod^k \mathbb{A}_j x \in S(|K|, u_0^\alpha)\}$,
4. $S(G_{k+1}, \psi_{k+1}) = \{x \in S(G_k, \psi_k); \ T_{k(i)} \mathbb{A}_j x \leq r_{k(i)}, \ \forall i \in \mathcal{I}_k, \ \forall j \in \mathcal{J}\}$,
5. $S(G_{k+1}, \psi_{k+1}) = \{x \in S(G_k, \psi_k); \ \mathbb{A}_j x \in S(G_k, \psi_k), \forall j \in \mathcal{J}\}$.

Also by construction, the following lemma can be stated.

**Lemma 2.2** *If there exists an index $k^\star$ such that $\mathcal{J}_{k^\star i}$ is empty for all $i \in \mathcal{I}_{k^\star}$ then one gets $S(G_k, \psi_k) = S(G_{k^\star}, \psi_{k^\star}), \forall k \geq k^\star$.*

*Remark 2.16* It should be noticed at this point that:

- $S(G_{\tilde{k}}, \psi_{\tilde{k}})$, corresponds to the set of initial states for which the corresponding trajectories of the polytopic system $\dot{x}(t) = \sum_{j=1}^{2^m} \lambda_j \mathbb{A}_j x(t)$ remain in $S(|K|, u_0^\alpha)$, for $k = 0$ to $k = \tilde{k}$, i.e. for $\tilde{k}$ steps.
- $S(G_{k^\star}, \psi_{k^\star})$ corresponds to the set of initial conditions for which the trajectories of the polytopic system $\dot{x}(t) = \sum_{j=1}^{2^m} \lambda_j \mathbb{A}_j x(t)$ remain in $S(|K|, u_0^\alpha)$, for all $k \geq 0$.

Before presenting conditions for the existence of such an index $k^\star$, let us give the following useful lemma.

**Lemma 2.3** *If pair $(K, A)$ is observable then pairs $(K, \mathbb{A}_j)$, $j = 1, \ldots, 2^m$, are observable.*

*Proof* Suppose that pair $(K, A)$ is observable, but that $(K, \mathbb{A}_j)$ is not observable. Then there exists a vector $x$, $x \neq 0$, such that

$$\begin{bmatrix} K \\ K \mathbb{A}_j \\ \vdots \\ K \mathbb{A}_j^{n-1} \end{bmatrix} x = 0$$

Recall that $\mathbb{A}_j = A + B \Gamma_j(\alpha_l) K$, therefore it follows that this vector $x$ satisfies

$$\begin{bmatrix} K \\ K A \\ \vdots \\ K A^{n-1} \end{bmatrix} x = 0$$

That is a contradiction with the hypothesis that the pair $(K, A)$ is observable. $\square$

## 2.6 Discrete-Time Case

### 2.6.2.2 Maximal Positively Invariant Set in $S(|K|, u_0^\alpha)$

By using results from previous sections, we are now ready to state the following result.

**Proposition 2.20** *If pair $(K, A)$ is observable and if there exists a symmetric positive definite matrix $P \in \Re^{n \times n}$ verifying*

$$\mathbb{A}'_j P \mathbb{A}_j - P < 0, \quad j = 1, \dots, 2^m \tag{2.152}$$

*then*:

1. *There exists an index $k^\star$ such that $S(G_k, \psi_k) = S(G_{k^\star}, \psi_{k^\star})$, $\forall k \geq k^\star$.*
2. *The set $S(G_{k^\star}, \psi_{k^\star})$ is positively invariant for the polytopic system*

$$x(k+1) = \sum_{j=1}^{2^m} \lambda_j(x) \mathbb{A}_j x(k) \tag{2.153}$$

3. *The set $S(G_{k^\star}, \psi_{k^\star})$ is the maximal positively invariant set for the polytopic system (2.153) contained in $S(G_0, \psi_0) = S(|K|, u_0^\alpha)$.*

*Proof* 1. Define the polyhedral set $S(Q_j, q_j) = \{x \in \Re^n; Q_j x \leq q_j\}$ where

$$Q_j = \begin{bmatrix} K \\ K \mathbb{A}_j \\ \vdots \\ K \mathbb{A}_j^{n-1} \\ -K \\ -K \mathbb{A}_j \\ \vdots \\ -K \mathbb{A}_j^{n-1} \end{bmatrix} ; \quad q_j = \begin{bmatrix} u_0^\alpha \\ u_0^\alpha \\ \vdots \\ u_0^\alpha \\ u_0^\alpha \\ u_0^\alpha \\ \vdots \\ u_0^\alpha \end{bmatrix}$$

From Lemma 2.3, it follows that the set $S(Q_j, q_j)$ is bounded, $\forall j = 1, \dots, 2^m$. Moreover, by construction one gets $S(G_{n-1}, \psi_{n-1}) \subseteq \bigcap_{j=1}^{2^m} S(Q_j, q_j)$, which implies that there exists an index $\bar{k} \leq n - 1$ such that $S(G_{\bar{k}}, \psi_{\bar{k}})$ is bounded.

Suppose that the positive scalar $\eta$ is chosen in order to satisfy $S(G_{\bar{k}}, \psi_{\bar{k}}) \subset \mathcal{E}(P, \eta)$. Furthermore, since matrix $P$ satisfies (2.152) then, by considering the quadratic function $V(x(k)) = x(k)' P x(k)$, it follows that $V(x(k+1)) - V(x(k)) < 0$ along the trajectories of the polytopic system (2.153). Hence the ellipsoid $\mathcal{E}(P, \eta)$ is a contractive set for the polytopic system (2.153). Thus, there exist an integer $M \geq 0$ and a positive scalar $\sigma_0$, $0 < \sigma_0 < 1$ such that

$$\left( \prod^M \mathbb{A}_j \right) S(G_{\bar{k}}, \psi_{\bar{k}}) \subset \left( \prod^M \mathbb{A}_j \right) \mathcal{E}(P, \eta) \subset \mathcal{E}(P, \sigma_0^{-M} \eta) \subset S(|K|, u_0^\alpha) \tag{2.154}$$

where $(\prod^M \mathbb{A}_j) S(G_{\bar{k}}, \psi_{\bar{k}})$ denotes the set of points obtained from the application of $(\prod^M \mathbb{A}_j)$ to any $x \in S(G_{\bar{k}}, \psi_{\bar{k}})$.

116    2  Stability Analysis and Stabilization—Polytopic Representation Approach

Consider $k^* = \bar{k} + M$ and suppose now that $x \in S(G_{k^*}, \psi_{k^*})$. Since $S(G_{k^*}, \psi_{k^*}) \subset S(G_{\bar{k}}, \psi_{\bar{k}})$ and $k^* + 1 > M$, from (2.154) we conclude that

$$\prod_{}^{k^*+1} \mathbb{A}_j x \in S(|K|, u_0^\alpha) \quad \Leftrightarrow \quad \left| K \prod_{}^{k^*+1} \mathbb{A}_j x \right| \preceq u_0^\alpha \tag{2.155}$$

On the other hand, note that by construction, each line of $T_{k^*}$ and each component of $r_{k^*}$ are given as $\pm K_{(i)} \prod^{k^*} \mathbb{A}_j$ and $\pm u_{0(i)}^\alpha$, respectively, considering some $i$, $i = 1, \ldots, m$ and some specific product of $k^*$ matrices $\mathbb{A}_j$. Hence, if (2.155) holds, it follows that

$$T_{k^*} \mathbb{A}_j x \preceq r_{k^*}, \quad \forall j \in \mathcal{J}$$

which means that $\tilde{\mathcal{I}}_{k^*} = \emptyset$, and thus

$$S(G_k, \psi_k) = S(G_{k^*}, \psi_{k^*}), \quad \forall k > k^*$$

2. Suppose that $x(k) \in S(G_{k^*}, \psi_{k^*})$. From item 1 and by construction it follows that $G_{k^*} \mathbb{A}_j x(k) \preceq \psi_{k^*}, \forall j \in \mathcal{J}$. Hence, by convexity, one obtains $G_{k^*} \sum_{j=1}^{2^m} \lambda_j(x) \times \mathbb{A}_j x(k) \preceq \psi_{k^*}$, with $0 \leq \lambda_j(x) \leq 1$ and $\sum_{j=1}^{2^m} \lambda_j(x) = 1$. Therefore, it follows that $G_{k^*} x(k+1) \preceq \psi_{k^*}$, i.e. $x(k+1) \in S(G_{k^*}, \psi_{k^*})$, which proves the positive invariance of the set $S(G_{k^*}, \psi_{k^*})$.

3. Suppose that $S(G_{k^*}, \psi_{k^*})$ is not the maximal positively invariant set for the polytopic system (2.153) contained in $S(|K|, u_0^\alpha)$. Denote by $\mathcal{M}$ this maximal set. Suppose now that $x \in \mathcal{M}$, but $x \notin S(G_{k^*}, \psi_{k^*})$. Since, $\mathcal{M}$ is positively invariant with respect to the polytopic system (2.153), it follows that

$$\prod_{}^{k} \mathbb{A}_j x \in S(|K|, u_0^\alpha) \quad \Leftrightarrow \quad \left| K \prod_{}^{k} \mathbb{A}_j x \right| \preceq u_0^\alpha, \quad \forall k \tag{2.156}$$

On the other hand since $\mathcal{M} \subseteq S(|K|, u_0^\alpha)$, it follows that $x \in S(|K|, u_0^\alpha)$. Hence, from (2.156) the Property 2.1 (item 3) and Remark 2.16, we conclude that $x \in S(G_{k^*}, \psi_{k^*})$, which is a contradiction. $\qquad\square$

One can also express the following corollary to Proposition 2.20.

**Corollary 2.2** *If the pair* $(K, A)$ *is observable and if there exists a symmetric positive definite matrix* $P \in \Re^{n \times n}$ *verifying relation* (2.152) *then*:

1. *The set* $S(G_{k^*}, \psi_{k^*})$ *is positively invariant for the saturated system* (2.140).
2. *The set* $S(G_{k^*}, \psi_{k^*})$ *is a region of asymptotic stability for system* (2.140). *Moreover, the quadratic function* $V(x(k)) = x(k)' P x(k)$ *is such that* $V(x(k+1)) - V(x(k)) < 0, \forall x(k) \in S(G_{k^*}, \psi_{k^*})$.

*Proof* 1. Recall that the polytopic model (2.153) represents the saturated system (2.140) for $x(k) \in S(|K|, u_0^\alpha)$. Thus since by construction $S(G_{k^*}, \psi_{k^*}) \subset S(|K|, u_0^\alpha)$ and since $S(G_{k^*}, \psi_{k^*})$ is positively invariant for the polytopic model (2.153), it is also a positively invariant set for the saturated system (2.140).

2.6 Discrete-Time Case                                                             117

2. By hypothesis, matrix $P$ satisfies relation (2.152), which implies that $V(x(k+1)) - V(x(k)) < 0$, $\forall x(k) \in S(G_{k^*}, \psi_{k^*})$, with respect to the trajectories of the polytopic system (2.153). Since $S(G_{k^*}, \psi_{k^*})$ is a positively invariant set for the polytopic system and it is included in the region $S(|K|, u_0^\alpha)$, all the trajectories of (2.153) initialized in $S(G_{k^*}, \psi_{k^*})$ converge asymptotically to the origin and never leave $S(|K|, u_0^\alpha)$. Since all the trajectories of the saturated system in $S(|K|, u_0^\alpha)$ are trajectories of the polytopic system (2.153), it follows that all the trajectories of (2.140), initialized in $S(G_{k^*}, \psi_{k^*})$ converge asymptotically to the origin. $\qquad\square$

*Remark 2.17* If we define $S(T_{k+1}, r_{k+1})_{i,j} = \{x \in \Re^n; T_{k(i)} \mathbb{A}_j x \leq \sigma r_{k(i)}\}$, with $0 < \sigma < 1$, the obtained polyhedron $S(G_{k^*}, \psi_{k^*})$ is $\sigma$-contractive with respect to the saturated system [130, 348]. This means that $\forall x(k) \in S(G_{k^*}, \psi_{k^*})$, $G_{k^*} x(k+1) \preceq \sigma \psi_{k^*}$. In this case the function

$$V(x(t)) = \max_i \left\{ \frac{G_{k^*(i)} x(t)}{\psi_{k^*(i)}} \right\}$$

is a strictly decreasing Lyapunov function for the saturated system in $S(G_{k^*}, \psi_{k^*})$ (see Chap. 4 for details).

From the results above, the implementation of an algorithm to compute $S(G_k^*, \psi_k^*)$ is straightforward. Basically, it suffices to solve and evaluate the linear programs (2.151) and repeat the procedure iteratively until the set $\tilde{\mathcal{I}}_k$ is empty [130, 348].

## Algorithm 2.2

1. Set $G_0 = \left[ \begin{smallmatrix} K \\ -K \end{smallmatrix} \right]$, $\psi_0 = \left[ \begin{smallmatrix} u_0^\alpha \\ u_0^\alpha \end{smallmatrix} \right]$, $T_0 = G_0$, $r_0 = \psi_0$, $m_0 = 2m$, $k = 0$.
2. Solve $LP_k(i, j)$ for all $i \in \mathcal{I}_k$, $j \in \mathcal{J}$.
3. Determine $\tilde{\mathcal{I}}_k$. Set $l = 1$.
4. If $\tilde{\mathcal{I}}_k = \emptyset$, then $S(G_k, \psi_k)$ is the maximal admissible set and Stop. Otherwise, go to Step 5.
5. For $i = 1$ to $m_k$: if $j \in \mathcal{J}_{ki} \neq \emptyset$ then $T_{k+1(l)} = T_{k(i)} \mathbb{A}_j$; $r_{k+1(l)} = \sigma r_{k(i)}$; $l = l+1$.
6. Set $G_{k+1} = \left[ \begin{smallmatrix} G_k \\ T_{k+1} \end{smallmatrix} \right]$, $\psi_{k+1} = \left[ \begin{smallmatrix} \psi_k \\ r_{k+1} \end{smallmatrix} \right]$. Denote $m_{k+1}$ the number of rows of $T_{k+1}$. $k = k+1$. Go to Step 2.

*Remark 2.18* In Proposition 2.20 and Corollary 2.2, the vector $\alpha_l$ is supposed to be given. Nevertheless, in order to obtain a larger region $S(|K|, u_0^\alpha)$ and therefore a larger region $S(G_{k^*}, \psi_{k^*})$, it is interesting to find $\alpha_l$ with components as small as possible, for which relation (2.152) is satisfied.

In practice, we can compute first an ellipsoidal RAS using polytopic model I, considering one of the criteria discussed in Sect. 2.2.5. Then, considering the set $S(|K|, u_0^\alpha)$ associated to the ellipsoidal RAS, we compute $S(G_{k^*}, \psi_{k^*})$. Note that the set $S(G_{k^*}, \psi_{k^*})$ will always include the ellipsoidal RAS, because it is the maximal invariant set contained in $S(|K|, u_0^\alpha)$.

118 2 Stability Analysis and Stabilization—Polytopic Representation Approach

*Remark 2.19* The $(K, A)$-observability assumption in Proposition 2.20 and Corollary 2.2 is necessary for guaranteeing that for a certain index $\bar{k}$, the resulting set $S(G_{\bar{k}}, \psi_{\bar{k}})$ is bounded. Such an assumption can be dropped when the set $S(|K|, u_0^\alpha) = S(G_0, \psi_0)$ is naturally bounded.

*Remark 2.20* A more abstracted presentation of these results, based on the concept of one-step set [34, 214], can be found in [238].

*Example 2.9* Consider system (2.140) described by the following data [348]:

$$A = \begin{bmatrix} 0.8 & 0.5 \\ -0.4 & 1.2 \end{bmatrix}; \qquad B = \begin{bmatrix} 0 \\ 1 \end{bmatrix}$$

$$K = \begin{bmatrix} 0.2888 & -1.8350 \end{bmatrix}; \qquad u_0 = 7$$

Note that matrix $A$ is unstable because its spectrum is $\sigma(A) = \{1 \pm j0.4\}$.

Considering $\sigma_0 = 0.999$, the minimal $\alpha_l$ for which it is possible to satisfy $\mathbb{A}'_j P \mathbb{A}_j - \sigma_0 P < 0$ is equal to 0.1661. In this case, one obtains

$$P = \begin{bmatrix} 0.00133 & 0.0000 \\ 0.0000 & 0.00198 \end{bmatrix}$$

Hence it follows that $\mathcal{E}(P, 1) \subset S(|K|, u_0^\alpha)$ and so we conclude that $\mathcal{E}(P, 1)$ is a RAS for the saturated closed-loop system.

Consider now $\sigma = 0.998$ (see Remark 2.17), we compute the maximal $\sigma$-invariant set $S(G_{k^*}, \psi_{k^*})$ contained in $S(|K|, u_0^\alpha)$ by applying Algorithm 2.2. One obtains

$$G_{k^*} = \begin{bmatrix} 0.2888 & -1.835 \\ -0.2888 & 1.835 \\ 0.877 & -1.498 \\ -0.877 & 1.498 \\ 0.4351 & 1.31 \\ -0.4351 & -1.31 \\ 1.229 & -0.9028 \\ -1.229 & 0.9028 \\ 0.8682 & 1.39 \\ -0.8682 & -1.39 \\ 1.301 & -0.1936 \\ -1.301 & 0.1936 \\ 0.2053 & 1.678 \\ -0.2053 & -1.678 \\ 1.084 & 1.188 \\ -1.084 & -1.188 \\ 0.4488 & 1.605 \\ -0.4488 & -1.605 \end{bmatrix}; \qquad \psi_{k^*} = \begin{bmatrix} 42.14 \\ 42.14 \\ 42.06 \\ 42.06 \\ 42.06 \\ 42.06 \\ 41.97 \\ 41.97 \\ 41.97 \\ 41.97 \\ 41.89 \\ 41.89 \\ 41.89 \\ 41.89 \\ 41.89 \\ 41.89 \\ 41.81 \\ 41.81 \end{bmatrix}$$

Figure 2.12 shows the ellipsoidal region $\mathcal{E}(P, 1)$, the final polyhedral region $S(G_{k^*}, \psi_{k^*})$ and the region of linearity of the system.

**Fig. 2.12** Example 2.9— ellipsoidal and polyhedral RAS and the region of linearity: (**a**) $S(|K|, u_0)$; (**b**) $\mathcal{E}(P, 1)$; (**c**) $S(G_{k^*}, \psi_{k^*})$

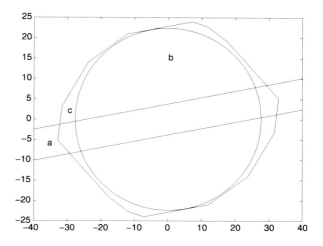

## 2.6.3 Extensions

### 2.6.3.1 External Stability

Basically, the conditions for the external stability and stabilization can be derived considering similar approaches to the ones adopted in Sects. 2.3 and 2.4.1.5 as follows.

*Amplitude Bounded External Signals* In this case we should ensure that

$$\begin{cases} \forall x(k); & x(k)'Px(k) \leq \eta^{-1} \\ \forall w(k); & w(k)'Rw(k) \leq \delta^{-1} \end{cases} \Rightarrow \quad x(k+1)'Px(k+1) \leq \eta^{-1} \quad (2.157)$$

and, in addition, ensure that $\mathcal{E}(P, \eta)$ is contained in $S(|K|, u_0^\alpha)$, $S(|H|, u_0)$ or $\bigcap_{j=2}^{2m} S(|H_j|, u_0)$, depending on whether polytopic model I, II or III is used.
Applying the S-procedure, it follows that (2.157) is verified if

$$x(k+1)'Px(k+1) - \eta^{-1} + \tau_1\left(\eta^{-1} - x(k)'Px(k)\right)$$
$$+ \tau_2\left(\delta^{-1} - w(k)'Rw(k)\right) < 0, \quad \tau_1 > 0, \ \tau_2 > 0 \quad (2.158)$$

Hence, to derive analysis or synthesis conditions it suffices to replace the polytopic model dynamics (2.141), (2.142) or (2.143) in (2.158). In [192], a slightly different procedure to show the $W$-invariance of $\mathcal{E}(P, \eta^{-1})$ is presented.

*Energy Bounded External Signals* In this case, it is supposed that

$$\sum_{k=0}^{\infty} w(k)'Rw(k) \leq \delta^{-1}$$

As discussed in Sect. 2.3.2, provided that $x(0) \in \mathcal{E}(P, \beta)$, the trajectory boundedness is obtained if the following inequality hold along the trajectories of the polytopic model

$$\Delta V\bigl(x(k)\bigr) - w(k)'Rw(k) < 0 \quad (2.159)$$

and, in addition, $\mathcal{E}(P, \mu)$, with $\mu^{-1} = \delta^{-1} + \beta^{-1}$, is contained in $S(|K|, u_0^{\alpha})$, $S(|H|, u_0)$ or $\bigcap_{j=2}^{2^m} S(|H_j|, u_0)$, depending on whether polytopic model I, II or III is considered.

Note that the verification of (2.159) along the trajectories of system the polytopic system (2.141), (2.142) or (2.143) implies that

$$V\bigl(x(T)\bigr) \leq \delta^{-1} + V\bigl(x(0)\bigr) = \delta^{-1} + \beta^{-1} = \mu^{-1}$$

If an $\mathcal{L}_2$-gain constraint is considered, then (2.159) should be replaced by

$$\Delta V(x) + \frac{1}{\gamma} z(k)' z(k) - w(k)' R w(k) < 0$$

#### 2.6.3.2 Stabilization

As in the continuous-time case, in order to obtain state feedback synthesis conditions, it suffices to perform the variable transformation $Y = K W$ in the conditions of Propositions 2.17, 2.18 or 2.19. For more details the reader can refer to [137], for the polytopic approach I or [192] for the polytopic approach II. The same optimization problems to compute the control law considered in Sect. 2.4.1 can therefore be applied.

For the discrete-time counterpart of the results considering the time-varying dynamic output feedback controller presented in Sect. 2.4.3, the reader can refer to [146] and [410], considering the polytopic approaches I and II, respectively. In those papers, it is also shown results concerning the external stabilization considering energy bounded disturbances.


## 2.7 Conclusion

In this chapter we presented several conditions that allow to analyse the stability as well as to compute stabilizing control laws for linear systems with saturating inputs. These conditions have been derived by using some polytopic differential inclusions for the saturated closed-loop system. It should be highlighted that these polytopic differential inclusions model the behavior of the actual closed-loop system only locally. Hence, the stability and stabilization conditions are basically obtained in a local (or regional) context. In fact, to each polytopic differential inclusion we can associate a region of validity of the model. In this case, considering the Lyapunov theory, the basic points to ensure the asymptotic stability are the following:

- determine a Lyapunov function $V(x)$ ensuring the asymptotic stability of the polytopic model;
- determine a level set of $V(x)$ inside the region of validity of the polytopic model.

Then, we can conclude that this level set is a region of asymptotic stability for the saturated closed-loop system. Regarding the external stability, the basic idea consists in ensuring that:

## 2.7 Conclusion

- the trajectories of the polytopic model do not leave a level set associated to a Lyapunov function;
- the level set is inside the region of validity of the polytopic model;
- the level set is also a region of asymptotic stability when the disturbance is vanishing.

The presented results mainly focused on the use of quadratic Lyapunov functions. In this context, the stability and stabilization conditions can be expressed in the form of linear or bilinear matrix inequalities. This is particularly interesting since convex optimization problems can be formulated to address analysis and synthesis problems. For instance, in the analysis context, it is possible to maximize ellipsoidal regions of asymptotic stability (i.e. to find good estimates of the region of attraction) or to maximize estimates of the maximal exogenous signals for which the input-to-state stability is ensured. On the other hand, it is possible to compute stabilizing control laws to maximize the associated region of stability (i.e. implicitly leading to a maximized region of attraction) or the tolerance to disturbances, or even to improve time-domain performance and disturbance rejection with a guaranteed region of stability. In particular, considering the discrete-time case, once an ellipsoidal region of asymptotic stability is determined, it is always possible to compute polyhedral region of asymptotic stability containing the ellipsoidal one. This polyhedral region is also included in the region of validity of the polytopic model and can be determined by linear programming techniques.

It should be pointed out that the conditions derived throughout the chapter are only sufficient. This comes from the fact that all the trajectories of the saturated system are trajectories of the polytopic models, but the converse is not true. Hence an inherent degree of conservatism is introduced in the conditions by the modeling itself. Furthermore, some other sources of conservatism are introduced when dealing with the external stability/stabilization problems, which can lead to conservative bounds on the maximal allowed disturbances and on the $\mathcal{L}_2$-gain between the exogenous inputs and the regulated outputs. Although conservative, the conditions are very useful to determine "analytical descriptions" of sets of admissible initial conditions as well as admissible exogenous signals associated to a "guaranteed safe behavior" for the closed-loop system. This is particularly useful in systems with many states, where it is complicated or impossible to characterize the region of attraction of the closed-loop system by intensive simulations.

The main purpose of this chapter has been to provide basic tools to apply polytopic differential inclusions to the analysis and synthesis of linear systems subject to input saturation. From the presented basic results many other problems and extensions have been addressed in the literature considering the same polytopic framework. For instance, among others, we can cite:

- time-delay systems [55, 98, 349, 354];
- rate saturation [16, 144, 417];
- switched and LPV systems [24, 399];
- model predictive control [54, 199, 238].

# Chapter 3
# Stability Analysis and Stabilization—Sector Nonlinearity Model Approach

## 3.1 Introduction

This chapter addresses two complementary problems: stability analysis and control design problems for linear control systems with input saturations using the sector nonlinearity model for representing the saturated closed-loop system. For the sake of simplicity and without loss of generality, the results are mainly developed considering state feedback control laws. Furthermore, the considered models are continuous-time systems. Some remarks about other control laws (like for example output dynamic controllers [119, 120]) and uncertain or discrete-time systems are provided at the end of the chapter (see Sects. 3.4.2, 3.4.3, 3.5 and 3.6).

As studied in Chap. 2, consider the continuous-time closed-loop system:

$$\dot{x}(t) = Ax(t) + B\operatorname{sat}\big(Kx(t)\big) + B_w w(t) \tag{3.1}$$

where $x \in \Re^n$, $\operatorname{sat}(Kx) \in \Re^m$ and $w \in \Re^q$. Matrices $A$, $B$, $B_w$ and $K$ are supposed to be real constant matrices of appropriate dimensions. The saturation bounds are symmetric and given by $u_0$, which is a vector with positive components. $w(t)$ is a vector-valued external signal, which can represent, for instance, a disturbance acting on the system.

Let us define the dead-zone nonlinearity $\phi(Kx(t))$ by

$$\phi\big(Kx(t)\big) = \operatorname{sat}\big(Kx(t)\big) - Kx(t) \tag{3.2}$$

or equivalently

$$\phi(K_{(i)}x) = \begin{cases} u_{0(i)} - K_{(i)}x & \text{if } K_{(i)}x > u_{0(i)} \\ 0 & \text{if } -u_{0(i)} \leq K_{(i)}x \leq u_{0(i)}, \quad \forall i = 1, \ldots, m \\ -u_{0(i)} - K_{(i)}x & \text{if } K_{(i)}x < -u_{0(i)} \end{cases} \tag{3.3}$$

From this definition, the closed-loop system (3.1) can be written as

$$\dot{x}(t) = (A + BK)x(t) + B\phi\big(Kx(t)\big) + B_w w(t) \tag{3.4}$$

In the stability analysis context (see Sect. 1.6.1), the feedback gain $K$ is supposed to be a given real constant matrix of appropriate dimensions computed from

S. Tarbouriech et al., *Stability and Stabilization of Linear Systems with Saturating Actuators*, DOI 10.1007/978-0-85729-941-3_3, © Springer-Verlag London Limited 2011

adequate design techniques (LQ control, pole placement, ... ). In the control design context (see Sect. 1.6.2), the gain $K$ has to be determined in order to satisfy some requirements in terms of stability, performance or robustness. As seen in Chap. 1, system (3.1), or equivalently system (3.4), can be viewed as a linear system in connection with a nonlinearity satisfying some sector condition. Hence, based on sector nonlinearity properties, the objective is both to derive constructive conditions ensuring the regional (local) stability of the closed-loop system and, from these conditions, to provide some efficient numerical algorithms for determining regions of stability as large as possible for the system. For this, we consider ellipsoidal regions of stability, that is, we consider sets related to quadratic Lyapunov functions (see also Appendix A). Some extensions related to the use of Lure Lyapunov functions are also considered.

The outline of the chapter is as follows. The problems to be solved are the same as those defined in Chap. 2 and therefore are not described again here. Section 3.2 presents some results regarding the stability analysis both in a regional (local) and a global contexts. The results are mainly based on the use of a quadratic Lyapunov function. The Lure Lyapunov function is also evoked. Some optimization and numerical aspects are briefly discussed. In Sect. 3.3, we derive some results to address the external stability problems. In particular, we consider two classes of disturbances: amplitude bounded and energy bounded ones. The results are mainly based on the use of a quadratic Lyapunov function, the Lure Lyapunov function being just mentioned. In Sect. 3.4, the problem of control design is addressed in both contexts of internal and external stabilization. The synthesis of state feedback gain is handled in Sect. 3.4.1, whereas the cases of observer-based feedback and dynamic output feedback stabilization are respectively studied in Sects. 3.4.2 and 3.4.3. Section 3.5 deals with the presence of model uncertainties. Norm-bounded, polytopic uncertainties and time-varying parameters are considered. Section 3.6 is briefly devoted to the discrete-time case. Nested saturations, nested nonlinearities and nonlinear and/or hybrid systems are discussed in Sect. 3.7. Finally, some concluding remarks end the chapter.

## 3.2 Asymptotic Stability Analysis

In this section, we focus on the internal asymptotic stability of system (3.1), or equivalently, of system (3.4). Hence, we assume that $w(t) = 0$. To ease the presentation of the results below let us note that system (3.1) with $w(t) = 0$ reads

$$\dot{x}(t) = Ax(t) + B \operatorname{sat}\big(Kx(t)\big) \tag{3.5}$$

or equivalently from (3.4)

$$\dot{x}(t) = (A + BK)x(t) + B\phi\big(Kx(t)\big) \tag{3.6}$$

In the sequel, two types of Lyapunov functions are considered: quadratic and Lure Lyapunov functions. All the results are provided using the generalized sector condition (see Sect. 1.7.2.2 and in particular Lemma 1.6 of Chap. 1).

## 3.2 Asymptotic Stability Analysis

### 3.2.1 Quadratic Lyapunov Function

In this section, we consider a quadratic Lyapunov function defined as

$$V(x) = x'Px, \quad P = P' > 0 \tag{3.7}$$

The following proposition allows to solve the stability analysis problem.

**Proposition 3.1** *If there exist a symmetric positive definite matrix $W \in \mathfrak{R}^{n \times n}$, a positive diagonal matrix $S \in \mathfrak{R}^{m \times m}$, a matrix $Z \in \mathfrak{R}^{m \times n}$ satisfying:*

$$\begin{bmatrix} W(A+BK)' + (A+BK)W & BS - Z' \\ SB' - Z & -2S \end{bmatrix} < 0 \tag{3.8}$$

$$\begin{bmatrix} W & WK'_{(i)} - Z'_{(i)} \\ K_{(i)}W - Z_{(i)} & u_{0(i)}^2 \end{bmatrix} \geq 0, \quad i = 1, \dots, m \tag{3.9}$$

*then the ellipsoid $\mathcal{E}(P, 1)$, with $P = W^{-1}$, is a region of asymptotic stability (RAS) for the saturated system (3.5).*

*Proof* Consider the quadratic Lyapunov function $V(x)$ defined in (3.7). Lemma 1.6 in Chap. 1 applies by choosing, as in [139], $v = Kx$ and $\omega = Gx$. For any $x$ belonging to the resulting set, defined in (1.63), which reads

$$S(|K - G|, u_0) = \{x \in \mathfrak{R}^n; \ -u_0 \preceq (K - G)x \preceq u_0\} \tag{3.10}$$

the sector nonlinearity $\phi(Kx)$, defined in (3.2), satisfies the inequality (1.64) (or equivalently inequality (1.65)), recalled below:

$$2\phi(Kx)'T(\phi(Kx) + Gx) \leq 0 \tag{3.11}$$

By noting $P = W^{-1}$ and using the change of variables $GW = Z$, the satisfaction of relation (3.9) means that the ellipsoid $\mathcal{E}(P, 1)$ is included in the polyhedral set $S(|K - G|, u_0)$.

To prove the asymptotic stability of system (3.5), the following condition:

$$\dot{V}(x) = \dot{x}'Px + x'P\dot{x} < 0$$

has to be satisfied along the trajectories of system (3.5). For this, we use the sector condition (3.11), which implies that

$$\dot{V}(x) \leq \dot{V}(x) - 2\phi(Kx)'T(\phi(Kx) + Gx), \quad \forall x \in \mathcal{E}(P, 1)$$

where $T$ is a positive diagonal matrix. Thus, the right term of the previous inequality reads

$$\begin{bmatrix} x' & \phi(Kx)' \end{bmatrix} \begin{bmatrix} (A+BK)'P + P(A+BK) & PB - G'T \\ B'P - TG & -2T \end{bmatrix} \begin{bmatrix} x \\ \phi(Kx) \end{bmatrix}$$

By left- and right-multiplying the above matrix by $\text{Diag}(P^{-1}, T^{-1})$, by denoting $W = P^{-1}$ and $S = T^{-1}$ and by using the change of variables $GW = Z$, one obtains

126    3  Stability Analysis and Stabilization—Sector Nonlinearity Model Approach

the matrix of the left-hand term of relation (3.8). Hence, if relation (3.8) is satisfied then $\dot{V}(x) \leq \dot{V}(x) - 2\phi(Kx)'T(\phi(Kx) + Gx) < 0$.

Then, provided that relations (3.8) and (3.9) are satisfied, one guarantees that, for any $x \in \mathcal{E}(P, 1)$, the asymptotic stability of the closed-loop system (3.5) is ensured.    $\square$

*Remark 3.1*  Another manner to apply the sector condition (1.64) may be considered. It consists in choosing $v = Kx$ and $\omega = Kx + Gx$. In this case the conditions of Proposition 3.1 are quite similar and read

$$\begin{bmatrix} W(A + BK)' + (A + BK)W & BS - WK' - Z' \\ SB' - KW - Z & -2S \end{bmatrix} < 0 \qquad (3.12)$$

$$\begin{bmatrix} W & Z'_{(i)} \\ Z_{(i)} & u^2_{0(i)} \end{bmatrix} \geq 0, \quad i = 1, \dots, m \qquad (3.13)$$

This is the way which will be followed in Chaps. 7 and 8 to provide the conditions for anti-windup design.

*Remark 3.2*  If there exists a positive scalar $\mu$ such that relation (3.9) and

$$\begin{bmatrix} W(A + BK)' + (A + BK)W + \mu W & BS - Z' \\ SB' - Z & -2S \end{bmatrix} < 0 \qquad (3.14)$$

are satisfied, then it follows that $\mathcal{E}(P, 1)$, with $P = W^{-1}$, is a $\mu$-contractive RAS for the closed-loop system (3.5). In other words, one verifies $\dot{V}(x) \leq -\mu V(x) < 0$, i.e. $V(x(t)) \leq e^{-\mu t} V(x(0))$, with $V(x) = x'Px$, $P = W^{-1}$, along the trajectories of the closed-loop system (3.5). In this case, it means that the trajectories initialized in $\mathcal{E}(P, 1)$ exponentially converge to the origin with a decay rate $\mu$.

*Remark 3.3*  In the context of the classical sector condition approach, by using Lemma 1.5 of Chap. 1, a proposition similar to Proposition 3.1 can be obtained. To do this, one has

- To replace the decision variable $Z$ by $\Lambda KW$ in relations (3.8) and (3.9).
- To replace the term $u^2_{0(i)}$ by $\eta u^2_{0(i)}$ in (3.9), where $\eta$ is a positive scalar and a new decision variable.
- To add the condition $0 < \Lambda_{(i,i)} \leq 1, i = 1, \dots, m$.

Therefore relations (3.8) and (3.9) are modified to

$$\begin{bmatrix} W(A + BK)' + (A + BK)W & BS - WK'\Lambda \\ SB' - \Lambda KW & -2S \end{bmatrix} < 0 \qquad (3.15)$$

$$\begin{bmatrix} W & (1 - \Lambda_{(i,i)})WK'_{(i)} \\ (1 - \Lambda_{(i,i)})K_{(i)}W & \eta u^2_{0(i)} \end{bmatrix} \geq 0, \quad i = 1, \dots, m \qquad (3.16)$$

$$0 < \Lambda_{(i,i)} \leq 1, \quad i = 1, \dots, m \qquad (3.17)$$

where the decision variables are $W$, $S$, $\Lambda$ and $\eta$. The region of asymptotic stability for system (3.5) is then $\mathcal{E}(P, \eta) = \{x \in \Re^n; \ x'Px \leq \eta^{-1}\}$, with $P = W^{-1}$.

3.2 Asymptotic Stability Analysis

*Remark 3.4* Observe that all the solutions obtained with $G = \Lambda K$ are also solutions of Proposition 3.1. One can conclude that Proposition 3.1 provides more generic and less conservative conditions than the classical approach. Moreover, it is important to note that relations (3.8) and (3.9) are LMIs in the decision variables $W$, $S$ and $Z$. At the opposite, in the context of the classical approach, as above mentioned, the resulting conditions are BMIs in the decision variables $W$ and $\Lambda$. See, for example, [183, 280, 379].

*Remark 3.5* In Proposition 3.1, the fact to consider $\eta u_{0(i)}^2$ instead of $u_{0(i)}^2$ in relation (3.9) does not add any new degree of freedom and $\eta$ can be considered as 1 without loss of generality. Actually, to understand this, it suffices to consider $\bar{W} = \eta^{-1}W$, $\bar{Z} = \eta^{-1}Z$ and $\bar{S} = \eta^{-1}S$. At the opposite, in the framework of the classical approach, however, it is not possible to eliminate $\eta$ by means of an appropriate change of variables due to the product between $\Lambda$ and $W$. Hence, the case $\eta = 1$ may be strongly conservative.

The previous results present regional (local) stability conditions for the closed-loop system (3.5). The global analysis problem is typically of somewhat lower complexity than the local stability analysis problem because, in the global case, the region of attraction (and therefore the region of stability) is the whole state space. Then, global stability conditions do not require any description of the region of stability of the saturated closed-loop system. For stable linear plants (i.e. $\mathfrak{Re}\lambda_i(A) < 0$, $\forall i$), it is possible to develop conditions which go beyond local guarantees and provide stability for all $x \in \mathfrak{R}^n$. The global stability of the closed-loop system (3.5) can then be considered as follows.

**Proposition 3.2** *If there exist a symmetric positive definite matrix $W \in \mathfrak{R}^{n \times n}$ and a positive diagonal matrix $S \in \mathfrak{R}^{m \times m}$ satisfying*

$$\begin{bmatrix} W(A + BK)' + (A + BK)W & BS - WK' \\ SB' - KW & -2S \end{bmatrix} < 0 \tag{3.18}$$

*then the origin is globally asymptotically stable for the saturated system* (3.5).

*Proof* It suffices to consider $G = K$. In that case, the sector condition (1.64) with $v = Kx$ is globally satisfied, that is, it is satisfied for any $x \in \mathfrak{R}^n$. Then $\phi(Kx)'S^{-1}(\phi(Kx) + Kx) \leq 0$, $\forall x \in \mathfrak{R}^n$, where $S$ is a diagonal positive definite matrix in $\mathfrak{R}^{m \times m}$. The satisfaction of relation (3.18) means that

$$\dot{V}(x) \leq \dot{V}(x) - 2\phi(Kx)'S^{-1}(\phi(Kx) + Kx) < 0, \quad \forall x \in \mathfrak{R}^n$$

The stability follows immediately from the above inequality which obviously implies that $\dot{V}(x) < 0$ along any closed-loop trajectory of the saturated system (3.5). $\qquad\square$

*Remark 3.6* The global case can also be addressed in the framework of the classical sector approach. In this case, it suffices to choose $\Lambda = I_m$ and then relation (3.15) reduces to (3.18). Hence, one retrieves the result of Proposition 3.2.

128            3 Stability Analysis and Stabilization—Sector Nonlinearity Model Approach

In order to ensure the global asymptotic stability of the system (3.5), the hypothesis of a diagonal multiplier $T$ in the sector condition can be relaxed. For this, consider the following lemma [359].

**Lemma 3.1** *The nonlinearity* $\phi(v) = \mathrm{sat}(v) - v$ *satisfies the inequality*

$$\phi(v)'T\big(\phi(v) + v\big) \leq 0 \tag{3.19}$$

*for any vector* $v \in \Re^m$ *and any matrix* $T \in \Re^{m \times m}$ *such that*

$$u_{0(i)}T_{(i,i)} \geq \sum_{i \neq j, j=1}^{m} u_{0(j)}|T_{(i,j)}|, \quad i = 1, \ldots, m \tag{3.20}$$

*Proof* From the definition of $\phi(v)$, it follows that

$$\phi(v)'T\big(\phi(v) + v\big)$$
$$= \sum_{i=1}^{m} \phi(v_{(i)})\left[ T_{(i,i)}\big(\phi(v_{(i)}) + v_{(i)}\big) + \sum_{i \neq j, j=1}^{m} T_{(i,j)}\big(\phi(v_{(j)}) + v_{(j)}\big)\right] \tag{3.21}$$

Consider now the three cases below.

Case 1: $\phi(v_{(i)}) > 0$. In this case $\phi(v_{(i)}) + v_{(i)} = -u_{0(i)}$ and therefore, $[T_{(i,i)}(\phi(v_{(i)}) + v_{(i)}) + \sum_{i \neq j, j=1}^{m} T_{(i,j)}(\phi(v_{(j)}) + v_{(j)})] \leq -u_{0(i)}T_{(i,i)} + \sum_{i \neq j, j=1}^{m} u_{0(j)}|T_{(i,j)}| \leq 0$. Hence if (3.20) is satisfied one has $\phi(v_{(i)})[T_{(i,i)}(\phi(v_{(i)}) + v_{(i)}) + \sum_{i \neq j, j=1}^{m} T_{(i,j)}(\phi(v_{(j)}) + v_{(j)})] \leq 0$.

Case 2: $\phi(v_{(i)}) < 0$. In this case $\phi(v_{(i)}) + v_{(i)} = u_{0(i)}$ and therefore, $[T_{(i,i)}(\phi(v_{(i)}) + v_{(i)}) + \sum_{i \neq j, j=1}^{m} T_{(i,j)}(\phi(v_{(j)}) + v_{(j)})] \geq u_{0(i)}T_{(i,i)} - \sum_{i \neq j, j=1}^{m} u_{0(j)}|T_{(i,j)}| \geq 0$. Hence if (3.20) is satisfied one has $\phi(v_{(i)})[T_{(i,i)}(\phi(v_{(i)}) + v_{(i)}) + \sum_{i \neq j, j=1}^{m} T_{(i,j)} \times (\phi(v_{(j)}) + v_{(j)})] \leq 0$.

Case 3: $\phi(v_{(i)}) = 0$. One has $\phi(v_{(i)})[T_{(i,i)}(\phi(v_{(i)}) + v_{(i)}) + \sum_{i \neq j, j=1}^{m} T_{(i,j)}(\phi(v_{(j)}) + v_{(j)})] = 0$.

Thus, from the three cases above and from (3.21), we can conclude that $\phi(v)'T \times (\phi(v) + v) \leq 0$. $\qquad\square$

The result of Lemma 3.1 can also be obtained by using some results based on integral quadratic constraints (IQC), as stated in [79, 229] for systems presenting repeated saturations.

Hence, by using Lemma 3.1 another condition for the global asymptotic stability can be stated.

**Proposition 3.3** *If there exist two symmetric positive definite matrices* $P \in \Re^{n \times n}$ *and* $T \in \Re^{m \times m}$ *satisfying*

## 3.2 Asymptotic Stability Analysis

$$\begin{bmatrix} (A+BK)'P + P(A+BK) & PB - K'T \\ B'P - TK & -2T \end{bmatrix} < 0 \qquad (3.22)$$

$$u_{0(i)}T_{(i,i)} \geq \sum_{j=1, j\neq i}^{m} u_{0(j)}|T_{(i,j)}|, \quad i = 1, \ldots, m \qquad (3.23)$$

then the origin is globally asymptotically stable for the saturated system (3.5).

*Proof* By considering Lemma 3.1, one wants to verify that

$$\dot{V}(x) \leq \dot{V}(x) - 2\phi(Kx)'T\big(\phi(Kx) + Kx\big) < 0$$

for any $x \in \mathfrak{R}^n$ and any matrix $T \in \mathfrak{R}^{m\times m}$ such that relation (3.20), or equivalently relation (3.23), holds. The satisfaction of relation (3.22) ensures that $\dot{V}(x) \leq \dot{V}(x) - 2\phi(Kx)'T(\phi(Kx) + Kx) < 0$. The stability immediately follows from the above inequality which obviously implies that $\dot{V}(x) < 0$ along any closed-loop trajectory. $\qquad \square$

*Remark 3.7* If the control bounds are normalized to $u_{0(i)} = 1, i = 1, \ldots, m$, condition (3.20) reads $T_{(i,i)} \geq \sum_{j=1, j\neq i}^{m} |T_{(i,j)}|$, which corresponds to the one given in [79] considering an IQC approach. Furthermore, in order to implement (3.23) in an LMI framework we can consider $T_{(i,j)} = T_{(i,j)}^{+} - T_{(i,j)}^{-}$, for $i \neq j$, with $T_{(j,i)}^{+} = T_{(i,j)}^{+} \geq 0$ and $T_{(j,i)}^{-} = T_{(i,j)}^{-} \geq 0$ and the following constraints [79]:

$$u_{0(i)}T_{(i,i)} \geq \sum_{j=1, j\neq i}^{m} u_{0(j)}\big(T_{(i,j)}^{+} + T_{(i,j)}^{-}\big), \quad i = 1, \ldots, m$$

### 3.2.2 Lure Lyapunov Function

In this section, we consider a Lure Lyapunov function defined by [215]

$$V(x) = x'Px - 2\sum_{i=1}^{m} \int_{0}^{K_{(i)}x(t)} \phi_{(i)}(\sigma)N_{(i,i)}\,d\sigma \qquad (3.24)$$

with $P = P' > 0$ and $N$ a diagonal positive definite matrix. From the definition of the sector nonlinearity $\phi(\cdot)$ given in (3.2), one can easily prove that the function $V(x)$ above defined is positive definite provided that $x \in S(|K - G|, u_0)$ (see the definition of this polyhedral set in (3.10)).

As in the quadratic case, conditions may be derived to solve the stability analysis problem with respect to the closed-loop system (3.5), by using Lure function (3.24).

**Proposition 3.4** *If there exist a symmetric positive definite matrix $P \in \mathfrak{R}^{n\times n}$, two diagonal positive definite matrices $N \in \mathfrak{R}^{m\times m}$ and $T \in \mathfrak{R}^{m\times m}$, a matrix $Z \in \mathfrak{R}^{m\times n}$ and a positive vector $\sigma \in \mathfrak{R}^{m}$ (i.e., $\sigma_{(i)} > 0, i = 1, \ldots, m$) satisfying:*

$$\begin{bmatrix} (A+BK)'P + P(A+BK) & PB - (A+BK)'K'N - Z' \\ B'P - NK(A+BK) - Z & -2T - NKB - B'K'N \end{bmatrix} < 0 \quad (3.25)$$

$$\begin{bmatrix} P & T_{(i,i)}K'_{(i)} - Z'_{(i)} \\ T_{(i,i)}K_{(i)} - Z_{(i)} & \sigma_{(i)}u^2_{0(i)} \end{bmatrix} \geq 0, \quad i = 1, \dots, m \quad (3.26)$$

then the region $S(V, \eta) = \{x \in \Re^n; \ V(x) \leq \eta^{-1}\}$, with $\eta \geq \max_i \frac{\sigma_{(i)}}{T^2_{(i,i)}}$ and $V(x)$ defined in (3.24), is a region of asymptotic stability (RAS) for the saturated system (3.5).

*Proof* Consider the Lure function defined in (3.24). This proof follows the same reasoning as that one of Proposition 3.1. By computing the time-derivative of the Lure function along the trajectories of the closed-loop system (3.5) or (3.6), one gets:

$$\dot{V}(x) = \dot{x}'Px + x'P\dot{x} - 2\phi(Kx)'NK\dot{x}$$

As in the proof of Proposition 3.1, we want to use the following inequality

$$\dot{V}(x) \leq \dot{V}(x) - 2\phi(Kx)'T\big(\phi(Kx) + Gx\big) < 0 \quad (3.27)$$

To do this, let us study the meaning of relation (3.26) and its link with the inclusion of the region $S(V, \eta) = \{x \in \Re^n; \ V(x) \leq \eta^{-1}\}$ in the region of validity $S(|K - G|, u_0)$ of the sector condition. From the definition of the Lure function, we can prove by computing the integral term of the Lure function that the following inequality holds [142, 183]:

$$x'Px \leq V(x) \leq x'\big(P + K'NK\big)x \quad (3.28)$$

Such a double inequality (3.28) corresponds to the following double inclusion:

$$\mathcal{E}\big(P + K'NK, \eta\big) \subset \big\{x \in \Re^n; \ V(x) \leq \eta^{-1}\big\} \subset \mathcal{E}(P, \eta) \quad (3.29)$$

Hence, to prove that the region $S(V, \eta) = \{x \in \Re^n; \ V(x) \leq \eta^{-1}\}$ is included in $S(|K - G|, u_0)$, it suffices to prove that the ellipsoid $\mathcal{E}(P, \eta)$ is included in $S(|K - G|, u_0)$. Such an inclusion corresponds to the satisfaction of the following inequality:

$$\begin{bmatrix} P & K'_{(i)} - G'_{(i)} \\ K_{(i)} - G_{(i)} & \eta u^2_{0(i)} \end{bmatrix} \geq 0, \quad i = 1, \dots, m \quad (3.30)$$

Therefore by post- and pre-multiplying inequality (3.30) by $\mathrm{Diag}(I_n, T_{(i,i)})$, one gets

$$\begin{bmatrix} P & T_{(i,i)}K'_{(i)} - T_{(i,i)}G'_{(i)} \\ T_{(i,i)}K_{(i)} - T_{(i,i)}G_{(i)} & T^2_{(i,i)}\eta u^2_{0(i)} \end{bmatrix} \geq 0, \quad i = 1, \dots, m \quad (3.31)$$

Hence, by setting $Z = TG$ and considering that $T^2_{(i,i)}\eta \geq \sigma_{(i)}$, $i = 1, \dots, m$, or equivalently, that $\eta \geq \max_i \frac{\sigma_{(i)}}{T^2_{(i,i)}}$, it follows that the satisfaction of relation (3.26) implies the satisfaction of relation (3.31), and then of (3.30). Thus, the satisfaction of (3.26) ensures the inclusion of $S(V, \eta) = \{x \in \Re^n; \ V(x) \leq \eta^{-1}\}$ in

## 3.2 Asymptotic Stability Analysis

$S(|K - G|, u_0)$.

As in the proof of Proposition 3.1, we want to satisfy $\dot{V}(x) - 2\phi(Kx)'T \times (\phi(Kx) + Gx) < 0$, for any $x$ in $S(V, \eta)$. The right-hand term of the inequality above reads

$$\begin{bmatrix} x' & \phi(Kx)' \end{bmatrix} \begin{bmatrix} (A + BK)'P + P(A + BK) & PB - (A + BK)'K'N - G'T \\ B'P - NK(A + BK) - TG & -2T - NKB - B'K'N \end{bmatrix}$$
$$\times \begin{bmatrix} x \\ \phi(Kx) \end{bmatrix}$$

With the change of variable $Z = TG$, it follows that the satisfaction of relation (3.25) guarantees that $\dot{V}(x) - 2\phi(Kx)'T(\phi(Kx) + Gx) < 0$ and therefore $\dot{V}(x) < 0$. We can conclude that the region $S(V, \eta) = \{x \in \Re^n; V(x) \le \eta^{-1}\}$, with $\eta \ge \max_i \frac{\sigma_{(i)}}{T_{(i,i)}^2}$ and $V(x)$ defined in (3.24), is a region of asymptotic stability for the saturated system (3.5). $\qquad\square$

**Remark 3.8** It has to be noted that the conditions (3.25) and (3.26) are LMIs in the decision variables $P$, $N$, $T$, $Z$ and $\sigma$.

**Remark 3.9** By considering $N = 0$, relations of Proposition 3.1 can be retrieved by using adequate change of variables and some congruence transformations.

**Remark 3.10** The region $S(V, \eta)$ issued from Proposition 3.4 and Lure function is not manageable, and only over and inner bounds provided by the double inclusion (3.29) may be easily pictured.

**Remark 3.11** As in Remark 3.3, the relations of Proposition 3.4 can be written in the context of the classical sector condition approach. To do this, one replaces the matrix $G$ by $\Lambda K$ and then one has

- To replace the decision variable $Z$ by $UK$ in relations (3.25) and (3.26), where $U$ is a new decision variable corresponding to a diagonal positive definite matrix.
- To add the condition $0 < \Lambda_{(i,i)} \le 1, i = 1, \ldots, m$.

Therefore relations (3.25) and (3.26) are modified to

$$\begin{bmatrix} (A + BK)'P + P(A + BK) & PB - (A + BK)'K'N - K'U \\ B'P - NK(A + BK) - UK & -2T - NKB - B'K'N \end{bmatrix} < 0 \quad (3.32)$$

$$\begin{bmatrix} P & (T_{(i,i)} - U_{(i,i)})K'_{(i)} \\ (T_{(i,i)} - U_{(i,i)})K_{(i)} & \sigma_{(i)}u_{0(i)}^2 \end{bmatrix} \ge 0, \quad i = 1, \ldots, m \quad (3.33)$$

$$0 < U_{(i,i)} \le T_{(i,i)}, \quad i = 1, \ldots, m \quad (3.34)$$

The decision variables are $P$, $T$, $N$, $U$ and $\sigma$. The domain of asymptotic stability for the system (3.5) is $S(V, \eta) = \{x \in \Re^n; V(x) \le \eta^{-1}\}$, with $\eta \ge \max_i \frac{\sigma_{(i)}}{T_{(i,i)}^2}$.

As in the quadratic case with Proposition 3.2, the global stability can be studied when the open-loop matrix $A$ is Hurwitz.

132  3 Stability Analysis and Stabilization—Sector Nonlinearity Model Approach

**Proposition 3.5** *If there exist a symmetric positive definite matrix $P \in \mathfrak{R}^{n \times n}$, two diagonal positive definite matrices $N \in \mathfrak{R}^{m \times m}$ and $T \in \mathfrak{R}^{m \times m}$ satisfying*

$$\begin{bmatrix} (A + BK)'P + P(A + BK) & PB - (A + BK)'K'N - K'T \\ B'P - NK(A + BK) - TK & -2T - NKB - B'K'N \end{bmatrix} < 0 \quad (3.35)$$

*then the origin is globally asymptotically stable for the saturated system* (3.5).

*Remark 3.12* By considering $N = 0$ in relation (3.35), one retrieves the result of Proposition 3.2.

*Remark 3.13* By using Lemma 3.1, a proposition similar to Proposition 3.3 can be derived in the Lure case. Then one has to take into account the condition (3.23) on $T$ in addition to relation (3.35).

### *3.2.3 Computational Burden*

Let us suggest some comments regarding the numerical complexity of the proposed results (for both quadratic and Lure cases). In particular, we can compare the number of variables and the number of lines in the LMIs to be solved considering the different propositions of this current chapter, but also Chap. 2 considering the polytopic approach to model the saturated system in a regional (local) or global context. Indeed, when the complexity analysis of a numerical method based on matrix inequalities is addressed, the numbers of lines and of variables of the considered LMIs play a crucial role. The computational complexity grows when these numbers become larger: see, for example the complexity analysis of interior-point method [101], or the complexity analysis for a primal method (Theorem 5.1 in [80]) or for a dual method (Theorem 5.2 in [80]). LMI conditions can be solved in polynomial time for instance by specialized algorithms as in [101], with complexity proportional to $\mathcal{C} = \mathcal{D}^3 \mathcal{L}$ (where $\mathcal{D}$ is the number of decision variables and $\mathcal{L}$ is the number of lines). Of course, other LMI solvers may perform differently and other indices, such as the CPU time, may be considered. Interesting reader may refer to [80, 101].

Considering the regional context, the number of decision variables and of lines in the LMIs are compared in Table 3.1 for three alternative propositions which all solve the same stability analysis problem for the saturated system (3.5).

One can observe that the numerical complexity associated to Proposition 3.1 increases more slowly than that one associated to Proposition 3.4 (in terms of decision variables). Proposition 2.2 involves less lines than Propositions 3.1 and 3.4 when $m = 1$. However, the number of lines increases much more quickly with increasing $m$ for Proposition 2.2 than for Propositions 3.1 and 3.4.

Similarly in the global context, the number of variables and the number of lines in the LMIs to be solved considering Propositions 3.2, 3.3 and 3.5 are compared in Table 3.2. It suggests that there is not much difference between the propositions unless $m$ is significantly increased.

## 3.2 Asymptotic Stability Analysis

**Table 3.1** Number of decision variables and of lines in Propositions 3.1, 3.4 and 2.2

|  | Decision variables ($\mathcal{D}$) | Lines ($\mathcal{L}$) |
| --- | --- | --- |
| Proposition 3.1 | $n(n+1)/2 + m(n+1)$ | $n + m + m(n+1)$ |
| Proposition 3.4 | $n(n+1)/2 + m(n+3)$ | $n + m + m(n+1)$ |
| Proposition 2.2 (Polytopic Approach II) | $n(n+1)/2 + mn$ | $n2^m + m(n+1)$ |

**Table 3.2** Number of decision variables and of lines in Propositions 3.2, 3.3 and 3.5

|  | Decision variables ($\mathcal{D}$) | Lines ($\mathcal{L}$) |
| --- | --- | --- |
| Proposition 3.2 | $n(n+1)/2 + m$ | $n + m$ |
| Proposition 3.3 | $n(n+1)/2 + m(m+1)/2$ | $n + 2m$ |
| Proposition 3.5 | $n(n+1)/2 + 2m$ | $n + m$ |

## 3.2.4 Optimization Problems

### 3.2.4.1 Quadratic Case

The conditions of Propositions 3.1 or 3.2 are linear in the decision variables and therefore can be easily checked in terms of feasibility. Moreover, as in the Chap. 2, the implicit objective is to maximize the set $\mathcal{E}(P, \eta)$. Therefore, different criteria can be associated to the different ways to measure the size of the set $\mathcal{E}(P, \eta)$: see Sect. 2.2.5.1.

The following simple algorithm can be used in order to exhibit solution to the stability analysis problem.

### Algorithm 3.1

1. Choose a convex criterion $f(\mathcal{E}(P, 1))$.
2. Solve for $W$, $S$ and $Z$ the following convex optimization problem

$$\min \ f\big(\mathcal{E}(P, 1)\big)$$

$$\text{subject to} \quad \text{inequalities (3.8)–(3.9)}$$

Algorithm 3.1 provides a solution to the local stability analysis problem for the closed-loop system (3.5).

Regarding the global stability of the closed-loop system, it suffices to search a feasible solution to relation (3.18) of Proposition 3.2 or to relations (3.22) and (3.23) of Proposition 3.3.

134 3 Stability Analysis and Stabilization—Sector Nonlinearity Model Approach

### 3.2.4.2 Lure Case

It is important to note that due to the inclusion

$$\mathcal{E}(P + K'NK, \eta) \subset \{x \in \Re^n; \; V(x) \leq \eta^{-1}\} \subset \mathcal{E}(P, \eta)$$

mentioned in the proof of Proposition 3.4, a way to maximize the region $\{x \in \Re^n; \; V(x) \leq \eta^{-1}\}$, may consist of maximizing the size of the ellipsoid $\mathcal{E}(P + K'NK, \eta)$.

Hence, we can consider the following algorithm.

### Algorithm 3.2

1. Choose a convex criterion $f(\mathcal{E}(P + K'NK, \eta))$.
2. Solve for $P$, $T$, $N$, $Z$ and $\sigma$ the following convex optimization problem:

$$\min \; f\big(\mathcal{E}(P + K'NK, \eta)\big)$$

$$\text{subject to} \quad \text{inequalities (3.25)–(3.26)}$$

*Remark 3.14* For example, due to the definition of $\eta$, namely $\eta \geq \max_i \frac{\sigma_{(i)}}{T_{(i,i)}^2}$, one can choose for $f(\mathcal{E}(P + K'NK, \eta))$ the following function:

$$f\big(\mathcal{E}(P + K'NK, \eta)\big) = \beta_1 \, \text{trace}\big(P + K'NK\big) + \sum_{i=1}^{m} (\beta_{2i}\sigma_{(i)} - \beta_{3i} T_{(i,i)})$$

where the positive scalars $\beta_1$, $\beta_{2i}$ and $\beta_{3i}$, $i = 1, \ldots, m$, are given weighting parameters. Moreover, a more mature approach based on a linearization algorithm, as detailed in [88], would allow us to tackle the presence of both $\sigma_{(i)}$ and $T_{(i,i)}^2$ in the conditions.

Note that, in the global context, relation (3.35) of Proposition 3.5 has just to be tested in terms of feasibility to find $P$, $N$ and $T$.

*Example 3.1* Consider the system studied in [218], already used in Chap. 2 (Example 2.1), recalled here:

$$A = \begin{bmatrix} 0.1 & -0.1 \\ 0.1 & -3 \end{bmatrix}; \qquad B = \begin{bmatrix} 5 & 0 \\ 0 & 1 \end{bmatrix}$$

$$K = \begin{bmatrix} -0.7283 & -0.0338 \\ -0.0135 & -1.3583 \end{bmatrix}; \qquad u_0 = \begin{bmatrix} 5 \\ 2 \end{bmatrix}$$

Algorithm 3.1 may be applied in the quadratic case to solve the stability analysis problem. Figure 3.1 depicts the obtained ellipsoids of closed-loop stability considering two different criteria:

- minimization of $\text{trace}(W^{-1})$ (dashed line);
- maximization in the directions formed by the vertices of the unit square box (solid line).

## 3.2 Asymptotic Stability Analysis

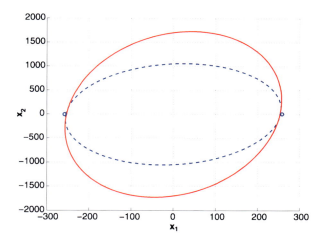

**Fig. 3.1** Example 3.1—regions of stability, using sector nonlinearities approaches with different criteria

This figure may be compared with Fig. 2.1 obtained using differential inclusions. Again, it shows that the estimation of the region of attraction comes very close to the unstable equilibrium points at

$$x_{eq} = \pm \begin{bmatrix} 257.931 \\ 7.931 \end{bmatrix}$$

To go further, this analysis of the state-feedback closed-loop control may be directly extended to the analysis of the saturated system interconnected with a dynamic output feedback described by the quadruplet $(A_c, B_c, C_c, D_c)$. The closed-loop augmented system is then described as

$$\dot{x}(t) = \mathbb{A}x(t) + \mathbb{B}\operatorname{sat}(\mathbb{K}x(t))$$

where

$$x = \begin{bmatrix} x_p \\ x_c \end{bmatrix}; \qquad \mathbb{A} = \begin{bmatrix} A & 0 \\ B_c C & A_c \end{bmatrix}; \qquad \mathbb{B} = \begin{bmatrix} B \\ 0 \end{bmatrix}; \qquad \mathbb{K} = \begin{bmatrix} D_c C & C_c \end{bmatrix}$$

and $x_p$ and $x_c$ are the state of the plant and the state of the dynamic controller, respectively.

Algorithm 3.1 may be applied directly to the augmented system. Considering the system above, the output $y_p = x_p$, and the dynamic feedback described by

$$A_c = \begin{bmatrix} -171.2 & 27.2 \\ -68 & -626.8 \end{bmatrix}; \qquad B_c = \begin{bmatrix} -598.2 & 5.539 \\ -4.567 & 149.8 \end{bmatrix}$$

$$C_c = \begin{bmatrix} 0.146 & 0.088 \\ -6.821 & -5.67 \end{bmatrix}; \qquad D_c = \begin{bmatrix} 0 & 0 \\ 0 & 0 \end{bmatrix}$$

the ellipsoidal set solution to the optimization problem considering the maximization in the directions formed by the vertices of the unit square box in the plant state is plotted (dashed-line ellipsoid) on Fig. 3.2. The figure only presents the projection of the ellipsoidal set of dimension 4 in the subspace of the plant state, considering

**Fig. 3.2** Example 3.1—regions of stability considering a state-feedback controller (*solid-line ellipsoid*) and a dynamic-feedback controller (*dashed-line ellipsoid*). Zoom close to the parasitic equilibrium points

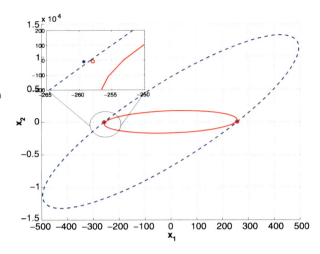

that the controller state is equal to 0 (we are comparing sets of admissible initial conditions, for which it is considered that the control system is initially at rest). The region plotted corresponds to the set of $x_p$ such that $x = \begin{bmatrix} x_p \\ 0 \end{bmatrix} \in \mathcal{E}(P, 1)$. It is compared to the ellipsoidal set previously obtained, with the same optimization criterion, considering the state-feedback case (solid ellipsoid). The nonlinear equilibrium points are given by

$$x_{\text{eq}} = \begin{bmatrix} -259.3103 & 259.3103 \\ -9.3103 & 9.3103 \\ 890.3703 & -890.3703 \\ -96.9298 & 96.9298 \end{bmatrix}$$

It can also be noted that they have been slightly moved with the dynamic controller (blue stars) with respect to those associated to the state feedback case (red circle).

## 3.3 External Stability

As in Chap. 2, the external stability analysis problems may be addressed. In particular, we consider two kinds of disturbances: amplitude bounded and energy bounded ones. The results are mainly based on the use of a quadratic Lyapunov function. The Lure Lyapunov function is briefly evoked as an extension of the quadratic results.

### 3.3.1 Amplitude Bounded Exogenous Signals

Consider the exogenous signal $w(t)$ belonging to the following set:

$$\mathcal{W} = \left\{ w \in \Re^q ; \ w'Rw \leq \delta^{-1} \right\} \tag{3.36}$$

## 3.3 External Stability

with $R = R' > 0$ and $\delta > 0$. In this case, the vector $w(t)$ is bounded by a quadratic norm which reflects amplitude bounds on this vector. Note that the classical $\mathcal{L}_\infty$-norm constraints can be straightforwardly considered.

Let us consider the quadratic Lyapunov function defined in (3.7). The following proposition allows one to solve the external stability analysis problem.

**Proposition 3.6** *If there exist a symmetric positive definite matrix $W \in \Re^{n \times n}$, a positive diagonal matrix $S \in \Re^{m \times m}$, a matrix $Z \in \Re^{m \times n}$, three positive scalars $\tau_1$, $\tau_2$ and $\eta$ satisfying:*

$$\begin{bmatrix} W(A + BK)' + (A + BK)W + \tau_1 W & BS - Z' & B_w \\ SB' - Z & -2S & 0 \\ B'_w & 0 & -\tau_2 R \end{bmatrix} < 0 \quad (3.37)$$

$$\begin{bmatrix} W & WK'_{(i)} - Z'_{(i)} \\ K_{(i)}W - Z_{(i)} & \eta u^2_{0(i)} \end{bmatrix} \geq 0, \quad i = 1, \ldots, m \quad (3.38)$$

$$-\tau_1 \delta + \tau_2 \eta < 0 \quad (3.39)$$

*then*

1. *for $w = 0$, the ellipsoid $\mathcal{E}(P, \eta)$, with $P = W^{-1}$, is a region of asymptotic stability (RAS) for the saturated system* (3.1);
2. *for any $w \in \mathcal{W}$ and $x(0) \in \mathcal{E}(P, \eta)$, the trajectories of the saturated system* (3.1) *do not leave the ellipsoid $\mathcal{E}(P, \eta)$.*

*Proof* The same procedure as in Chap. 2 is used. We want to prove that $\dot{V}(x) < 0$ along the trajectories of the saturated system (3.1) for any $x$ such that $x'Px > \eta^{-1}$ and for any $w \in \mathcal{W}$, or equivalently such that $w'Rw \leq \delta^{-1}$. This is a way to guarantee the invariance of the ellipsoid $\mathcal{E}(P, \eta)$, with $P = W^{-1}$ with respect to the trajectories of system (3.1). In other words, we have to verify by using the S-procedure (see Appendix C):

$$\dot{V}(x) + \tau_1 \left(x'Px - \eta^{-1}\right) + \tau_2 \left(\delta^{-1} - w'Rw\right) < 0 \quad (3.40)$$

Thus, relation (3.40) is verified if $\dot{V}(x) + \tau_1 x'Px - \tau_2 w'Rw < 0$ and $-\tau_1 \eta^{-1} + \tau_2 \delta^{-1} < 0$.

As in the proof of Proposition 3.1, we use the inequality

$$\dot{V}(x) + \tau_1 x'Px - \tau_2 w'Rw$$
$$\leq \dot{V}(x) + \tau_1 x'Px - \tau_2 w'Rw - 2\phi(Kx)'T\left(\phi(Kx) + Gx\right) < 0$$

By setting $P = W^{-1}$, $T = S^{-1}$ and $Z = GW$, the left-hand term of the inequality above reads

$$
\begin{bmatrix} W^{-1}x \\ S^{-1}\phi(Kx) \\ w \end{bmatrix}' \begin{bmatrix} W(A+BK)' + (A+BK)W + \tau_1 W & BS - Z' & B_w \\ SB' - Z & -2S & 0 \\ B'_w & 0 & -\tau_2 R \end{bmatrix}
$$
$$
\times \begin{bmatrix} W^{-1}x \\ S^{-1}\phi(Kx) \\ w \end{bmatrix}
$$

Thus, if relations (3.37) and (3.39) are satisfied then it follows that condition (3.40) holds.

The satisfaction of relation (3.38) guarantees that the ellipsoid $\mathcal{E}(P,\eta)$ is included in $S(|K - G|, u_0)$.

Note that if relation (3.37) is satisfied, it follows that condition (3.14) is satisfied with $\mu = \tau_1$. Hence, one can conclude that for $w = 0$, the ellipsoid $\mathcal{E}(P,\eta)$ is a region of asymptotic stability for the saturated system (3.1) since one gets $\dot{V}(x) < 0$ along the trajectories of the saturated system (3.1). Furthermore, for any $w \in \mathcal{W}$, the trajectories of the saturated system (3.1) do not leave the ellipsoid $\mathcal{E}(P,\eta)$ thanks to the satisfaction of condition (3.40). $\qquad\square$

*Remark 3.15* A particular case is considered when the disturbance $w$ directly affects the input. Thus, the saturated system (3.1) can be written as

$$
\dot{x}(t) = Ax(t) + B \operatorname{sat}(Kx(t) + K_w w(t)) \tag{3.41}
$$

The same technique employed in Proposition 3.6 can be used and in particular, by using adequately Lemma 1.6 (see also Remark 3.1) with $v = Kx + K_w w$ and $\omega = Kx + K_w w + Gx$, we want to verify the following inequality:

$$
\dot{V}(x) + \tau_1 x' P x - \tau_2 w' R w
$$
$$
\leq \dot{V}(x) + \tau_1 x' P x - \tau_2 w' R w
$$
$$
- 2\phi(Kx + K_w w)'T(\phi(Kx + K_w w) + Kx + K_w w + Gx) < 0
$$

Hence, by setting $P = W^{-1}$, $T = S^{-1}$ and $Z = GW$, the relations (3.37), (3.38) and (3.39) are modified to

$$
\begin{bmatrix} W(A+BK)' + (A+BK)W + \tau_1 W & BS - WK' - Z' & 0 \\ SB' - KW - Z & -2S & -K_w \\ 0 & -K'_w & -\tau_2 R \end{bmatrix} < 0 \tag{3.42}
$$

$$
\begin{bmatrix} W & Z'_{(i)} \\ Z_{(i)} & \eta u^2_{0(i)} \end{bmatrix} \geq 0, \quad i = 1, \dots, m \tag{3.43}
$$

$$
-\tau_1 \delta + \tau_2 \eta < 0 \tag{3.44}
$$

As in Sect. 3.2, assuming that the matrix $A$ is Hurwitz, we can adapt Proposition 3.6 to deal with the global case.

**Proposition 3.7** *If there exist a symmetric positive definite matrix $W \in \Re^{n \times n}$, a positive diagonal matrix $S \in \Re^{m \times m}$, three positive scalars $\tau_1$, $\tau_2$ and $\eta$ satisfying:*

## 3.3 External Stability 139

$$\begin{bmatrix} W(A+BK)'+(A+BK)W+\tau_1 W & BS-WK' & B_w \\ SB'-KW & -2S & 0 \\ B'_w & 0 & -\tau_2 R \end{bmatrix} < 0 \quad (3.45)$$

$$-\tau_1\delta + \tau_2\eta < 0 \quad (3.46)$$

*then*

1. *for* $w = 0$, *the origin is globally asymptotically stable for the saturated system* (3.1);
2. *for any* $w \in \mathcal{W}$, *the trajectories of the saturated system* (3.1) *do not leave the ellipsoid* $\mathcal{E}(P, \eta)$, *with* $P = W^{-1}$, *for any* $x(0) \in \mathcal{E}(P, \eta)$;
3. *for any* $w \in \mathcal{W}$ *and any* $x(0) \in \Re^n\backslash\mathcal{E}(P, \eta)$, *the trajectories of the saturated system* (3.1) *converge toward the set* $\mathcal{E}(P, \eta)$.

*Proof* It suffices to consider $G = K$, or equivalently $Z = KW$ in the relations of Proposition 3.6. In this case, the sector condition of Lemma 1.6 is globally satisfied and therefore relation (3.38) becomes useless. Furthermore, in order to prove that $\dot{V}(x) < 0$ along the trajectories of the saturated system (3.1) for any $x$ such that $x'Px > \eta^{-1}$ and for any $w \in \mathcal{W}$, or equivalently $w'Rw \leq \delta^{-1}$, relations (3.37) and (3.39) are modified as (3.45) and (3.46). Hence, by invoking similar arguments as in Proposition 3.6, we can conclude that for any $w \in \mathcal{W}$, the trajectories of the saturated system (3.1) do not leave the ellipsoid $\mathcal{E}(P, \eta)$, with $P = W^{-1}$, for any $x(0) \in \mathcal{E}(P, \eta)$ (invariance of the set $\mathcal{E}(P, \eta)$). Moreover the fact to verify $\dot{V}(x) < 0$ for any $w \in \mathcal{W}$ and any $x(0) \in \Re^n\backslash\mathcal{E}(P, \eta)$ (i.e. $x'Px > \eta^{-1}$) along the trajectories of the saturated system (3.1) means that for any $w \in \mathcal{W}$ and any $x(0) \in \Re^n\backslash\mathcal{E}(P, \eta)$ the closed-loop trajectories converge toward the set $\mathcal{E}(P, \eta)$. That corresponds to the notion of ultimate boundedness in [215] (see also the survey of Sontag [328], p. 14). $\square$

Note that item 3 of Proposition 3.7 holds for any positive $\delta$ satisfying (3.46). Recall that the smaller is $\delta$, the greater is the admissible disturbance. From (3.46), it follows that $\eta < \frac{\tau_1\delta}{\tau_2}$, and then the smaller is $\delta$, the larger is the region $\mathcal{E}(P, \eta)$.

*Remark 3.16* Lure Lyapunov function can also be used to solve the external stability analysis problem in the context of amplitude bounded exogenous signals. Let us consider $V(x)$ as defined in (3.24). The technique of Proposition 3.6 is adapted by using the inequality (3.27) and therefore verifying the following inequality:

$$\dot{V}(x) + \tau_1 V(x) - \tau_2 w'Rw \leq \dot{V}(x) + \tau_1 x'\big(P + K'NK\big)x - \tau_2 w'Rw$$
$$- 2\phi(Kx)'T\big(\phi(Kx) + Gx\big) < 0$$

The term $\dot{V}(x) + \tau_1 x'(P + K'NK)x - \tau_2 w'Rw - 2\phi(Kx)'T(\phi(Kx) + Gx) < 0$ of the above inequality may be written as

$$
\begin{bmatrix} x \\ \phi(Kx) \\ w \end{bmatrix}'
$$

$$
\times \begin{bmatrix} (A+BK)'P + P(A+BK) + \tau_1(P+K'NK) & & \\ \star & & \\ \star & & \end{bmatrix}
$$

$$
\begin{bmatrix} PB - (A+BK)'K'N - G'T & PB_w \\ -2T - NKB - B'K'N & -NKB_w \\ \star & -\tau_2 R \end{bmatrix}
$$

$$
\times \begin{bmatrix} x \\ \phi(Kx) \\ w \end{bmatrix} < 0 \tag{3.47}
$$

Hence, by invoking similar arguments as in Proposition 3.4, relations (3.37), (3.38) and (3.39) are modified to

$$
\begin{bmatrix} (A+BK)'P + P(A+BK) + \tau_1(P+K'NK) & & \\ \star & & \\ \star & & \\ PB - (A+BK)'K'N - G'T & PB_w \\ -2T - NKB - B'K'N & -NKB_w \\ \star & -\tau_2 R \end{bmatrix} < 0 \tag{3.48}
$$

$$
\begin{bmatrix} P & K'_{(i)} - G'_{(i)} \\ K_{(i)} - G_{(i)} & \eta u^2_{0(i)} \end{bmatrix} \geq 0, \quad i = 1, \ldots, m \tag{3.49}
$$

$$
-\tau_1 \delta + \tau_2 \eta < 0 \tag{3.50}
$$

where the symmetric positive definite matrix $P \in \Re^{n \times n}$, two diagonal positive matrices $N \in \Re^{m \times m}$ and $T \in \Re^{m \times m}$, a matrix $G \in \Re^{m \times n}$, positive scalars $\eta$, $\tau_1$ and $\tau_2$ are the decision variables.

Moreover, it is important to point out that if we choose $N = 0$ by using the same change of variables as in Proposition 3.6 (i.e. $W = P^{-1}$, $Z = GW$ and $S = T^{-1}$) and the same congruence transformation, relations (3.48), (3.49) and (3.50) give the relations of Proposition 3.6.

*Remark 3.17* In the global context, Proposition 3.7 can be adapted by considering a Lure Lyapunov function. Hence, by considering $G = K$, relations (3.45) and (3.46) are modified as

$$
\begin{bmatrix} (A+BK)'P + P(A+BK) + \tau_1(P+K'NK) & & \\ \star & & \\ \star & & \\ PB - (A+BK)'K'N - K'T & PB_w \\ -2T - NKB - B'K'N & -NKB_w \\ \star & -\tau_2 R \end{bmatrix} < 0 \tag{3.51}
$$

$$
-\tau_1 \delta + \tau_2 \eta < 0 \tag{3.52}
$$

3.3 External Stability

Furthermore, note that if we choose $N = 0$ by using the same change of variables as in Proposition 3.7 (i.e., $W = P^{-1}$ and $S = T^{-1}$) and the same congruence transformation, relations (3.51) and (3.52) are the relations of Proposition 3.7.

### 3.3.2 Energy Bounded Exogenous Signals

As in Chap. 2, let us consider that the exogenous signal $w(t)$ is energy bounded, i.e. it belongs to the following set of functions:

$$\mathcal{W} = \left\{ w : [0, \infty) \to \mathfrak{R}^q; \int_0^\infty w(\tau)' R w(\tau) \, d\tau \leq \delta^{-1} \right\} \tag{3.53}$$

for some $\delta > 0$ and symmetric positive definite matrix $R$. In this case, the energy of $w(t)$ is bounded by $\delta^{-1}$. When $R = I_q$, definition (3.53) corresponds to the square of the $\mathcal{L}_2$-norm of the signal $w(t)$, i.e

$$\|w\|_2 = \sqrt{\int_0^\infty w(\tau)' w(\tau) \, d\tau}$$

and therefore $\mathcal{W}$ represents a set of $\mathcal{L}_2$-bounded disturbances.

Let us consider the quadratic Lyapunov function defined in (3.7). The following proposition allows to solve the external stability analysis problem involving energy bounded exogenous signals.

**Proposition 3.8** *If there exist a symmetric positive definite matrix $W \in \mathfrak{R}^{n \times n}$, a positive diagonal matrix $S \in \mathfrak{R}^{m \times m}$, a matrix $Z \in \mathfrak{R}^{m \times n}$ and a positive scalar $\eta$ satisfying:*

$$\begin{bmatrix} W(A + BK)' + (A + BK)W & BS - Z' & B_w \\ SB' - Z & -2S & 0 \\ B_w' & 0 & -R \end{bmatrix} < 0 \tag{3.54}$$

$$\begin{bmatrix} W & WK_{(i)}' - Z_{(i)}' \\ K_{(i)}W - Z_{(i)} & \eta u_{0(i)}^2 \end{bmatrix} \geq 0, \quad i = 1, \ldots, m \tag{3.55}$$

$$-\delta + \eta < 0 \tag{3.56}$$

*then*

1. *for $w = 0$, the ellipsoid $\mathcal{E}(P, \eta)$, with $P = W^{-1}$, is a region of asymptotic stability (RAS) for the saturated system (3.1);*
2. *for any $w \in \mathcal{W}$, the trajectories of the saturated system (3.1) do not leave the ellipsoid $\mathcal{E}(P, \eta)$, $\forall x(0) \in \mathcal{E}(P, \beta)$, with $0 < \beta^{-1} \leq \eta^{-1} - \delta^{-1}$.*

*Proof* The same procedure as in Chap. 2 is used. Then, we want to prove that $\dot{V}(x) - w'Rw < 0$ along the trajectories of the saturated system (3.1). As in the proof of Proposition 3.1, we use the following inequality:

$$\dot{V}(x) - w'Rw \leq \dot{V}(x) - w'Rw - 2\phi(Kx)'T(\phi(Kx) + Gx) < 0 \tag{3.57}$$

By setting $P = W^{-1}$, $T = S^{-1}$ and $Z = GW$, the term $\dot{V}(x) - w'Rw - 2\phi(Kx)'T(\phi(Kx) + Gx) < 0$ reads

$$\begin{bmatrix} W^{-1}x \\ S^{-1}\phi(Kx) \\ w \end{bmatrix}' \begin{bmatrix} W(A+BK)' + (A+BK)W & BS - Z' & B_w \\ SB' - Z & -2S & 0 \\ B'_w & 0 & -R \end{bmatrix}$$
$$\times \begin{bmatrix} W^{-1}x \\ S^{-1}\phi(Kx) \\ w \end{bmatrix} < 0$$

Hence, if relation (3.54) is satisfied then one gets $\dot{V}(x) - w'Rw < 0$. By integrating this inequality, it follows: $V(x(T)) - V(x(0)) - \int_0^T w(t)'Rw(t)\,dt < 0$, $\forall T > 0$. In other words, $\forall x(0) \in \mathcal{E}(P, \beta)$ and $\forall w(t) \in \mathcal{W}$ as defined in (3.53), it follows that:

- $V(x(T)) < \|w\|_2^2 + V(x(0)) \leq \delta^{-1} + \beta^{-1}$, $\forall T > 0$, i.e. the trajectories of the system do not leave the set $\mathcal{E}(P, (\delta^{-1} + \beta^{-1})^{-1})$;
- if $w(t) = 0$, $\forall t > t_1 \geq 0$, then $\dot{V}(x(t)) < 0$, which ensures that $x(t) \to 0$ as $t \to \infty$.

The satisfaction of relation (3.55) guarantees that the ellipsoid $\mathcal{E}(P, \eta)$ is included in $S(|K - G|, u_0)$ and therefore the sector condition is verified.

Hence, one can conclude that for $w = 0$, the ellipsoid $\mathcal{E}(P, \eta)$ is a region of asymptotic stability since one gets $\dot{V}(x) < 0$ along the trajectories of the saturated system (3.1). Furthermore, for any $w \in \mathcal{W}$, as defined in (3.53), the trajectories of the saturated system (3.1) do not leave the ellipsoid $\mathcal{E}(P, \eta)$ for any initial condition $x(0)$ belonging to $\mathcal{E}(P, \beta)$, thanks to the satisfaction of conditions (3.54), (3.55) and (3.56). $\qquad\square$

*Remark 3.18* As in Chap. 2, it is important to note that the set $\mathcal{E}(P, \eta)$ obtained in Proposition 3.8 is not an invariant set with respect to the trajectories of the saturated system (3.1). Actually, if $x(0) \in \mathcal{E}(P, \eta)$ and $x(0) \notin \mathcal{E}(P, \beta)$, there is no guarantee that the trajectories will remain confined in $\mathcal{E}(P, \eta)$, $\forall w \in \mathcal{W}$, as defined in (3.53). At the opposite, the ellipsoid $\mathcal{E}(P, \eta)$ obtained in Proposition 3.6 for the amplitude bounded exogenous signal case is an invariant set with respect to the trajectories of the saturated system (3.1).

*Remark 3.19* The alternative case discussed in Remark 3.15 can also be tackled. Considering that the disturbance $w$ directly affects the input, and therefore that the saturated system (3.1) is defined as in (3.41), the technique employed in Proposition 3.8 can also be used. By using adequately Lemma 1.6 (see also Remark 3.1) with $v = Kx + K_w w$ and $\omega = Kx + K_w w + Gx$, we want to verify the following inequality:

$$\dot{V}(x) - w'Rw \leq \dot{V}(x) - w'Rw - 2\phi(Kx + K_w w)'T(\phi(Kx + K_w w) + \omega) < 0$$

Hence, by setting $P = W^{-1}$, $T = S^{-1}$ and $Z = GW$, relations (3.54), (3.55) and (3.56) are modified as follows:

## 3.3 External Stability

$$\begin{bmatrix} W(A+BK)'+(A+BK)W & BS-WK'-Z' & 0 \\ SB'-KW-Z & -2S & -K_w \\ 0 & -K'_w & -R \end{bmatrix} < 0 \quad (3.58)$$

$$\begin{bmatrix} W & Z'_{(i)} \\ Z_{(i)} & \eta u^2_{0(i)} \end{bmatrix} \ge 0, \quad i=1,\ldots,m \quad (3.59)$$

$$-\delta + \eta < 0 \quad (3.60)$$

In the case where the matrix $A$ is Hurwitz, Proposition 3.8 may be extended to the global case.

**Proposition 3.9** *If there exist a symmetric positive definite matrix $W \in \Re^{n \times n}$ and a positive diagonal matrix $S \in \Re^{m \times m}$ satisfying:*

$$\begin{bmatrix} W(A+BK)'+(A+BK)W & BS-WK' & B_w \\ SB'-KW & -2S & 0 \\ B'_w & 0 & -R \end{bmatrix} < 0 \quad (3.61)$$

*then*

1. *for $w=0$, the origin is globally asymptotically stable for the saturated system (3.1);*
2. *for any $w \in \mathcal{W}$, the trajectories of the saturated system (3.1) remain bounded for any initial condition $x(0) \in \Re^n$ as follows:*

$$x(T)'Px(T) \le x(0)'Px(0) + \delta^{-1}, \quad \forall T > 0$$

*Proof* It suffices to consider $G=K$, or equivalently $Z=KW$ in the relations of Proposition 3.8. In this case, the sector condition of Lemma 1.6 is globally satisfied and therefore relation (3.55) becomes useless. Furthermore, in order to prove that $\dot{V}(x) - w'Rw < 0$ along the trajectories of the saturated system (3.1) for any $x \in \Re^n$ and for any $w \in \mathcal{W}$, relation (3.54) is modified as (3.61). Relation (3.56) becomes useless. $\square$

Note that the result of Proposition 3.9 is independent of the value of $\delta$. Actually, once condition (3.61) is verified the boundedness of the closed-loop trajectories is ensured for any $w$ such that $\int_0^\infty w(\tau)'Rw(\tau)\,d\tau \le \delta^{-1}$. In the particular case $R=I$, it follows that the closed-loop trajectories are bounded for any $w \in \mathcal{L}_2$.

*Remark 3.20* Proposition 3.8 can be modified by considering a Lure Lyapunov function. Hence, let us consider $V(x)$ as defined in (3.24). The technique of Proposition 3.8 is then adapted by using inequality (3.27) and therefore verifying the following inequality:

$$\dot{V}(x) - w'Rw \le \dot{V}(x) - w'Rw - 2\phi(Kx)'T\big(\phi(Kx) + Gx\big) < 0$$

The term of the above inequality, i.e. $\dot{V}(x) - w'Rw - 2\phi(Kx)'T(\phi(Kx)+Gx) < 0$, reads as

$$
\left[\begin{array}{c} x \\ \phi(Kx) \\ w \end{array}\right]'
$$

$$
\times \left[\begin{array}{ccc} (A+BK)'P+P(A+BK) & PB-(A+BK)'K'N-G'T & PB_w \\ \star & -2T-NKB-B'K'N & -NKB_w \\ \star & \star & -R \end{array}\right]
$$

$$
\times \left[\begin{array}{c} x \\ \phi(Kx) \\ w \end{array}\right] < 0 \tag{3.62}
$$

Hence, by invoking similar arguments as in Proposition 3.4, the relations (3.54), (3.55) and (3.56) are modified to

$$
\left[\begin{array}{ccc} (A+BK)'P+P(A+BK) & PB-(A+BK)'K'N-G'T & PB_w \\ \star & -2T-NKB-B'K'N & -NKB_w \\ \star & \star & -R \end{array}\right] < 0 \tag{3.63}
$$

$$
\left[\begin{array}{cc} P & K'_{(i)}-G'_{(i)} \\ K_{(i)}-G_{(i)} & \eta u_{0(i)}^2 \end{array}\right] \geq 0, \quad i=1,\ldots,m \tag{3.64}
$$

$$
-\delta + \eta < 0 \tag{3.65}
$$

where the symmetric positive definite matrix $P \in \Re^{n \times n}$, two diagonal positive matrices $N \in \Re^{m \times m}$ and $T \in \Re^{m \times m}$, a matrix $G \in \Re^{m \times n}$, and a positive scalar $\eta$ are the decision variables. By using (3.28) and (3.29), one can ensure that the trajectories of the closed-loop system (3.1) do not leave the set $S(V, \eta)$, $\forall x(0) \in \mathcal{E}(P+K'NK, \beta)$, with $0 < \beta^{-1} \leq \eta^{-1} - \delta^{-1}$.

Moreover, it is important to point out that if we choose $N = 0$, by using the same change of variables as in Proposition 3.6 (i.e., $W = P^{-1}$, $Z = GW$ and $S = T^{-1}$) and the same congruence transformation, one retrieves the relations of Proposition 3.8.

*Remark 3.21* In the global context, Proposition 3.9 can be adapted by considering a Lure Lyapunov function. Hence, by considering $G = K$, the relation (3.61) becomes

$$
\left[\begin{array}{ccc} (A+BK)'P+P(A+BK) & PB-(A+BK)'K'N-K'T & PB_w \\ \star & -2T-NKB-B'K'N & -NKB_w \\ \star & \star & -R \end{array}\right] < 0 \tag{3.66}
$$

In this case, it follows that

$$
x(T)'Px(T) \leq V\big(x(T)\big)
$$

$$
\leq V\big(x(0)\big) + \delta^{-1} \leq x(0)'\big(P+K'NK\big)x(0) + \delta^{-1}, \quad \forall T > 0
$$

By choosing $N = 0$ by using the same change of variables as in Proposition 3.7 (i.e., $W = P^{-1}$ and $S = T^{-1}$) and the same congruence transformation, one retrieves the relation of Proposition 3.9.

3.3 External Stability

*Remark 3.22* It is possible to analyze the external stability for systems subject to exogenous signals which are both amplitude and energy bounded. An interesting class of such exogenous signals includes the signals generated by stable autonomous systems [27, 305]. To deal with both amplitude and energy bounded exogenous signals, the results stated in Propositions 3.6 and 3.8 could be combined. In order to reduce the conservatism, an alternative approach would be to use Finsler's lemma [81].

### 3.3.3 Optimization Issues

Similarly to Sect. 3.2.4, some comments regarding potential optimization problems associated to Propositions 3.6 and 3.8 are suggested.

#### 3.3.3.1 Amplitude Bounded Exogenous Signals

Relations in Proposition 3.6 are linear in the decision variables $W$, $S$, $Z$, $\eta$ provided that the parameters $\tau_1$ and $\tau_2$ are fixed. Although their numerical value may have an impact on the obtained solution, very often, it is rather easy to find an a priori choice which allows to obtain an admissible solution. Otherwise, one can perform a search on a grid defined by parameters $\tau_1$ and $\tau_2$. Moreover, $R$ and $\delta$ can be supposed either a priori given or decision variables. The following simple algorithms can be used in order to exhibit solution to the maximization of the size of $\mathcal{E}(P, \eta)$ or to the maximization of the size of $\mathcal{W}$.

**Algorithm 3.3** Maximization of the size of $\mathcal{E}(P, \eta)$
1. Given $R$ and $\delta$, choose a convex criterion $f(\mathcal{E}(P, \eta))$.
2. Fix $\tau_1$ and $\tau_2$.
3. Solve for $W$, $S$, $Z$, $\eta$ the following convex optimization problem

$$\min \ f\big(\mathcal{E}(P, \eta)\big)$$

subject to   inequalities (3.37)–(3.39)

**Algorithm 3.4** Maximization of the size of $\mathcal{W}$
1. Choose a convex criterion $f(\mathcal{W})$.
2. Fix $\tau_1$ and $\tau_2$.
3. Solve for $W$, $S$, $Z$, $R$, $\eta$, $\delta$ the following convex optimization problem

$$\min \ f(\mathcal{W})$$

subject to   inequalities (3.37)–(3.39)

Algorithms 3.3 and 3.4 provide local stability solutions for the closed-loop system (3.1). When regarding the global stability of the closed-loop system, it suffices to search for a feasible solution to relation (3.18) of Proposition 3.7.

### 3.3.3.2 Energy Bounded Exogenous Signals

Relations of Proposition 3.8 are linear in the decision variables. Hence, as in Chap. 2 LMI-based optimization problems can be proposed to address the following criteria:

- maximization of the $\mathcal{L}_2$-bound on the admissible disturbances (disturbance tolerance maximization which corresponds to minimize $\delta$);
- maximization of the region where the asymptotic stability of the closed-loop system is ensured (maximization of the region of attraction).

Some details regarding the associated algorithms will be given in the stabilization framework (Sect. 3.4).

## 3.4 Stabilization

### 3.4.1 State Feedback Stabilization

In this section, we propose conditions to design a state feedback gain $K \in \mathfrak{R}^{m \times n}$ such that the saturated system (3.5) is regionally asymptotically stable or such that the external stability of the saturated system (3.1) is ensured.

#### 3.4.1.1 Asymptotic Stabilization

Basically, we search for a gain $K$ in order to ensure the asymptotic stability for a given set of admissible initial conditions, or in order to maximize the region of attraction of the origin. This problem can be indirectly addressed by computing the gain $K$ in order to maximize an ellipsoidal estimate $\mathcal{E}(P, 1)$ of the region of attraction.

To do this, we can extend the results of Proposition 3.1 as follows.

**Proposition 3.10** *If there exist a symmetric positive definite matrix $W \in \mathfrak{R}^{n \times n}$, a positive diagonal matrix $S \in \mathfrak{R}^{m \times m}$, two matrices $Y \in \mathfrak{R}^{m \times n}$, $Z \in \mathfrak{R}^{m \times n}$ satisfying:*

$$\begin{bmatrix} WA' + AW + BY + Y'B' & BS - Z' \\ SB' - Z & -2S \end{bmatrix} < 0 \qquad (3.67)$$

$$\begin{bmatrix} W & Y'_{(i)} - Z'_{(i)} \\ Y_{(i)} - Z_{(i)} & u^2_{0(i)} \end{bmatrix} \geq 0, \quad i = 1, \ldots, m \qquad (3.68)$$

*then the gain $K = YW^{-1}$ guarantees the asymptotic stability for the saturated system (3.5) in the ellipsoid $\mathcal{E}(P, 1)$, with $P = W^{-1}$.*

## 3.4 Stabilization

*Proof* The proof is quite similar to that one of Proposition 3.1. The classical change of variable $Y = KW$ is used in order to linearize the product $KW$, introducing a new decision variable $Y$. □

As in Sect. 3.2.1, solutions to the global case can be addressed. Hence, for asymptotically stable linear plants (i.e. $\Re e\lambda_i(A) < 0$, $\forall i$) the global asymptotic stabilization of the closed-loop system can be considered as follows.

**Proposition 3.11** *If there exist a symmetric positive definite matrix* $W \in \Re^{n \times n}$, *a positive diagonal matrix* $S \in \Re^{m \times m}$ *and a matrix* $Y \in \Re^{m \times n}$ *satisfying*

$$\begin{bmatrix} WA' + Y'B' + AW + BY & BS - Y' \\ SB' - Y & -2S \end{bmatrix} < 0 \tag{3.69}$$

*then the saturated system* (3.5) *is globally asymptotically stabilizable by the state feedback* $K = YW^{-1}$.

*Proof* It suffices to consider $G = K$. In this case, the sector condition of Lemma 1.6 is globally satisfied, that is, it is satisfied for any $x \in \Re^n$, or equivalently $\phi(Kx)'S^{-1}(\phi(Kx) + Kx) \leq 0$, $\forall x \in \Re^n$, where $S$ is a diagonal positive definite matrix in $\Re^{m \times m}$. By considering the same change of variables as in Proposition 3.10, the satisfaction of relation (3.18) means that $\dot{V}(x) \leq \dot{V}(x) - 2\phi(Kx)'S^{-1}(\phi(Kx) + Kx) < 0$, with $K = YW^{-1}$. The stability immediately follows from the above inequality which implies that $\dot{V}(x) < 0$ along any closed-loop trajectory. □

*Remark 3.23* The use of Proposition 3.3 involving non-diagonal multipliers $T$ for control design purpose in a global context appears to be a difficult task due to the fact that we need to use the inverse of the multiplier $T$, namely $S$. Note, however, that in the particular case $m = 2$, we can reformulate the condition (3.23). By noting that for $T = \begin{bmatrix} T_{(1,1)} & T_{(1,2)} \\ T_{(1,2)} & T_{(2,2)} \end{bmatrix}$, one gets $S = T^{-1} = \frac{1}{(T_{(1,1)}T_{(2,2)} - T_{(1,2)}^2)} \begin{bmatrix} T_{(2,2)} & -T_{(1,2)} \\ -T_{(1,2)} & T_{(1,1)} \end{bmatrix}$, relation (3.23) reads as a function of $S$ as follows:

$$u_{0(1)}S_{(2,2)} \geq u_{0(2)}|S_{(1,2)}|$$

$$u_{0(2)}S_{(1,1)} \geq u_{0(1)}|S_{(1,2)}|$$

Then the relations of Proposition 3.3 are modified to

$$\begin{bmatrix} WA' + AW + BY + Y'B' & BS - Y' \\ SB' - Y & -2S \end{bmatrix} < 0 \tag{3.70}$$

$$u_{0(1)}S_{(2,2)} \geq u_{0(2)}|S_{(1,2)}|$$
$$u_{0(2)}S_{(1,1)} \geq u_{0(1)}|S_{(1,2)}| \tag{3.71}$$

where the decision variables are two symmetric positive definite matrices $W \in \Re^{n \times n}$, $S \in \Re^{m \times m}$ and a matrix $Y \in \Re^{m \times n}$.

*Remark 3.24* The extension of Proposition 3.4 manipulating Lure functions for control design purpose is a difficult task due to the presence of several products between decision variables.

**Optimization Issues** Note that Proposition 3.10 provides LMIs in the decision variables $W$, $S$, $Y$ and $Z$. Hence, the conditions can be very easily tested in terms of feasibility. On the other hand, as in Chap. 2, we may be interested in computing $K$ in order to guarantee stability for a given set of admissible initial states, or even to maximize an estimate of the region of attraction. Hence, all the optimization problems presented in Sect. 2.4.1 can be formulated considering the conditions stated in Proposition 3.10. Furthermore, it is possible to compute global stabilizing gains by using the result of Proposition 3.11.

Similarly, as studied in Chap. 2, some performance requirements can be considered in the state feedback design procedure. Hence, several kinds of performance can be tackled, like time-domain performance improvement including rate of convergence and damping. The comments provided in Sect. 2.4.1.2 remain valid in the current case. In other words, the idea is to consider different performance criteria depending on the region of operation: for example a pole placement criterion can be associated to the region of linearity (that one in which only the linear behavior is allowed) and another one (like speed of convergence) can be associated to the region in which the nonlinear behavior is taken into account.

### 3.4.1.2 External Stabilization

Considering the linearizing variable change $Y = KW$, stabilization conditions for the case where the system is subject to the action of additive exogenous signals can be straightforwardly derived. Hence, by considering the saturated system (3.1), both cases of amplitude and energy bounded exogenous signals are investigated extending Propositions 3.6 and 3.8.

**Amplitude Bounded Exogenous Signals** As in the context of Sect. 3.3.1, let us consider that the exogenous signal $w(t)$ belongs to $\mathcal{W}$, as defined in (3.36).

By using the quadratic Lyapunov function defined in (3.7), the following proposition allows to solve the external stabilization problem.

**Proposition 3.12** *If there exist a symmetric positive definite matrix $W \in \Re^{n \times n}$, a positive diagonal matrix $S \in \Re^{m \times m}$, two matrices $Y \in \Re^{m \times n}$, $Z \in \Re^{m \times n}$, three positive scalars $\tau_1$, $\tau_2$ and $\eta$ satisfying*

$$\begin{bmatrix} WA' + Y'B' + AW + BY + \tau_1 W & BS - Z' & B_w \\ SB' - Z & -2S & 0 \\ B_w' & 0 & -\tau_2 R \end{bmatrix} < 0 \quad (3.72)$$

$$\begin{bmatrix} W & Y_{(i)}' - Z_{(i)}' \\ Y_{(i)} - Z_{(i)} & \eta u_{0(i)}^2 \end{bmatrix} \geq 0, \quad i = 1, \dots, m \quad (3.73)$$

$$-\tau_1 \delta + \tau_2 \eta < 0 \quad (3.74)$$

*then the gain $K = YW^{-1}$ is such that*

1. *for $w = 0$, the ellipsoid $\mathcal{E}(P, \eta)$, with $P = W^{-1}$, is a region of asymptotic stability (RAS) for the saturated system (3.1);*

## 3.4 Stabilization

2. *for any $w \in \mathcal{W}$ and $x(0) \in \mathcal{E}(P, \eta)$, the trajectories of the saturated system (3.1) do not leave the ellipsoid $\mathcal{E}(P, \eta)$.*

**Energy Bounded Exogenous Signals**  As in the context of Sect. 3.3.2, let us consider that the exogenous signal $w(t)$ is energy bounded, i.e. it belongs to $\mathcal{W}$ defined in (3.53) for some $\delta > 0$ and symmetric positive definite matrix $R$.

By considering the quadratic Lyapunov function defined in (3.7), the following proposition allows to solve the external stabilization problem in the current case.

**Proposition 3.13**  *If there exist a symmetric positive definite matrix $W \in \mathfrak{R}^{n \times n}$, a positive diagonal matrix $S \in \mathfrak{R}^{m \times m}$, two matrices $Y \in \mathfrak{R}^{m \times n}$, $Z \in \mathfrak{R}^{m \times n}$, a positive scalar $\eta$ satisfying:*

$$\begin{bmatrix} WA' + Y'B' + AW + BY & BS - Z' & B_w \\ SB' - Z & -2S & 0 \\ B_w' & 0 & -R \end{bmatrix} < 0 \tag{3.75}$$

$$\begin{bmatrix} W & Y_{(i)}' - Z_{(i)}' \\ Y_{(i)} - Z_{(i)} & \eta u_{0(i)}^2 \end{bmatrix} \geq 0, \quad i = 1, \dots, m \tag{3.76}$$

$$-\delta + \eta < 0 \tag{3.77}$$

*then the gain $K = YW^{-1}$ is such that*

1. *for $w = 0$, the ellipsoid $\mathcal{E}(P, \eta)$, with $P = W^{-1}$, is a region of asymptotic stability (RAS) for the saturated system (3.1);*
2. *for any $w \in \mathcal{W}$, the trajectories of the saturated system (3.1) do not leave the ellipsoid $\mathcal{E}(P, \eta)$, $\forall x(0) \in \mathcal{E}(P, \beta)$, with $0 < \beta^{-1} \leq \eta^{-1} - \delta^{-1}$.*

By defining the regulated output $z(t)$ as

$$z(t) = C_z x(t) \tag{3.78}$$

we can consider an additional control objective regarding the $\mathcal{L}_2$-gain of the saturated system (3.1) between the admissible disturbance $w$ and the controlled output $z$. Hence, we want to ensure some level (in the $\mathcal{L}_2$ sense) of disturbance rejection. This objective can be achieved by defining a function [45, 215]

$$\mathcal{J}(t) = \dot{V}(x(t)) - w(t)'w(t) + \frac{1}{\gamma} z(t)'z(t)$$

with $V(x(t)) = x(t)'Px(t)$. Hence, if $\mathcal{J}(t) < 0$, one obtains

$$\int_0^T \mathcal{J}(t)\,dt = V(x(T)) - V(x(0)) - \int_0^T w(t)'w(t)\,dt$$
$$+ \frac{1}{\gamma} \int_0^T z(t)'z(t)\,dt < 0, \quad \forall T > 0$$

and it is possible to conclude that:

- $V(x(T)) < V(x(0)) + \|w\|_2^2 \leq \delta^{-1}$, $\forall T > 0$, i.e. the trajectories of the system (3.1) and (3.78) are bounded for $w(t)$ such that $\|w\|_2^2 \leq \delta^{-1}$;

150　　　　　3　Stability Analysis and Stabilization—Sector Nonlinearity Model Approach

- for $T \to \infty$, $\|z\|_2^2 < \gamma(\|w\|_2^2 + V(x(0)))$;
- if $w(t) = 0$, $\forall t \geq t_1 \geq 0$, then $\dot{V}(x(t)) < -\frac{1}{\gamma}z(t)'z(t) < 0$, which ensures that in this case $x(t) \to 0$.

In this context, the following proposition can be stated.

**Proposition 3.14** *If there exist a symmetric positive definite matrix* $W \in \Re^{n \times n}$, *a positive diagonal matrix* $S \in \Re^{m \times m}$, *two matrices* $Y \in \Re^{m \times n}$, $Z \in \Re^{m \times n}$, *positive scalars* $\eta$ *and* $\gamma$ *satisfying:*

$$\begin{bmatrix} WA' + Y'B' + AW + BY & BS - Z' & B_w & WC_z' \\ SB' - Z & -2S & 0 & 0 \\ B_w' & 0 & -I & 0 \\ C_z W & 0 & 0 & -\gamma I \end{bmatrix} < 0 \qquad (3.79)$$

$$\begin{bmatrix} W & Y_{(i)}' - Z_{(i)}' \\ Y_{(i)} - Z_{(i)} & \eta u_{0(i)}^2 \end{bmatrix} \geq 0, \quad i = 1, \dots, m \qquad (3.80)$$

$$-\delta + \eta < 0 \qquad (3.81)$$

*then the gain* $K = YW^{-1}$ *is such that*

1. *for* $w = 0$, *then* $\dot{V}(x) < -\frac{1}{\gamma}z'z < 0$ *and therefore the ellipsoid* $\mathcal{E}(P, \eta)$, *with* $P = W^{-1}$, *is a region of asymptotic stability (RAS) for the saturated system* (3.1);
2. *for any* $w \in \mathcal{W}$, *with* $R = I$,
   (a) *the trajectories of the saturated system* (3.1) *do not leave the ellipsoid* $\mathcal{E}(P, \eta)$, $\forall x(0) \in \mathcal{E}(P, \beta)$, *with* $0 < \beta^{-1} \leq \eta^{-1} - \delta^{-1}$;
   (b) *one gets:* $\|z\|_2^2 < \gamma(\|w\|_2^2 + V(x(0))) = \gamma(\|w\|_2^2 + x(0)'Px(0))$.

*Proof* By left- and right-multiplying relation (3.79) by $\text{Diag}(W^{-1}, S^{-1}, I, I) = \text{Diag}(P, T, I, I)$, by applying Schur's complement twice and with the change of variables $K = YW^{-1} = YP$, $G = ZP$ one obtains

$$\begin{bmatrix} (A + BK)'P + P(A + BK) + \frac{1}{\gamma}C_z'C_z + PB_w B_w'P & PB - G'T \\ B'P - TG & -2T \end{bmatrix} < 0$$

This implies that

$$\mathcal{J}(t) = \dot{V}\big(x(t)\big) - w(t)'w(t) + \frac{1}{\gamma}z(t)'z(t)$$

$$\leq \mathcal{J}(t) - 2\phi(Kx)'T\big(\phi(Kx) + Gx\big) < 0$$

with $V(x(t)) = x(t)'Px(t)$ along the trajectories of the saturated system (3.1), provided that the ellipsoid $\mathcal{E}(P, \eta)$ is included in $S(|K - G|, u_0)$. This inclusion is ensured thanks to the satisfaction of relation (3.80). Furthermore, one verifies that the trajectories of the saturated system (3.1) do not leave the ellipsoid $\mathcal{E}(P, \eta)$, $\forall x(0) \in \mathcal{E}(P, \beta)$, with $0 < \beta^{-1} \leq \eta^{-1} - \delta^{-1}$ thanks to the satisfaction of relation (3.81). $\qquad \square$

# 3.4 Stabilization 151

*Remark 3.25* In the case $x(0) = 0$, the conditions of Proposition 3.14 ensure the $\mathcal{L}_2$ input-to-state stability of the closed-loop system (3.1). Thus, in this case, the $\mathcal{L}_2$-gain of the system is $\frac{1}{\sqrt{\gamma}}$, since one gets $\|z\|_2^2 < \gamma \|w\|_2^2$.

*Remark 3.26* Another interesting approach, proposed in [62], resides in the use of a multiplier approach, obtained from the use of Finsler's Lemma, to give a solution to the $\mathcal{L}_2$-stabilization problem. In particular, such an approach is an interesting alternative solution for multi-objective problems, since it is then possible to associate different Lyapunov matrices with different objectives. Furthermore, in this case, the Lyapunov matrix $P$ appears isolated in the conditions and therefore the resulting state feedback gain $K$ does not explicitly depend on it. See also [81] for more details regarding the use of Finsler's Lemma.

*Remark 3.27* Proposition 3.14 based on the definition of the regulated output $z(t)$ given in (3.78) could be easily extended by considering that $z$ is defined by

$$\begin{aligned} z(t) &= C_z x(t) + D_z u(t) + D_w w(t) \\ &= C_z x(t) + D_z \operatorname{sat}(K x(t)) + D_w w(t) \\ &= (C_z + D_z K) x(t) + D_z \phi(K x(t)) + D_w w(t) \end{aligned} \tag{3.82}$$

This case is addressed in details to deal with the anti-windup compensator design problem in Chap. 7.

**Optimization Issues** In the context of external stabilization, similarly to Chap. 2, several optimization convex problems can be proposed by using Propositions 3.12, 3.13, 3.14. Hence, the maximization of the disturbance tolerance consists in maximizing the bound on the disturbance, for which the closed-loop trajectories remain bounded: this consists in minimizing $\delta$. The maximization of the disturbance rejection consists in minimizing the upper bound for the $\mathcal{L}_2$-gain from $w(t)$ to $z(t)$: one then wants to minimize $\gamma$.

*Example 3.2* Consider the example of the balancing pointer discussed in Chap. 1, but now considering in plus an exogenous signal which affects its dynamics. System (3.1) is now defined by the following data:

$$A = \begin{bmatrix} 0 & 1 \\ 1 & 0 \end{bmatrix}; \qquad B = \begin{bmatrix} 0 \\ -1 \end{bmatrix}; \qquad B_w = \begin{bmatrix} 1 \\ 1 \end{bmatrix}; \qquad u_0 = 5$$

Let us first consider that the system is affected by an amplitude bounded disturbance ($R = 1$, $\delta = 0.5$, corresponding to a maximum amplitude equal to 1.4142). Considering for the optimization problem the maximization of the disturbance tolerance (minimize $\delta$), for the nominal controller $K = [\, 13\ 7\,]$, the minimal value of $\delta$ is $\delta = 0.8648$. Implicitly, it signifies that the analysis problem for $\delta = 0.5$ is unfeasible, i.e. that it cannot be evaluated a set of safe initial conditions for such a disturbance. It may be checked on Fig. 3.3 that there exist initial conditions selected inside the ellipsoidal set associated to the case without disturbance, which diverge when a disturbance of amplitude 1.4142 is applied during 5 seconds.

**Fig. 3.3** Example 3.2—nominal controller. Ellipsoidal region of stability without disturbance. Trajectories of the closed-loop system in presence of a disturbance of amplitude 1.4142 during 5 seconds

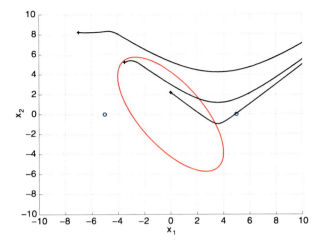

**Fig. 3.4** Example 3.2—designed controller. Ellipsoidal region of stability with disturbance. Trajectories of the closed-loop system in presence of a disturbance of amplitude 1.4142 during 5 seconds

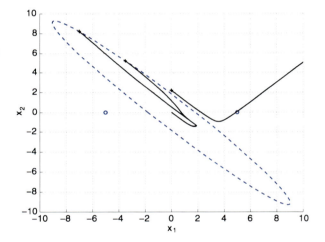

On the other hand, the design of a state feedback may be investigated to evaluate the largest set of safe initial conditions in presence of the disturbance associated to $\delta = 0.5$. In that case, we consider an optimization problem related to the size of the ellipsoidal domain. This is done by considering the maximization of the set in the direction $[1\ 1]'$, considering also a pole placement requirement in a disk centered in $-5$ and of radius 4, and conditioning of matrix $W$ (to avoid an increase in the non-useful infinite directions). The control gain solution of the problem is $K = [6.1433\ 6.1299]$. The ellipsoidal set solution to this synthesis problem is plotted in Fig. 3.4. Considering the same disturbance (amplitude 1.4142 during 5 seconds) and initial conditions as previously, it may be verified that the trajectories initialized inside the ellipsoidal set issued from the synthesis problem converge toward 0. Moreover, the disturbance affects the domain of stability especially in the direction of the equilibrium points. This is confirmed by the trajectory initialized

## 3.4 Stabilization

outside the set issued for the synthesis problem, but close to the set, which cannot be conveyed toward 0 when facing the disturbance.

*Example 3.3* The following example intends to illustrate various aspects related to global stability analysis. Let us consider the following open-loop stable example borrowed to [139]

$$A = \begin{bmatrix} -0.1 & 0 \\ 0 & -0.1 \end{bmatrix}; \qquad B = \begin{bmatrix} 1.5 & 4 \\ 1.2 & 3 \end{bmatrix}; \qquad C = \begin{bmatrix} 1 & 0 \\ 0 & 1 \end{bmatrix}; \qquad u_0 = \begin{bmatrix} 1 \\ 1 \end{bmatrix}$$

augmented with an additive energy bounded disturbance $w$ and a controlled output $z$ through

$$B_w = \begin{bmatrix} 1 \\ 1 \end{bmatrix}; \qquad C_z = \begin{bmatrix} 1 & 1 \end{bmatrix}$$

Let us first consider the output dynamic controller given by

$$A_c^1 = \begin{bmatrix} -1 & 0.1 \\ -0.1 & -1.05 \end{bmatrix}; \qquad B_c^1 = \begin{bmatrix} 0.9 & -0.1 \\ 0.1 & 0.9 \end{bmatrix}$$
$$C_c^1 = \begin{bmatrix} 0 & -0.6667 \\ 0 & 0.25 \end{bmatrix}; \qquad D_c^1 = \begin{bmatrix} 0 & 0 \\ 0 & 0 \end{bmatrix}$$

The closed-loop spectrum associated to this stable controller is

$$\{-0.1; \ -0.15; \ -1 \pm 0.1i\}$$

The global stability analysis may be investigated ($A$ and $A_c$ are asymptotically stable) by using the LMI condition (3.61) of Proposition 3.9. Unfortunately, for this controller, the associated LMI constraint is unfeasible. This is easily confirmed by the presence of four parasitic equilibrium points at the positions:

$$x_{eq} = \pm \begin{bmatrix} -3.8747 \\ -2.1560 \\ -3.4604 \\ -1.8875 \end{bmatrix} \quad \text{and} \quad \pm \begin{bmatrix} -25.0000 \\ -18.0000 \\ -22.2689 \\ -15.6887 \end{bmatrix}$$

For simulation purposes, it is considered an exponentially decreasing disturbance $w(t) = 10e^{-t}$. The closed-loop trajectory of the system, initialized in 0 and in presence of an exponentially decreasing disturbance, using this first controller is plotted on Fig. 3.5. It may be checked that once the disturbance has vanished, the closed-loop system converges toward a parasitic equilibrium instead of toward the origin of the state space.

A second attempt of output dynamic controller is considered, for which the closed-loop spectrum is $\{-0.5040; \ -0.5; \ -0.5; \ -0.4960\}$:

$$A_c^2 = \begin{bmatrix} -0.9 & 0 \\ 0 & -0.9 \end{bmatrix}; \qquad B_c^2 = \begin{bmatrix} 0.4 & 0 \\ 0 & -0.4 \end{bmatrix}$$
$$C_c^2 = \begin{bmatrix} 4 & 5.3333 \\ -1.6 & -2 \end{bmatrix}; \qquad D_c^2 = \begin{bmatrix} 0 & 0 \\ 0 & 0 \end{bmatrix}$$

**Fig. 3.5** Example 3.3—controller ($A_c^1$, $B_c^1$, $C_c^1$, $C_c^1$). The trajectory of the closed-loop system in presence of an energy bounded disturbance converges toward a parasitic equilibrium point

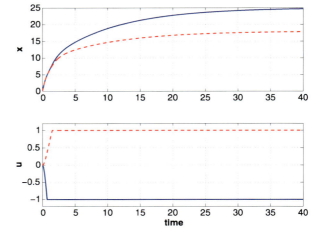

**Fig. 3.6** Example 3.3—controller ($A_c^2$, $B_c^2$, $C_c^2$, $C_c^2$). The trajectory of the closed-loop system in presence of a energy bounded disturbance converges toward the origin of the state space once the disturbance has vanished

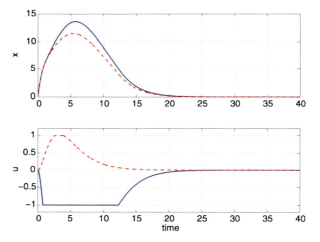

The LMI condition of Proposition 3.9 is feasible, which guarantees the global asymptotic stability of the closed-loop system with this second controller. This can be checked on Fig. 3.6, which shows that, once the additive energy bounded disturbance has vanished, the closed-loop system comes back to the origin of the state space. Any initial condition will also be brought to the origin after some transient period.

Let us now comment a third case, which illustrates that the global asymptotic stability condition given in Proposition 3.9 is only a sufficient condition. The controller defined by the following data:

$$A_c^3 = \begin{bmatrix} -1.9 & 0 \\ 0 & -1.55 \end{bmatrix}; \qquad B_c^3 = \begin{bmatrix} 1.4 & 0 \\ 0 & 1.3 \end{bmatrix}$$

# 3.4 Stabilization

$$C_c^3 = \begin{bmatrix} 4 & -2 \\ -1.6 & 0.75 \end{bmatrix}; \qquad D_c^3 = \begin{bmatrix} 0 & 0 \\ 0 & 0 \end{bmatrix}$$

which places the poles of the closed-loop system in $\{-1.5; -1.4; -0.5; -0.25\}$ cannot be guaranteed to globally asymptotically stabilize the system of interest (condition (3.61) is found unfeasible). This controller does not generate, however, any parasitic equilibrium, and it has not been possible to find any initial condition or additive disturbance for which the closed-loop trajectory does not converge to the origin.

## 3.4.2 Observer-Based Feedback Stabilization

We now suppose that the state of the system is not fully available for measurement. Then we suppose that the measured output of the system is defined by

$$y = Cx \tag{3.83}$$

with $y \in \Re^p$ and $C \in \Re^{p \times n}$. In this case, an estimated state feedback can be considered. In other words, system (3.1) is modified as

$$\dot{x}(t) = Ax(t) + B\,\mathrm{sat}\big(K\hat{x}(t)\big) + B_w w(t) \tag{3.84}$$

where $\hat{x}$ is the estimated state. To simplify the presentation, we consider the case without additive perturbation, i.e., $w = 0$.

Hence, in the stabilization problem context, we are interested in computing a state observer in order to ensure the stability of the closed-loop system under an estimated state feedback, while characterizing the region of stability. It is important to point out that the region of stability is now associated to the augmented state (composed by the state of the plant plus the state of the observer) taking into account that the dynamics of the plant state and that one of the observer are coupled.

Similarly to Chap. 2, we discuss this issue and present a solution to the problem of computing a state-observer-based control law, considering explicitly the control saturation effects. To do this, we consider a full-order Luenberger state observer [64, 277] given by

$$\dot{\hat{x}}(t) = A\hat{x}(t) + Bu(t) - LC\big(x(t) - \hat{x}(t)\big) \tag{3.85}$$

where $\hat{x}(t) \in \Re^n$ is the estimate of the state and $L \in \Re^{n \times p}$ is the observer gain. The applied control law is now defined by

$$u(t) = \mathrm{sat}\big(K\hat{x}(t)\big) \tag{3.86}$$

Two problems can be of interest: (i) The state feedback gain $K$ is supposed known (for example computed from previous sections) and in this case only $L$ needs to be computed; (ii) Both gains $K$ and $L$ have to be calculated.

Let us define the estimation error $e$ as follows:

$$e = x - \hat{x} \tag{3.87}$$

156    3 Stability Analysis and Stabilization—Sector Nonlinearity Model Approach

The closed-loop system in function of $x$ and $e$ can be written as

$$\dot{x}(t) = (A + BK)x(t) - BKe(t) + B\phi\big(Kx(t) - Ke(t)\big)$$
$$\dot{e}(t) = (A + LC)e(t) \tag{3.88}$$

By considering the augmented state vector $\xi = [x' \ e']' \in \Re^{2n}$, system (3.88) reads

$$\dot{\xi}(t) = \mathcal{A}\xi(t) + \mathcal{B}\phi\big(\mathcal{K}\xi(t)\big) \tag{3.89}$$

with

$$\mathcal{A} = \begin{bmatrix} A + BK & -BK \\ 0 & A + LC \end{bmatrix}; \qquad \mathcal{B} = \begin{bmatrix} B \\ 0 \end{bmatrix}; \qquad \mathcal{K} = [\, K \ -K \,] \tag{3.90}$$

Note that the gains $K$ and $L$ are such that matrices $A + BK$ and $A + LC$ are Hurwitz. The following result allowing one to characterize a region of asymptotic stability for the closed-loop system (3.88) or (3.89) can be stated.

**Proposition 3.15** *If there exist two symmetric positive definite matrices $W \in \Re^{n \times n}$, $P_2 \in \Re^{n \times n}$, a positive diagonal matrix $S \in \Re^{m \times m}$, matrices $K \in \Re^{m \times n}$, $Z_1 \in \Re^{m \times n}$, $G_2 \in \Re^{m \times n}$, $U \in \Re^{n \times p}$ and a positive scalar $\eta$ satisfying*

$$\begin{bmatrix} W(A+BK)' + (A+BK)W & BS - Z_1' & -BK \\ SB' - Z_1 & -2S & -G_2 \\ -K'B' & -G_2' & A'P_2 + P_2A + UC + C'U' \end{bmatrix} < 0 \tag{3.91}$$

$$\begin{bmatrix} W & 0 & WK_{(i)}' - Z_{1(i)}' \\ 0 & P_2 & -K_{(i)}' - G_{2(i)}' \\ K_{(i)}W - Z_{1(i)} & -K_{(i)} - G_{2(i)} & \eta u_{0(i)}^2 \end{bmatrix} \geq 0, \quad i = 1, \dots, m \tag{3.92}$$

*then the gains $K$ and $L = P_2^{-1}U$ are such that the ellipsoid $\mathcal{E}(\mathrm{Diag}(P_1, P_2), \eta) = \{(x, e) \in \Re^n \times \Re^n; \ x'P_1x + e'P_2e \leq \eta^{-1}\}$ is a region of asymptotic stability (RAS) for the closed-loop system* (3.88).

*Proof* The proof is quite similar to that one of Proposition 3.1. The quadratic Lyapunov function $\xi' \mathrm{Diag}(P_1, P_2)\xi = x'P_1x + e'P_2e$ is considered. Lemma 1.6 in Chap. 1 applies by choosing, $v = \mathcal{K}\xi = Kx - Ke$ and $\omega = \mathcal{G}\xi = G_1x + G_2e$. By noting $P_1 = W^{-1}$ and using the change of variables $G_1W = Z_1$, the satisfaction of relation (3.92) means that the ellipsoid $\mathcal{E}(\mathrm{Diag}(P_1, P_2), \eta) = \{(x, e) \in \Re^n \times \Re^n; \ x'P_1x + e'P_2e \leq \eta^{-1}\}$ is included in the polyhedral set $S(|\mathcal{K} - \mathcal{G}|, u_0)$.

By left- and right-multiplying relation (3.91) by $\mathrm{Diag}(W^{-1}, S^{-1}, I)$ with the change of variables $L = P_2^{-1}U$ and $S = T^{-1}$, one obtains

$$\begin{bmatrix} (A+BK)'P_1 + P_1(A+BK) & P_1B - G_1'T & -P_1BK \\ B'P_1 - TG_1 & -2T & -TG_2 \\ -K'B'P_1 & -G_2'T & (A+LC)'P_2 + P_2(A+LC) \end{bmatrix}$$
$$< 0$$

That means that $\dot{V}(\xi) \leq \dot{V}(\xi) - 2\phi(\mathcal{K}\xi(t))'T(\phi(\mathcal{K}\xi(t)) + \mathcal{G}\xi(t)) < 0$, $\forall \xi \in \mathcal{E}(\mathrm{Diag}(P_1, P_2), \eta)$, along the trajectories of the closed-loop system (3.88). $\qquad \square$

3.4 Stabilization 157

*Remark 3.28* In Proposition 3.15 the gains $K$ and $L$ cannot be calculated simultaneously with a linear problem. On the other hand, if the gain $K$ is given the relations of Proposition 3.15 are linear in the decision variables.

*Remark 3.29* In Propositions 3.15, the quadratic Lyapunov function used is $\xi' P \xi = \xi' \text{Diag}(P_1, P_2) \xi = x' P_1 x + e' P_2 e$. By considering an approach based on the use of Finsler's Lemma [81], it is possible to relax this diagonal form for the $P$ matrix and therefore to reduce the possible conservatism associated to quadratic Lyapunov functions: see, for example, [279, 281].

An alternative solution may be studied by using a slightly different paradigm based on the notion of input-to-state stability (ISS: see for example the survey of Sontag [328], Theorem 5, p. 16). In this case, the closed-loop system is defined in function of $\hat{x}$ and $e$ by

$$\dot{\hat{x}}(t) = (A + BK)\hat{x}(t) + B\phi(K\hat{x}(t)) - LCe(t) \tag{3.93}$$

$$\dot{e}(t) = (A + LC)e(t) \tag{3.94}$$

which reads by considering the augmented state vector $\bar{\xi} = [\hat{x}' \ e']' \in \Re^{2n}$:

$$\dot{\bar{\xi}}(t) = \bar{A}\bar{\xi}(t) + \bar{B}\phi(\bar{K}\bar{\xi}(t)) \tag{3.95}$$

with

$$\bar{A} = \begin{bmatrix} A + BK & -LC \\ 0 & A + LC \end{bmatrix}; \qquad \bar{B} = \begin{bmatrix} B \\ 0 \end{bmatrix}; \qquad \bar{K} = \begin{bmatrix} K & 0 \end{bmatrix} \tag{3.96}$$

*Remark 3.30* Since, by construction, $A + LC$ is Hurwitz, the error system (3.94) is exponentially decreasing for any initial value $e(0)$. Then one gets $\|e(t)\| \le \kappa e^{-\alpha t}$ for all $t \ge 0$ and for some positive scalars $\alpha$ and $\kappa$ depending of $A + LC$ and $e(0)$, or still $\int_0^\infty e(t)'e(t)\,dt \le \frac{\kappa^2}{2\alpha}$.

According to Remark 3.30, let us give a simple lemma to rely the upper bound of $\int_0^\infty e(t)'e(t)\,dt$ to $A + LC$ and $e(0)$.

**Lemma 3.2** *If there exist a symmetric positive definite matrix $P_2 \in \Re^{n \times n}$, a matrix $L \in \Re^{n \times p}$ and a positive scalar $\alpha$ satisfying:*

$$(A + LC)'P_2 + P_2(A + LC) + 2\alpha P_2 < 0 \tag{3.97}$$

*then one gets*

$$\int_0^\infty e(t)'e(t)\,dt \le \frac{e(0)'P_2e(0)}{2\alpha\lambda_{\min}(P_2)} \tag{3.98}$$

Based on the notion of input-to-state stability, we can state the following result.

158        3 Stability Analysis and Stabilization—Sector Nonlinearity Model Approach

**Proposition 3.16** *Given any positive scalar $\rho$, if there exist two symmetric positive definite matrices $W \in \Re^{n \times n}$, $P_2 \in \Re^{n \times n}$, a positive diagonal matrix $S \in \Re^{m \times m}$, three matrices $K \in \Re^{m \times n}$, $L \in \Re^{n \times p}$, $Z \in \Re^{m \times n}$, positive scalars $\theta$, $\eta$, $\alpha$ and $\gamma$ satisfying*

$$\begin{bmatrix} W(A+BK)' + (A+BK)W + \theta W & BS - Z' & -LC \\ SB' - Z & -2S & 0 \\ -C'L' & 0 & -\gamma I \end{bmatrix} < 0 \quad (3.99)$$

$$\begin{bmatrix} W & WK'_{(i)} - Z'_{(i)} \\ K_{(i)}W - Z_{(i)} & \eta u^2_{0(i)} \end{bmatrix} \geq 0, \quad i = 1, \ldots, m \quad (3.100)$$

$$(A+LC)'P_2 + P_2(A+LC) + 2\alpha P_2 < 0 \quad (3.101)$$

$$\gamma \eta - \delta < 0 \quad (3.102)$$

*with $\delta = 2\alpha \lambda_{\min}(P_2)\rho$, then the gains $K$ and $L$ are such that*

1. *$e$ converges toward zero and then $\hat{x}$ converges toward zero;*
2. *for any $e(0)$ such that $e(0)'P_2 e(0) \leq \rho^{-1}$, the trajectories of the system (3.93) do not leave the ellipsoid $\mathcal{E}(P_1, \eta)$, with $P_1 = W^{-1}$, for $\hat{x}(0) = 0$;*
3. *for any $e(0)$ such that $e(0)'P_2 e(0) \leq \rho^{-1}$, one gets*

$$\int_0^\infty \hat{x}(t)' P_1 \hat{x}(t)\, dt \leq \frac{\gamma}{\theta} \int_0^\infty e(t)'e(t)\, dt \leq \frac{\gamma}{2\alpha \lambda_{\min}(P_2)\rho\theta}$$

4. *the region of asymptotic stability (RAS) for the closed-loop system (3.95) is included in $\mathcal{E}(P_1, \eta) \times \mathcal{E}(P_2, \rho)$.*

*Proof* If relation (3.101) is verified then matrix $A + LC$ is Hurwitz and Remark 3.30 applies. Thus, by choosing any positive scalar $\rho$, the ellipsoid $\mathcal{E}(P_2, \rho) = \{e \in \Re^n; \ e'P_2 e \leq \rho^{-1}\}$ is an invariant and contractive set with respect to the trajectories of system (3.94).

Let us consider the quadratic Lyapunov function $V(\hat{x}) = \hat{x}'P_1\hat{x}$, $P_1 = P_1' > 0$, with $P_1 = W^{-1}$. Lemma 1.6 in Chap. 1 applies by choosing, $v = \mathcal{K}\bar{\xi} = K\hat{x}$ and $\omega = \mathcal{G}\bar{\xi} = G\hat{x}$. By using the change of variables $GW = Z$, the satisfaction of relation (3.100) means that the ellipsoid $\mathcal{E}(P_1, \eta) = \{\hat{x} \in \Re^n; \ \hat{x}'P_1\hat{x} \leq \eta^{-1}\}$ is included in the polyhedral set $S(|K - G|, u_0)$. Then one gets by noting that $S = T^{-1}$:

$$\dot{V}(\hat{x}) \leq \dot{V}(\hat{x}) - 2\phi(K\hat{x})'T\big(\phi(K\hat{x}) + G\hat{x}\big)$$

for any $\hat{x} \in \mathcal{E}(P_1, \eta)$.

By considering the quadratic Lyapunov function $V(\hat{x}) = \hat{x}'P_1\hat{x}$ and by using the ISS condition of [328], we want to satisfy

$$\dot{V}(\hat{x}) \leq -\theta V(\hat{x}) + \gamma e'e$$

along the trajectories of system (3.93). Hence, provided that the ellipsoid $\mathcal{E}(P_1, \eta)$ is included in $S(|K - G|, u_0)$, one wants to verify

$$\dot{V}(\hat{x}) \leq \dot{V}(\hat{x}) - 2\phi(K\hat{x})'T\big(\phi(K\hat{x}) + G\hat{x}\big) \leq -\theta V(\hat{x}) + \gamma e'e$$

## 3.4 Stabilization

By left- and right-multiplying relation (3.99) by $\text{Diag}(W^{-1}, S^{-1}, I) = \text{Diag}(P_1, T, I)$, one obtains

$$\begin{bmatrix} (A+BK)'P_1 + P_1(A+BK) + \theta P_1 & P_1 B - G'T & -P_1 LC \\ B'P_1 - TG & -2T & 0 \\ -C'L'P_1 & 0 & -\gamma I \end{bmatrix} < 0$$

That corresponds to satisfying

$$\hat{x}'\big((A+BK)'P_1 + P_1(A+BK) + \theta P_1\big)\hat{x} + 2\hat{x}'P_1 B\phi(K\hat{x}) - 2\hat{x}'P_1 LCe \\ - 2\phi(K\hat{x})'T\big(\phi(K\hat{x}) + G\hat{x}\big) - \gamma e'e < 0$$

which can be also written as

$$\dot{V}(\hat{x}) - 2\phi(K\hat{x})'T\big(\phi(K\hat{x}) + G\hat{x}\big) + \theta V(\hat{x}) - \gamma e'e < 0$$

Hence, if relations (3.99), (3.100), (3.101) and (3.102) are verified, it follows that

$$\dot{V}(\hat{x}) \leq \dot{V}(\hat{x}) - 2\phi(K\hat{x})'S^{-1}\big(\phi(K\hat{x}) + G\hat{x}\big) \leq -\theta V(\hat{x}) + \gamma e'e \qquad (3.103)$$

By integrating the inequality (3.103) along the trajectories of system (3.93) with $\hat{x}(0) = 0$, it follows that

$$\theta \int_0^T V\big(\hat{x}(t)\big)\, dt \leq \theta \int_0^T V\big(\hat{x}(t)\big)\, dt + V\big(\hat{x}(T)\big) \leq \gamma \int_0^T e(t)'e(t)\, dt, \quad \forall T > 0 \tag{3.104}$$

and also

$$V\big(\hat{x}(T)\big) \leq \theta \int_0^T V\big(\hat{x}(t)\big)\, dt + V\big(\hat{x}(T)\big) \leq \gamma \int_0^T e(t)'e(t)\, dt, \quad \forall T > 0 \tag{3.105}$$

The inequality (3.104) allows to verify that for any $e(0)$ such that $e(0)'P_2 e(0) \leq \rho^{-1}$, one gets

$$\int_0^\infty \hat{x}(t)'P_1\hat{x}(t)\, dt \leq \frac{\gamma}{\theta} \int_0^\infty e(t)'e(t)\, dt \leq \frac{\gamma}{2\alpha \lambda_{\min}(P_2)\rho\theta}$$

The inequality (3.105) allows to verify that for $\hat{x}(0) = 0$ and any $e(0) \in \mathcal{E}(P_2, \rho)$ one gets from Lemma 3.2

$$V\big(\hat{x}(T)\big) \leq \gamma \frac{e(0)'P_2 e(0)}{2\alpha \lambda_{\min}(P_2)} \leq \gamma \frac{1}{2\alpha \lambda_{\min}(P_2)\rho}$$

That means that the trajectories of system (3.93) do not leave the ellipsoid $\mathcal{E}(P_1, \eta)$, for $\hat{x}(0) = 0$ and any $e(0) \in \mathcal{E}(P_2, \rho)$ thanks to the satisfaction of relation (3.102). One can therefore conclude that the items of Proposition 3.16 hold. □

*Remark 3.31* In Proposition 3.16 the gains $K$ and $L$ cannot be computed simultaneously. If the gain $L$ is given the relations of Proposition 3.16 are linear in the decision variable $Y$ (which corresponds to the classical change of variable $Y = KW$). The converse is, however, not so trivial. Thus, according to Remark 3.28, Propositions 3.15 and 3.16 are complementary.

160              3 Stability Analysis and Stabilization—Sector Nonlinearity Model Approach

Then, if we consider that the gain $K$ is known, for example allowing to stabilize the closed-loop system (3.5) for any $x(0) \in \mathcal{X}_0 \subseteq \Re^n$, the idea is to compute the observer gain $L$ in order to ensure that the closed-loop trajectories of system (3.89) or (3.95) converge to the origin for any $x(0) \in \mathcal{X}_0$. In this case, one can consider that $\hat{x}(0) = 0$. By considering that the set $\mathcal{X}_0$ is defined by its vertices $v_i$, $i = 1, \ldots, n_v$, such a solution can be achieved through the following algorithm, associated to Proposition 3.15:

**Algorithm 3.5**

1. Initialize $K$ (for example issued from state feedback problem).
2. Given $K$ compute $W$, $S$, $Z_1$, $G_2$, $P_2$, $U$, $\eta$, $\mu$ solutions to

$$\min \ \mu + \eta$$

$$\text{subject to} \quad \text{inequalities (3.91), (3.92)} \tag{3.106}$$

$$\begin{bmatrix} \mu & v_i' \\ v_i & W \end{bmatrix} \geq 0, \quad i = 1, \ldots, n_v$$

which gives $L = P_2^{-1} U$.

On other hand, when $L$ is fixed, the following algorithm, originating with Proposition 3.16, may be used.

**Algorithm 3.6**

1. Initialize $L$ such that relation (3.101) holds for some symmetric positive definite matrix $P_2$ and positive scalar $\alpha$.
2. Given $L$, compute $W$, $Y$, $Z$, $S$, $\eta$, $\theta$, $\gamma$, $\mu$ solutions to

$$\min \ \mu + \eta$$

$$\text{subject to} \quad \begin{bmatrix} WA + Y'B' + AW + BY + \theta W & BS - Z' & -LC \\ SB' - Z & -2S & 0 \\ -C'L' & 0 & -\gamma I \end{bmatrix} < 0$$

$$\begin{bmatrix} W & WK_{(i)}' - Z_{(i)}' \\ K_{(i)}W - Z_{(i)} & \eta u_{0(i)}^2 \end{bmatrix} \geq 0, \quad i = 1, \ldots, m$$

$$\begin{bmatrix} \mu & v_i' \\ v_i & W \end{bmatrix} \geq 0, \quad i = 1, \ldots, n_v$$

$$\gamma \eta - \delta < 0$$

$$\tag{3.107}$$

which gives $K = YW^{-1}$.

*Remark 3.32* The global case can also be considered when the open-loop matrix $A$ is Hurwitz. To do this, one can use the results developed in [328, 329]. The relations of Proposition 3.16 are then modified as follows by considering $G = K$, or equivalently, $Z = KW$:

3.4 Stabilization 161

$$\begin{bmatrix} W(A+BK)' + (A+BK)W + \theta W & BS - WK' & -LC \\ SB' - KW & -2S & 0 \\ -C'L & 0 & -\gamma I \end{bmatrix} < 0 \qquad (3.108)$$

$$(A+LC)'P_2 + P_2(A+LC) < 0 \qquad (3.109)$$

Hence, one can conclude that every trajectory of system (3.93), for which $e(t) \to 0$ as $t \to \infty$, converges to the origin [338].

### 3.4.3 Dynamic Output Feedback Stabilization

The problem considered in this section regards the stabilization of the system through a dynamic output feedback compensator. To reach this goal, let us redefine system (3.1) for dynamic output feedback compensator purpose with a slightly modified notation:

$$\dot{x}_p(t) = Ax_p(t) + B\,\mathrm{sat}\big(y_c(t)\big) + B_w w(t)$$
$$y_p(t) = Cx_p(t) \qquad (3.110)$$

where $x_p \in \mathfrak{R}^n$ is the state and $y_p \in \mathfrak{R}^p$ is the measured output; $y_c$ represents the output of the controller entering the saturation block (which corresponds in Sect. 3.4.1 to a static state feedback gain); $w$ is a vector-valued external signal, which can represent, for instance, a disturbance acting on the system. Pairs $(A, B)$ and $(C, A)$ are supposed to be stabilizable and detectable, respectively.

The dynamic output feedback compensator we want to design is described by

$$\dot{x}_c(t) = A_c x_c(t) + B_c y_p(t)$$
$$y_c(t) = C_c x_c(t) + D_c y_p(t) \qquad (3.111)$$

where $x_c \in \mathfrak{R}^{n_c}$ is the controller state, $y_p$ is the controller input and $y_c \in \mathfrak{R}^m$ is the controller output. $A_c$, $B_c$, $C_c$, $D_c$ are matrices of appropriate dimensions, yet to be determined.

*Remark 3.33* Note that the observer case (3.85)–(3.86) studied in the previous Sect. 3.4.2 is a particular case of the dynamic controller (3.111) with

$$x_c = \hat{x}; \qquad A_c = A + BK + LC; \qquad B_c = -LC; \qquad C_c = K; \qquad D_c = 0$$

and an additional term due to the saturation affecting the dynamics $\dot{x}_c$ (i.e. the term $B(\mathrm{sat}(y_c) - y_c)$).

The connection between system (3.110) and the controller (3.111), leads to the following closed-loop system:

$$\dot{x}_p(t) = Ax_p(t) + B\,\mathrm{sat}\big(C_c x_c(t) + D_c Cx_p(t)\big) + B_w w(t)$$
$$\dot{x}_c(t) = A_c x_c(t) + B_c Cx_p(t) \qquad (3.112)$$

162    3 Stability Analysis and Stabilization—Sector Nonlinearity Model Approach

Let us define the following augmented vector:

$$x = \begin{bmatrix} x_p \\ x_c \end{bmatrix} \in \Re^{n+n_c} \tag{3.113}$$

and the matrices

$$\mathbb{A} = \begin{bmatrix} A + BD_cC & BC_c \\ B_cC & A_c \end{bmatrix}; \qquad \mathbb{B} = \begin{bmatrix} B \\ 0 \end{bmatrix}$$

$$\mathbb{B}_w = \begin{bmatrix} B_w \\ 0 \end{bmatrix}; \qquad \mathbb{K} = \begin{bmatrix} D_cC & C_c \end{bmatrix} \tag{3.114}$$

From (3.113) and (3.114), the closed-loop system reads

$$\dot{x}(t) = \mathbb{A}x(t) + \mathbb{B}\phi\big(y_c(t)\big) + \mathbb{B}_w w(t)$$

$$y_c(t) = \mathbb{K}x(t) \tag{3.115}$$

with $\phi(y_c) = \mathrm{sat}(y_c) - y_c$.

As in the state feedback case, the design of the controller can be oriented in order to maximize the disturbance tolerance, the disturbance rejection or the region where the asymptotic stability of the closed-loop system is ensured.

In the sequel, we provide some results to address the design problem of dynamic controller, that is, the determination of matrices $A_c$, $B_c$, $C_c$, $D_c$, in both cases of amplitude bounded and energy bounded signals for $w(t)$. Moreover, we focus on the design of a full-order dynamic output controller. Hence, the results developed in the sequel hold for $n_c = n$, and therefore the state $x$ of the closed-loop system (3.115) is an element of $\Re^{2n}$.

**Amplitude Bounded Exogenous Signals**    As in the context of Sect. 3.3.1, let us consider that the exogenous signal $w$ belongs to $\mathcal{W}$, as defined in (3.36).

By considering the quadratic Lyapunov function defined in (3.7), the following proposition allows to solve the external output feedback stabilization problem for the system facing amplitude bounded exogenous signals.

**Proposition 3.17**    *If there exist two symmetric positive definite matrices $X \in \Re^{n \times n}$, $Y \in \Re^{n \times n}$, a positive diagonal matrix $S \in \Re^{m \times m}$, matrices $L \in \Re^{m \times n}$, $F \in \Re^{n \times p}$, $W \in \Re^{n \times n}$, $Z \in \Re^{m \times n}$, $Z_1 \in \Re^{m \times n}$, $D_c \in \Re^{m \times p}$, three positive scalars $\tau_1$, $\tau_2$ and $\eta$ satisfying*

$$\begin{bmatrix} AY + YA' + BL + L'B' + \tau_1 Y & W + \tau_1 I & BS - Z' & B_w \\ W' + \tau_1 I & A'X + XA + C'F' + FC + \tau_1 X & XBS - Z_1' & XB_w \\ SB' - Z & SB'X - Z_1 & -2S & 0 \\ B_w' & B_w'X & 0 & -\tau_2 R \end{bmatrix}$$
$$< 0 \tag{3.116}$$

$$\begin{bmatrix} Y & I & L'_{(i)} - Z'_{(i)} \\ I & X & C'D'_{c(i)} - Z'_{1(i)} \\ L_{(i)} - Z_{(i)} & D_{c(i)}C - Z_{1(i)} & \eta u_{0(i)}^2 \end{bmatrix} \geq 0, \quad i = 1, \dots, m \tag{3.117}$$

$$-\tau_1 \delta + \tau_2 \eta < 0 \tag{3.118}$$

*then the controller defined by the matrices*

$$D_c$$
$$C_c = (L - D_c C Y)(V')^{-1}$$
$$B_c = U^{-1}(F - X B D_c) \tag{3.119}$$
$$A_c = U^{-1}(W' - (A + B D_c C)' - X A Y - X B L - U B_c C Y)(V')^{-1}$$

*where matrices $U$ and $V$ verify $U V' = I - X Y$, and matrix $P$ is defined by*

$$P = \begin{bmatrix} X & U \\ U' & \hat{X} \end{bmatrix} \tag{3.120}$$

*with $\hat{X} \in \Re^{n \times n}$ a positive definite symmetric matrix, is such that*

1. *for $w = 0$, the ellipsoid $\mathcal{E}(P, \eta)$ is a region of asymptotic stability (RAS) for the saturated system (3.115);*
2. *for any $w \in \mathcal{W}$ and $x(0) \in \mathcal{E}(P, \eta)$, the trajectories of the saturated system (3.115) do not leave the ellipsoid $\mathcal{E}(P, \eta)$.*

*Proof* As in the proof of Proposition 3.6, we want to satisfy the inequality

$$\dot{V}(x) + \tau_1 x' P x - \tau_2 w' R w$$
$$\leq \dot{V}(x) + \tau_1 x' P x - \tau_2 w' R w - 2\phi(\mathbb{K}x)' T (\phi(\mathbb{K}x) + Gx) < 0$$

along the trajectories of system (3.115). Following the same reasoning as that one of the proof of Proposition 3.6, this inequality is satisfied if relation (3.118) and the following ones hold:

$$\begin{bmatrix} (\mathbb{A} + \mathbb{B}\mathbb{K})' P + P(\mathbb{A} + \mathbb{B}\mathbb{K}) + \tau_1 P & P\mathbb{B} - G'T & P\mathbb{B}_w \\ \mathbb{B}'P - TG & -2T & 0 \\ \mathbb{B}'_w P & 0 & -\tau_2 R \end{bmatrix} < 0 \tag{3.121}$$

$$\begin{bmatrix} P & \mathbb{K}'_{(i)} - G'_{(i)} \\ \mathbb{K}_{(i)} - G_{(i)} & \eta u_{0(i)}^2 \end{bmatrix} \geq 0, \quad i = 1, \dots, m \tag{3.122}$$

Consider the following structures for matrices $P$ and $P^{-1}$:

$$P = \begin{bmatrix} X & U \\ U' & \hat{X} \end{bmatrix}; \qquad P^{-1} = \begin{bmatrix} Y & V \\ V' & \hat{Y} \end{bmatrix} \tag{3.123}$$

where $X \in \Re^{n \times n}$, $Y \in \Re^{n \times n}$, $\hat{X} \in \Re^{n \times n}$, $\hat{Y} \in \Re^{n \times n}$ are positive definite symmetric matrices and $U, V$ matrices of appropriate dimensions. Thus, it follows that

$$X Y + U V' = I; \qquad U'V + \hat{X}\hat{Y} = I$$
$$U'Y + \hat{X}V' = 0; \qquad X V + U\hat{Y} = 0$$

Define now the matrix (see [314])

$$J = \begin{bmatrix} Y & V \\ I & 0 \end{bmatrix}$$

Note that it is always possible to compute square and nonsingular matrices $V$ and $U$ verifying the equation $UV' = I - XY$. This fact ensures that $J$ is nonsingular.

By pre- and post-multiplying (3.121) respectively by $\mathrm{Diag}(J, S, I)$ and $\mathrm{Diag}(J', S, I)$ with $S = T^{-1}$, one gets

$$\begin{bmatrix} J((\mathbb{A} + \mathbb{B}\mathbb{K})'P + P(\mathbb{A} + \mathbb{B}\mathbb{K}) + \tau_1 P)J' & JP\mathbb{B}S - JG' & JP\mathbb{B}_w \\ S\mathbb{B}'PJ' - GJ' & -2S & 0 \\ \mathbb{B}'_w PJ' & 0 & -\tau_2 R \end{bmatrix} < 0$$

$$(3.124)$$

Partitioning $G = [\, G_1 \; G_2 \,]$, and considering the change of variables

$$L = D_c C Y + C_c V'; \qquad F = X B D_c + U B_c$$
$$W = V A'_c U' + A + B D_c C + Y A' X + L' B' X + Y C' B'_c U'$$
$$Z = G_1 Y + G_2 V'; \qquad Z_1 = G_1$$

it follows that

$$J\mathbb{A}'PJ' = \begin{bmatrix} YA' + L'B' & W - (A + BD_cC) \\ (A + BD_cC)' & A'X + C'F' \end{bmatrix}$$

$$JP\mathbb{B}S = \begin{bmatrix} BS \\ XBS \end{bmatrix}; \qquad JG' = \begin{bmatrix} Z' \\ Z'_1 \end{bmatrix}$$

$$JP\mathbb{B}_w = \begin{bmatrix} B_w \\ XB_w \end{bmatrix}; \qquad JPJ' = \begin{bmatrix} Y & I \\ I & X \end{bmatrix}; \qquad J\mathbb{K}' = \begin{bmatrix} L' \\ C'D'_c \end{bmatrix}$$

Hence, relation (3.116) is directly obtained from (3.124).

By the same way, relation (3.117) is obtained by pre- and post-multiplying inequality (3.122) respectively by $\mathrm{Diag}(J, 1)$ and $\mathrm{Diag}(J', 1)$, and by using the change of variables above defined.

The proof is completed by using the same arguments as in the proof of Proposition 3.6. $\qquad\square$

*Remark 3.34* The results of Proposition 3.17 can be adapted in order to provide global stabilizing conditions, applicable when the open-loop system is asymptotically stable. In this case, it suffices to consider $Z = L$ and $Z_1 = D_c C$. Thus, according to Proposition 3.7, the conditions to be satisfied are the following:

$$\begin{bmatrix} AY + YA' + BL + L'B' + \tau_1 Y & W + \tau_1 I & BS - L' & B_w \\ W' + \tau_1 I & A'X + XA + C'F' + FC + \tau_1 X & XBS - C'D'_c & XB_w \\ SB' - L & SB'X - D_cC & -2S & 0 \\ B'_w & B'_w X & 0 & -\tau_2 R \end{bmatrix}$$
$$< 0 \qquad\qquad (3.125)$$

$$\begin{bmatrix} Y & I \\ I & X \end{bmatrix} > 0 \qquad\qquad (3.126)$$

$$-\tau_1 \delta + \tau_2 \eta < 0 \qquad\qquad (3.127)$$

Note that, in this case, since the condition (3.117) is no more considered, we have to add the condition (3.126) to guarantee that $I - XY$ is nonsingular.

## 3.4 Stabilization

It has to be noted that in relation (3.116) there exist several nonlinearities: $\tau_1 X$, $\tau_1 Y$ and $XBS$ (and its symmetric $SB'X$). The first two nonlinearities are not complicated to manage since $\tau_1$ can be fixed through a one-dimensional sweep on $\tau_1 \in (0, \infty)$. The last one can be more complicated to manage although $S \in \Re^{m \times m}$ is a diagonal positive definite matrix. However, a simple way to remove this nonlinearity consists in modifying the controller as in [120, 147] by adding a static anti-windup part to the dynamic controller.[1] The modified controller to be synthesized is then described by

$$\dot{x}_c(t) = A_c x_c(t) + B_c C x_p(t) + E_c \phi(y_c)$$
$$y_c(t) = C_c x_c(t) + D_c C x_p(t) \tag{3.128}$$

where now $A_c$, $B_c$, $C_c$, $D_c$ and $E_c$ are matrices of appropriate dimensions to be determined. In the closed-loop system (3.115) only the $\mathbb{B}$ matrix is then modified to

$$\mathbb{B}_{\text{new}} = \begin{bmatrix} B \\ E_c \end{bmatrix} \tag{3.129}$$

*Remark 3.35* According to Remark 3.33, note that the observer-based controller (3.85)–(3.86) studied in Sect. 3.4.2 can be viewed as a particular case of the dynamic controller (3.128) with

$$x_c = \hat{x}; \qquad A_c = A + BK + LC; \qquad B_c = -LC$$
$$C_c = K; \qquad D_c = 0; \qquad E_c = B$$

Then by considering the closed-loop system (3.115) with (3.129), the following proposition extending Proposition 3.17 can be stated.

**Proposition 3.18** *If there exist two symmetric positive definite matrices $X \in \Re^{n \times n}$, $Y \in \Re^{n \times n}$, a positive diagonal matrix $S \in \Re^{m \times m}$, matrices $Q \in \Re^{n \times m}$, $L \in \Re^{m \times n}$, $F \in \Re^{n \times p}$, $W \in \Re^{n \times n}$, $Z \in \Re^{m \times n}$, $Z_1 \in \Re^{m \times n}$, $D_c \in \Re^{m \times p}$, three positive scalars $\tau_1$, $\tau_2$ and $\eta$ satisfying:*

$$\begin{bmatrix} AY + YA' + BL + L'B' + \tau_1 Y & W + \tau_1 I & BS - Z' & B_w \\ W' + \tau_1 I & A'X + XA + C'F' + FC + \tau_1 X & Q - Z_1' & XB_w \\ SB' - Z & Q' - Z_1 & -2S & 0 \\ B_w' & B_w'X & 0 & -\tau_2 R \end{bmatrix}$$
$$< 0 \tag{3.130}$$

$$\begin{bmatrix} Y & I & L_{(i)}' - Z_{(i)}' \\ I & X & C'D_{c(i)}' - Z_{1(i)}' \\ L_{(i)} - Z_{(i)} & D_{c(i)}C - Z_{1(i)} & \eta u_{0(i)}^2 \end{bmatrix} \geq 0, \quad i = 1, \ldots, m \tag{3.131}$$

$$-\tau_1 \delta + \tau_2 \eta < 0 \tag{3.132}$$

---

[1] Anti-windup strategies will be studied and discussed in Chaps. 6, 7 and 8.

*then the controller defined by the following matrices*

$$E_c = U^{-1}(Q - XBS)S^{-1}$$

$$D_c$$

$$C_c = (L - D_c C Y)(V')^{-1} \tag{3.133}$$

$$B_c = U^{-1}(F - XBD_c)$$

$$A_c = U^{-1}(W' - (A + BD_cC)' - XAY - XBL - UB_cCY)(V')^{-1}$$

*where matrices $U$ and $V$ verify $UV' = I - XY$ and matrix $P$ is defined in (3.120), is such that*

1. *for $w = 0$, the ellipsoid $\mathcal{E}(P, \eta)$ is a region of asymptotic stability (RAS) for the saturated system (3.115) with (3.129);*
2. *for any $w \in \mathcal{W}$ and $x(0) \in \mathcal{E}(P, \eta)$, the trajectories of the saturated system (3.115) with (3.129) do not leave the ellipsoid $\mathcal{E}(P, \eta)$.*

*Proof* The same arguments as in the proof of Proposition 3.17 can be invoked. The main difference is in the calculation of the term (see the definition of $\mathbb{B}_{\text{new}}$ in (3.129))

$$J P \mathbb{B}_{\text{new}} S = \begin{bmatrix} BS \\ XBS + UE_cS \end{bmatrix}$$

and the change of variables: $Q = XBS + UE_cS$. $\qquad\square$

*Remark 3.36* As in Remark 3.34, the results of Proposition 3.18 can be adapted in order to provide global stabilizing conditions, applicable when the open-loop system is asymptotically stable. In this case, by considering $Z = L$ and $Z_1 = D_cC$, the conditions to be satisfied are relations (3.126), (3.127) and the following:

$$\begin{bmatrix} AY + YA' + BL + L'B' + \tau_1 Y & W + \tau_1 I & BS - L' & B_w \\ W' + \tau_1 I & A'X + XA + C'F' + FC + \tau_1 X & Q - C'D_c' & XB_w \\ SB' - L & Q' - D_cC & -2S & 0 \\ B_w' & B_w'X & 0 & -\tau_2 R \end{bmatrix} < 0 \tag{3.134}$$

**Energy Bounded Exogenous Signals** As in the context of Sect. 3.3.2, let us consider that the exogenous signal $w(t)$ is energy bounded, i.e. it belongs to $\mathcal{W}$ defined in (3.53) for some $\delta > 0$ and symmetric positive definite matrix $R$.

By considering the quadratic Lyapunov function defined in (3.7) and by using similar arguments as in Proposition 3.17, the following proposition allows us to solve the external stabilization problem in the current case.

**Proposition 3.19** *If there exist two symmetric positive definite matrices $X \in \mathfrak{R}^{n \times n}$, $Y \in \mathfrak{R}^{n \times n}$, a positive diagonal matrix $S \in \mathfrak{R}^{m \times m}$, matrices $L \in \mathfrak{R}^{m \times n}$, $F \in \mathfrak{R}^{n \times p}$, $W \in \mathfrak{R}^{n \times n}$, $Z \in \mathfrak{R}^{m \times n}$, $Z_1 \in \mathfrak{R}^{m \times n}$, $D_c \in \mathfrak{R}^{m \times p}$, a positive scalar $\eta$ satisfying*

## 3.4 Stabilization

$$
\begin{bmatrix}
AY + YA' + BL + L'B' & W & BS - Z' & B_w \\
W' & A'X + XA + C'F' + FC & XBS - Z'_1 & XB_w \\
SB' - Z & SB'X - Z_1 & -2S & 0 \\
B'_w & B'_w X & 0 & -R
\end{bmatrix}
$$
$$< 0 \tag{3.135}$$

$$
\begin{bmatrix}
Y & I & L'_{(i)} - Z'_{(i)} \\
I & X & C'D'_{c(i)} - Z'_{1(i)} \\
L_{(i)} - Z_{(i)} & D_{c(i)}C - Z_{1(i)} & \eta u^2_{0(i)}
\end{bmatrix} \geq 0, \quad i = 1, \ldots, m \tag{3.136}
$$

$$-\delta + \eta < 0 \tag{3.137}$$

*then the controller defined by the following matrices*

$$D_c$$

$$
\begin{aligned}
C_c &= (L - D_c CY)(V')^{-1} \\
B_c &= U^{-1}(F - XBD_c) \\
A_c &= U^{-1}\big(W' - (A + BD_cC)' - XAY - XBL - UB_cCY\big)(V')^{-1}
\end{aligned} \tag{3.138}
$$

*where matrices $U$ and $V$ verify $UV' = I - XY$ and matrix $P$ is defined in (3.120), is such that*

1. *for $w = 0$, the ellipsoid $\mathcal{E}(P, \eta)$ is a region of asymptotic stability (RAS) for the saturated system (3.115);*
2. *for any $w \in \mathcal{W}$, the trajectories of the saturated system (3.115) do not leave the ellipsoid $\mathcal{E}(P, \eta)$, $\forall x(0) \in \mathcal{E}(P, \beta)$, with $0 < \beta^{-1} \leq \eta^{-1} - \delta^{-1}$.*

*Proof* The same procedure as in the proof of Proposition 3.8 is used. In other words, we want to prove that $\dot{V}(x) - w'Rw \leq \dot{V}(x) - w'Rw - 2\phi(\mathbb{K}x)'T(\phi(\mathbb{K}x) + Gx) < 0$ along the trajectories of the saturated system (3.115). The same arguments as in the proof of Proposition 3.17 are used to exhibit relations (3.135), (3.136) and (3.137). $\qquad\square$

By using the modified controller (3.128), the following proposition extending Proposition 3.19 can be stated, with respect to the closed-loop system (3.115)–(3.129).

**Proposition 3.20** *If there exist two symmetric positive definite matrices $X \in \mathfrak{R}^{n \times n}$, $Y \in \mathfrak{R}^{n \times n}$, a positive diagonal matrix $S \in \mathfrak{R}^{m \times m}$, matrices $Q \in \mathfrak{R}^{n \times m}$, $L \in \mathfrak{R}^{m \times n}$, $F \in \mathfrak{R}^{n \times p}$, $W \in \mathfrak{R}^{n \times n}$, $Z \in \mathfrak{R}^{m \times n}$, $Z_1 \in \mathfrak{R}^{m \times n}$, $D_c \in \mathfrak{R}^{m \times p}$, a positive scalar $\eta$ satisfying*

$$
\begin{bmatrix}
AY + YA' + BL + L'B' & W & BS - Z' & B_w \\
W' & A'X + XA + C'F' + FC & Q - Z'_1 & XB_w \\
SB' - Z & Q' - Z_1 & -2S & 0 \\
B'_w & B'_w X & 0 & -R
\end{bmatrix} < 0
$$
$$\tag{3.139}$$

$$\begin{bmatrix} Y & I & L'_{(i)} - Z'_{(i)} \\ I & X & C'D'_{c(i)} - Z'_{1(i)} \\ L_{(i)} - Z_{(i)} & D_{c(i)}C - Z_{1(i)} & \eta u^2_{0(i)} \end{bmatrix} \geq 0, \quad i = 1, \dots, m \qquad (3.140)$$

$$-\delta + \eta < 0 \qquad (3.141)$$

*then the controller defined by the following matrices*

$$E_c = U^{-1}(Q - XBS)S^{-1}$$

$$D_c$$

$$C_c = (L - D_c CY)(V')^{-1} \qquad (3.142)$$

$$B_c = U^{-1}(F - XBD_c)$$

$$A_c = U^{-1}\big(W' - (A + BD_cC)' - XAY - XBL - UB_cCY\big)(V')^{-1}$$

*where matrices $U$ and $V$ verify $UV' = I - XY$ and matrix $P$ is defined in (3.120), is such that*

1. *for $w = 0$, the ellipsoid $\mathcal{E}(P, \eta)$ is a region of asymptotic stability (RAS) for the saturated system (3.115) with (3.129);*
2. *for any $w \in \mathcal{W}$, the trajectories of the saturated system (3.115) with (3.129) do not leave the ellipsoid $\mathcal{E}(P, \eta)$, $\forall x(0) \in \mathcal{E}(P, \beta)$, with $0 < \beta^{-1} \leq \eta^{-1} - \delta^{-1}$.*

**Remark 3.37** As in Remarks 3.34 and 3.36, the results of Propositions 3.19 and 3.20 can be adapted in order to provide global stabilizing conditions, applicable when the open-loop system is asymptotically stable by considering $Z = L$ and $Z_1 = D_c C$.

**Remark 3.38** By defining a regulated output $z(t)$, for example as defined in (3.78), we can consider an additional control objective regarding the $\mathcal{L}_2$-gain between the admissible disturbance $w$ and this regulated output $z$. In this context, Proposition 3.14 can be adapted by using the same developments as in Propositions 3.19 and 3.20.

**Remark 3.39** All the results developed in Sect. 3.4.3, based on the definition of the measured output $y_p$ in (3.110) can be extended by considering that $y_p$ is defined by

$$y_p = Cx_p + D\,\mathrm{sat}\big(y_c(t)\big) + D_w w(t) \qquad (3.143)$$

This case is considered in Part III.

**Remark 3.40** The results developed in this section can be easily adapted to cope with actuators presenting both magnitude and rate saturations. In this case, the basic idea consists in introducing a set of $m$ input/output saturating integrators in the controller. The saturating outputs of the integrators are injected in the actuators input. On the other hand, the saturating inputs of the integrators are fed by the output of a dynamic stabilizing linear compensator. This nonlinear structure ensures that the control signal delivered to the actuator input is always within its amplitude and rate limits. The simultaneous design of the dynamic stabilizing compensator and some static anti-windup loops is carried out by similar conditions to the ones developed in this section, but considering, in this case, an augmented system composed by the states of the plant and the integrators. This approach has been applied for instance in [209, 282], where classical sector conditions were considered, leading to BMI synthesis results. The generalized sector condition is applied in [147] and [20], leading to LMI synthesis results. In particular, in [147] and [20] the problems of enlarging the basin of attraction, disturbance tolerance or rejection are explicitly addressed. The same idea can be extended to limit the amplitude of any derivative order of the control signal (see for instance [19]).

**Optimization Issues**   Conditions proposed in the two previous paragraphs can be used to find the matrices of the dynamic controller considering several optimization problems, as the maximization of the disturbance tolerance (i.e., maximize the bound on the disturbance), the maximization of the disturbance rejection (i.e., for an a priori given bound on the $\mathcal{L}_2$-norm of the admissible disturbances, minimize the upper bound for the $\mathcal{L}_2$-gain), or still the maximization of the size of the region of stability.

Let us consider this last case (which consists in considering $w = 0$). For the sake of simplicity, in this case, we set $\eta = 1$. A way to indirectly maximize a measure of the size of the set $\mathcal{E}(P, 1)$ is to minimize the trace of $P$ which can be written as

$$\text{trace}(P) = \text{trace}(X) + \text{trace}(\hat{X})$$

From the definition of $P$ and $P^{-1}$, we can deduce that

$$\hat{X} = U'\left(X - Y^{-1}\right)^{-1}U$$

and then minimizing trace($P$) can be done by minimizing

$$\text{trace}(X) + \rho$$

where $\rho$ is a positive scalar such that $\hat{X} \leq \rho I$, or equivalently, according to the expression of $\hat{X}$, such that

$$\begin{bmatrix} \rho I & U' & 0 \\ U & X & I \\ 0 & I & Y \end{bmatrix} \geq 0 \tag{3.144}$$

It is important to note that inequality (3.144) depends on matrix $U$. Actually, matrix $U$ is a degree of freedom related to the controller realization. The inequality (3.144) connects matrices $X$, $Y$ and $U$ and its role is to select, among all the possible solutions, the one enlarging the region of stability with respect to the trace criterion previously defined. In order to ensure that matrix $U$ is nonsingular, the following constraint can be added:

$$U + U' > 0 \tag{3.145}$$

170 3 Stability Analysis and Stabilization—Sector Nonlinearity Model Approach

**Table 3.3** Example 3.145—minimization of $\gamma$ for a given $\delta$

| $\delta$ | 0.321 | 0.33 | 0.34 | 0.35 | 0.4 | 0.5 | 0.7 | 0.8 | 1.0 | 5.0 | 10.0 |
|---|---|---|---|---|---|---|---|---|---|---|---|
| $\gamma$ | 318.0160 | 30.1425 | 14.2622 | 9.0318 | 2.7269 | 0.8816 | 0.2722 | 0.1841 | 0.1004 | 0.0027 | 0.0006 |

*Example 3.4* Once again, the balancing pointer example (see Chap. 1) is considered in presence of an additional exogenous signal which affects its dynamics. System (3.1) is defined by the following data, considering only the first state as available for measurement:

$$A = \begin{bmatrix} 0 & 1 \\ 1 & 0 \end{bmatrix}; \qquad B = \begin{bmatrix} 0 \\ -1 \end{bmatrix}; \qquad B_w = \begin{bmatrix} 1 \\ 1 \end{bmatrix}; \qquad u_0 = 5$$

$$C = \begin{bmatrix} 1 & 0 \end{bmatrix}; \qquad D = 0; \qquad C_z = \begin{bmatrix} 1 & 1 \end{bmatrix}$$

and is affected by an energy bounded exogenous signal for which the set $\mathcal{W}$ is defined according to (3.53) with $R = 1$ and bound $\delta^{-1}$.

The first optimization problem to be considered is to find the maximal energy bounded admissible disturbance $w(t)$, i.e. the minimal value of $\delta$ for which a dynamic output controller (involving the additional anti-windup term $E_c$) which ensures that the trajectories remain bounded may be computed. This optimization procedure is stated as minimization of $\delta$ under conditions (3.139), (3.140) and (3.141). The solution of this optimization problem is $\delta = 0.32$.

A second optimization problem may be to evaluate, for given bounds $\delta$ of the admissible disturbance, dynamic output feedback controllers which minimize the upper bound for the $\mathcal{L}_2$-gain $\gamma$ from $w(t)$ to $z(t)$ (see Remark 3.38). Yet without any considerations about the size of the region of stability or about the dynamics of the controller and closed-loop system, Table 3.3 summarizes the optimal $\gamma$ for various disturbance bounds $\delta$. It may be noticed that the bound of the $\mathcal{L}_2$-gain drastically increases as soon as $\delta$ decreases toward its feasibility limit. Note also that the results match those obtained by using polytopic model II (see Example 2.7, state-feedback stabilization).

Consider now the problem of computing a controller which maximizes the region of stability (minimization of trace($P$) such as described above), in the case without disturbance ($w = 0$). In order to avoid ill conditioning problems and to ensure some performance of the linear behavior, an additional constraint is considered to place the eigenvalues of the closed-loop system in a vertical strip in the complex plane [68] delimited by $-0.5$ and $-100$. The stabilizing controller solution to the optimization problem is

$$A_c^1 = \begin{bmatrix} -40.1494 & 88.0838 \\ -117.5713 & -122.1316 \end{bmatrix}; \qquad B_c^1 = \begin{bmatrix} -521.6256 \\ -349.3032 \end{bmatrix}$$

$$C_c^1 = \begin{bmatrix} -13.4748 & -34.7899 \end{bmatrix}; \qquad D_c^1 = \begin{bmatrix} 25.9528 \end{bmatrix}; \qquad E_c^1 = \begin{bmatrix} -5.5967 \\ 0.5379 \end{bmatrix}$$

The cut of the associated ellipsoidal region of stability in the plane defined by the state of the system $x_p$ is plotted in solid line ($w = 0$, synthesis) in Fig. 3.7. It may

## 3.5 Uncertain Systems

**Fig. 3.7** Example 3.145—cut of the regions of stability in the system space ($x_p$). Cases without disturbance ($w = 0$) and with disturbance ($\delta = 1$), synthesis and analysis (with and without the anti-windup term)

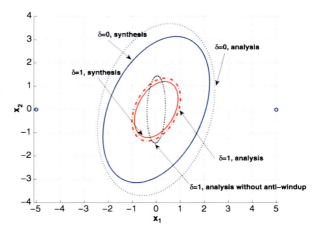

be verified that the optimization criterion makes sense since the analysis of this controller provides a similar ellipsoidal set (dashed-line ellipsoid ($w = 0$, analysis)). Note that the analysis of this controller but without the anti-windup scheme gives the same analysis ellipsoid, which means that, in that case, the anti-windup term mainly appears as an extra degree of freedom in the numerical conditions. Furthermore, its presence allows one to overcome the bilinearities in the conditions.

Further, in presence of an additive disturbance bounded by $\delta = 1$, the optimization problem may be stated as follows: $\min \gamma + \mu$, where $\frac{1}{\sqrt{\mu}}$ represents the scaling factor of a shape set of initial states $\mathcal{X}_0$. The solution of this optimization problem allows to exhibit the controller:

$$A_c^2 = \begin{bmatrix} -61.3435 & 24.3063 \\ -20.3029 & -77.4251 \end{bmatrix}; \qquad B_c^2 = \begin{bmatrix} -797.3229 \\ 276.3576 \end{bmatrix}$$

$$C_c^2 = \begin{bmatrix} 3.9024 & -8.9195 \end{bmatrix}; \qquad D_c^2 = \begin{bmatrix} 98.8622 \end{bmatrix}; \qquad E_c^2 = \begin{bmatrix} -6.9409 \\ 4.1551 \end{bmatrix}$$

for which $\gamma = 1.4711$, and the cut in the system state space of the ellipsoidal region of asymptotic stability is plotted in solid red line ($\delta = 1$, synthesis) in Fig. 3.7. The analysis of this controller allows to provide a slightly larger region of stability ($\delta = 1$, analysis), but to the detriment of the $\mathcal{L}_2$-gain which is degraded to $\gamma = 1.8938$. Note also the influence of the anti-windup term in that case, since the analysis of the controller without this term results in a significantly smaller region of stability ($\delta = 1$, analysis without anti-windup) and a larger bound $\gamma = 3.6623$.

## 3.5 Uncertain Systems

In the previous sections of the chapter, results dealing with saturating inputs have been exploited for systems whose matrices are supposed to be perfectly known, that is, without being affected by parametric uncertainties or by time-varying parameters.

172        3  Stability Analysis and Stabilization—Sector Nonlinearity Model Approach

Thus throughout this section, we consider the uncertain case, that is, system (3.5) with uncertain parameter $F \in \mathcal{F}$. More precisely, we consider that matrices $A$ and $B$ in the closed-loop system are replaced by $A(F)$ and $B(F)$. To ease the developments, only the case $w = 0$, and therefore system (3.5), is considered. Depending on the description of the set $\mathcal{F}$, we are interested in considering two different types of uncertainty (see also Appendix B), namely:

- Norm-bounded uncertainty, where

$$A(F) = A_0 + DF(t)E_1 \quad \text{and} \quad B(F) = B_0 + DF(t)E_2$$

$$\text{with } F(t)'F(t) \le I_r, \ \forall t \tag{3.146}$$

for given constant real matrices $D$, $E_1$ and $E_2$ of appropriate dimensions.

- Polytopic uncertainty where matrices $A(F)$ and $B(F)$ belong to matrix polytopes

$$A(F) \in \mathcal{D}_A = \left\{ A(F) \in \mathfrak{R}^{n \times n}; \ A(F) = \sum_{i=1}^{N_A} \lambda_i A_i, \ \sum_{i=1}^{N_A} \lambda_i = 1, \ \lambda_i \ge 0 \right\}$$

$$B(F) \in \mathcal{D}_B = \left\{ B(F) \in \mathfrak{R}^{n \times m}; \ B(F) = \sum_{k=1}^{N_B} \mu_k B_k, \ \sum_{k=1}^{N_B} \mu_k = 1, \ \mu_k \ge 0 \right\}$$

$$\tag{3.147}$$

Considering the uncertainty representations above, analogous results to those of previous Sects. 3.2, 3.3 and 3.4 can be derived. In the sequel, we present the results of stability analysis by considering the quadratic Lyapunov function defined in (3.7), applied to the system

$$\dot{x}(t) = A(F)x(t) + B(F)\text{sat}\big(Kx(t)\big) = \big(A(F) + B(F)K\big)x(t) + B(F)\phi\big(Kx(t)\big) \tag{3.148}$$

with uncertainty described by (3.146) or (3.147).

**Norm-Bounded Uncertainty**    The following proposition allows us to solve the stability analysis problem for system (3.148) with norm-bounded uncertainty described by (3.146).

**Proposition 3.21** *If there exist a symmetric positive definite matrix* $W \in \mathfrak{R}^{n \times n}$, *a positive diagonal matrix* $S \in \mathfrak{R}^{m \times m}$, *a matrix* $Z \in \mathfrak{R}^{m \times n}$ *and a positive scalar* $\varepsilon$ *satisfying*:

$$\begin{bmatrix} W(A_0 + B_0K)' + (A_0 + B_0K)W + \varepsilon DD' & B_0S - Z' & W(E_1 + E_2K)' \\ SB_0' - Z & -2S & SE_2' \\ (E_1 + E_2K)W & E_2S & -\varepsilon I \end{bmatrix} < 0 \tag{3.149}$$

$$\begin{bmatrix} W & WK_{(i)}' - Z_{(i)}' \\ K_{(i)}W - Z_{(i)} & u_{0(i)}^2 \end{bmatrix} \ge 0, \quad i = 1, \dots, m \tag{3.150}$$

*then the ellipsoid* $\mathcal{E}(P, 1)$, *with* $P = W^{-1}$, *is a region of asymptotic stability (RAS) for the saturated system* (3.148).

## 3.5 Uncertain Systems

*Proof* The proof follows the same lines as that one of Proposition 3.1. Define $\dot{V}(x) - 2\phi(Kx)'T(\phi(Kx) + Gx) = \mathcal{L}(x)$. Hence, one gets along the trajectories of the saturated system (3.148)

$$
\begin{aligned}
\mathcal{L}(x) = {} & x'\big((A_0 + B_0 K)'P + P(A_0 + B_0 K)\big)x + 2x'P B_0\phi(Kx) \\
& - 2\phi(Kx)'T\big(\phi(Kx) + Gx\big) + 2x'PDF\big((E_1 + E_2 K)x + E_2\phi(Kx)\big)
\end{aligned}
$$

By using the property on norm-bounded uncertainty (as described in Appendix B), it follows:

$$
\begin{aligned}
\mathcal{L}(x) \leq {} & x'\big((A_0 + B_0 K)'P + P(A_0 + B_0 K)\big)x + 2x'P B_0\phi(Kx) \\
& - 2\phi(Kx)'T\big(\phi(Kx) + Gx\big) + \varepsilon x'PDD'Px \\
& + \varepsilon^{-1}\big((E_1 + E_2 K)x + E_2\phi(Kx)\big)'\big((E_1 + E_2 K)x + E_2\phi(Kx)\big)
\end{aligned}
$$

The right-hand term of this inequality reads

$$
\begin{bmatrix} x' & \phi(Kx)' \end{bmatrix} \left( M_0 + \varepsilon^{-1} \begin{bmatrix} (E_1 + E_2 K)' \\ E_2' \end{bmatrix} \begin{bmatrix} E_1 + E_2 K & E_2 \end{bmatrix} \right) \begin{bmatrix} x \\ \phi(Kx) \end{bmatrix}
$$

with

$$
M_0 = \begin{bmatrix} (A_0 + B_0 K)'P + P(A_0 + B_0 K) + \varepsilon PDD'P & P B_0 - G'T \\ B_0'P - TG & -2T \end{bmatrix}
$$

By using the Schur's complement the matrix

$$
M = M_0 + \varepsilon^{-1} \begin{bmatrix} (E_1 + E_2 K)' \\ E_2' \end{bmatrix} \begin{bmatrix} E_1 + E_2 K & E_2 \end{bmatrix}
$$

can be written as

$$
M = \begin{bmatrix} (A_0 + B_0 K)'P + P(A_0 + B_0 K) + \varepsilon PDD'P & P B_0 - G'T & (E_1 + E_2 K)' \\ B_0'P - TG & -2T & E_2' \\ (E_1 + E_2 K) & E_2 & -\varepsilon I \end{bmatrix}
$$

By left- and right-multiplying $M$ by $\mathrm{Diag}(P^{-1}, T^{-1}, I)$, by denoting $W = P^{-1}$ and $S = T^{-1}$ and by using the change of variables $GW = Z$, one obtains condition (3.149). Hence, if relation (3.149) is satisfied then one gets $\dot{V}(x) \leq \mathcal{L}(x) < 0$, as soon as $\mathcal{E}(P, 1)$ is included in $S(|K - G|, u_0)$, i.e. as soon as (3.150) is satisfied. $\qquad\square$

**Polytopic Uncertainty** By the same way, in the context of polytopic uncertainty as described in (3.147), one can state the following proposition.

**Proposition 3.22** *If there exist a symmetric positive definite matrix $W \in \Re^{n \times n}$, a positive diagonal matrix $S \in \Re^{m \times m}$ and a matrix $Z \in \Re^{m \times n}$:*

$$
\begin{bmatrix} W(A_i + B_k K)' + (A_i + B_k K)W & B_k S - Z' \\ S B_k' - Z & -2S \end{bmatrix} < 0
$$
$$
i = 1, \ldots, N_A, \ k = 1, \ldots, N_B \tag{3.151}
$$

$$
\begin{bmatrix} W & W K_{(j)}' - Z_{(j)}' \\ K_{(j)} W - Z_{(j)} & u_{0(j)}^2 \end{bmatrix} \geq 0, \quad j = 1, \ldots, m \tag{3.152}
$$

174            3 Stability Analysis and Stabilization—Sector Nonlinearity Model Approach

*then the ellipsoid $\mathcal{E}(P, 1)$, with $P = W^{-1}$, is a region of asymptotic stability (RAS) for the saturated system* (3.148).

*Proof* To prove that $\dot{V}(x) < 0$ with respect to the trajectories of the saturated system (3.148) with polytopic uncertainty as described in (3.147), it suffices by convexity to prove that $\dot{V}(x) < 0$ for all the vertices $A_i$, $i = 1, \ldots, N_A$, and $B_k$, $k = 1, \ldots, N_B$ (see also Appendix B). Keeping this in mind, the proof follows exactly the same steps as that one of Proposition 3.1.        $\square$

All the points previously addressed in this chapter (global stability analysis, external stability analysis, stabilization and external stabilization problems) can be carried out by considering both types of uncertainties on matrices $A$ and $B$.

**Gain Scheduling Approach**     Moreover, when the system is affected by time-varying parameters and these parameters are available (measured or estimated) in real-time for feedback, it is known that gain-scheduled strategies can provide better performance than robust controllers (i.e. parameter-independent controllers): see, for instance, [237, 306, 313] and references herein. Note that LMI-based conditions have been used to design gain-scheduled controllers: see, for example, [83, 263]. In some cases, the bounds on the rate of variation of the parameters are taken into account, but when the parameters are allowed to vary arbitrarily fast, and also when no prior information on these bounds is given, quadratic stability (i.e. a Lyapunov function with a fixed matrix) [45] is maybe the only known methodology that allows to transform the synthesis of gain-scheduled controllers into convex LMI optimization problems. Such LMI-based conditions, as several other conditions to design gain-scheduled controllers, rely on Lyapunov functions which ensure the stability of the closed-loop system for the entire domain of the parameters but, in general, without considering the limits of actuators.

Let us consider the system defined by

$$\dot{x}(t) = A(\alpha(t))x(t) + B(\alpha(t))\,\text{sat}(K(\alpha(t))x(t))$$
$$= (A(\alpha(t)) + B(\alpha(t))K(\alpha(t)))x(t) + B(\alpha(t))\phi(K(\alpha(t))x(t)) \quad (3.153)$$

where $\alpha(t) \in \Re^N$ is the vector of time-varying parameters, belonging to the unit simplex $\mathcal{S}$ for almost every $t \geq 0$. The set $\mathcal{S}$ is defined by

$$\mathcal{S} = \left\{ \alpha(t) \in \Re^N; \ \sum_{j=1}^{N} \alpha_j(t) = 1, \ \alpha_j(t) \geq 0, \ j = 1, \ldots, N \right\} \quad (3.154)$$

$\alpha(t)$ is supposed to be available in real-time and is used in the saturated state feedback control law $u(t) = \text{sat}(K(\alpha(t))x(t))$, $K(\alpha(t)) \in \Re^{m \times n}$.

The time-derivative of $\alpha(t)$, $\dot{\alpha}(t)$, is supposed unknown and is allowed to take even arbitrarily large values. Matrices $A(\alpha(t)) \in \Re^{n \times n}$ and $B(\alpha(t)) \in \Re^{n \times m}$ belong to the polytope

$$\mathcal{P} = \left\{ (A, B)(\alpha(t)); \ (A, B)(\alpha(t)) = \sum_{j=1}^{N} \alpha_j(t)(A, B)_j, \ \alpha(t) \in \mathcal{S} \right\} \quad (3.155)$$

3.5 Uncertain Systems                                                                     175

A sufficient condition to design a locally stabilizing state feedback gain, based on parameter-dependent matrix variables, is presented in the next proposition.

**Proposition 3.23** *If there exist a symmetric positive definite matrix* $W \in \Re^{n \times n}$, *parameter-dependent matrices* $Z(\alpha) \in \Re^{m \times n}$, $Y(\alpha) \in \Re^{m \times n}$, *a diagonal parameter-dependent positive matrix* $S(\alpha) \in \Re^{m \times m}$ *and a real positive scalar* $\eta$ *such that,* $\forall \alpha \in \mathcal{S}$,

$$\begin{bmatrix} \mathcal{Y}(\alpha) & B(\alpha)S(\alpha) - Z(\alpha)' \\ S(\alpha)B(\alpha)' - Z(\alpha) & -2S(\alpha) \end{bmatrix} < 0 \qquad (3.156)$$

$$\begin{bmatrix} W & Y_{(i)}(\alpha)' - Z_{(i)}(\alpha)' \\ Y_{(i)}(\alpha) - Z_{(i)}(\alpha) & \eta u_{0(i)}^2 \end{bmatrix} \geq 0, \quad i = 1, \dots, m \quad (3.157)$$

*with* $\mathcal{Y}(\alpha)$ *given by*

$$\mathcal{Y}(\alpha) = A(\alpha)W + WA(\alpha)' + B(\alpha)Y(\alpha) + Y(\alpha)'B(\alpha)'$$

*then the parameter-dependent control gain* $K(\alpha) = Y(\alpha)W^{-1}$ *ensures the asymptotic stability of the closed-loop system,* $\forall \alpha \in \mathcal{S}$, *in the ellipsoidal region* $\mathcal{E}(P, \eta) = \{x \in \Re^n; \ x'Px \leq \eta^{-1}\}$, *with* $P = W^{-1}$.

The proof of Proposition 3.23 is based on the use of an extension of Lemma 1.6 with parameter-dependent matrices, already presented in [264] for $K(\alpha) = K$ (constant) and stated below.

**Lemma 3.3** *For all* $x \in S(|K(\alpha) - G(\alpha)|, u_0)$ *defined by*

$$S(|K(\alpha) - G(\alpha)|, u_0) = \{x \in \Re^n; \ -u_0 \preceq (K(\alpha) - G(\alpha))x \preceq u_0\} \qquad (3.158)$$

*and for all* $\alpha \in \mathcal{S}$, *the following condition is satisfied:*

$$\phi(K(\alpha)x)'T(\alpha)(\phi(K(\alpha)x) + G(\alpha)x) \leq 0 \qquad (3.159)$$

*for any positive definite diagonal matrix* $T(\alpha) \in \Re^{m \times m}$.

*Proof* The proof of Proposition 3.23 follows the same lines as that one of Theorem 2 in [265]. Hence, the proof can be developed by using a quadratic Lyapunov function $V(x) = x'W^{-1}x$, $W = W' > 0$, by taking into account the variable transformations $Y(\alpha) = K(\alpha)W$, $Z(\alpha) = G(\alpha)W$ and by considering the extension of Lemma 1.6 with parameter-dependent matrices stated in Lemma 3.3.  $\square$

*Remark 3.41* The conditions in Proposition 3.23 are parameter-dependent LMIs with time-varying parameters $\alpha(t)$ belonging to $\mathcal{S}$ for almost every $t \geq 0$. As a consequence, if there exist a matrix $W$ and parameter-dependent matrices $Y(\alpha)$, $Z(\alpha)$ and $S(\alpha)$ verifying conditions of Proposition 3.23 then, without loss of generality, there also exist a matrix $W$ and homogeneous polynomial parameter-dependent (HPPD) matrices $Y_g(\alpha)$, $Z_g(\alpha)$ and $S_g(\alpha)$, of arbitrary degree $g$, verifying conditions of Proposition 3.23 [40].

176 3 Stability Analysis and Stabilization—Sector Nonlinearity Model Approach

Let us first give some definitions concerning HPPD matrices.

**Definition 3.1** A HPPD matrix $Z_g(\alpha)$ of degree $g$ can be generally written as

$$Z_g(\alpha) = \sum_{k \in \mathcal{K}(g)} \alpha_1^{k_1} \alpha_2^{k_2} \cdots \alpha_N^{k_N} Z_k, \quad k = k_1 k_2 \cdots k_N \tag{3.160}$$

where $\alpha_1^{k_1} \alpha_2^{k_2} \cdots \alpha_N^{k_N}$, $\alpha \in \mathcal{S}$, $k_i \in \mathcal{N}$, $i = 1, \ldots, N$ are the monomials, and $Z_k \in \mathfrak{R}^{n \times n}$, $\forall k \in \mathcal{K}(g)$ are matrix-valued coefficients to be determined. By definition, $\mathcal{K}(g)$ is the set of $N$-tuples obtained as all possible combinations of nonnegative integers $k_i$, $i = 1, \ldots, N$, such that $k_1 + k_2 + \cdots + k_N = g$. The number of elements in $\mathcal{K}(g)$ is given by

$$J(g) = \frac{(N + g - 1)!}{g!(N - 1)!} \tag{3.161}$$

where $N$ is the number of vertices in the polytope $\mathcal{P}$ defined in (3.155). For $N$-tuples $k, k'$ one writes $k \succeq k'$ if $k_i \geq k'_i$, $i = 1, \ldots, N$. Usual operations of summation $k + k'$ and subtraction $k - k'$ (whenever $k \succeq k'$) are defined componentwise. Consider also the following definitions for the $N$-tuple $e_i$ and the coefficients $\pi(k)$:

$$e_i = 0 \cdots 0 \underbrace{1}_{i\text{th}} 0 \cdots 0, \qquad \pi(k) = (k_1!)(k_2!) \cdots (k_N!)$$

The following proposition generalizes Proposition 3.23 by using Definition 3.1.

**Proposition 3.24** *There exist a symmetric positive definite matrix* $W \in \mathfrak{R}^{n \times n}$, *HPPD matrices* $Z_g(\alpha) \in \mathfrak{R}^{m \times n}$, $Y_g(\alpha) \in \mathfrak{R}^{m \times n}$, *a diagonal positive definite HPPD matrix* $S_g(\alpha) \in \mathfrak{R}^{m \times m}$ *of arbitrary degree* $g$ *and a real positive scalar* $\eta$ *satisfying* (3.156) *and* (3.157) *if and only if there exist a symmetric positive definite matrix* $W \in \mathfrak{R}^{n \times n}$, *matrices* $Y_k \in \mathfrak{R}^{m \times n}$, $Z_k \in \mathfrak{R}^{m \times n}$, *diagonal positive matrices* $S_k \in \mathfrak{R}^{m \times m}$, $k \in \mathcal{K}(g)$, *a real positive scalar* $\eta$ *and a sufficiently large* $d \in \mathcal{N}$ *satisfying*

$$\sum_{\substack{k' \in \mathcal{K}(d) \\ k \succeq k'}} \sum_{\substack{i \in \{1, \ldots, N\} \\ k_i > k'_i}} \frac{d!}{\pi(k')} \left[ \begin{array}{cc} \frac{g!(k_i - k'_i)}{\pi(k-k')}(A_i W + W A'_i) + B_i Y_{k-k'-e_i} + Y'_{k-k'-e_i} B'_i \\ \star \end{array} \right.$$

$$\left. \begin{array}{c} B_i S_{k-k'-e_i} - Z'_{k-k'-e_i} \\ -2S_{k-k'-e_i} \end{array} \right] < 0$$

$$\forall k \in \mathcal{K}(g + d + 1) \tag{3.162}$$

$$\sum_{\substack{k' \in \mathcal{K}(d) \\ k \succeq k'}} \frac{d!}{\pi(k')} \left[ \begin{array}{cc} \frac{g!}{\pi(k-k')} W & (Y_{k-k'})'_{(\ell)} - (Z_{k-k'})'_{(\ell)} \\ \star & \frac{g!}{\pi(k-k')} \eta u^2_{0(\ell)} \end{array} \right] \geq 0,$$

$$\ell = 1, \ldots, m, \ \forall k \in \mathcal{K}(g + d) \tag{3.163}$$

*In this case,* $K_g(\alpha)$, *given by* $K_g(\alpha) = Y_g(\alpha) W^{-1}$, *is a HPPD control gain that ensures the asymptotic stability of the closed-loop system in* $\mathcal{E}(P, \eta) = \{x \in \mathfrak{R}^n; x'Px \leq \eta^{-1}\}$ *with* $P = W^{-1}$.

3.5 Uncertain Systems 177

*Proof* Notice that the left-hand term in (3.156) can be rewritten as

$$T(\alpha) = \sum_{k \in \mathcal{K}(g+d+1)} \alpha_1^{k_1} \alpha_2^{k_2} \cdots \alpha_N^{k_N} \mathcal{T}_k$$

where $\mathcal{T}_k$ corresponds to the matrix in (3.162). It is clear that if $\mathcal{T}_k < 0$, then $T(\alpha) < 0$, $\forall \alpha \in S$. This also ensures that $S_g(\alpha) > 0$, $\forall \alpha \in S$. Conversely, based on the extension of Pólya's theorem in [275, Theorem 2], one has that if $T(\alpha) < 0$, $\forall \alpha \in S$, then there exists a sufficiently large $d \in \mathcal{N}$ such that all the matrix-valued coefficients $\mathcal{T}_k$ in (3.162) are negative definite. Similar steps can be used to show that (3.163) is sufficient and necessary (for sufficiently large $d \in \mathcal{N}$) to ensure that (3.157) is feasible $\forall \alpha \in S$, for a given degree $g$. The control gain is also a HPPD matrix, given by $K_g(\alpha) = Y_g(\alpha)W^{-1}$. □

*Example 3.5* Let us illustrate the previous Propositions 3.23 and 3.24. Consider system (3.153) with four vertices randomly generated for matrices $A$ and $B$ given by

$$A_1 = \begin{bmatrix} 0.67 & 0.26 \\ 0.09 & 0.14 \end{bmatrix}; \qquad A_2 = \begin{bmatrix} 0.28 & 0.71 \\ 0.78 & 0.60 \end{bmatrix}$$

$$A_3 = \begin{bmatrix} 0.04 & 0.23 \\ 0.85 & 0.50 \end{bmatrix}; \qquad A_4 = \begin{bmatrix} 0.03 & 0.40 \\ 0.65 & 0.77 \end{bmatrix}$$

$$B_1 = \begin{bmatrix} 0.38 & 0.41 \\ 0.89 & 0.38 \end{bmatrix}; \qquad B_2 = \begin{bmatrix} 0.09 & 0.92 \\ 0.98 & 0.46 \end{bmatrix}$$

$$B_3 = \begin{bmatrix} 0.56 & 0.99 \\ 0.49 & 0.03 \end{bmatrix}; \qquad B_4 = \begin{bmatrix} 0.26 & 0.12 \\ 0.33 & 0.75 \end{bmatrix}$$

This system has two actuators, with magnitude saturation levels given by $u_{0(i)} = 10$, $i = 1, 2$. The aim here is to compute stabilizing state feedback scheduled controllers and estimates of the region of attraction for this system. The associate conditions (3.162) and (3.163) are not feasible for $g = 0$ (i.e. for robust control gains). On the other hand, the conditions are feasible for $g \geq 1$, providing HPPD stabilizing controllers.

Consider the convex optimization problem:

$$\min \sigma$$
$$\text{subject to} \quad \begin{bmatrix} \sigma I & I \\ I & W \end{bmatrix} \geq 0, \ (3.162), \ (3.163) \tag{3.164}$$

with $\eta = 1$, in order to maximize the estimates of the region of attraction of the system. The optimal value $\sigma^\star$ of the optimization problem (3.164) is given in Table 3.4 for different values of $d$ and $g$.

It is important to note the improvement obtained (measured by the reduction on $\sigma^*$) as one increases $d$ and $g$. For a fixed $g$, the increase on $d$ signifies that the LMI constraints (3.162) and (3.163) are relaxed, without increasing the number of scalar variables. The increase on $g$ implies that more variables and LMIs are used, allowing to reduce the conservatism of the solutions. The results in Table 3.4

178   3 Stability Analysis and Stabilization—Sector Nonlinearity Model Approach

**Table 3.4** Example 3.5—estimates ($\sigma^*$) of the domain of attraction

| $g$ | $d = 0$ | $d = 1$ | $d = 2$ |
|---|---|---|---|
| 0 | unfeasible | unfeasible | unfeasible |
| 1 | 0.0774 | 0.0755 | 0.0694 |
| 2 | 0.0640 | 0.0633 | 0.0595 |
| 3 | 0.0565 | 0.0561 | 0.0537 |
| 4 | 0.0526 | 0.0524 | 0.0512 |
| 5 | 0.0507 | 0.0506 | 0.0500 |

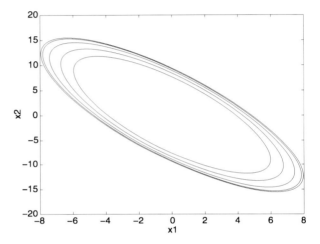

**Fig. 3.8** Example 3.5—estimates of the domain of attraction for $d = 0$, from $g = 1$ (*smallest ellipsoid*) to $g = 5$ (*largest ellipsoid*)

illustrate the efficiency of the proposed conditions to design HPPD controllers of arbitrary degree, which can progressively enlarge the region of stable operation. This is also depicted on Fig. 3.8.

## 3.6 Discrete-Time Case

In this section, we show that the discrete-time case can be carried out by similar arguments as in the continuous-time case. To simplify let us consider a discrete-time system without additive disturbance described by

$$x(k+1) = Ax(k) + B\,\text{sat}\big(Kx(k)\big) \qquad (3.165)$$

Thus, by defining the dead-zone nonlinearity $\phi(Kx(k)) = \text{sat}(Kx(k)) - Kx(k)$ the closed-loop system equivalently reads

$$x(k+1) = (A + BK)x(k) + B\phi\big(Kx(k)\big) \qquad (3.166)$$

In this section, we just present the basic ideas by addressing the stability analysis problem but all the problems addressed in the previous sections can be considered

## 3.6 Discrete-Time Case

in a straightforward manner. Hence, consider a quadratic Lyapunov function defined as

$$V\big(x(k)\big) = x(k)'Px(k), \quad P = P' > 0$$

In the case of discrete-time system (3.165), the following proposition can be stated.

**Proposition 3.25** *If there exist a symmetric positive definite matrix $W \in \mathfrak{R}^{n \times n}$, a positive diagonal matrix $S \in \mathfrak{R}^{m \times m}$ and a matrix $Z \in \mathfrak{R}^{m \times n}$ satisfying:*

$$\begin{bmatrix} -W & -Z' & W(A+BK)' \\ -Z & -2S & SB' \\ (A+BK)W & BS & -W \end{bmatrix} < 0 \qquad (3.167)$$

$$\begin{bmatrix} W & WK'_{(i)} - Z'_{(i)} \\ K_{(i)}W - Z_{(i)} & u_{0(i)}^2 \end{bmatrix} \geq 0, \quad i = 1, \ldots, m \qquad (3.168)$$

*then the ellipsoid $\mathcal{E}(P, 1)$, with $P = W^{-1}$, is a region of asymptotic stability (RAS) for the saturated system (3.165).*

*Proof* The proof follows the same way as that one of Proposition 3.1. Let us consider the quadratic Lyapunov function $V(x(k)) = x(k)'Px(k)$, $P = P' > 0$. Indeed, one wants to verify along the trajectories of the saturated system (3.165)):

$$\Delta V(x) \leq \Delta V(x) - 2\phi(Kx)'T\big(\phi(Kx) + Gx\big)$$
$$= V\big(x(k+1)\big) - V\big(x(k)\big) - 2\phi(Kx)'T\big(\phi(Kx) + Gx\big) < 0$$

provided that the ellipsoid $\mathcal{E}(P, 1)$ is included in $S(|K - G|, u_0)$, i.e. provided that (3.168) is satisfied. Defining $\mathcal{L}(x) = \Delta V(x) - 2\phi(Kx)'T(\phi(Kx) + Gx)$, it follows that

$$\mathcal{L}(x) = \begin{bmatrix} x' & \phi(Kx)' \end{bmatrix} M \begin{bmatrix} x \\ \phi(Kx) \end{bmatrix}$$

with

$$M = \begin{bmatrix} (A+BK)'P(A+BK) - P & (A+BK)'PB - G'T \\ B'P(A+BK) - TG & -2T + B'PB \end{bmatrix}$$

By using Schur's complement, by left- and right-multiplying the resulting matrix $M$ by $\mathrm{Diag}(P^{-1}, T^{-1}, I)$, denoting $W = P^{-1}$ and $S = T^{-1}$, and by using the change of variables $GW = Z$, one obtains condition (3.167). Hence, if conditions (3.167) and (3.168) are satisfied then $\Delta V(x(k)) < 0$. $\qquad\square$

As in the continuous-time system case (see Proposition 3.2), the global stabilization of the closed-loop system can be examined via the following proposition provided that all the eigenvalues of matrix $A$ are inside the unit circle.

180         3 Stability Analysis and Stabilization—Sector Nonlinearity Model Approach

**Proposition 3.26** *If there exist a symmetric positive definite matrix* $W \in \mathfrak{R}^{n \times n}$ *and a positive diagonal matrix* $S \in \mathfrak{R}^{m \times m}$ *satisfying:*

$$\begin{bmatrix} -W & -WK' & W(A+BK)' \\ -KW & -2S & SB' \\ (A+BK)W & BS & -W \end{bmatrix} < 0 \qquad (3.169)$$

*then the origin is globally asymptotically stable for the saturated system* (3.165).

*Proof* It directly follows from Proposition 3.25 by setting $G = K$. $\qquad\qquad\square$

## 3.7 Extensions

Several extensions can be considered: the presence of nested saturations including, for example, the case of magnitude and rate saturations, the presence of saturation in the output, the presence of nested nonlinearities or the presence of delays in the system. Furthermore, the nonlinear and/or hybrid natures of the system are also very important features to tackle.

### 3.7.1 Nested Saturations

An important class of systems to be studied consists in systems presenting nested saturations. In particular, such a structure frequently appears in aerospace control systems (e.g., launcher and aircraft control) where actuators are generally limited both in magnitude and rate (dynamics) (see [233] and the reference therein). Furthermore, the presence of both sensor and actuator amplitude limitations can also lead to a closed-loop system presenting nested saturations. This is for example the case in linear systems controlled by dynamic output feedback controllers in the presence of such saturating sensors and actuators. On the other hand, analysis and design methodologies for systems presenting nested saturations can be useful to address stability issues of more general classes of nonlinear systems. For instance, the use of nested saturations becomes very interesting when one uses forwarding techniques for cascade systems with linear part [319, 365].

Linear systems subject to nested saturations can be generically described by

$$\dot{x} = A_p x + B_p \operatorname{sat}_p \left( A_{p-1} x + B_{p-1} \operatorname{sat}_{p-1} \left( A_{p-2} x \right. \right.$$
$$\left. \left. + \cdots + \left( A_1 x + B_1 \operatorname{sat}_1(Cx) \right) \cdots \right) \right) \qquad (3.170)$$

where $x \in \mathfrak{R}^n$ is the state of the system. For all $j = 1, \ldots, p$, $A_j$, $B_j$ and $C$ are matrices of appropriate dimensions (possibly having different dimensions depending on the index $j$). Furthermore, $\operatorname{sat}_j$ is a componentwise saturation map $\mathfrak{R}^{m_j} \to \mathfrak{R}^{m_j}$ defined by

$$\left( \operatorname{sat}_j(v) \right)_{(i)} = \operatorname{sat}_j(v_{(i)}) = \operatorname{sign}(v_{(i)}) \min \left( u_{j(i)}, |v_{(i)}| \right), \quad \forall i = 1, \ldots, m_j$$

3.7 Extensions 181

where $u_{j(i)}$, denotes the bound on the $i$th component of the $j$th saturation function. In contrast with [16], accordingly to the different $m_j$, more generic matrices $B_j$, $A_j$ and $C$ can be considered. In particular, note also that the matrices $B_j$ do not need to be diagonal.

Several results dealing with system (3.170) have been developed to address the stability analysis in a regional (local) or global context, the design of stabilizing controllers or still the external stability: see, for example, [16, 359, 361, 417] and references therein. The results in [417] are based on a polytopic representation of the saturation generalizing the approach in [1], whereas Lemma 1.6 is used in [359].

On the other hand, in practice both sensors and actuators present amplitude constraints (as is the case, for instance, in aerospace applications and vibration control). Then techniques considering the stabilization taking into account simultaneously actuator and sensor saturation are of major interest. However, if few results are available for systems with sensor saturation (see, for example, [57, 206, 222, 227, 241, 347]), still less results concern the case of systems with both actuator and sensor limitations. In [95], adaptive integral control design for linear systems with actuator and sensor nonlinearities is addressed by considering the asymptotic stability of the open-loop. In [127], different methods of analysis and synthesis are presented, in particular relative to PID controllers. Nevertheless, the methods proposed in [95] and [127] do not offer a systematic way to deal with the control design problem. In [120], results concerning the problem of controlling linear systems with both sensors and actuators subject to saturation, through dynamic output feedback, have been presented using similar tools to those described in Sect. 3.4.3.

### *3.7.2 Nested Nonlinearities*

Beyond saturation nonlinearities, non-smooth nonlinearities, such as hysteresis, backlash or dead-zone may also occur in real process control, due to physical, technological or safety constraints, imperfections, or even inherent characteristics of considered controlled systems. Hence, a wide variety of practical systems and devices, like, for example, servo systems, flexible systems, may be represented by the interconnection of a non-smooth nonlinear operator with a plant and an actuator/controller device. In this context, it is important to be able to take into account these types of nonlinearities for analysis or synthesis purposes. Furthermore, these non-smooth nonlinearities can appear jointly with saturation ones. The occurrence of nested nonlinearities of different kind can result from the nature of the system considered or from the structure of the control law investigated.

Let us give some examples of such nested nonlinearities. Hence, we can mention two cases of nested nonlinearities due to the nature of the system: saturation plus backlash in [364] or saturation plus quantization in [350, 351]. In [364], the problem of stability analysis for a certain class of nonlinear systems resulting from nested backlash and saturation operators is carried out. Due to the presence of the backlash in the loop of the control system, global stability should be understood as convergence to a set of equilibrium points (maybe not reduced to the origin). As mentioned previously, the nested nonlinearities also appear due to the control law chosen. This is the case in [63], which considers a linear system affected by a state-dependent nonlinearity belonging to a general class of sectors and subject to amplitude saturation in the input. The two nested nonlinearities are a saturation nonlinearity and a cone bounded nonlinearity, the control law investigated for stabilization purposes depending both on the state and on the cone bounded nonlinearity.

### 3.7.3 Nonlinear and/or Hybrid Systems

An interesting class of nonlinear systems which has attracted increasing attention in the literature is that one of quadratic systems in which the system state equation features a quadratic-type nonlinearity. One of the interesting characteristics of such systems is that the region of attraction of an equilibrium point is the whole state space for just some particular cases. For the large majority of cases such systems have more than one equilibrium point and, furthermore, the geometry of the region of attraction for these equilibria is quite complex. It is typically of interest, therefore, to estimate their region of attraction [6, 65] and to investigate their behavior in presence of input saturation [381]. Tractable conditions to characterize an estimation of the basin of attraction of the origin through an ellipsoidal region have been provided. By the same way, constructive conditions to determine the largest invariant ellipsoid within the region of attraction of the origin through the design of a stabilizing state-feedback control law has been developed. Note that in the general case it is difficult to predict which nonlinearity (quadratic terms or saturation) is more critical for the optimization of the estimation of the region of attraction. The generalization of such results by considering polynomial nonlinear systems can be handled by using more complex Lyapunov functions (as for example those of degree higher than 2) [156, 157, 200, 380]. The case of rational systems and the use of rationale Lyapunov functions is considered in [75] by using an extended version of Lemma 1.6.

Another interesting class of systems is that one of hybrid systems in a large sense, i.e. including switched systems, impulsive systems, reset systems, ... Hybrid systems are dynamical systems exhibiting heterogeneous interactions between logic and differential or difference dynamics, or still mixing both continuous-time and discrete-time dynamics. The importance of hybrid systems in the field of stability analysis and control design has been growing in the last decades, due in particular to the increasing application of digital devices for real control systems. For example, communication networks, automotive industry, chemical processes, biological systems are fields in which hybrid systems as modeling framework are more and more present and useful. Furthermore, the interest for this class of systems resides in the fact that adding some hybrid loops may be fruitful in comparison to linear controllers. See e.g. [17, 91] for some academic examples, and consider [43, 48, 160, 412] for specific control applications where hybrid controllers are used. Even if the stability analysis may become complicated, the performance improvement of hybrid strategies with respect to more classical controllers (i.e. regular linear or nonlinear controllers) is now proved (see, e.g., [72, 128, 293, 294, 408], for different uses of hybrid controllers). However, such a class of feedback laws may introduce—possibly unrealistic—large transition input values, which constitutes a motivation to study hybrid systems subject to input saturation. A first work studying the stability issue of a nonlinear systems resulting from a First Order Reset Element (FORE) and a saturation in the input has been proposed in [249] by using an adaptation of Lemma 1.6. Some extensions in the same context have been proposed in [93] by using a geometrical characterization of the condition of contractiveness of an ellipsoid using convexity related properties. The value of the saturated function is then shown to be contained into a set determined by a finite number of vertices, easily computable.

## 3.8 Conclusion

This chapter was dedicated to present conditions, mainly LMI-based, to address both asymptotic stability analysis and control design problems. By modeling the saturation term through a generalized sector condition and by using quadratic Lyapunov functions, tractable conditions were exhibited in order to characterize the region of stability of the saturated closed-loop system, to deal with the external stability in presence of additive exogenous signals or to synthesize stabilizing control laws. Several extensions have also been discussed pointing out that numerous forthcoming issues need to be deeply studied. The elements provided in this chapter pave the way for these future developments.

# Chapter 4
# Analysis via the Regions of Saturation Model

## 4.1 Introduction

This chapter considers the problem of stability analysis of linear control systems with input saturations using the regions of saturation model introduced in Sect. 1.7.3. For a sake of simplicity and without loss of generality, the results are developed considering state feedback control laws.

As pointed out in Sect. 1.7.3, considering the closed-loop system

$$\dot{x}(t) = Ax(t) + B \operatorname{sat}(Kx(t)) \tag{4.1}$$

the state space can be divided in $3^m$ regions of saturation

$$S(R_j, d_j) = \{x \in \Re^n : R_j x \preceq d_j\}, \quad j = 1, \dots, 3^m \tag{4.2}$$

with $R_j \in \Re^{l_j \times n}$ and $d_j \in \Re^{l_j}$ defined from the matrix $K$ and the control limits $u_{\max}$ and $u_{\min}$. Inside each region $S(R_j, d_j)$, the behavior of system (4.1) can be represented by an affine system:

$$\dot{x}(t) = \bar{A}_j x(t) + p_j \tag{4.3}$$

where $\bar{A}_j = A + B \operatorname{diag}(1_m - |\xi_j|)K$ and $p_j = Bu(\xi_j)$, with $\xi_{j(i)}$ taking the value $1, 0$ or $-1$.

Based on this piecewise affine model for system (4.1), in this chapter, conditions to ensure the regional (local) stability of the closed-loop system are presented. From these conditions, numerical algorithms for determining regions of asymptotic stability as large as possible are provided. In particular, we focus on two kinds of regions of stability: polyhedral and ellipsoidal ones. As we will see, these regions are in fact level surfaces of polyhedral (also called Minkowski functionals) and quadratic Lyapunov functions.

Considering the polyhedral regions, the stability conditions and the numerical algorithms can be cast in a linear programming framework. On the other hand, concerning the ellipsoidal regions, the stability conditions are stated in terms of bilinear matrix inequalities. The numerical algorithms are then based on LMI relaxation schemes.

S. Tarbouriech et al., *Stability and Stabilization of Linear Systems with Saturating Actuators*, DOI 10.1007/978-0-85729-941-3_4, © Springer-Verlag London Limited 2011

The chapter is organized as follows. In Sect. 4.2 the determination of polyhedral regions of stability is addressed. The stability analysis based on ellipsoidal regions is presented in Sect. 4.3. The chapter ends with the presentation of some extensions regarding discrete-time systems and unbounded regions of stability.

## 4.2 Polyhedral Regions of Stability

### 4.2.1 Positive Invariance

Consider a polyhedral set $S(G, w)$ defined in the state space[1] by

$$S(G, w) = \left\{ x \in \Re^n; \; Gx \preceq w \right\}$$

with $G \in \Re^{g \times n}$ and $w \in \Re^g$. int $S(G, w)$ and $\partial S(G, w)$ denote the interior and the boundary of $S(G, w)$, respectively. The $i$th face of $S(G, w)$, denoted by $\partial_i S(G, w)$, is described by

$$\partial_i S(G, w) = \left\{ x \in S(G, w); \; G_{(i)} x = w_{(i)} \right\}$$

The following lemma states a generic condition to ensure the positive invariance of a polyhedron $S(G, w)$ with respect to the saturated system (4.1).

**Lemma 4.1** *The set $S(G, w)$ is positively invariant with respect to system* (4.1) *if*

$$G_{(i)} \dot{x}(t) < 0, \quad \forall x(t) \in \partial_i S(G, w), \; \forall i = 1, \ldots, g \tag{4.4}$$

*Proof* Since the state trajectory is continuous, for the positive invariance of $S(G, w)$ it is sufficient to guarantee the admissibility of any infinitesimal motion starting from any point on the boundary of this domain [58]. Hence, if at instant $t$, $x(t)$ belongs to the face $\partial_i S(G, w)$ it suffices that

$$G_{(i)} x(t + \tau) < G_{(i)} x(t)$$

with $\tau \to 0$.

From a Taylor expansion we have

$$x(t + \tau) = x(t) + \tau \frac{\dot{x}(t)}{1!} + \tau^2 \frac{x^{(2)}(t)}{2!} + \cdots + \tau^n \frac{x^{(n)}(t)}{n!} + \mathcal{O}_{n+1}$$

with $\mathcal{O}_{n+1}$ containing all the terms of order greater or equal to $n + 1$. Hence, if $G_{(i)} \dot{x}(t) < 0$, for a small enough $\tau > 0$ it follows that

$$G_{(i)} \left( \tau \frac{\dot{x}(t)}{1!} + \tau^2 \frac{x^{(2)}(t)}{2!} + \cdots + \tau^n \frac{x^{(n)}(t)}{n!} + \mathcal{O}_{n+1} \right) < 0$$

---

[1]Recall that the symbols "$\preceq$" and "$\succeq$" denote componentwise inequalities.

## 4.2 Polyhedral Regions of Stability

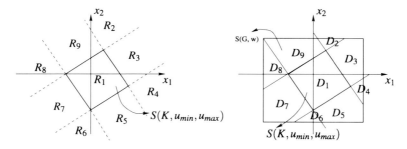

**Fig. 4.1** Representation of $S(R_j, d_j)$ and $S(D_j, s_j)$ for $n = m = 2$

which allows us to conclude that $G_{(i)}x(t + \tau) < G_{(i)}x(t)$. Since condition (4.4) should be valid $\forall i$, we conclude that $x(t + \tau) \in \text{int } S(G, w)$, for any motion starting on the boundary of $S(G, w)$, which concludes the proof. □

Now, based on the piecewise affine representation (4.3) of the closed-loop system (4.1), it is possible to derive an algebraic condition that allows to conclude that (4.4) holds. With this aim, consider the following definition.

**Definition 4.1** Let $\mathcal{J}$ be the set of indices $j$ such that $S(R_j, d_j) \cap S(G, w) \neq \emptyset$. The regions $S(R_j, d_j)$ such that $j \in \mathcal{J}$ are called *target regions*.

For each target region, define a polyhedral set

$$S(D_j, s_j) = S(R_j, d_j) \cap S(G, w) = \{x \in \Re^n; \ D_j x \leq s_j\} \quad (4.5)$$

where $D_j = \begin{bmatrix} \bar{G}_j \\ R_j \end{bmatrix}$, $s_j = \begin{bmatrix} \bar{w}_j \\ d_j \end{bmatrix}$, with $\bar{G}_j \in \Re^{g_j \times n}$ and $\bar{w}_j \in \Re^{g_j}$ corresponding to the faces of $S(G, w)$ that have a nonempty intersection with $S(R_j, s_j)$. Figure 4.1 depicts the regions $S(R_j, d_j)$ (notation $R_j$) and $S(D_j, s_j)$ (notation $D_j$) for a second order system with two control inputs.

The following Proposition provides a sufficient algebraic test to conclude about the positive invariance of a set $S(G, w)$ with respect to the closed-loop system (4.1).

**Proposition 4.1** *The set $S(G, w)$ is positively invariant with respect to system* (4.1) *if, for each target region $j$, there exists a matrix $H_j \in \Re^{g_j \times (g_j + l_j)}$, with $H_{j(i,l)} \geq 0$ if $i \neq l$, such that*

$$H_j D_j = \bar{G}_j \bar{A}_j \quad (4.6)$$

$$H_j s_j < -\bar{G}_j p_j \quad (4.7)$$

188           4 Analysis via the Regions of Saturation Model

*Proof* For each target region of saturation $j$, consider the following set of linear programs:

$$LP_{j,i}: \begin{cases} y_{j(i)} = \min_{H_{j(i)}} H_{j(i)}s_j + \bar{G}_{j(i)}p_j \\ \quad\quad \text{subject to} \quad H_{j(i)}D_j = \bar{G}_{j(i)}\bar{A}_j \\ \quad\quad\quad\quad H_{j(i,l)} \geq 0, \quad \text{if } l \neq i \end{cases} \quad (4.8)$$

where $H_{j(i)} \in \Re^{1 \times (g_j + l_j)}$. Let $H^*_{j(i)}$ be the optimal solution of $LP_{j,i}$. The satisfaction of relations (4.6) and (4.7) implies that

$$H^*_{j(i)}s_j + \bar{G}_{j(i)}p_j < 0 \quad (4.9)$$

By duality (see for example [252]), each of the programs $LP_{j,i}$ is equivalent to

$$DLP_{j,i}: \begin{cases} y_{j(i)} = \max_{x} \bar{G}_{j(i)}\bar{A}_j x + \bar{G}_{j(i)}p_j \\ \quad\quad \text{subject to} \quad \bar{G}_{j(i)}x = \bar{w}_{j(i)} \\ \quad\quad\quad\quad \bar{G}_{j(l)}x \leq \bar{w}_{j(l)}, \quad l = 1, \dots, g_j; \ l \neq i \\ \quad\quad\quad\quad R_j x \preceq d_j \end{cases} \quad (4.10)$$

Let $x^*_{j(i)}$ be the optimal solution of $DLP_{j,i}$. From duality, it also follows that

$$\bar{G}_{j(i)}\bar{A}_j x^*_{j(i)} + \bar{G}_{j(i)}p_j = H^*_{j(i)}s_j + \bar{G}_{j(i)}p_j < 0 \quad (4.11)$$

Hence, from (4.10), $\forall x \in (\partial_i S(G, w) \cap S(D_j, s_j)) = \{x \in \Re^n; \ R_j x \leq d_j, \ \bar{G}_{j(i)}x = \bar{w}_{j(i)}, \ \bar{G}_{j(l)}x \leq \bar{w}_{j(l)}, \ \forall l \neq i, \ \forall i = 1, \dots, g_j, \ \forall j \in \mathcal{J}\}$, it follows that

$$\bar{G}_{j(i)}(\bar{A}_j x + p_j) < 0 \quad (4.12)$$

Therefore, the satisfaction of (4.6) and (4.7), for all target regions $j$, with $H_{j(i,l)} \geq 0$ if $i \neq l$, implies that

$$G_{(i)}\dot{x}(t) < 0, \quad \forall x(t) \in \partial_i S(G, w), \ \forall i = 1, \dots, g$$

which, from Lemma 4.1, ensures the positive invariance of $S(G, w)$ with respect to system (4.1). $\qquad\square$

*Remark 4.1* Note that the characterization of positively invariant sets for continuous-time systems can be studied by using Nagumo's theorem [270] and the notion of tangent cone [12, 44, 71]. The tangent cone for convex sets is closed while in Lemma 4.1 we consider only its interior [71]. This is the reason why condition (4.4) is only a sufficient condition. Nonetheless, by using similar arguments as those developed in Chap. 4 of [39], and then based on a different procedure from that one used in Lemma 4.1, one can prove that

$$G_{(i)}\dot{x}(t) \leq 0, \quad \forall x(t) \in \partial_i S(G, w)$$

is a necessary and sufficient condition for the positive invariance of the set $S(G, w)$ with respect to system (4.1). In this case, replacing (4.7) by

$$H_j s_j \preceq -\bar{G}_j p_j$$

4.2 Polyhedral Regions of Stability 189

Proposition 4.1 can be re-stated to provide a necessary and sufficient condition.

We have chosen to use a strict inequality in Lemma 4.1 because similar arguments to those developed in the proof of this lemma will be used to derive the results in the sequel.

*Remark 4.2* In the absence of control bounds, the closed-loop system is represented globally by

$$\dot{x}(t) = (A + BK)x(t) \tag{4.13}$$

In this case, the classical strict polyhedral positive invariance relations [31, 58]:

$$HG = G(A + BK) \tag{4.14}$$

$$Hw \prec 0 \tag{4.15}$$

with $H_{(i,l)} \geq 0$ if $i \neq l$, are a particular case of relations (4.6) and (4.7). Actually, the only target region is $S(G, w)$. In this case $D_j = G$, $s_j = w$, $\bar{A}_j = (A + BK)$ and $p_j = 0$. We can also conclude that if $S(G, w) \subseteq S(K, u_{\min}, u_{\max})$, conditions in Proposition 4.1 reduce to (4.14) and (4.15).

## 4.2.2 Contractivity—Compact Case

The conditions stated in Proposition 4.1 guarantee that $\forall x(0) \in S(G, w)$, the corresponding trajectories of the saturated system (4.1) are confined in $S(G, w)$, $\forall t \geq 0$. However, this property is not sufficient to ensure the convergence of the trajectories to the origin. Of course, in the case where the positive invariant set $S(G, w)$ is contained in $S(K, u_{\min}, u_{\max})$, this set is also a region where the asymptotic stability is guaranteed if and only if all the eigenvalues of $(A + BK)$ are in the open left half-plane. However, this is only a necessary condition if $S(G, w) \not\subset S(K, u_{\min}, u_{\max})$. In this case, since the behavior of the system is nonlinear, the possible existence of limit cycles and/or parasitic equilibrium points inside $S(G, w)$ has to be considered. Hence, before concluding that the polyhedron $S(G, w)$ is also a region of asymptotic stability for system (4.1), it is necessary to eliminate these possibilities. The verification of the existence of equilibrium points, different from the origin, inside $S(G, w)$ is trivial. Nevertheless, in general, it is not easy to verify if there exist limit cycles inside $S(G, w)$.

In this case, a conservative, but relatively easy way, of ensuring the asymptotic stability of the system inside a positively invariant set consists in guaranteeing the contractivity of this set.

**Definition 4.2** The polyhedron $S(G, w)$ is said to be contractive with respect to system (4.1), if the following implication holds $\forall \alpha \in (0, 1]$:

$$x(t) \in \partial S(G, \alpha w) \quad \Rightarrow \quad x(t + \tau) \in \text{int}\, S(G, \alpha w), \quad \forall \tau > 0$$

Similarly to the case of contractive ellipsoids, which are associated to quadratic functions, the contractivity in this case can be characterized by using a Minkowski functional $\mathcal{V}(x(t))$ associated to $S(G, w)$, which is defined by [35, 217, 340]

$$\mathcal{V}\big(x(t)\big) = \max_i \left\{ \frac{G_{(i)}x(t)}{w_{(i)}} \right\} \tag{4.16}$$

The upper right Dini derivative of $\mathcal{V}(x(t))$ is defined as follows [232]:

$$D^+\mathcal{V}\big(x(t)\big) = \lim_{\tau \to 0^+} \sup \frac{\mathcal{V}(x(t+\tau)) - \mathcal{V}(x(t))}{\tau} \tag{4.17}$$

Since $\mathcal{V}(x(t))$ is a continuous function, $D^+\mathcal{V}(x(t))$ is well-defined and its existence is guaranteed.

Thus, if $S(G, w)$ is compact the following lemmas can be stated.

**Lemma 4.2** *A compact polyhedron $S(G, w)$ is contractive with respect to system (4.1) if and only if*

$$D^+\mathcal{V}\big(x(t)\big) < 0 \tag{4.18}$$

$\forall x(t) \in S(G, w), \ x(t) \neq 0, \ \forall t \geq 0.$

*Proof Sufficiency:* Suppose that $x(t) \in \partial S(G, \alpha w)$, $\alpha \in (0, 1]$. In this case, $\mathcal{V}(x(t)) = \alpha$. Hence, if (4.18) holds, it follows that $\mathcal{V}(x(t + \tau)) < \alpha$, that is, $x(t + \tau) \in \mathrm{int}\, S(G, \alpha w)$ where $\tau$ is a positive infinitesimal time. Since this is valid for all $x(t) \in S(G, w)$ contractivity is proven according to Definition 4.2.

*Necessity:* Suppose that $S(G, w)$ is contractive and $D^+\mathcal{V}(x(t)) \geq 0$ for some $x(t) \in \partial S(G, \alpha w)$, $\alpha \in (0, 1]$. In this case, $\mathcal{V}(x(t + \tau)) \geq \alpha$ and it follows that $x(t + \tau) \notin \mathrm{int}\, S(G, w)$, which contradicts the assumption that $S(G, w)$ is contractive. $\qquad\square$

**Lemma 4.3** *A compact polyhedron $S(G, w)$ is contractive with respect to system (4.1) if and only if*

$$G_{(i)}\dot{x}(t) < 0 \tag{4.19}$$

$\forall x(t) \in \partial_i S(G, w\alpha), \ \forall \alpha \in (0, 1], \ \forall i = 1, \ldots, g, \forall t \geq 0.$

*Proof* From Lemma 4.2 the contractivity of $S(G, w)$ is equivalent to satisfying

$$D^+\mathcal{V}\big(x(t)\big) = \lim_{\tau \to 0^+} \frac{\max_i\{\frac{G_{(i)}x(t+\tau)}{w_{(i)}}\} - \max_i\{\frac{G_{(i)}x(t)}{w_{(i)}}\}}{\tau} < 0 \tag{4.20}$$

Since $x(t)$ is a continuous function and by supposing that at instant $t$, $x(t) \in \partial_i S(G, \alpha w)$ it follows that (4.20) is equivalent to

$$D^+\mathcal{V}\big(x(t)\big) = \lim_{\tau \to 0^+} \frac{\frac{G_{(i)}x(t+\tau)}{w_{(i)}} - \frac{G_{(i)}x(t)}{w_{(i)}}}{\tau} < 0 \tag{4.21}$$

By expanding $x(t + \tau)$ in Taylor's series one obtains

## 4.2 Polyhedral Regions of Stability

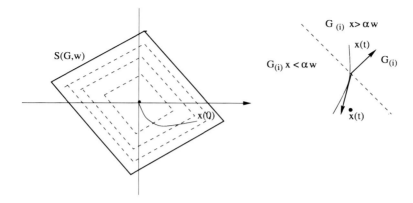

**Fig. 4.2** Polyhedral contractivity

$$\frac{G_{(i)}x(t+\tau)}{w_{(i)}} - \frac{G_{(i)}x(t)}{w_{(i)}} = \frac{\tau G_{(i)}\dot{x}(t)}{w_{(i)}1!} + \tau^2 \frac{G_{(i)}x^{(2)}(t)}{w_{(i)}2!} + \cdots$$
$$+ \tau^n \frac{G_{(i)}x^{(n)}(t)}{w_{(i)}n!} + \cdots \quad (4.22)$$

From (4.22), it follows that

$$D^+\mathcal{V}(x(t)) = \lim_{\tau \to 0^+} \frac{\tau G_{(i)}\dot{x}(t)}{\tau w_{(i)}}$$
$$+ \lim_{\tau \to 0^+} \left( \frac{\tau^2 G_{(i)}x^{(2)}(t)}{\tau w_{(i)}2!} + \cdots + \frac{\tau^n G_{(i)}x^{(n)}(t)}{\tau w_{(i)}n!} + \cdots \right) = \frac{G_{(i)}\dot{x}(t)}{w_{(i)}}$$
$$(4.23)$$

whence we can conclude that $D^+\mathcal{V}(x(t)) < 0$ if and only if $G_{(i)}\dot{x}(t) < 0$.

Since this reasoning can be applied $\forall x(t) \in \partial_i S(G, w\alpha)$, $\forall \alpha \in (0, 1]$, $\forall i = 1, \ldots, g$, $\forall t \geq 0$, it follows that condition (4.19) is equivalent to have $D^+\mathcal{V}(x(t)) < 0$, $\forall x(t) \in S(G, w)$, $\forall t \geq 0$ which proves the lemma. □

Figure 4.2 gives a graphical interpretation of the Lemmas above. From Lemma 4.2, the satisfaction of (4.18) means that the state trajectories enter, at each instant, inner homothetic sets to $S(G, \alpha w)$. In other words, if at instant $t$, $x(t) \in \partial_i S(G, \alpha w)$, $\alpha \in (0, 1])$ then at instant $t + \tau$, with an infinitesimal time $\tau$, $x(t + \tau) \in \text{int } S(G, \alpha w)$. From Lemma 4.3, this fact means that the angle between the normal to the face $\partial_i S(G, \alpha w)$ (i.e. the gradient of $\mathcal{V}(x(t))$) and $\dot{x}$ is larger than 90°.

Associated to the face $\partial_i S(G, w)$, define the following polyhedral cone:

$$\mathcal{K}_i(G, w) = \left\{ x \in \mathfrak{R}^n; \; \left( \frac{G_{(k)}}{w_{(k)}} - \frac{G_{(i)}}{w_{(i)}} \right) x \leq 0, \; \forall k \neq i \right\} \quad (4.24)$$

**Fig. 4.3** The cone $\mathcal{K}_i(G, w)$

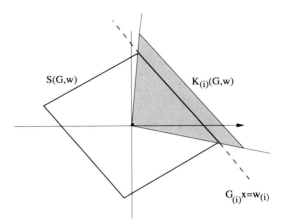

Note that if $x \in \mathcal{K}_i(G, w)$, it follows that $\frac{G_{(k)}x}{w_{(k)}} \leq \frac{G_{(i)}x}{w_{(i)}}$, $\forall k \neq i$, i.e. $x \in \partial_i S(G, \alpha w)$, for some $\alpha \in (0, 1]$. Figure 4.3 depicts the cone $\mathcal{K}_i(G, w)$.

We can now state the main result of this section.

Let $\mathcal{J}$ be the set of indices $j$ such that $S(G, w) \cap S(R_j, d_j) \neq \emptyset$. For each $j \in \mathcal{J}$, define $\mathcal{I}_j$ as the set of indices $i$ such that the cone $\mathcal{K}_i(G, w)$ has a nonempty intersection with the region $S(R_j, d_j)$, i.e.

$$\mathcal{I}_j = \{i;\ \mathcal{K}_i(G, w) \cap S(R_j, d_j) \neq \emptyset\}$$

**Proposition 4.2** *Consider the following linear programs for $j \in \mathcal{J}$ and $i \in \mathcal{I}_j$:*

$$LP_{j,i} = \begin{cases} y_{j(i)} = \max_x G_{(i)} \bar{A}_j x + G_{(i)} p_j \\ \text{subject to} \quad x \in \left(\mathcal{K}_i(G, w) \cap S(R_j, d_j) \cap S(G, w)\right) \end{cases} \quad (4.25)$$

*Define $y_j = \max\{y_{j(i)};\ i \in \mathcal{I}_j\}$. A compact polyhedron $S(G, w)$ is contractive with respect to system (4.1) if and only if the following conditions hold:*

(i) $y_j < 0$ *for each $j \in \mathcal{J}$ such that $S(R_j, d_j) \neq S(K, u_{\min}, u_{\max})$;*
(ii) $y_j = 0$ *for $j$ such that $S(R_j, d_j) = S(K, u_{\min}, u_{\max})$ and, in this case, the optimal solution of each linear program $LP_{j,i}$ is unique and obtained for $x = 0$.*

*Proof Sufficiency:* For all $x(t) \in S(G, w)$, $x(t) \neq 0$, it follows that $x(t)$ belongs to at least one face of $S(G, \alpha w)$, $0 < \alpha \leq 1$. In other words, $x(t) \in \partial_i S(G, \alpha w)$ and it follows that $x(t) \in \mathcal{K}_i(G, w)$. Moreover $x(t)$ belongs to some region of saturation $S(R_j, d_j)$, $j \in \mathcal{J}$. Consider that $x(t) \in S(R_j, d_j) \neq S(K, u_{\min}, u_{\max})$. Hence, if *(i)* holds, it follows that

$$G_{(i)} \bar{A}_j x(t) + G_{(i)} p_j < 0$$

i.e., $G_{(i)} \dot{x}(t) < 0$. On the other hand, if $x(t) \in S(R_j, d_j) = S(K, u_{\min}, u_{\max})$, one has $p_j = 0$ and, if (ii) holds, it follows that

$$G_{(i)} \bar{A}_j x(t) < 0, \quad \forall x(t) \neq 0$$

4.2 Polyhedral Regions of Stability 193

which also implies that $G_{(i)}\dot{x}(t) < 0$. Since this reasoning can be applied $\forall x(t) \in S(G, w)$, from Lemma 4.3 the contractivity of $S(G, w)$ is guaranteed if conditions (i) and (ii) are verified.

*Necessity:* Suppose that $S(G, w)$ is contractive and condition (i) or (ii) is not verified. Then, for some $i \in \mathcal{I}_j, j \in \mathcal{J}$, there exists $x(t) \in (\mathcal{K}_i(G, w) \cap S(R_j, d_j) \cap S(G, w))$, $x(t) \neq 0$, such that $G_{(i)}(\bar{A}_j x(t) + p_j) \geq 0$. In this case $x(t) \in \partial_i S(G, \alpha w)$, for some $0 < \alpha \leq 1$, and it follows that $G_{(i)}\dot{x}(t) \geq 0$. Since $x(t) \in S(G, w)$, this contradicts the assumption that $S(G, w)$ is contractive and thus the necessity of the condition is proved. $\square$

The basic idea of Proposition 4.2 consists in analyzing the closed-loop system trajectories in some sub-domains of $S(G, w)$. These sub-domains are delimited by a polyhedral cone $\mathcal{K}_i(G, w)$ and a region of saturation $S(R_j, d_j)$. In each one of these sub-domains we verify if for every $x(t) \neq 0$ belonging to the domain one obtains $D^+ V(x(t)) < 0$. This test is accomplished by solving linear programs like (4.25). Hence, since $S(G, w)$ is supposed to be compact, if conditions (i) and (ii) are verified, we can conclude that $D^+ V(x(t)) < 0, \forall x(t) \in S(G, w), x(t) \neq 0$.

Note that if $S(G, w)$ is compact $\lim_{\alpha \to 0} S(G, \alpha w) = \{0\}$. Thus, the contractivity of $S(G, w)$ implies the asymptotic convergence to the origin of all the trajectories emanating from $S(G, w)$. In fact, since $S(G, w)$ is supposed to be compact and contains the origin, the function $V(x(t))$ defined by (4.16) is a strictly decreasing Lyapunov function for system (4.1) in $S(G, w)$. From these considerations, the following corollary can be stated.

**Corollary 4.1** *If $S(G, w)$ is compact and conditions of Proposition 4.2 holds, then*

(i) *the polyhedral function $V(x(t)) = \max_i \{ \frac{G_{(i)}x(t)}{w_{(i)}} \}$ is a strictly decreasing Lyapunov function for system (4.1) in $S(G, w)$;*
(ii) *$S(G, w)$ is a region of asymptotic stability (RAS) for system (4.1).*

### *4.2.3 Determination of Stability Regions*

In this section, we consider that a contractive compact polyhedron $S(G, w) \subseteq S(K, u_{\min}, u_{\max})$ with respect to system (4.13) was computed by one of the methods proposed in the literature (see for instance [38, 58, 383]). We present now an algorithm to evaluate the maximum scaling factor, $\delta_{\max}$, for which the property of contractivity of $S(G, \delta_{\max} w)$ with respect to the saturated system (4.1) is preserved. The idea is to expand the original set $S(G, w)$ over the region of nonlinear behavior of the closed-loop system. In this case, note that the condition (ii) of Proposition 4.2 is automatically verified.

**Algorithm 4.1**

*Step 0* Initialize: $\delta = \delta_0$. Choose a computational accuracy.

194                                    4  Analysis via the Regions of Saturation Model

*Step 1* Determine $\mathcal{J}$ with respect to $S(G, \delta w)$. For each $j \in \mathcal{J}$ solve the following linear programs $\forall i \in \mathcal{I}_j$:

$$y_{j(i)} = \max_x G_{(i)} \bar{A}_j x + G_{(i)} p_j$$

$$\text{subject to} \begin{cases} \left(\dfrac{G_{(k)}}{w_{(k)}} - \dfrac{G_{(i)}}{w_{(i)}}\right) x \leq 0, & \forall k \neq i \\ R_j x \preceq d_j \\ Gx \preceq \delta w \end{cases} \tag{4.26}$$

*Step 2* If conditions (i) and (ii) of Proposition 4.2 hold, goto Step 4. Otherwise goto Step 3.
*Step 3* Decrease $\delta$ and return to Step 1.
*Step 4* If the difference between the $\delta$ of this iteration and the previous iterations is larger than the chosen accuracy, increase $\delta$ and return to Step 1. Otherwise stop: $\delta_{\max} = \delta$.

From Corollary 4.1, the obtained set $S(G, \delta_{\max} w)$ is a region of asymptotic stability for the saturated system (4.1). The proposed algorithm can be viewed as a tool to generate polyhedral regions of local stability and thus to approximate the region of attraction of the origin for system (4.1). This approximation can be improved by considering, for example, the union of different polyhedra obtained by the application of Algorithm 4.1. In this case, the final domain may be non-convex.

*Example 4.1*  Consider system (4.1) described by the following matrices [218]:

$$A = \begin{bmatrix} 0.1 & -0.1 \\ 0.1 & -3 \end{bmatrix}; \qquad B = \begin{bmatrix} 5 & 0 \\ 0 & 1 \end{bmatrix}$$

$$u_{\min} = u_{\max} = \begin{bmatrix} 5 & 2 \end{bmatrix}'$$

A stabilizing state feedback matrix $K$ and a positively invariant set $S(G, w) \subset S(K, u_{\min}, u_{\max})$ are given by

$$K = \begin{bmatrix} -0.7283 & -0.0338 \\ -0.0135 & -1.3583 \end{bmatrix}$$

$$G = \begin{bmatrix} 1 & 0 \\ 0 & 1 \\ -1 & 0 \\ 0 & -1 \end{bmatrix}; \qquad w = \begin{bmatrix} 1 \\ 1 \\ 1 \\ 1 \end{bmatrix}$$

As presented in Sect. 1.7.3, the matrix $K$ and the control constraints define nine regions of saturation. Since the polyhedra $S(G, w)$ and $S(K, u_{\min}, u_{\max})$ are symmetric, we can analyze only five of these regions:

*Region 1* ($\xi_1 = [0\ 0]' \Leftrightarrow$ Reg. linearity):

$$\bar{A}_1 = A + BK = \begin{bmatrix} -3.5415 & -0.2690 \\ 0.0865 & -4.3583 \end{bmatrix}; \qquad p_1 = \begin{bmatrix} 0 \\ 0 \end{bmatrix}$$

## 4.2 Polyhedral Regions of Stability

$$R_1 = \begin{bmatrix} K \\ -K \end{bmatrix}; \qquad d_1 = \begin{bmatrix} 5 \\ 2 \\ 5 \\ 2 \end{bmatrix}$$

*Region 2* $(\xi_2 = [0 \; -1]')$:

$$\bar{A}_2 = \begin{bmatrix} -3.5415 & -0.2690 \\ 0.1 & -3 \end{bmatrix}; \qquad p_2 = \begin{bmatrix} 0 \\ -2 \end{bmatrix}$$

$$R_2 = \begin{bmatrix} -0.7283 & -0.0338 \\ 0.7283 & 0.0338 \\ -0.0135 & -1.3583 \end{bmatrix}; \qquad d_2 = \begin{bmatrix} 5 \\ 5 \\ -2 \end{bmatrix}$$

*Region 3* $(\xi_3 = [-1 \; -1]')$:

$$\bar{A}_3 = A; \qquad p_3 = \begin{bmatrix} -25 \\ -2 \end{bmatrix}$$

$$R_3 = \begin{bmatrix} -0.7283 & -0.0338 \\ -0.0135 & -1.3583 \end{bmatrix}; \qquad d_3 = \begin{bmatrix} -5 \\ -2 \end{bmatrix}$$

*Region 4* $(\xi_4 = [-1 \; 0]')$:

$$\bar{A}_4 = \begin{bmatrix} 0.1 & -0.1 \\ 0.0865 & -4.3583 \end{bmatrix}; \qquad p_4 = \begin{bmatrix} -25 \\ 0 \end{bmatrix}$$

$$R_4 = \begin{bmatrix} -0.7283 & -0.0338 \\ -0.0135 & -1.3583 \\ 0.0135 & 1.3583 \end{bmatrix}; \qquad d_4 = \begin{bmatrix} -5 \\ 2 \\ 2 \end{bmatrix}$$

*Region 5* $(\xi_5 = [1 \; -1]')$:

$$\bar{A}_5 = A; \qquad p_5 = \begin{bmatrix} 25 \\ -2 \end{bmatrix}$$

$$R_5 = \begin{bmatrix} 0.7283 & 0.0338 \\ -0.0135 & -1.3583 \end{bmatrix}; \qquad d_5 = \begin{bmatrix} -5 \\ -2 \end{bmatrix}$$

By applying Algorithm 4.1 one obtains $\delta_{max} = 124.99$ (*accuracy* $= 0.01$). This value gives the maximal homothetical set of $S(G, w)$ which is positively invariant and contractive with respect to the closed-loop saturated system (4.1). Since $S(G, \delta_{max} w)$ is bounded, it is a region of asymptotic stability for system (4.1). It is worth noticing that the maximal homothetical set of $S(G, w)$ contained in the region of linearity $S(K, u_{min}, u_{max})$ is obtained for $\delta = 1.45$. Figure 4.4 depicts $S(G, \delta_{max} w)$ and the regions of saturation.

In Fig. 4.5, several trajectories for initial conditions taken on the boundary of $S(G, \delta_{max} w)$ are shown. As expected, all the trajectories lie in $S(G, \delta_{max} w)$ and converge asymptotically to the origin.

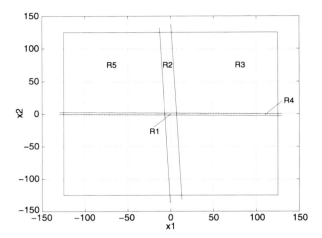

**Fig. 4.4** Example 4.1— $S(G, \delta_{\max} w)$ and the regions of saturation

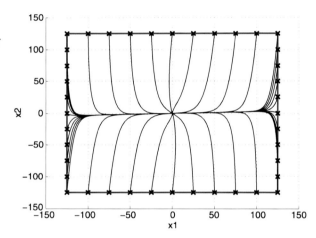

**Fig. 4.5** Example 4.1— trajectories for initial conditions on the boundary of $S(G, \delta_{\max} w)$

## 4.3 Ellipsoidal Regions

In this section, some conditions to ensure the contractivity of an ellipsoidal set for system (4.1) are presented. Similarly to the results presented for polyhedral sets, the strategy to derive these conditions consists in analyzing the behavior of a quadratic function along the trajectories of the system inside each one of the regions of saturation. By definition, we consider $S(R_1, d_1) = S(|K|, u_0)$.

Consider an ellipsoidal set generically described as

$$\mathcal{E}(P, c^{-1}) = \left\{ x \in \mathfrak{R}^n;\ x'Px \leq c \right\} \quad (4.27)$$

where $P = P' > 0$ and $c > 0$.

As seen in Sect. 2.2.1, the set $\mathcal{E}(P, c^{-1})$ is *contractive* with respect to system (4.1) if the function $V(x) = x'Px$ is strictly decreasing along the trajectories of system (4.1) in $\mathcal{E}(P, c^{-1}) - \{0\}$. In particular, if $\mathcal{E}(P, c^{-1})$ is contractive, then it

## 4.3 Ellipsoidal Regions

is a region of asymptotic stability for the closed-loop system. Hence, the following proposition provides the basis for the determination of contractive ellipsoids, using the regions of saturation description. Since $\mathcal{E}(P, c^{-1})$ is symmetric with respect to the origin, we consider that the control bounds are also symmetric, i.e. $u_{\max} = u_{\min} = u_0$.

**Proposition 4.3** *The function $V(x) = x'Px$ is a strictly decreasing Lyapunov function for the saturated system* (4.1) *in $\mathcal{E}(P, c^{-1})$ if and only if the following conditions hold*:

$$x'\left[P(A + BK) + (A + BK)'P\right]x < 0, \quad \forall x \in S(|K|, u_0) \cap \mathcal{E}(P, c^{-1}), \ x \neq 0 \tag{4.28}$$

$$x'P(\bar{A}_j x + p_j) + (\bar{A}_j x + p_j)'Px < 0, \quad \forall x \in S(R_j, d_j) \cap \mathcal{E}(P, c^{-1}),$$
$$\forall j, \ j = 2, \dots, 3^m, \ such \ that \ S(R_j, d_j) \cap \mathcal{E}(P, c^{-1}) \neq \emptyset \tag{4.29}$$

*Proof* It follows directly by considering $\dot{V}(x) < 0$ and $\dot{x}$ given by (4.3). $\qquad\square$

Although Proposition 4.3 provides a necessary and sufficient condition for a set $\mathcal{E}(P, c^{-1})$ to be contractive, it still lacks of practical benefit because the conditions (4.28) and (4.29) are not easily testable. In the sequel we present two conditions that, despite being only sufficient conditions for the satisfaction of (4.28) and (4.29), are numerically tractable.

### 4.3.1 Test Condition 1

Basically, the proposition below corresponds to a generalization, to multi-input systems, of the results proposed in [96].

**Proposition 4.4** *If there exist a symmetric positive definite matrix $P \in \mathfrak{R}^{n \times n}$, a positive scalar $c$ and non-negative scalars $\kappa_j$ and $\tau_{j(i)}, i = 1, \dots, l_j$, satisfying the conditions*

$$P(A + BK) + (A + BK)'P < 0 \tag{4.30}$$

$$\begin{bmatrix} P\bar{A}_j + \bar{A}_j'P - \kappa_j P & Pp_j - 0.5R_j'T_j' \\ p_j'P - 0.5T_j R_j & \kappa_j c + T_j d_j \end{bmatrix} < 0$$
$$\forall j, \ j = 2, \dots, 3^m, \ such \ that \ S(R_j, d_j) \cap \mathcal{E}(P, c^{-1}) \neq \emptyset \tag{4.31}$$

*with $T_j = [\tau_{j(1)} \ \cdots \ \tau_{j(l_j)}] \in \mathfrak{R}^{1 \times l_j}$, then the set $\mathcal{E}(P, c^{-1})$ is a region of asymptotic stability (RAS) for the saturated system* (4.1).

*Proof* Relation (4.30) implies that relation (4.28) is satisfied.

198  4  Analysis via the Regions of Saturation Model

For all regions such that $S(R_j, d_j) \cap \mathcal{E}(P, c^{-1}) \neq \emptyset$ it follows that

$$\begin{cases} x'Px - c \leq 0 \\ R_{j(i)}x - d_{j(i)} \leq 0, \quad i = 1, \ldots, l_j \end{cases}$$

Hence, considering the time-derivative of the Lyapunov function $V(x) = x'Px$ along the trajectories of system (4.1) and applying the S-procedure (see [45]), if there exist non-negative scalars $\kappa_j$ and $\tau_{j(i)}, i = 1, \ldots, l_j$, such that $\forall x, x \neq 0$

$$x'(P\bar{A}_j + \bar{A}'_j P)x + x'Pp_j + p'_j Px - \kappa_j(x'Px - c)$$

$$-\sum_{i=1}^{l_j} \tau_{j(i)}(R_{j(i)}x - d_{j(i)}) < 0 \tag{4.32}$$

or, equivalently

$$\begin{bmatrix} x' & 1 \end{bmatrix} \begin{bmatrix} P\bar{A}_j + \bar{A}'_j P - \kappa_j P & Pp_j - 0.5\sum_{i=1}^{l_j} \tau_{j(i)} R'_{j(i)} \\ p'_j P - 0.5\sum_{i=1}^{l_j} \tau_{j(i)} R_{j(i)} & \kappa_j c + \sum_{i=1}^{l_j} \tau_{j(i)} d_{j(i)} \end{bmatrix} \begin{bmatrix} x \\ 1 \end{bmatrix} < 0 \tag{4.33}$$

then (4.29) is satisfied.

It follows that a sufficient condition for the satisfaction of (4.33), and, in consequence, for the satisfaction of (4.29) is given by (4.31), which completes the proof. □

The result of Proposition 4.4 allows to verify whether a given ellipsoidal set $\mathcal{E}(P, c^{-1})$ is contractive or not. In such a case, the condition (4.31) is just an LMI feasibility test. Some variants related to the application of Proposition 4.4 are as follows.

- Given a given matrix $P$, compute the maximal $c$ for which the set $\mathcal{E}(P, c^{-1})$ is a RAS for the saturated system (4.1). This can be accomplished by interactively increasing $c$ and testing the condition (4.31). In this case, since $P$ is given and $c$ is fixed, the test corresponds to solve an LMI feasibility problem on the decision variables $T_j$ and $\kappa_j$.
- The conditions (4.30) and (4.31) can also be used to find a contractive set $\mathcal{E}(P, c^{-1})$ for system (4.1), which is a fundamental problem in general. However, the condition (4.31) becomes a BMI since $P$, $c$ and $\kappa_j$ are decision variables. In this case, the following algorithm based on the solution of optimization problems with LMI constraints can be applied to compute $\mathcal{E}(P, c^{-1})$.

**Algorithm 4.2**

*Step 1* Choose $\kappa_j = \kappa, \forall j = 1, \ldots, 3^m$.
*Step 2* Set $c = 1$. Fix $\kappa_j$, $j = 1, \ldots, 3^m$, obtained in the previous step and search for $P$ and $T_j$ by optimizing a criterion on the size of $\mathcal{E}(P, c^{-1})$ subject to the LMI conditions (4.30) and (4.31).

## 4.3 Ellipsoidal Regions

*Step 3* Fix $P$ obtained in Step 2. Maximize $c$ subject to conditions (4.30) and (4.31), with $\kappa_j$ and $T_j$, $j = 1, \ldots, 3^m$ as free variables.

*Step 4* If no further improvement is obtained regarding the size of $\mathcal{E}(P, c^{-1})$ stop. Otherwise, go to Step 2.

In Step 2, note that because the scalars $\kappa_i$ are given, the condition (4.31) is an LMI. As optimization criteria on the size $\mathcal{E}(P, c^{-1})$, we can consider the ones presented in Sect. 2.2.5.1. The Steps 2 and 3 of Algorithm 4.2 are performed iteratively until a desired precision in the size criterion for $\mathcal{E}(P, c^{-1})$ is achieved. Note that the $(P, \kappa_j, T_j)$ obtained in Step 2 are a feasible solution for Step 3 with $c = 1$. It is important to remark that, since $P$ and $T_j$ are decision variables, $c$ can be taken as 1 without loss of generality. Conversely $(P, c, \kappa_j, T_j)$ obtained in Step 3 is a feasible solution for Step 2 by setting $P \leftarrow c^{-1}P$ and $T_j \leftarrow c^{-1}T_j$. Hence the convergence of the algorithm is always ensured. The maximization of $c$ in Step 3 can be accomplished by iteratively increasing $c$ and testing (4.30) and (4.31) as an LMI feasibility problem.

*Remark 4.3* The computational burden can be reduced in the implementation of inequalities (4.30) and (4.31) by removing all regions of saturation that are symmetric to any other, since the satisfaction of (4.30) and (4.31) in one region of saturation also implies its satisfaction in the region symmetric to it.

*Remark 4.4* The condition (4.29) has been turned into condition (4.31), which can be verified by an LMI test or, in the worst case, by a BMI test. In this procedure, however, some conservatism is introduced due to the following facts:

1. The use of the S-procedure in step (4.32). Indeed, the S-procedure is only a sufficient condition in this case because there is more than a single constraint involved [45].
2. The LMI test (4.31). It implies that

$$\begin{bmatrix} x' & \xi \end{bmatrix} \begin{bmatrix} P\bar{A}_j + \bar{A}'_j P - \kappa_j P & Pp_j - 0.5R'_j T'_j \\ p'_j P - 0.5T_j R_j & \kappa_j c + T_j d_j(i) \end{bmatrix} \begin{bmatrix} x' \\ \xi \end{bmatrix} < 0 \qquad (4.34)$$

   for all $(x, \xi) \neq 0$, while it would be enough to check the case where $\xi = 1$.
3. The application of an LMI-based relaxation algorithm. This fact implies that we are not certain to find a solution of the problem even if it exists. Moreover, whenever a solution is found, there is no guarantee that it is actually the optimal one.
4. It is clear that the contractive set $\mathcal{E}(P, c^{-1})$ does not necessarily intersect all the regions of saturation. Thus, only the regions that intersect $\mathcal{E}(P, c^{-1})$ need to be tested. However, if the set $\mathcal{E}(P, c^{-1})$ is being synthesized, it is not possible to determine, a priori, whether the searched ellipsoid will intersect or not some of the regions of saturation. Then, in Algorithm 4.2, the test of (4.31) is performed for all regions of saturation and it can happen that condition (4.31) is unnecessarily verified in some region $j$.

## 4.3.2 Test Condition 2

We derive now a condition mainly inspired by the results presented in [202] for generic hybrid systems.

**Proposition 4.5** *If there exist a symmetric positive definite matrix* $P \in \Re^{n \times n}$, *a positive scalar* $c$, *non-negative scalars* $\kappa_j$ *and symmetric matrices* $M_j \in \Re^{l_j \times l_j}$, $j = 2, \dots, 3^m$, *with non-negative entries satisfying the following conditions*

$$(A + BK)'P + P(A + BK) < 0 \tag{4.35}$$

$$\begin{bmatrix} P\bar{A}_j + \bar{A}'_j P + R'_j M_j R_j - \kappa_j P & Pp_j - R'_j M_j d_j \\ p'_j P - d'_j M_j R_j & \kappa_j c + d'_j M_j d_j \end{bmatrix} < 0$$

$$\forall j, \ j = 2, \dots, 3^m, \ such \ that \ S(R_j, d_j) \cap \mathcal{E}(P, c^{-1}) \neq \emptyset \tag{4.36}$$

*then the set* $\mathcal{E}(P, c^{-1})$ *is a region of asymptotic stability (RAS) for the saturated system* (4.1).

*Proof* Relation (4.35) implies that relation (4.28) is satisfied.

Condition (4.29) can be rewritten as

$$[x' \quad 1] \begin{bmatrix} P & 0 \\ 0 & 0 \end{bmatrix} \begin{bmatrix} \bar{A}_j & p_j \\ 0 & 0 \end{bmatrix} \begin{bmatrix} x \\ 1 \end{bmatrix} + [x' \quad 1] \begin{bmatrix} \bar{A}'_j & 0 \\ p'_j & 0 \end{bmatrix} \begin{bmatrix} P & 0 \\ 0 & 0 \end{bmatrix} \begin{bmatrix} x \\ 1 \end{bmatrix} < 0 \tag{4.37}$$

$\forall j = 2, \dots, 3^m, \forall x$ such that

$$\begin{cases} x'Px - c \leq 0 \\ R_{j(i)}x - d_{j(i)} \leq 0, \quad i = 1, \dots, l_j \end{cases}$$

Let now $M_j \in \Re^{l_j \times l_j}$ be a symmetric matrix with non-negative entries and let $\kappa_j$ be a non-negative scalar. It follows that

$$[x' \quad 1] \begin{bmatrix} -R'_j \\ d'_j \end{bmatrix} M_j [-R_j \quad d_j] \begin{bmatrix} x \\ 1 \end{bmatrix} \geq 0 \quad \forall x \ such \ that \ R_j x - d_j \preceq 0 \tag{4.38}$$

$$\kappa_j [x' \quad 1] \begin{bmatrix} -P & 0 \\ 0 & c \end{bmatrix} \begin{bmatrix} x \\ 1 \end{bmatrix} \geq 0 \quad \forall x \ such \ that \ x'Px - c \leq 0 \tag{4.39}$$

Using now the S-procedure, it follows that a sufficient condition for the satisfaction of condition (4.29) is that, for some symmetric matrix $M_j \in \Re^{l_j \times l_j}$ with non-negative entries and a non-negative scalar $\kappa_j$, the following inequality is verified $\forall x \neq 0$:

$$[x' \quad 1] \left( \begin{bmatrix} P & 0 \\ 0 & 0 \end{bmatrix} \begin{bmatrix} \bar{A}_j & p_j \\ 0 & 0 \end{bmatrix} + \begin{bmatrix} \bar{A}'_j & 0 \\ p'_j & 0 \end{bmatrix} \begin{bmatrix} P & 0 \\ 0 & 0 \end{bmatrix} \right.$$

$$\left. + \begin{bmatrix} -R'_j \\ d'_j \end{bmatrix} M_j [-R_j \quad d_j] + \kappa_j \begin{bmatrix} -P & 0 \\ 0 & c \end{bmatrix} \right) \begin{bmatrix} x \\ 1 \end{bmatrix} < 0 \tag{4.40}$$

## 4.3 Ellipsoidal Regions

On the other hand, a sufficient condition for the satisfaction of (4.40), and, in consequence, for the satisfaction of (4.29) is given by

$$
\begin{bmatrix} P & 0 \\ 0 & 0 \end{bmatrix} \begin{bmatrix} \bar{A}_j & p_j \\ 0 & 0 \end{bmatrix} + \begin{bmatrix} \bar{A}'_j & 0 \\ p'_j & 0 \end{bmatrix} \begin{bmatrix} P & 0 \\ 0 & 0 \end{bmatrix}
$$
$$
+ \begin{bmatrix} -R'_j \\ d'_j \end{bmatrix} M_j \begin{bmatrix} -R_j & d_j \end{bmatrix} + \kappa_j \begin{bmatrix} -P & 0 \\ 0 & c \end{bmatrix} < 0 \tag{4.41}
$$

which is equivalent to (4.36). $\qquad\square$

As can be concluded from the proofs, Propositions 4.4 and 4.5 basically differ in the strategy the S-procedure is handled. In Proposition 4.5 the constraints are transformed into quadratic forms (4.38)–(4.39) before being included in the conditions. Due to the similarity in the development of the two Propositions, all the remarks made about Proposition 4.4, including the relaxation algorithm, apply to Proposition 4.5.

*Remark 4.5* In the single input case, $M_j$ is a positive scalar. It follows that $d'_j M_j d_j$ is always positive. Thus, inequality (4.36) never admits a feasible solution. In order to avoid this problem, we can consider a modified form for (4.38), which yields the following equation as a replacement for (4.36):

$$
[x' \quad 1] \left( \begin{bmatrix} P & 0 \\ 0 & 0 \end{bmatrix} \begin{bmatrix} \bar{A}_j & p_j \\ 0 & 0 \end{bmatrix} + \begin{bmatrix} \bar{A}'_j & 0 \\ p'_j & 0 \end{bmatrix} \begin{bmatrix} P & 0 \\ 0 & 0 \end{bmatrix} \right.
$$
$$
\left. + \begin{bmatrix} -\bar{R}'_j \\ \bar{d}'_j \end{bmatrix} M_j \begin{bmatrix} -\bar{R}_j & \bar{d}_j \end{bmatrix} + \kappa_j \begin{bmatrix} -P & 0 \\ 0 & c \end{bmatrix} \right) \begin{bmatrix} x \\ 1 \end{bmatrix} < 0 \tag{4.42}
$$

with $\bar{R}_j = \begin{bmatrix} R_j \\ 0 \end{bmatrix}$ and $\bar{d}_j = \begin{bmatrix} d_j \\ 1 \end{bmatrix}$. In this case $M_j \in \Re^{2\times 2}$ and $\bar{d}'_j M_j \bar{d}_j$ can assume negative values depending on the $M_j$ entries.

*Example 4.2* Consider the same system as treated in Examples 4.1 and 2.1.

Using test condition 1 and considering as criterion the maximization of the minor axis of the ellipsoidal region (i.e. minimization of the larger eigenvalue of $P$), Algorithm 4.2 gives the following values

$$
\kappa = 0.25; \qquad P = 10^{-3} \begin{bmatrix} 0.5886 & 0.0023 \\ 0.0023 & 0.2800 \end{bmatrix}; \qquad c = 1
$$

On the other hand, no feasible solution is found with test condition 2.

Figure 4.6 depicts the ellipsoidal region of asymptotic stability obtained in Example 2.1 (considering the polytopic model approaches) and the current one obtained with Test condition 1.

Note that the set obtained with the regions of saturation representation and the test condition 1 condition is significantly smaller than the sets obtained with the polytopic approaches. This reflects the high level of conservatism introduced in the procedure to obtain a tractable condition as commented in Remark 4.4.

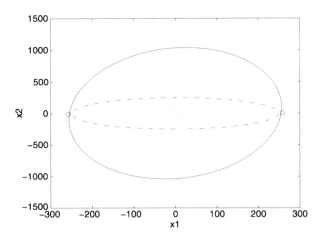

**Fig. 4.6** Example 4.2—RAS: test condition 1 (···); polytopic approach I (–·–); polytopic approach II (—)

## 4.4 Discrete-Time Case

Consider the discrete-time system described by

$$x(k+1) = Ax(k) + B\,\text{sat}(Kx(k)) \tag{4.43}$$

Results concerning the positive invariance and the contractivity of a polyhedral set $S(G, w)$ with respect to the discrete-time system (4.43) can be stated similarly to those stated for the continuous-time system presented in the previous section. In this case, if $x(k) \in S(R_j, d_j)$, it follows that

$$x(k+1) = \bar{A}_j x(k) + p_j \tag{4.44}$$

As we will see in the sequel, necessary and sufficient conditions for the positive invariance and contractivity can also be formulated in terms of the test of some linear programs. The main differences in the conditions come from the definitions of positive invariance and contractivity for discrete-time systems.

### 4.4.1 Positive Invariance

A necessary and sufficient condition for the positive invariance of a set $S(G, w)$ with respect to the discrete-time system (4.43) consists in verifying the implication

$$\forall x(k) \in S(G, w) \quad \Rightarrow \quad x(k+1) \in S(G, w) \tag{4.45}$$

with $x(k+1)$ given by (4.43). Note that, for continuous-time systems, it is sufficient to focus on the behavior of the system only on the boundary of the set $S(G, w)$, which is not the case here.

In order to state conditions for testing the implication (4.45) consider the polyhedral sets $S(D_j, s_j)$ defined in (4.5) and the set $\mathcal{J} = \{j\,;\; S(D_j, s_j) \neq \emptyset\}$.

A necessary and sufficient condition to guarantee the positive invariance of $S(G, w)$ with respect to system (4.43) can now be stated.

4.4 Discrete-Time Case 203

**Proposition 4.6** $S(G, w)$ *is positively invariant with respect to system* (4.43) *if and only if, for each region* $S(D_j, s_j)$, $j \in \mathcal{J}$, *there exists a non-negative matrix* $H_j \in \Re^{g \times (g_j + l_j)}$ *such that*

$$H_j D_j = G \bar{A}_j \qquad (4.46)$$

$$H_j s_j \preceq w - G p_j \qquad (4.47)$$

*Proof* It is similar to the proof of Proposition 4.1. For more details see [132]. □

### 4.4.2 Contractivity—Compact Case

**Definition 4.3** The set $S(G, w)$ is contractive with respect to system (4.43) if $\forall x(k) \in \partial S(G, \lambda_k w)$, $\lambda_k \in \Re$, $0 < \lambda_k \leq 1$, there exists $\lambda_{k+1}$, $0 \leq \lambda_{k+1} < \lambda_k$ such that $x(k+1) \in \partial S(G, \lambda_{k+1} w)$ for all integer $k \geq 0$.

According to Definition 4.3, the following proposition, quite similar to Proposition 4.2, can be stated.

**Proposition 4.7** *Consider the description of system* (4.43) *in the region of saturation* $j$, $j \in \mathcal{J}$, *given by* (4.44). *For each* $i \in \mathcal{I}_j$ *define the following linear programs for* $l = 1, \ldots, g$:

$$LP_{j(l,i)} \begin{cases} y_{j(l,i)} = \max_x \left( G_{(l)} \bar{A}_j - \dfrac{w_{(l)}}{w_{(i)}} G_{(i)} \right) x + G_{(l)} p_j \\ \text{subject to} \quad x \in \left( \mathcal{K}_i(G, w) \cap S(R_j, d_j) \cap S(G, w) \right) \end{cases} \qquad (4.48)$$

*Define* $y_j = \max\{y_{j(l,i)}; \ i \in \mathcal{I}_j, \ l = 1, \ldots, g\}$. *The polyhedron* $S(G, w)$ *is contractive with respect to system* (4.43) *if and only if the following conditions hold*:

(i) $y_j < 0$ *for each* $j \in \mathcal{J}$ *such that* $S(R_j, d_j) \neq S(K, u_{\min}, u_{\max})$;
(ii) $y_j = 0$ *for* $j$ *such that* $S(R_j, d_j) = S(K, u_{\min}, u_{\max})$ *and, in this case, the optimal solution to each linear program* $LP_{j(l,i)}$ *is unique and obtained for* $x = 0$.

*Proof* It is similar to the proof of Proposition 4.2. For more details see [136]. □

*Remark 4.6* An interesting extension of the results presented in this section is given in [258]. The idea is to consider piecewise affine (PWA) Lyapunov functions instead of Minkowski functionals. Basically, considering a polyhedral set $S(G, 1_g)$, the PWA function is a generalization of a Minkowski functional, obtained by introducing scalars $c_i$, $i = 1, \ldots, l_g$, by

$$\mathcal{V}(x(k)) = \max_i G_{(i)} x(k) + c_i$$

Hence, using linear programming techniques and modeling of the closed-loop system by regions of saturation, it is possible to obtain contractive sets associated to

204                             4   Analysis via the Regions of Saturation Model

the PWA function. These sets are also polyhedral, but do not have the same shape as $S(G, 1_g)$. This fact implies that better shaped and larger polyhedral region of asymptotic stability (RAS) can in general be obtained.

## 4.5 Unbounded Sets of Stability

### 4.5.1 Unbounded Polyhedra

From Proposition 4.2, we can conclude about the stability in $S(G, w)$ only if it is a compact set. However, in some cases, the condition of compactness can be relaxed as shown in this section. For this, consider the following assumption.

**Assumption 4.1** Consider that $S(G, w)$ is unbounded and there exists $0 < \alpha \le 1$ such that $S(G, \alpha w) \subseteq S(K, u_{\min}, u_{\max})$.

Note that the directions in which $S(G, w)$ is unbounded correspond to the null space of matrix $G$. Hence, Assumption 4.1 implies that $\text{Ker}(G) \subseteq \text{Ker}(K)$.

It is easy to see that a necessary condition to the contractivity of $S(G, w)$ is that the set $S(G, \alpha w) \subseteq S(K, u_{\min}, u_{\max})$, for some appropriated positive constant $\alpha$, must be positively invariant with respect to the linear system $\dot{x}(t) = (A + BK)x(t)$. On the other hand, it is shown in [59] that a necessary condition for this is that $\text{Ker}(G)$ is $(A + BK)$-invariant. Since, by assumption, $\text{Ker}(G) \subseteq \text{Ker}(K)$, we can conclude that a necessary condition for the contractivity of $S(G, w)$ in this case is the $A$-invariance of $\text{Ker}(G)$.

Consider the case that $\text{rank}(G) = g < n$, then $\text{Ker}(G)$ has dimension $g - r$. Define now a matrix $Q = [\, Q_o \; Q_r \,]$, $Q \in \Re^{n \times n}$, where:

- $Q_0 \in \Re^{n \times (n-g)}$ is a basis for the subspace $\mathcal{S}_0 = \text{Ker}(G)$,
- $Q_r$ is a basis for the complementary subspace of $\mathcal{S}_0$ denoted by $\mathcal{S}_r$.

Define now the similarity transformation

$$x = [\, Q_0 \quad Q_r \,] \begin{bmatrix} z_0 \\ z_r \end{bmatrix}$$

with $z_0 \in \Re^{n-g}$ and $z_r \in \Re^g$.

Suppose now that $\text{Ker}(G)$ is $A$-*invariant*. In this case, the representation of the linear operator $A$ in the basis given by the columns of $Q$ takes the following form:

$$Q^{-1} A Q = \begin{bmatrix} R_0 & R_2 \\ 0 & R_r \end{bmatrix}$$

with $R_0 \in \Re^{(n-g) \times (n-g)}$ and $R_r \in \Re^{g \times g}$. Moreover, since $\text{Ker}(G) \subseteq \text{Ker}(K)$ and $Q_0$ defines a basis for $\text{Ker}(G)$ it follows that $K Q_0 = 0$.

Hence, system (4.1) can be rewritten, in the basis defined by the columns of matrix $Q$, by

4.5 Unbounded Sets of Stability

$$\dot{z}_0(t) = R_0 z_0(t) + R_2 z_r(t) + B_0 \,\text{sat}\big(K_r z_r(t)\big) \tag{4.49}$$

$$\dot{z}_r(t) = R_r z_r(t) + B_r \,\text{sat}\big(K_r z_r(t)\big) \tag{4.50}$$

where $K_r = K Q_r$.

The projection of the polyhedron $S(G, w)$ in $S_r$ along $S_0$ is given by

$$S(G_r, w) = \big\{ z_r \in \Re^g : G_r z_r \le w \big\} \tag{4.51}$$

with $G_r = G Q_r$.

**Proposition 4.8** *The unbounded polyhedron $S(G, w)$, under Assumption* 4.1, *is a contractive set for system* (4.1) *if and only if $S(G_r, w_r)$ is a contractive set for system* (4.50).

*Proof* Recalling that

$$Gx = G \begin{bmatrix} Q_0 & Q_r \end{bmatrix} \begin{bmatrix} z_0 \\ z_r \end{bmatrix} = \begin{bmatrix} 0 & G_r \end{bmatrix} \begin{bmatrix} z_0 \\ z_r \end{bmatrix} = G_r z_r$$

then the proof directly follows. $\qquad\square$

**Proposition 4.9** *Consider that $S(G, w)$ is an unbounded polyhedron satisfying Assumption* 4.1. *If $S(G_r, w_r)$ is contractive for system* (4.50) *and all the eigenvalues of matrix $R_0$ have negative real-part then $S(G, w)$ is a region of asymptotic stability (RAS) for system* (4.1).

*Proof* Since $S(G_r, w)$ is a compact set, if $S(G_r, w)$ is a contractive set it is also a domain of local asymptotic stability for the reduced-order system (4.50). It follows that $z_r(t) \to 0$, as $t \to \infty$, $\forall z_r(0) \in S(G_r. w)$. Since all the eigenvalues of $R_0$ have negative real-part, from Proposition 1 in [403] we can conclude that $z_0(t) \to 0$, as $t \to \infty$, $\forall z_r(0) \in S(G_r, w)$, $\forall z_0(0) \in \Re^{n-g}$. Since $Q_0 z_0(t) \in \text{Ker}(G)$, $\forall t \ge 0$, it follows that $\forall x(0) \in S(G, w)$, the corresponding trajectory converges asymptotically to the origin. $\qquad\square$

Hence, provided that Assumption 4.1 is verified, the contractivity of an unbounded polyhedron $S(G, w)$ can be analyzed by using reduced-order system (4.50). In this case, Algorithm 4.1 can be applied by considering the polyhedron $S(G_r, w)$ and system (4.50).

*Remark 4.7* Note that if Assumption 4.1 is verified, it means that pair $(G, A)$ is non observable. In this case, the unobservable subspace of $(G, A)$ corresponds to $\text{Ker}(G)$. Moreover, the eigenvalues of $A$ associated to $\text{Ker}(G)$ are the eigenvalues of $R_0$. Hence, from Proposition 4.9, a necessary condition for the contractivity of an unbounded polyhedra $S(G, w)$ satisfying Assumption 4.1 is in fact the detectability of pair $(G, A)$.

*Remark 4.8* Considering (4.49)–(4.50), if Assumption 4.1 is satisfied it follows that the motion of the system inside the region of linearity is given by

$$\begin{bmatrix} \dot{z}_0(t) \\ \dot{z}_r(t) \end{bmatrix} = \begin{bmatrix} R_0 & R_2 + B_0 K_r \\ 0 & R_r + B_r K_r \end{bmatrix} \begin{bmatrix} z_0(t) \\ z_r(t) \end{bmatrix}$$

Therefore, we conclude that having simultaneously $\text{Ker}(G)$ $A$-invariant and $\text{Ker}(G) \subseteq \text{Ker}(K)$ implies that matrix $K$ provides a partial pole placement. In this case, the eigenvalues associated to the subspace defined by $\text{Ker}(G)$ are also eigenvalues of $(A + BK)$. Note that the eigenvalues of $R_0$ are in fact eigenvalues of $A$ and also of $(A + BK)$. Actually, the A-invariance of $\text{Ker}(G)$ is equivalent to the existence of a matrix $H$ such that the equation $HG = G(A + BK)$ is satisfied. In particular, it is shown in [58, 59] that when $\text{Ker}(G) \subseteq \text{Ker}(K)$ this equation can be interpreted as a canonical projection equation and it is equivalent to a partial pole placement.

*Remark 4.9* The construction of polyhedra $S(G, w)$ and the design of a gain $K$ satisfying Assumption 4.1 is addressed in [169]. In this context, a method using a reduced-order system is proposed in [61].

### 4.5.2 Unbounded Ellipsoidal Sets

The conditions provided by Proposition 4.4 and 4.5 consider positive definite matrices $P$ and, in this case, the associated ellipsoidal set $\mathcal{E}(P, c^{-1})$ is compact. When the matrix $P$ is semi-positive definite, i.e. $P = P' \geq 0$, the set $\mathcal{E}(P, c^{-1})$ is unbounded in the directions associated to $\text{Ker}(P)$ and takes the form of an hyper cylinder with ellipsoidal section. In this case, using similar arguments to those in Sect. 4.5.1, it is possible to show that under certain additional conditions, the stability of system (4.1) in a set $\mathcal{E}(P, c^{-1})$ with $P = P' \geq 0$ can be ensured. With this aim, consider the following assumption.

**Assumption 4.2** The matrix $P = P' \geq 0$ is such that

(i) $\text{Ker}(P) \subset \text{Ker}(K)$;
(ii) $\text{Ker}(P)$ is an $A$-invariant subspace.

Let now $Q = [Q_o \ Q_r]$, $Q \in \Re^{n \times n}$, where $Q_0 \in \Re^{n \times (n-g)}$ is an orthonormal basis for the subspace $\mathcal{S}_0 = \text{Ker}(P)$ and $Q_r$ is a basis for the complementary subspace of $\mathcal{S}_0$ denoted by $\mathcal{S}_r$. In this basis, system (4.1) is given by equation (4.49)–(4.50).

Suppose that matrix $P$ defining an unbounded ellipsoidal set $\mathcal{E}(P, c^{-1})$ satisfies Assumption 4.2. Then it follows that the projection of this set in the subspace $\mathcal{S}_r$ along the subspace $\mathcal{S}_o$ is given by the ellipsoidal compact set:

$$\mathcal{E}_r\left(P_r, c^{-1}\right) = \left\{ z_r \in \Re^r; \ z_r' P_r z_r \leq c \right\} \tag{4.52}$$

where $P_r = Q_r' P Q_r > 0$.

## 4.6 Conclusion

**Proposition 4.10** *The unbounded ellipsoid $\mathcal{E}(P, c^{-1})$, under Assumption 4.2, is a contractive set for system (4.1) if and only if $\mathcal{E}_r(P_r, c^{-1})$ is a contractive set for system (4.50).*

*Proof* It mimics that one of Proposition 4.8. $\qquad\qquad\square$

**Proposition 4.11** *Consider that $\mathcal{E}(P, c^{-1})$ is unbounded with $P$ satisfying Assumption 4.2. If $\mathcal{E}_r(P_r, c^{-1})$ is contractive for system (4.50) and all the eigenvalues of matrix $R_0$ have negative real-part, then $\mathcal{E}(P, c^{-1})$ is a region of asymptotic stability (RAS) for system (4.1).*

*Proof* It mimics that one of Proposition 4.9. $\qquad\qquad\square$

Hence, we can conclude that for an unbounded ellipsoidal set $\mathcal{E}(P, c^{-1})$ satisfying Assumption 4.2, it suffices to apply the test conditions stated in Sect. 4.3 with respect to the reduced-order system (4.50).

## 4.6 Conclusion

In this chapter, some results allowing the computation of regions of asymptotic stability have been stated considering the piecewise affine modeling of the saturated system (regions of saturation model).

First, conditions to ensure the positive invariance and the contractivity of polyhedral sets with respect to the saturated closed-loop system have been derived using linear programming arguments. It has been shown that these polyhedral sets are associated to polyhedral Lyapunov functions, also called Minkowsky functionals. Indeed, the level sets of these functions are polyhedra. Hence, once the contractivity conditions are verified it is ensured that the trajectories enter an inner scaled polyhedral set at each instant, i.e., the Lyapunov polyhedral function is strictly decreasing. The key point here is that the contractivity condition is a necessary and sufficient one. Hence, it is possible to determine the largest scaled polyhedron, with a given shape, for which the contractivity of the trajectories is ensured for any trajectory initialized into it. The contractivity conditions basically resume to test linear programming problems associated to each region of saturation that has a nonempty intersection with the considered polyhedron.

Differently from the determination of ellipsoidal regions of asymptotic stability (RAS) using, for instance an LMI framework, where the matrix $P$ is free and so is the Lyapunov function, the shape of the polyhedral set should be fixed a priori. This can be seen as a drawback, since a bad choice of the polyhedron shape can lead to a RAS that does not fit well in the actual region of attraction. On the other hand, the determination of polyhedral invariant and contractive sets for systems presenting control constraints has a special interest in the guarantee of stability in model predictive control (MPC) strategies. In this case, for numerical issues, it is suitable to consider polyhedral invariant terminal sets. Furthermore, the larger is the terminal set, the larger is the region of attraction for the system under the predictive control law. More details about the application of contractive polyhedra for systems with saturating controls in MPC can be found in [238].

On the other hand, conditions to determine ellipsoidal regions of asymptotic stability from the regions of saturation modeling have been presented. Although the modeling itself does not introduce any conservatism, the determination of tractable conditions, formulated as matrix inequalities, involves some conservatism mainly related to the application of the S-procedure. Furthermore, when matrix $P$ is a free variable two problems arise: the conditions are in fact BMIs (and so potentially conservative LMI relaxations schemes are considered to compute solutions) and the tests should be performed in all regions of saturation since the shape and the size of the ellipsoidal set is not known a priori.

Regarding synthesis issues, the main drawback of the modeling of the regions of saturation is that the regions of saturation actually depend on the control law to be synthesized, which makes the problem very complicated (and virtually impossible) to solve.

# Chapter 5
# Synthesis via a Parameterized ARE Approach or a Parameterized LMI Approach

## 5.1 Introduction

In the context of linear systems, optimal control is one of the major topics which captured the interest of many researches. Several books developing theoretical aspects and presenting important applications were published [11, 47, 230]. From a practical point of view, the interest for such a problem resides in its ability to take into account some kind of compromise between control effort and desired system response through a quadratic criterion involving two weighting matrices associated with the control and state. When dealing with state feedback control, in the infinite horizon case, the control is derived through the positive definite solution of the well-known algebraic Riccati equation [230] (see Appendix C).

In this setting, the control effort is expressed in terms of energy but not in terms of amplitude, meaning that from an optimal energy point of view, the obtained control can be satisfactory but its amplitude may be out of the limits of actuators. Taking directly the control limits in the derivation of the optimal control law is not, however an easy task [11]. In the context of regulation problems, when the state is near the origin, intuitively we can suspect that the control necessary to drive the system to the equilibrium point is, in general, under the actuator limits (linear behavior). But when the initial state is far from the origin, the limits of actuators can be attained leading to undesirable effects. As mentioned above, control effort in the criterion is taken into account through a weighting matrix which constitutes a degree of freedom and which can be used in order to maintain the control under the actuator limits. In the case where norm-bounded uncertainty affects the model of the system, optimal control techniques can be extended to guaranteed cost designs whose important results are presented in the quadratic stability framework [15] in Appendix B. One of the main noticeable points is that classical Riccati equation is replaced by algebraic Riccati equation involved in $H_\infty$-control [216, 288] whose properties are close to those of the standard one. The main idea is to derive a nested family of ellipsoids in the system state space. Once initialized in the outermost ellipsoid of the previous family, system trajectories are steered to the origin by applying increasing linear gains without causing input saturation. In the innermost ellipsoid, a guaranteed cost

S. Tarbouriech et al., *Stability and Stabilization of Linear Systems with Saturating Actuators*, DOI 10.1007/978-0-85729-941-3_5, © Springer-Verlag London Limited 2011

can be applied to ensure performance and go beyond stabilization. This idea can be seen as extensions of works in [321, 335, 395] where only matched uncertainties are considered. As is classical now, it is possible to replace the Riccati equation by a set of LMIs and then derive an algorithm to design a control law with the same characteristics. Even if there is a connection between the two methods, they are quite different from a technical point of view. Each of them has a specific interest and possesses its own features. On the one hand, the ARE approach allows for entering into structural aspects of the problem and leads to an elegant derivation from a mathematical point of view. Unfortunately, from a computational side, an exact solution is in general unattainable and some approximations are necessary to derive a control law with the desired properties. On the other hand, the LMI approach is derived via simple manipulations, but the structural properties of the problem are implicitly absorbed in the LMI formulation. Computationally, it offers a greater flexibility which can be advantageously exploited for obtaining control laws with additional properties.

The techniques developed in the chapter are complementary with those developed in the previous ones. Actually, they propose conditions to take into account simultaneously the criteria of the size of the region of asymptotic stability and the performance requirements for the closed-loop system. Hence, as mentioned in Chap. 1 (see Sect. 1.6.2), the approach pursued can be ranged in the saturation avoidance class. The idea is to determine a nonlinear control $v(x)$ instead of linear one in the previous chapter, but ensuring that $\mathrm{sat}(v(x)) = v(x)$ for given values of $x$ belonging to some set.

This chapter is organized as follows. Section 5.2 presents the ARE approach. Section 5.3 develops the LMI approach and Sects. 5.4 and 5.5 propose some extensions which include the multi-objective control design where pole placement and guaranteed cost are combined and the disturbance rejection problem. When the multi-objective control is particularized to the case of discrete-time systems with pole placement in unitary disk, we obtain the counterpart of the ARE approach for discrete-time systems. Only state feedback control is considered in this chapter. The case of output feedback is more involved. However in the LMI approach, it is possible adding some adequate structural constraints, to deal in a simple way with static output feedback control. All the methods are illustrated by numerical examples.

## 5.2 The Parameterized ARE Approach

Consider the system

$$\dot{x}(t) = \big(A_0 + DF(t)E_1\big)x(t) + Bu(t) = \big(A_0 + DF(t)E_1\big)x(t) + B\,\mathrm{sat}(t) \quad (5.1)$$

where $x \in \Re^n$, $u \in \Re^m$ are, respectively, the state and the control. $A_0$, $B$ are constant appropriately dimensioned matrices. $DF(t)E_1$ is an uncertainty which affects the dynamical matrix $A_0$ with $D$ and $E_1$ constant matrices of appropriate dimensions

5.2 The Parameterized ARE Approach

and $F(t)$ is an uncertain matrix satisfying

$$F(t)'F(t) \le I_r$$

We assume now that the control takes values in the compact set:

$$\mathcal{U} = \left\{ u \in \mathfrak{R}^m; \ -u_{0(i)} \le u_{(i)} \le u_{0(i)}, \ u_{0(i)} > 0, \ i = 1, \dots, m \right\} \qquad (5.2)$$

For the sake of simplicity, the single-input case, i.e. $u \in [-u_0, u_0]$ and $B \in \mathfrak{R}^n$ is first presented. In this chapter, the underlying idea is to build a nonlinear control $v(x)$ instead of linear one in the previous chapters but ensuring that $\text{sat}(v(x)) = v(x)$ for given values of $x$ belonging to some set. In practice, this corresponds to adjoint the region of linearity to enclose the state $x$, such that the computed control $v(x)$ never violates the saturation bounds $u_0$.

## 5.2.1 Preliminaries

To clarify the numerous calculations which follow, the time dependence of the state $x$ is omitted throughout the chapter. For simplicity reasons, we consider the case where no uncertainty affects the input matrix $B$ and uncertainty matrix $F$ is constant. It is possible to extend the following results, with a little effort, to the case where uncertainty also affects matrix $B$ and where uncertainty matrix $F$ is time-varying. It is well known that a necessary and sufficient condition for quadratic stabilizability of system (5.1) by a linear state feedback is given in the following proposition [288], see Appendix B.

**Proposition 5.1** *The system (5.1) is quadratically stabilizable if and only if given $R \in \mathfrak{R}^{m \times m}, R = R' > 0, Q \in \mathfrak{R}^{n \times n}, Q = Q' > 0$, there exist $\varepsilon > 0$ and a positive definite symmetric matrix $P \in \mathfrak{R}^{n \times n}$ solutions to*

$$A_0'P + PA_0 + \varepsilon PDD'P - PBR^{-1}B'P + \varepsilon^{-1}E_1'E_1 + Q = 0 \qquad (5.3)$$

*If these conditions are met, a control gain is given by*

$$K = -R^{-1}B'P$$

Note that the existence of a positive definite symmetric matrix satisfying the conditions of Proposition 5.1 is independent of the choice of the $Q$ and $R$ matrices. See [288] and Appendix B for details. Associated with the system (5.1) is the cost function:

$$J = \int_0^\infty \left[ x'Qx + u'Ru \right] dt \qquad (5.4)$$

and we suppose that $x(0) = x_0$ is the initial state. Now define

$$E = \left\{ \varepsilon > 0 : \text{there exists } P(\varepsilon) = P'(\varepsilon) > 0 \text{ solution to } (5.3) \right\}$$

and the set

$$\mathcal{K} = \left\{ K = -R^{-1}B'P(\varepsilon) : \varepsilon \in E \right\}$$

It is shown in [289] that if $K \in \mathcal{K}$, the closed-loop uncertain system is quadratically stable and in addition, the corresponding value of the cost function satisfies (see Appendix B)

$$J \le x_0'P(\varepsilon)x_0, \quad \forall F : F'F \le I_r$$

Note that the above bound depends of the initial condition $x(0) = x_0$. To remove this dependence, we assume that $x_0$ is a zero mean random variable satisfying $E[x_0 x_0'] = I_n$, where $E[\cdot]$ denotes the expectation. In this case, the bound becomes

$$\bar{J} = E[J] \le \text{Trace}\left[P(\varepsilon)\right]$$

This result implies that a control, which minimizes the cost bound, can be obtained by choosing $\varepsilon > 0$ to minimize $\text{Trace}[P(\varepsilon)]$. Then we have to solve the following optimization problem:

$$J^* = \min_{\varepsilon \in E} \text{Trace}\left[P(\varepsilon)\right]$$

If one is able to solve this problem, a certain performance level is attained for the closed-loop system. Moreover, it is shown in [289] that $P(\varepsilon)$ is a convex function with respect to $\varepsilon$ over $E$. This property ensures that a global minimum is reachable by an one-line search algorithm, which consists in solving iteratively an algebraic Riccati equation. In the sequel, we denote

$$\varepsilon^* = \text{Arg}\left\{\min_{\varepsilon \in E} \text{Trace}\left[P(\varepsilon)\right]\right\} \tag{5.5}$$

Suppose that the guaranteed cost control problem has a solution. Then for $R$ and $Q \in \Re^{n \times n}$ defined by (5.4), there exist $\varepsilon^* > 0$ and $P = P' > 0$ solutions to:

$$A_0'P + PA_0 + \varepsilon^* PDD'P - PBR^{-1}B'P + \varepsilon^{*-1}E_1'E_1 + Q = 0 \tag{5.6}$$

The control is written

$$u = -R^{-1}B'Px \tag{5.7}$$

Taking $S = \varepsilon^*P$, (5.6) becomes

$$A_0'S + SA_0 + SDD'S - \varepsilon^{*-1}SBR^{-1}B'S + E_1'E_1 + \varepsilon^*Q = 0 \tag{5.8}$$

The idea is similar to the one presented in [335], but for uncertain system (5.1). In a first time, we parameterize (5.8) in the following way:

$$A_0'S(\tau) + S(\tau)A_0 + S(\tau)DD'S(\tau) - \varepsilon^{*-1}S(\tau)BR^{-1}B'S(\tau) + E_1'E_1 + \varepsilon^*Q/\tau = 0 \tag{5.9}$$

## 5.2 The Parameterized ARE Approach

with $1 \leq \tau < \infty$. For $\tau = 1$, we obtain (5.8) and the guaranteed cost control is deduced letting $\tau = 1$ in the control:

$$u = -\varepsilon^{*-1} R^{-1} B' S(\tau) x$$

The first step is to verify that if (5.9) has a positive definite symmetric solution for $\tau = 1$, it has also a solution for $\tau > 1$.

**Lemma 5.1** *Suppose that the guaranteed cost control problem has a solution. Then there exist $\varepsilon^* > 0$ and $S(1) = S(1)' > 0$ satisfying (5.9). Moreover, for all $\tau > 1$, (5.9) has a positive definite symmetric solution $S(\tau) = S(\tau)' > 0$.*

*Proof* For $\tau = 1$, $S(1) = S(1)' > 0$. Right- and left-multiplying (5.9) by $S^{-1}(1)$, it can be put in the following form:

$$-\begin{bmatrix} I_n & S^{-1}(1) \end{bmatrix} \mathcal{H}(1) \begin{bmatrix} I_n \\ S^{-1}(1) \end{bmatrix} = 0$$

with

$$\mathcal{H}(\tau) = \begin{bmatrix} \varepsilon^{*-1} B R^{-1} B' - D'D & -A_0' \\ -A_0 & -(E_1' E_1 + \varepsilon^* Q/\tau) \end{bmatrix}$$

If $\tau > 1$, then $Q/\tau < Q$ and:

$$\mathcal{H}(\tau) - \mathcal{H}(1) = \begin{bmatrix} 0 & 0 \\ 0 & -\varepsilon^* (Q/\tau - Q) \end{bmatrix} \geq 0$$

By Theorem 2.2 in [296], we have

$$S^{-1}(\tau) \geq S^{-1}(1) > 0$$

Then the Riccati equation (5.9) possesses a positive definite symmetric solution for $\tau > 1$. Moreover,

$$S(1) \geq S(\tau) > 0 \qquad \qquad \square$$

### 5.2.2 The Single-Input Case

Define the set:[1]

$$\mathcal{E}\left(S(\tau), c^{-1}(\tau)\right) = \left\{ x \in \mathfrak{R}^n; \ x' S(\tau) x \leq \frac{u_0^2 \varepsilon^{*2}}{R^{-1} B' S(\tau) B R^{-1}} = c(\tau) \right\} \qquad (5.10)$$

---

[1] Throughout this chapter, ellipsoidal sets are defined with the notation $c$ instead of $\eta^{-1}$ (see Notation) to simplify the presentation of the results.

214    5  Synthesis via a Parameterized ARE Approach or a Parameterized LMI

**Lemma 5.2** $\mathcal{E}(S(\tau), c^{-1}(\tau))$ *is the maximal ellipsoid defined by the quadratic function* $x'S(\tau)x$ *where the feedback* $u = -\varepsilon^{*-1}R^{-1}B'S(\tau)x$ *is bounded by* $u_0$.

*Proof* Writing the Lagrangian and expressing the necessary condition for an extremum for the following optimization problem (because the control constraints are symmetrical):

$$\begin{cases} \max_{x} x'S(\tau)x = c(\tau) \\ -\varepsilon^{*-1}R^{-1}B'S(\tau)x = u_0 \end{cases}$$

we obtain (if $\lambda$ denotes the Lagrange multiplier and $\bar{x}$ the optimal value of $x$):

$$\bar{x} = \frac{1}{2}\lambda\varepsilon^{*-1}BR^{-1}$$

Using the fact that

$$-\varepsilon^{*-1}R^{-1}B'S(\tau)\bar{x} - u_0 = 0$$

it follows that

$$\lambda = \frac{-2u_0}{\varepsilon^{*-2}R^{-1}B'S(\tau)BR^{-1}} \quad \text{and} \quad \bar{x} = \frac{-BR^{-1}u_0}{\varepsilon^{*-1}R^{-1}B'S(\tau)BR^{-1}}$$

and

$$\bar{x}'S(\tau)\bar{x} = \frac{u_0^2\varepsilon^{*2}}{R^{-1}B'S(\tau)BR^{-1}} = c(\tau) \qquad \square$$

In the sequel, we normalize the ellipsoids (5.10) defining the positive definite matrix $\mathcal{X}(\tau) = S(\tau)/c(\tau)$, $1 \le \tau < \infty$, we denote

$$\mathcal{E}(\mathcal{X}(\tau), 1) = \{x \in \mathfrak{R}^n; \ x'\mathcal{X}(\tau)x \le 1\} \tag{5.11}$$

the ellipsoidal set for a given $\tau$ and $\Xi(\tau)$ the associated $\tau$-parameterized family of ellipsoidal sets defined as

$$\Xi(\tau) = \{\mathcal{E}(\mathcal{X}(\tau_i), 1), \ 1 \le \tau_i \le \tau\} \tag{5.12}$$

$\Xi(1)$ is the inner (smaller) ellipsoidal set related to the guaranteed cost control.

**Lemma 5.3** $\frac{d\mathcal{X}(\tau)}{d\tau}$ *is negative definite.*

*Proof* Differentiating (5.9) with respect to $\tau$, we obtain

$$A_0'\frac{dS(\tau)}{d\tau} + \frac{dS(\tau)}{d\tau}A_0 + \frac{dS(\tau)}{d\tau}DD'S(\tau) + S(\tau)DD'\frac{dS(\tau)}{d\tau}$$

$$- \varepsilon^{*-1}\frac{dS(\tau)}{d\tau}BR^{-1}BS(\tau) - \varepsilon^{*-1}S(\tau)BR^{-1}B'\frac{dS(\tau)}{d\tau} = \varepsilon^*Q/\tau^2$$

## 5.2 The Parameterized ARE Approach

and:

$$\left(A_0 + DD'S(\tau) - \varepsilon^{*-1}BR^{-1}B'S(\tau)\right)'\frac{dS(\tau)}{d\tau}$$

$$+ \frac{dS(\tau)}{d\tau}\left(A_0 + DD'S(\tau) - \varepsilon^{*-1}BR^{-1}B'S(\tau)\right) = \varepsilon^* Q/\tau^2$$

Since $(A_0 + DD'S(\tau) - \varepsilon^{*-1}BR^{-1}B'S(\tau))$ is asymptotically stable [414], by Lyapunov arguments, it follows that

$$\frac{dS(\tau)}{d\tau} < 0$$

On the other hand:

$$\frac{d\mathcal{X}(\tau)}{d\tau} = \frac{\frac{dS(\tau)}{d\tau}c(\tau) - S(\tau)\frac{dc(\tau)}{d\tau}}{c(\tau)^2}$$

with

$$\frac{dc(\tau)}{d\tau} = -\frac{\varepsilon^{*2}u_0^2 R^{-1}B'\frac{dS(\tau)}{d\tau}BR^{-1}}{(R^{-1}B'S(\tau)BR^{-1})^2} > 0$$

then $\frac{d\mathcal{X}(\tau)}{d\tau} < 0$. $\qquad\square$

It is also possible to prove the following proposition which shows that a nested family of ellipsoids is obtained when $\tau$ varies [118].

**Proposition 5.2** *The $\tau$-parameterized family of ellipsoids $\Xi(\tau)$ is a nested family set, that is,*

$$\Xi(\tau_1) \subset \text{int } \Xi(\tau_2), \quad \text{whenever } \tau_1 < \tau_2$$

*with a maximal element $\Xi = \bigcup_\tau \Xi(\tau)$. This set $\Xi$ may not be an ellipsoidal set in the case of infinite directions, but a degraded ellipsoid (see Proposition 5.6).*

In order to derive the controller, we introduce the following definition.

**Definition 5.1** Given $x^* \in \Xi$ and $\tau^* \in [1, \infty)$ such that $x^* \in \partial\Xi(\tau^*)$ ($\partial\Xi(\tau)$ denoting the boundary of the set $\Xi(\tau)$), define $\tau(x^*) = \tau^*$ and if $x^* \in \Xi(1)$ define $\tau(x^*) = 1$.

Using the same arguments as in [335], the following result can be stated.

**Proposition 5.3** *$\tau$ is a continuous function from $\Xi$ to $[1, \infty)$ which is differentiable in $\Xi \setminus \Xi(1)$.*

216                    5   Synthesis via a Parameterized ARE Approach or a Parameterized LMI

We are now in position to introduce the controller. It is defined by

$$u(x) = \begin{cases} -\varepsilon^{*-1} R^{-1} B' S(\tau(x)) x & \text{if } x \in \Xi \setminus \Xi(1) \\ -\varepsilon^{*-1} R^{-1} B' S(1) x & \text{if } x \in \Xi(1) \end{cases} \tag{5.13}$$

The idea is to drive the system trajectories to the neighborhood of the origin $\Xi(1)$ and then to apply the control $-\varepsilon^{*-1} R^{-1} B' S(1) x$ which ensures asymptotic stability and a certain performance level. The following result is central.

**Proposition 5.4** *Suppose that the guaranteed cost control problem has a solution. Let $S(\tau)$ be the positive definite symmetric solution of the Riccati equation (5.9) and $\Xi(\tau)$ the $\tau$-parameterized family of ellipsoids defined from (5.12). Then the controller defined by (5.13) satisfies the constraint $-u_0 \leq u(x) \leq u_0$ and drives any point $x \in \Xi = \bigcup_\tau \Xi(\tau)$ to the origin.*

*Proof* In $\Xi(1)$ the system is asymptotically stable and constrained by $u_0$. To prove that the compact set $\Xi(1)$ is a finite attractor in $\Xi$ for the closed-loop system, it suffices to show that for all $x$ in the closure of $\Xi \setminus \Xi(1)$, the time derivative $\dot\tau$ given by

$$\dot\tau = \frac{-2x' S(\tau) \dot x}{x' \dot S(\tau) x - \frac{dc(\tau)}{d\tau}}$$

is negative along the closed-loop vector fields. We have

$$\begin{aligned} 2x' S(\tau) \dot x &= x' \big[ \big( A_0 + DFE_1 - \varepsilon^{*-1} BR^{-1} B' S(\tau) \big)' S(\tau) \\ &\quad + S(\tau) \big( A_0 + DFE_1 - \varepsilon^{*-1} BR^{-1} B' S(\tau) \big) \big] x \\ &\leq x' \big[ \big( A_0 - \varepsilon^{*-1} BR^{-1} B' S(\tau) \big)' S(\tau) + S(\tau) \big( A_0 - \varepsilon^{*-1} BR^{-1} B' S(\tau) \big) \\ &\quad + S(\tau) DD' S(\tau) + E_1' E_1 \big] x \\ &= -x' \big[ \varepsilon^* Q/\tau + \varepsilon^{*-1} BR^{-1} B' S(\tau) \big] x < 0 \end{aligned}$$

Noting that $\frac{dS(\tau)}{d\tau} < 0$ and $\frac{dc(\tau)}{d\tau} > 0$, it follows that $\dot\tau < 0$. □

## 5.2.3  The Multi-variable Case

In this section, we move to the multi-inputs case. Matrix $B$ is written as $B = [B_1 \ \ldots \ B_m]$, $B_i \in \mathfrak{R}^n$, $i = 1, \ldots, m$ and we take:

$$R = \text{Diag}(r_1, \ldots, r_m) > 0$$

Define also:

$$c_i(\tau) = \frac{u_{0(i)}^2 \varepsilon^{*2} r_i^2}{B_i' S(\tau) B_i}, \quad i = 1, \ldots, m$$

5.2 The Parameterized ARE Approach 217

and:

$$C(\tau) = \min_i c_i(\tau)$$

It is possible to extend the results of the previous section and prove the following proposition.

**Proposition 5.5** *Let $S(\tau)$ be the positive definite symmetric solution to (5.9). Define the function $\tau(x)$ in the following way:*

- *for $x \in \Xi(1) = \{x \in \mathfrak{R}^n : x'S(1)x \le C(1)\}$, $\tau(x) = 1$;*
- *for $x \in \Xi \setminus \Xi(1)$, $\tau(x)$ is the positive solution to the equation*

$$x'S(\tau)x - C(\tau) = 0$$

*then, the control $u(x)$ defined by*

$$u(x) = \begin{cases} -\varepsilon^{*-1}R^{-1}B'S(\tau(x))x & \text{if } x \in \Xi \setminus \Xi(1) \\ -\varepsilon^{*-1}R^{-1}B'S(1)x & \text{if } x \in \Xi(1) \end{cases} \tag{5.14}$$

*satisfies the constraints*

$$-u_{0(i)} \le u_{(i)}(x) \le u_{0(i)}, \quad i = 1, \dots, m$$

*and drives any point of $\Xi$ to the origin.*

Suppose now that $\tau \longrightarrow \infty$, then the Riccati equation (5.9) becomes

$$A_0'S(\infty) + S(\infty)A_0 + S(\infty)DD'S(\infty) + E_1'E_1 - \varepsilon^{*-1}S(\infty)BR^{-1}B'S(\infty) = 0 \tag{5.15}$$

It is well known that [414]:

– if $(E_1, A_0)$ is observable and if $(A_0, B)$ is stabilizable, $S(\infty)$ is positive definite;
– if $(E_1, A_0)$ is detectable and if $(A_0, B)$ is stabilizable, $S(\infty)$ is positive semi-definite.

Hence the set $\Xi$ is characterized in the following proposition.

**Proposition 5.6** *Suppose that the pair $(A_0, B)$ is stabilizable. Then $\Xi$ is defined by*

1. *If $(E_1, A_0)$ is observable:*

$$\Xi = \left\{ x \in \mathfrak{R}^n; \ \max_i \left[ \frac{B_i'S(\infty)B_i}{u_{0(i)}^2 \varepsilon^{*2}r_i^2} \right] x'S(\infty)x < 1 \right\}$$

2. *If $(E_1, A_0)$ is detectable, $\Xi$ is equal to the Cartesian product of $\mathfrak{R}^{\dim(\mathrm{Ker}\,S(\infty))}$ and a $k$-dimensional open ellipsoid ($k = n - \dim(\mathrm{Ker}\,S(\infty))$). This ellipsoid is*

*a linear transformation of the set:*

$$\varXi = \left\{ y \in \Re^k; \ \max_i \left[ \frac{B_i' S(\infty) B_i}{u_{0(i)}^2 \varepsilon^{*2} r_i^2} \right] y' S_k(\infty) y < 1 \right\}$$

*where*

$$M^{-1} S(\infty) M = \begin{bmatrix} 0 & 0 \\ 0 & S_k(\infty) \end{bmatrix}$$

*with M an appropriate change of coordinates.*

*Proof* The proof follows from the above arguments. $\qquad\square$

In some problems, it could be interesting to maximize the size of $\varXi$. A way to do this, is to select appropriately the parameter $\varepsilon$. Actually (5.15) has a positive definite or semi-definite symmetric solution for $\varepsilon \in (0, \bar{\varepsilon}[$. An approximation of this interval can be obtained by the following algorithm:

**Algorithm 5.1**

1. Initialize $\varepsilon$ to some starting value $\varepsilon_{\text{init}}$.
2. Solve Riccati equation (5.15).
3. If the solution is positive definite or semi-definite, increase $\varepsilon_{\text{init}}$ and go to step 2. If not decrease $\varepsilon_{\text{init}}$ and go to step 2.

The algorithm is stopped when $\bar{\varepsilon}$ is attained with some accuracy. Let denote $\varepsilon_{\text{app}}$ the approximation of $\bar{\varepsilon}$. For simplicity, assume that $r_i = r > 0$, $u_{0(i)} = u_0$, for all $i$. $\varXi$ is then defined by

$$\varXi = \left\{ y \in \Re^k; \ \max_i \left[ B_i' S(\infty) B_i \right] u_0^{-2} \varepsilon^{-2} r^{-2} y' S_k(\infty) y < 1 \right\} \times \Re^{\dim(\text{Ker } S(\infty))}$$

It is well known that the volume of the ellipsoid in the brackets is proportional to $\det(V^{-2}(\varepsilon))$ where

$$V(\varepsilon) = \max_i \left[ B_i' S(\infty) B_i \right] u_0^{-2} \varepsilon^{-2} r^{-2} S_k(\infty)$$

A way to maximize the size of $\varXi$ is to solve the following optimization problem

$$\begin{cases} \min_{\varepsilon \in (0, \varepsilon_{\text{app}}]} \sigma \\ V(\varepsilon) < \sigma I_k \\ \sigma > 0 \end{cases} \tag{5.16}$$

An algorithm which approximates the solution of (5.16) is the following.

## 5.2 The Parameterized ARE Approach 219

**Algorithm 5.2**

1. Using the algorithm described above find $\bar{\varepsilon}$. Compute $\varepsilon_1, \ldots, \varepsilon_N$ obtained by sampling $(0, \varepsilon_{\text{app}}]$.
2. Solve Riccati equation (5.15) for each $\varepsilon_j$, $j = 1, \ldots, N$.
3. Compute the eigenvalues of $V(\varepsilon_j)$ denoted $\sigma_1(\varepsilon_j), \ldots, \sigma_k(\varepsilon_j)$, for $j = 1, \ldots, N$.
4. The solution $\sigma$ of (5.16) can be approximated by

$$\sigma = \min_{j=1,\ldots,N} \max_{l=1,\ldots,k} \sigma_l(\varepsilon_j)$$

Hence, it is clear that $\varepsilon$ has to be chosen in order to satisfy a certain compromise between the two following requirements.

- The performance level ($\tau \longrightarrow 1$): see (5.5) which gives $\varepsilon_1^*$.
- The maximization of the volume of $\Xi$ ($\tau \longrightarrow \infty$): see (5.16) which gives $\varepsilon_2^*$.

### 5.2.4 Control Computation and Implementation

In order to derive the control, it is necessary to obtain the function $\tau(x)$ satisfying

$$x'S(\tau)x - C(\tau) = 0$$

Analytically, this is not possible in general. In practice, a numerical approximation of the solution is used. A family of ellipsoids is obtained for a set of values of $\tau$ obtained by sampling the interval $[1, \infty)$, these calculations being done off-line. The algorithm can be summarized as follows.

**Algorithm 5.3**

1. Solve the following optimization problem (optimal guaranteed cost):

$$J^* = \min_{\varepsilon \in E} \text{Trace}\left[P(\varepsilon)\right]$$

Let

$$\varepsilon^* = \text{Arg}\left\{\min_{\varepsilon \in E} \text{Trace}\left[P(\varepsilon)\right]\right\}$$

Choose $N$ values of $\tau$ such that $\tau_0 = 1 < \tau_1 < \tau_2 < \cdots < \tau_N < \infty$. For $\tau = \tau_N$, solve Riccati equation (5.9). We obtain the corresponding ellipsoid $\mathcal{E}(\mathcal{X}(\tau_N), 1)$ and the control $K(\tau_N)$. Set $\Xi = \mathcal{E}(\mathcal{X}(\tau_N), 1)$ and the $\tau_N$-parameterized family

$$\Xi(\tau_N) = \mathcal{E}\left(\mathcal{X}(\tau_N), 1\right), \qquad K = \left\{K(\tau_N)\right\}$$

i. Take $\tau = \tau_{N-i}$. Solve the Riccati equation (5.9). We obtain $\mathcal{E}(\mathcal{X}(\tau_{N-i}), 1)$ and $K(\tau_{N-i})$. Set

$$\Xi(\tau_N) = \left\{\Xi(\tau_N), \mathcal{E}\left(\mathcal{X}(\tau_{N-i}), 1\right)\right\}, \qquad K = \left\{K, K(\tau_{N-i})\right\}$$

If $i < N$ go to step $i + 1$, else stop.

220    5 Synthesis via a Parameterized ARE Approach or a Parameterized LMI

At the end of the algorithm, we obtain a nested family of ellipsoids $\mathcal{E}(\tau_N)$, and corresponding control gains. To apply the control, we measure $x(t)$ and identify the outer ellipsoid in $\mathcal{E}(\tau_N)$, which contains $x(t)$, and the corresponding control is applied. With this method, a piecewise control is obtained.

*Remark 5.1* An important point in Algorithm 5.3 is the choice of a sequence of parameters $\tau_i$ in step 0. By experience, a geometric sequence produces satisfying results [395], namely:

$$\tau_{N-i} = \tau_N (\Delta\tau)^i, \quad \tau_N \text{ being fixed}$$

where $\Delta\tau = \tau_N^{-1/N}$. As pointed out in [395], to help in the selection of an appropriate spacing factor $\Delta\tau$, we can use a function which measures the percentage change in the major axes of two consecutive nested ellipsoids.

*Example 5.1* Consider the model of a balancing pointer given by

$$\dot{x} = \begin{bmatrix} 0 & 1 \\ 1 & 0 \end{bmatrix} x + \begin{bmatrix} 0 \\ -1 \end{bmatrix} u$$

The weighting matrices are taken as

$$Q = I_2 \quad \text{and} \quad R = 1$$

and the control is constrained by $u_0 = 5$. For this example, we can compute analytically the solution of the ARE. The solution in that case reads

$$S(\tau) = \begin{bmatrix} 1 + \frac{1}{\tau} + \sqrt{1 + \frac{1}{\tau}} & 1 + \sqrt{1 + \frac{1}{\tau}} \\ 1 + \sqrt{1 + \frac{1}{\tau}} & 1 + \sqrt{1 + \frac{1}{\tau}} \end{bmatrix}$$

and the control gains are

$$K(\tau) = -\begin{bmatrix} 1 + \sqrt{1 + \frac{1}{\tau}} & 1 + \sqrt{1 + \frac{1}{\tau}} \end{bmatrix}$$

The guaranteed cost control is nothing but the LQ optimal control characterized by

$$K(1) = -\begin{bmatrix} 1 + \sqrt{2} & 1 + \sqrt{2} \end{bmatrix}$$

Algorithm 5.3 is implemented with $N = 6$ and $\tau_N = 10$. The sequence of values of $\tau$ is generated taking into account Remark 5.1. Algorithm 5.3 returns 7 switching surfaces shown in Fig. 5.1. When $\tau \to \infty$, we have

$$S(\infty) = \begin{bmatrix} 2 & 2 \\ 2 & 2 \end{bmatrix} = M^{-1} \begin{bmatrix} 0 & 0 \\ 0 & 4 \end{bmatrix} M$$

## 5.2 The Parameterized ARE Approach

**Fig. 5.1** Example 5.1— Ellipsoids for $1 \leq \tau \leq 10$

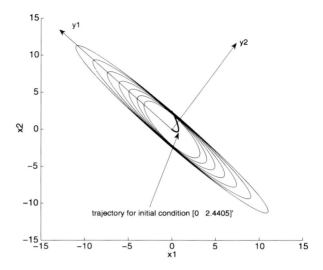

with

$$M = \begin{bmatrix} -\sqrt{2} & \sqrt{2} \\ \sqrt{2} & \sqrt{2} \end{bmatrix}$$

With the change of coordinate $x = My$, the ellipsoid degenerates when $\tau \to \infty$ to the set (see Proposition 5.6):

$$\Re \times \left\{ y_1 \in \Re; \ y_1^2 \leq \frac{25}{8} \right\}$$

Figure 5.1 also represents a trajectory for the initial condition $[0\ 2.4405]'$ and the axis $y_1$ and $y_2$ are also depicted which explicit the infinite direction ($y_1$) and the constrained one ($y_2$). In Fig. 5.2, we compare the guaranteed cost control when input control is not constrained and our piecewise-linear control designed to handle the constraints. Because the number of ellipsoids is restricted, we observe a chattering phenomenon which can be attenuated if the number of ellipsoids increases.

*Example 5.2* Consider the uncertain system (5.1) which models the dynamics of an helicopter in a vertical plane [315]. The corresponding linear model has four state variables, two inputs and reads

$$A_0 = \begin{bmatrix} -0.0336 & 0.0271 & 0.0188 & -0.4555 \\ 0.0482 & -1.0100 & 0.0024 & -4.0208 \\ 0.1002 & 0.2855 & -7070 & 1.3229 \\ 0 & 0 & 1 & 0 \end{bmatrix}$$

**Fig. 5.2** Example 5.1—nonlinear bounded control and LQ control

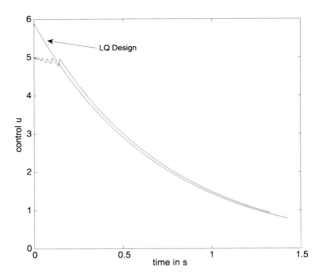

$$B = \begin{bmatrix} 0.4422 & 0.1761 \\ 3.0447 & -7.5922 \\ -5.5200 & 4.9900 \\ 0 & 0 \end{bmatrix}$$

$$D = \begin{bmatrix} 0 & 0 & 0 \\ 0 & 0 & 2.0673 \\ 0.2192 & 1.2031 & 0 \\ 0 & 0 & 0 \end{bmatrix}; \quad E_1 = \begin{bmatrix} 0 & 1 & 0 & 0 \\ 0 & 0 & 0 & 1 \\ 0 & 0 & 0 & 0 \end{bmatrix}; \quad u_0 = \begin{bmatrix} 1 \\ 1 \end{bmatrix}$$

We take for $R$ and $Q$

$$Q = I_4; \qquad R = I_2$$

First, the guaranteed cost control is designed solving the following problem:

$$\min_{\varepsilon} \ \text{trace}\big[P(\varepsilon)\big]$$

The optimal value corresponds to $\varepsilon^* = 0.752$. The guaranteed cost control is given by

$$u = \begin{bmatrix} -1.03 & 0.10 & 1.13 & 1.86 \\ 0.25 & 1.50 & -0.02 & -0.89 \end{bmatrix} x$$

Algorithm 5.3 is implemented with $\tau_N = 10^6$. Figure 5.3 shows the projections of switching surfaces in the different state space planes. Figure 5.4 compares the guaranteed cost control when saturation does not affect the input (dashed line) and our piecewise-linear control designed to avoid saturation (solid line) for the nominal system when we have the initial condition $x_0 = [-0.021 \ 0.92 \ -0.61 \ 0]'$.

## 5.2 The Parameterized ARE Approach

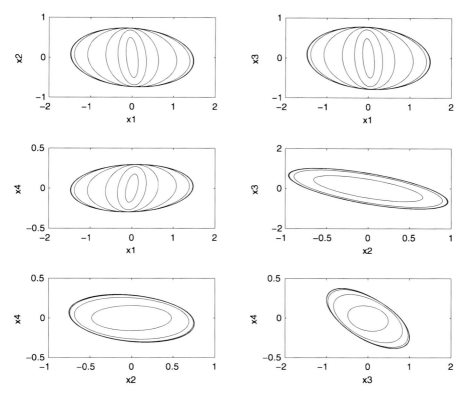

**Fig. 5.3** Example 5.2—eight ellipsoids for $\tau$ between 1 and $10^6$

**Fig. 5.4** Example 5.2—guaranteed cost and nonlinear controls

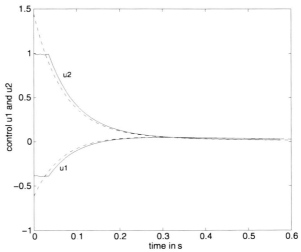

224      5   Synthesis via a Parameterized ARE Approach or a Parameterized LMI

## 5.3 A Parameterized LMI Approach

Another parameterization of the control gain matrix is possible through an LMI formulation. We have seen at the end of Sect. 5.2.3 that there are two conflicting objectives in our control problem: the stabilization of the largest domain of initial conditions and the best achievable performance around the origin. When the control is constrained, it is well known that the whole space cannot be stabilized unless the open-loop system is assumed to be stable (see Sect. 1.6.2). The following lemma aims at maximizing the size of the stabilizable initial condition set while ensuring robust stability under control constraints, i.e. the control is not allowed to saturate. The stabilizable initial condition set belongs to the region of linearity of the closed-loop system $R_L$ (cf. Chap. 1).

**Lemma 5.4** *Define $W^*$ and $Y^*$ as the optimal solution to the LMI determinant maximization problem:*

$$\min_{W,Y} \; \log\left(\det\left(W^{-1}\right)\right)$$

$$\begin{bmatrix} W & Y'_{(i)} \\ Y_{(i)} & u^2_{0(i)} \end{bmatrix} \geq 0, \quad i = 1, \dots, m \tag{5.17}$$

$$\begin{bmatrix} A_0 W + W A_0' + B Y + Y' B' + D' D & W E'_1 \\ E_1 W & -I \end{bmatrix} < 0$$

*Then the Lyapunov level set $\mathcal{E}(W^{*-1}, 1)$ is a maximum ellipsoid such that for any initial condition $x_0$ in $\mathcal{E}(W^{*-1}, 1)$, the linear state feedback control:*

$$u = K^* x = Y^* W^{*-1} x$$

*belongs to $\mathcal{U}$ and asymptotically stabilizes system (5.1).*

*Proof* See Sect. 7.2 in [45].        □

Note that problem (5.17) is a convex optimization problem which can easily be solved by efficient numerical tools. To introduce the LMI parameterization method, the following lemma proposes a counterpart to the ARE guaranteed cost approach presented in the previous section. It aims at addressing the problem of achieving the best performance around the origin with respect to the quadratic criterion (5.4) with a control belonging to $\mathcal{U}$.

**Proposition 5.7** *Define $W^*, Y^*$ and $\mu^*$ as the optimal solution to the LMI optimization problem:*

## 5.3 A Parameterized LMI Approach

$$\min_{W=W'>0, Y, \mu>0} \mu$$

$$\begin{bmatrix} \mu I_n & I_n \\ I_n & W \end{bmatrix} \geq 0$$

$$\begin{bmatrix} A_0 W + W A_0' + BY + Y'B + DD' & W E_1' & Y' & W \\ E_1 W & -I & 0 & 0 \\ Y & 0 & -R^{-1} & 0 \\ W & 0 & 0 & -Q^{-1} \end{bmatrix} < 0 \tag{5.18}$$

$$\begin{bmatrix} W & Y_{(i)}' \\ Y_{(i)} & u_{0(i)}^2 \end{bmatrix} \geq 0, \quad i = 1, \ldots, m$$

Then for any initial condition $x_0$ in the Lyapunov level set $\mathcal{E}(W^{*-1}, 1)$ the linear state feedback control

$$u = K^* x = Y^* W^{*-1} x$$

belongs to $\mathcal{U}$, asymptotically stabilizes system (5.1) and ensures a certain performance level characterized by $J \leq \text{Trace}[W^*] \leq n\mu^*$.

The first and second inequalities ensure that matrix $W$ is a guaranteed cost matrix (see Appendix B). The last inequality ensures that $u \in \mathcal{U}$. We are now in position for introducing the parameterized optimization problem which combines the design objectives of the two previous lemmas.

Let us introduce the LMI optimization problem:

$$\min_{W=W'>0, Y, \mu>0} (1-\alpha) \log\left(\det\left(W^{-1}\right)\right) + \alpha\mu$$

$$\begin{bmatrix} \mu I_n & I_n \\ I_n & W \end{bmatrix} \geq 0$$

$$\begin{bmatrix} A_0 W + W A_0' + BY + Y'B + DD' & W E_1' & \alpha Y' & \alpha W \\ E_1 W & -I & 0 & 0 \\ \alpha Y & 0 & -R^{-1} & 0 \\ \alpha W & 0 & 0 & -Q^{-1} \end{bmatrix} < 0 \tag{5.19}$$

$$\begin{bmatrix} W & Y_{(i)}' \\ Y_{(i)} & u_{0(i)}^2 \end{bmatrix} \geq 0, \quad i = 1, \ldots, m$$

$$0 \leq \alpha \leq 1$$

When $\alpha$ tends to 1, the problem above becomes problem (5.18) and when $\alpha$ tends to 0, it becomes problem (5.17). Here, the idea is to let $\alpha$ increase from 0 to 1 as a function of $x$ when $x$ approaches the origin. The parameterized gain

$$K(\alpha) = Y(\alpha)W(\alpha)^{-1}$$

varies from low-gain $K(0)$ to high-gain $K(1)$.

To ensure that the ellipsoid $\mathcal{E}(W(\alpha)^{-1}, 1)$ contains ellipsoid $\mathcal{E}(W(\beta)^{-1}, 1)$ when $0 \le \alpha \le \beta \le 1$, the constraints:

$$0 < W(\beta) < W(\alpha) \tag{5.20}$$

are inserted into problem (5.19). The following algorithm can be implemented for designing the control law.

**Algorithm 5.4**

0. Build an increasing sequence $\alpha_0 = 0 < \alpha_1 < \cdots < \alpha_N = 1$. Solve problem (5.18). Set $W(\alpha_0) = W^*$, $Y(\alpha_0) = Y^*$. Let $\varXi(0) = \{\mathcal{E}(W(\alpha_0)^{-1}, 1)\}$, $K = \{Y(\alpha_0)W(\alpha_0)^{-1}\}$ and $\beta = \alpha_0$.
i. Let $\alpha = \alpha_{N-i}$. Solve LMI problem (5.19) and (5.20) for $W(\alpha)$ and $Y(\alpha)$. Let $\varXi(0) = \{\varXi(0); \mathcal{E}(W(\alpha)^{-1}, 1)\}$, $K = \{K; Y(\alpha)W(\alpha)^{-1}\}$, $\beta = \alpha$.
   If $i < N$ go to step $i + 1$ else stop.

All the computations that are required to implement the control law generation can be performed off-line. Only on-line memory is needed to store the family of nested ellipsoids $\varXi(0)$ and piecewise-linear control $K$. The choice of a sequence of parameters $\alpha_i$ in step 0 can be done following similar arguments as for the ARE approach. In step 0, a good choice for a sequence parameters $\alpha_i$ could be a logarithmic sequence:

$$\alpha_i = \frac{\rho^{i/N} - 1}{\rho - 1} \tag{5.21}$$

for $i = 1, \ldots, N$ and $\rho > 1$. When $\rho$ approaches 1, the scalars $\alpha_i$ tend to be linearly equally spaced. To increase $\rho$ implies tighter ellipsoid nesting far from the origin. For more details see [173].

*Remark 5.2* There are two main reasons for which the control laws obtained from our parameterized ARE or LMI approaches differ from the one proposed in [321, 395]. On the one hand, our domain of attraction is the innermost ellipsoid where guaranteed cost is ensured, whereas it is the origin in [321, 395]. On the other hand, the control law in [321, 395] does not handle uncertainty.

*Remark 5.3* As pointed out at the beginning of the chapter, it is possible to extend the proposed approaches to the case where uncertainty also affects input matrix $B$. Concerning the ARE approach, this extension is not trivial because of the complexity of the corresponding ARE and the control expression. It induces a loss of monotonicity properties which makes the derivation of a family

5.3  A Parameterized LMI Approach

of nested ellipsoids more elaborated. For more details, the reader is referred to [173].

Concerning the LMI approach, this extension is quite trivial. If the input matrix is written

$$B = B_0 + DFE_2$$

with $E_2$ of appropriate dimensions, it suffices to replace (5.19) by

$$\min_{W=W'>0,Y,\mu>0} (1-\alpha) \log \left( \det (W^{-1}) \right) + \alpha\mu$$

$$\begin{bmatrix} \mu I_n & I_n \\ I_n & W \end{bmatrix} \geq 0$$

$$\begin{bmatrix} A_0 W + W A_0' + B_0 Y + Y' B_0 + DD' & W E_1' + Y' E_2' & \alpha Y' & \alpha W \\ E_1 W + E_2 Y & -I & 0 & 0 \\ \alpha Y & 0 & -R^{-1} & 0 \\ \alpha W & 0 & 0 & -Q^{-1} \end{bmatrix} < 0$$

$$\begin{bmatrix} W & Y_{(i)}' \\ Y_{(i)} & u_{0(i)}^2 \end{bmatrix} \geq 0, \quad i = 1, \dots, m$$

$$0 \leq \alpha \leq 1$$

$$(5.22)$$

*Remark 5.4* The maximization of the set of admissible initial conditions is taken into account into problem (5.19) through the term $\log \det W^{-1}$. In order to solve the problem with the help of the majority of the available numerical tools, we can transform this term adding the following constraint:

$$\begin{bmatrix} \gamma I_n & I_n \\ I_n & W \end{bmatrix} \geq 0 \qquad (5.23)$$

and replacing it by "$\gamma$" in the criterion. This constraint ensures that

$$\text{Trace}\left[ W^{-1} \right] \leq n\gamma$$

and minimizing $\gamma$ goes with the objective of maximizing the volume of the set of admissible initial conditions. In fact the new criterion writes: $(1 - \alpha)\gamma + \alpha\mu$. We may also remark that if the first constraint of problem (5.19) is satisfied together with (5.23), we have

$$\begin{bmatrix} ((1-\alpha)\gamma + \alpha\mu) I_n & I_n \\ I_n & W \end{bmatrix} \geq 0$$

and denoting $\theta = (1 - \alpha)\gamma + \alpha\mu$, an optimization problem whose solution goes with the objective of problem (5.19) can be written as

$$\min_{W=W'>0, Y, \theta>0} \theta$$

$$\begin{bmatrix} \theta I_n & I_n \\ I_n & W \end{bmatrix} \geq 0$$

$$\begin{bmatrix} A_0 W + W A_0' + BY + Y'B + DD' & W E_1' & \alpha Y' & \alpha W \\ E_1 W & -I_{f_1} & 0 & 0 \\ \alpha Y & 0 & -R^{-1} & 0 \\ \alpha W & 0 & 0 & -Q^{-1} \end{bmatrix} < 0 \quad (5.24)$$

$$\begin{bmatrix} W & Y_{(i)}' \\ Y_{(i)} & u_{0(i)}^2 \end{bmatrix} \geq 0, \quad i = 1, \ldots, m$$

$$0 \leq \alpha \leq 1$$

*Example 5.3* Consider the same system as in Example 5.2. We run Algorithm 5.4 replacing problem (5.19) by problem (5.24). The values of $\alpha$ are set thanks to expression (5.21). $\rho$ is equal to 20 and 11 ellipsoids are generated. With $R = I_m$ and $Q = I_n$, the LMI guaranteed cost control, obtained for $\alpha = 1$, reads

$$u = \begin{bmatrix} -0.34 & 0.29 & 0.46 & 0.84 \\ 0.09 & 0.50 & 0.07 & -0.50 \end{bmatrix} x$$

Figure 5.5 represents the projections of ellipsoids in the different state space planes. The volume ratio of the ellipsoids obtained through the LMI approach and the ARE approach is approximately 100 for the innermost ellipsoids where performance is ensured and 200 for the outermost ellipsoids of stabilizable initial conditions. This clearly illustrates the greater flexibility of the LMI approach over the ARE approach. However, the price to pay is a higher computational load [173]. However, all the computations are done off-line.

*Remark 5.5* If only a measured output is available for control purpose, namely:

$$y = Cx$$

where $y \in \Re^p$ and $C \in \Re^{p \times n}$ is the appropriately dimensioned output matrix supposed to be full rank, the state feedback can be replaced by an output feedback control, for example, by the static output control expressed as

$$u = Ly$$

where $L \in \Re^{m \times p}$ is the static output control gain matrix. In the context of the ARE approach, finding a solution is really difficult, if not impossible. In the context of

## 5.3 A Parameterized LMI Approach

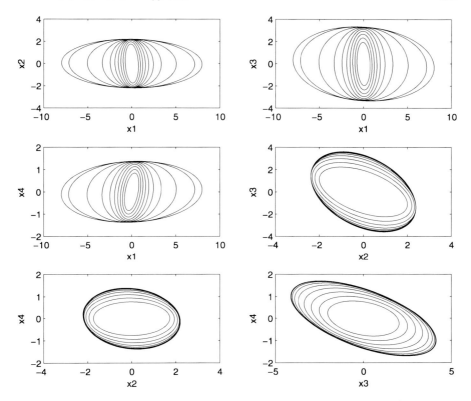

**Fig. 5.5** Example 5.3—eleven ellipsoids for $\alpha$ between 0 (the outer one) and 1 (the inner one)

LMI approach, it is possible to try to obtain a solution according to the following algorithm [116].

**Algorithm 5.5**

1. Select a matrix $N \in \Re^{(n-p) \times n}$ such that the matrix

$$M = \begin{bmatrix} C \\ N \end{bmatrix}$$

is invertible. This is possible because $C$ is full rank.

2. Perform the following change of state vector:

$$\bar{x} = Mx = \begin{bmatrix} y \\ z \end{bmatrix}$$

The new matrices of the state space equation are

$$\bar{A}_0 = MA_0 M^{-1}, \qquad \bar{B}_0 = MB_0, \qquad \bar{D} = MD$$

$$\bar{E}_1 = M^{-1}E_1, \qquad \bar{E}_2 = E_2$$

and the new form for matrix $C$, central for the sequel, is

$$\bar{C} = CM^{-1} = [I_p \quad 0]$$

The control is written

$$u = Ly = LCx = LCM^{-1}\bar{x} = L[I_p \quad 0]\bar{x} = [L \quad 0]\bar{x}$$

The previous control is nothing but a state feedback with a structured gain.

3. Implement Algorithm 5.4 with the new state space matrices, adding the following constraints:

$$W(\alpha) = \begin{bmatrix} W_1(\alpha) & 0 \\ 0 & W_2(\alpha) \end{bmatrix} \quad \text{and} \quad Y(\alpha) = [Y_1(\alpha) \quad 0]$$

where $W_1(\alpha) \in \Re^{p \times p}$, $W_2(\alpha) \in \Re^{(n-p) \times (n-p)}$ and $Y_1(\alpha) \in \Re^{m \times p}$. Then, the static output control gain is given by

$$L(\alpha) = Y_1(\alpha)W_1(\alpha)^{-1}$$

## 5.4 Multi-objective Control: Eigenvalues Placement and Guaranteed Cost

In this section, the results of the previous sections are extended to the case where in plus of a performance expressed through a guaranteed cost, we want to guarantee performance in terms of time constant response around the origin by eigenvalues placement. For reasons which will be made clear later, the region of placement is a disk. The case of uncertain continuous- and discrete-time systems is considered in the same framework. When the disk is the unitary disk and system discrete-time, the method is the counterpart of the previous one developed for continuous-time systems. The system under study is written

$$\delta[x] = (A_0 + DFE_1)x + Bu \tag{5.25}$$

with the same definition as (5.1) for all the matrices. $\delta$ is the derivation operator in the continuous-time case that is $\delta[x] = \dot{x}$ or the delay operator for the discrete one, $\delta[x(t)] = x(t+1)$. When no confusion is possible, the notation is the same as before.

### 5.4.1 Preliminaries

In [110], a concept similar to the quadratic stabilizability concept [15] was defined in order to solve the state feedback pole placement problem in a disk $D(r, \alpha)$ with

## 5.4  Multi-objective Control: Eigenvalues Placement and Guaranteed Cost

center $\alpha + j0$ and radius $r$. A disk for pole location offers a good compromise between mode damping and speed. For continuous-time systems, it suffices to include it, in a sector located in the left half complex plane, while for a discrete-time system, it satisfactory approximates a cardioid which is the discrete counterpart of a sector [110]. This concept referred as the d-quadratic stabilizability concept has proven to be efficient for solving some practical robust control problems [77]. In [110], a necessary and sufficient condition of d-quadratic stabilization for system (5.25) by a linear state feedback was derived (see Appendix B).

**Lemma 5.5** *The system (5.25) is d-quadratically stabilizable if and only if for all* $R \in \mathfrak{R}^{m \times m}$, $R = R' > 0$ *and* $Q \in \mathfrak{R}^{n \times n}$, $Q = Q' > 0$, *there exist* $\varepsilon > 0$ *and a positive definite matrix* $X \in \mathfrak{R}^{n \times n}$, $X = X' > 0$ *solution to*

$$A_r'\left(X^{-1} + B_r R^{-1} B_r' - \varepsilon D_r D_r'\right)^{-1} A_r - X + \varepsilon^{-1} E_r' E_r + Q = 0 \qquad (5.26)$$

*satisfying*

$$\varepsilon^{-1} 1 - D_r' X D_r > 0 \qquad (5.27)$$

*with*

$$A_r = \frac{A_0 - \alpha 1}{r}, \qquad B_r = \frac{B}{r}, \qquad E_r = \frac{E_1}{\sqrt{r}}, \qquad D_r = \frac{D}{\sqrt{r}}$$

*If such conditions are met, a control gain is given by*

$$K = -R^{-1} B_r' \left(X^{-1} + B_r R^{-1} B_r' - \varepsilon D_r D_r'\right)^{-1} A_r \qquad (5.28)$$

In [110], an algorithm is proposed to check the d-quadratic stabilizability. This algorithm converges in a finite number of steps and involves some elementary calculations such as solving a Riccati equation. We may note that the existence of a positive definite symmetric matrix $X$ is independent of the choice of the $Q$ and $R$ matrices. This degree of freedom can be used to combine pole location with guaranteed LQ cost specification [109]. Associated with the system (5.25) is the cost function:

$$J = \int_0^\infty \left[x' Q x + u' R u\right] dt, \qquad Q = Q' > 0, \ R = R' > 0,$$

for a continuous-time system

$$J = \sum_{t=0}^\infty \left[x' Q x + u' R u\right], \qquad Q = Q' > 0, \ R = R' > 0,$$

for a discrete-time system.

Now suppose that $x(0) = x_0$ is the state initial condition and define

$$E = \left\{\varepsilon > 0 : \text{there exists } X(\varepsilon) = X'(\varepsilon) > 0 \text{ solution to (5.26)}\right\}$$

and the set:

$$\mathcal{K} = \left\{ K = -R^{-1} B_r' \left( X^{-1} + B_r R^{-1} B_r' - \varepsilon D_r D_r' \right)^{-1} A_r : \varepsilon \in E \right\}$$

It is shown in [289] and [109] that if $K \in \mathcal{K}$, the closed-loop uncertain system is d-quadratically stable and in addition, the corresponding value of the cost function satisfies

$$J \le x_0' \frac{|\alpha| X(\varepsilon)}{r^2} x_0, \quad \forall F : F'F \le 1, \quad \text{for a continuous-time system}$$

$$J \le x_0' \frac{X(\varepsilon)}{r^2} x_0, \quad \forall F : F'F \le 1, \quad \text{for a discrete-time system}$$

Note that the above bounds depend on the initial condition $x(0) = x_0$. To remove this dependence, we assume as in Sect. 5.2 that $x_0$ is a zero mean random variable satisfying $E[x_0 x_0'] = 1$. In this case, the bounds become

$$\bar{J} = E[J] \le \operatorname{trace}\left[ \frac{|\alpha|}{r^2} X(\varepsilon) \right] = \bar{X}(\varepsilon), \quad \text{for a continuous-time system}$$

$$\bar{J} = E[J] \le \operatorname{trace}\left[ \frac{X(\varepsilon)}{r^2} \right] = \bar{X}(\varepsilon), \quad \text{for a discrete-time system}$$

This result implies that a control, which minimizes the cost bound, can be obtained by choosing $\varepsilon > 0$ minimizing $\bar{X}(\varepsilon)$. This control is known as a guaranteed cost control. Then we have to solve the following optimization problem:

$$J^* = \min_{\varepsilon \in E} \bar{X}(\varepsilon) \tag{5.29}$$

If one is able to solve this problem, a certain performance level is attained for the closed-loop system. Similarly to Sect. 5.2, it is shown in [109] that trace $X(\varepsilon)$ is a convex function with respect to $\varepsilon$ over $E$. For the sequel, we denote

$$\varepsilon_* = \operatorname{Arg}\left\{ \operatorname{Min} \bar{X}(\varepsilon) \right\}$$

The problem addressed can be stated as follows.

**Problem 5.1** Design a controller $u(x)$ and a set $\varXi$ such that

(i) $u(x) \in \mathcal{U}, \forall x \in \varXi$,
(ii) $\forall x_0 \in \varXi$, the resulting trajectories $x(t; x_0) \to 0$ as $t \to \infty$,
(iii) the closed-loop system is d-stable with a guaranteed cost in a subset of $\varXi$.

For simplicity, the single-input case is first addressed, i.e., $m = 1$ and $B \in \mathfrak{R}^n$.

5.4 Multi-objective Control: Eigenvalues Placement and Guaranteed Cost 233

### 5.4.2 The Single-Input Case

Suppose that the d-quadratic stabilization with a guaranteed cost problem has a solution. Then for $R = R' > 0$ and $Q = Q' > 0$, there exist $\varepsilon_* > 0$ and $X = X' > 0$ solutions to

$$A_r' \left( X^{-1} + B_r R^{-1} B_r' - \varepsilon_* D_r D_r' \right)^{-1} A_r - X + \varepsilon_*^{-1} E_r' E_r + Q = 0 \qquad (5.30)$$

and

$$\varepsilon_*^{-1} I - D_r' X D_r > 0 \qquad (5.31)$$

The control gain is obtained as

$$K = -R^{-1} B_r' \left( X^{-1} + B_r R^{-1} B_r' - \varepsilon_* D_r D_r' \right)^{-1} A_r \qquad (5.32)$$

The idea is to define a neighborhood of the origin where the control (5.32) is applied without saturation and where a performance level is attained through a closed-loop pole location and a guaranteed cost. Outside this neighborhood, a linear-like control with state dependent gains is derived from the solutions of a family of parameterized Riccati equations obtained in the following way:

$$A_r' \left( S(\tau)^{-1} + B_r R^{-1} B_r' - \varepsilon_* D_r D_r' \right)^{-1} A_r - S(\tau) + \varepsilon_*^{-1} E_r' E_r + Q/\tau = 0 \quad (5.33)$$

$$\varepsilon_*^{-1} I - D_r' S(\tau) D_r > 0 \qquad (5.34)$$

with $1 \leq \tau < \infty$. For $\tau = 1$, we obtain (5.30) and the corresponding control (5.32) is deduced letting $\tau = 1$ in:

$$K = -R^{-1} B_r' \left( S(\tau)^{-1} + B_r R^{-1} B_r' - \varepsilon_* D_r D_r' \right)^{-1} A_r \qquad (5.35)$$

We have the following important result.

**Lemma 5.6** *Suppose that the system* (5.25) *is d-quadratically stabilizable. Then there exist $\varepsilon_* > 0$ and $S(1) = S(1)' > 0$ satisfying* (5.30) *and* (5.31). *Then for all $\tau > 1$,* (5.33) *has a positive definite symmetric solution $S(\tau) = S(\tau)' > 0$ satisfying* (5.34).

This result follows from the fact that the existence of a positive definite symmetric matrix to solution (5.26), and therefore (5.33) is independent of the choice of the matrices $Q$ and $R$, see [109]. To simplify the notation, let us denote

$$\Delta = B_r R^{-1} B_r' - \varepsilon_* D_r D_r' \qquad (5.36)$$

and introduce the following parameterized ellipsoidal set:

$$\mathcal{E}\left( S(\tau), c(\tau)^{-1} \right) = \left\{ x \in \mathfrak{R}^n; \ x' S(\tau) x \leq c(\tau) = \frac{u_0^2}{R^{-1} B_r' (S(\tau)^{-1} + \Delta)^{-1} B_r R^{-1}} \right\}$$

Defining the positive definite symmetric matrix:

$$\mathcal{X}(\tau) = \frac{S(\tau)}{c(\tau)}$$

The previous ellipsoid can be normalized as follows:

$$\mathcal{E}\big(\mathcal{X}(\tau), 1\big) = \big\{x \in \Re^n;\ x'\mathcal{X}(\tau)x \le 1\big\} \tag{5.37}$$

and the associated $\tau$-parameterized family of ellipsoidal sets is defined by

$$\Xi(\tau) = \big\{\mathcal{E}\big(\mathcal{X}(\tau_i), 1\big), 1 \le \tau_i \le \tau\big\} \tag{5.38}$$

The following result can be stated.

**Lemma 5.7** $\frac{d\mathcal{X}(\tau)}{d\tau}$ *is negative definite and* $\frac{dc(\tau)}{d\tau} \ge 0$.

*Proof* Differentiating (5.33), we have

$$A_r'\big(S(\tau)^{-1} + \Delta\big)^{-1}S(\tau)^{-1}\frac{dS}{d\tau}S(\tau)^{-1}\big(S(\tau)^{-1} + \Delta\big)^{-1}A_r - \frac{dS}{d\tau} - \frac{Q}{\tau^2} = 0$$

since $S(\tau)^{-1}(S(\tau)^{-1} + \Delta)^{-1}A_r$ is a stable matrix [414], by Lyapunov arguments, we deduce that $\frac{dS}{d\tau} < 0$. On the other hand, one gets

$$\frac{d\mathcal{X}(\tau)}{d\tau} = \frac{\frac{dS(\tau)}{d\tau}c(\tau) - S(\tau)\frac{dc(\tau)}{d\tau}}{c(\tau)^2}$$

One can verify:

$$\frac{dc(\tau)}{d\tau} = -u_0^2 \frac{R^{-1}B'(S(\tau)^{-1} + \Delta)^{-1}S(\tau)^{-1}\frac{dS}{d\tau}S(\tau)^{-1}(S(\tau)^{-1} + \Delta)^{-1}BR^{-1}}{(R^{-1}B_r'(S(\tau)^{-1} + \Delta)^{-1}B_r R^{-1})^2} \ge 0 \tag{5.39}$$

It follows that

$$\frac{d\mathcal{X}(\tau)}{d\tau} < 0. \qquad \square$$

It is worth noticing that $c(\tau)$ is a monotonous increasing function of $\tau$ and the results in [4] can be applied.

**Proposition 5.8** [4] *We have the following statements.*

(i) $\Xi(\tau_1) \subset \operatorname{int} \Xi(\tau_2)$, *whenever* $\tau_1 < \tau_2$.
(ii) *The $\tau$-parameterized family of ellipsoids $\Xi(\tau)$ is a nested family set, that is,*

$$\Xi(\tau_1) \subset \operatorname{int} \Xi(\tau_2), \quad \text{whenever } \tau_1 < \tau_2$$

*with a maximal element $\Xi = \bigcup_\tau \Xi(\tau)$.*

5.4 Multi-objective Control: Eigenvalues Placement and Guaranteed Cost 235

(iii) $\frac{d\mathcal{X}(\tau)}{d\tau} < 0$ *if and only if* $\Xi(\tau_1) \subset \text{int } \Xi(\tau_2)$, *whenever* $\tau_1 < \tau_2$.

In order to formulate the control, we recall the following definition.

**Definition 5.2** Given $x^* \in \Xi$ and $\tau^* \in [1, \infty)$ such that $x^* \in \partial\Xi(\tau)$, then define $\tau(x^*) = \tau^*$. If $x^* \in \Xi(1)$ then define $\tau(x^*) = 1$.

**Proposition 5.9** [112] $\tau$ *is a continuous function from* $\Xi$ *to* $[1, \infty)$.

The $\tau$ function is implicitly defined as the positive solution to the algebraic equation (see Sect. 5.2)

$$x'S(\tau)x - c(\tau) = 0 \qquad (5.40)$$

We are now in position to define the controller. It is defined by

$$u(x) = \begin{cases} -R^{-1}B'_r(S(\tau)^{-1} + B_r R^{-1}B'_r - \varepsilon_* D_r D'_r)^{-1}A_r x & \text{if } x \in \Xi \setminus \Xi(1) \\ -R^{-1}B'_r(S(1)^{-1} + B_r R^{-1}B'_r - \varepsilon_* D_r D'_r)^{-1}A_r x & \text{if } x \in \Xi(1) \end{cases} \qquad (5.41)$$

The idea is to drive the system to the neighborhood of the origin applying the control $-R^{-1}B'_r(S(\tau)^{-1} + B_r R^{-1}B'_r - \varepsilon_* D_r D'_r)^{-1}A_r x$ then near the origin to apply the control $-R^{-1}B'_r(S(1)^{-1} + B_r R^{-1}B'_r - \varepsilon_* D_r D'_r)^{-1}A_r x$ which ensures asymptotic stability and a certain performance level by a closed-loop pole location and a guaranteed cost.

**Proposition 5.10** *Suppose that the system* (5.25) *is d-quadratically stabilizable. Let* $S(\tau)$ *be the positive definite symmetric solution to the Riccati equation* (5.33) *and* $\Xi(\tau)$ *be the* $\tau$-*parameterized family of ellipsoids defined by* (5.38), *then the controller defined by* (5.41) *satisfies the constraint* $-u_0 \le u(x) \le u_0$ *and drives any point* $x \in \Xi = \bigcup_\tau \Xi(\tau)$ *to the origin.*

*Proof* The first step is to show that control (5.41) is bounded by $u_0$. To show that, define the following optimization problem which allows one to find the maximal ellipsoid defined by the quadratic function $x'S(\tau)x$ where the feedback $-R^{-1}B'_r(S(\tau)^{-1} + B_r R^{-1}B'_r - \varepsilon_* D_r D'_r)^{-1}A_r x$ is bounded by $u_0$:

$$\begin{cases} \max_x x'S(\tau)x = \widetilde{c}(\tau) \\ \text{subject to} \quad \left|-R^{-1}B'_r(S(\tau)^{-1} + B_r R^{-1}B'_r - \varepsilon_* D_r D'_r)^{-1}A_r x\right| = u_0 \end{cases}$$

Applying the necessary condition for an extremum, after some simple calculations, we obtain

$$x'S(\tau)x = \widetilde{c}(\tau) = \frac{u_0^2}{R^{-1}B'_r(S(\tau)^{-1} + \Delta)^{-1}A_r S(\tau)^{-1}A'_r(S(\tau)^{-1} + \Delta)^{-1}B_r R^{-1}} \qquad (5.42)$$

where $\Delta$ is defined by (5.36). From (5.33), we can deduce that

$$\begin{bmatrix} -(S(\tau)^{-1} + \Delta) & A_r \\ A_r' & -S(\tau) \end{bmatrix} < 0$$

and then $A_r S(\tau)^{-1} A_r' < S(\tau)^{-1} + \Delta$. From this last inequality, we have $c(\tau) < \tilde{c}(\tau)$. We conclude that control (5.41) is bounded by $u_0$. We have to prove now that the compact set $\Xi(1)$ is a finite attractor in $\Xi$ for the closed-loop system. It suffices to show that for all $x$ in the closure of $\Xi \backslash \Xi(1)$, for a continuous-time system one gets $\dot{\tau} < 0$ and for a discrete-time system one gets $\tau(x(t+1)) - \tau(x(t)) < 0$. The Riccati equation (5.33) can be written as

$$(A_r + B_r K)' \big(S(\tau)^{-1} - \varepsilon_* D_r D_r'\big)^{-1} (A_r + B_r K)$$

$$- S(\tau) + \varepsilon_*^{-1} E_r' E_r + K' R K + Q/\tau = 0 \qquad (5.43)$$

and then we have [110]

$$\begin{bmatrix} -S(\tau)^{-1} & A_r + D_r F E_r + B_r K \\ (A_r + D_r F E_r + B_r K)' & -S(\tau) \end{bmatrix} < 0 \qquad (5.44)$$

*Continuous-Time System* The time derivative of $\tau$ is given by

$$\dot{\tau} = \frac{-2x' S(\tau) \dot{x}}{x' \frac{dS(\tau)}{d\tau} x - \frac{dc(\tau)}{d\tau}} \qquad (5.45)$$

We have

$$2x' S(\tau) \dot{x} = x' \big[ (A + DFE + BK)' S(\tau) + S(\tau)(A + DFE + BK) \big] x < 0$$

by Theorem 2 in [110]. Noting that $\frac{dS}{d\tau} < 0$ and $\frac{dc}{d\tau} \geq 0$, it follows that $\dot{\tau} < 0$.
*Discrete-Time System* First recall that $x' S(\tau) x - c(\tau) = 0$ can be written in a normalized form:

$$x' \mathcal{X}(\tau) x - 1 = 0$$

Then

$$x'(t+1) \mathcal{X}\big(\tau(t+1)\big) x(t+1) - x'(t) \mathcal{X}\big(\tau(t)\big) x(t) = 0$$

We have

$$x'(t+1) \mathcal{X}\big(\tau(t+1)\big) x(t+1) - x'(t) \mathcal{X}\big(\tau(t+1)\big) x(t)$$

$$+ x'(t) \mathcal{X}\big(\tau(t+1)\big) x(t) - x'(t) \mathcal{X}\big(\tau(t)\big) x(t) = 0$$

The sum of the two first terms being negative (Theorem 2 in [110]), we deduce that

$$x'(t) \big[ \mathcal{X}\big(\tau(t+1)\big) - \mathcal{X}\big(\tau(t)\big) \big] x(t) > 0$$

## 5.4 Multi-objective Control: Eigenvalues Placement and Guaranteed Cost

Furthermore, $c(\tau)$ being a monotonous increasing function of $\tau$, by Lemma 5.7, we deduce that

$$\tau(t+1) < \tau(t)$$

and thus this completes the proof of the proposition. $\qquad\square$

*Remark 5.6* The proposed method gives a solution to the problem stated in [4]. Actually, we extend the results of [4] to the case of uncertain systems. When $D = 0, E = 0, \alpha = 0$ and $r = 1$, the problem solved here is exactly the same as the one defined in [4]. The main difference resides in the definition of the family of ellipsoids. In [4], it is defined by

$$x'S(\tau)x \le \frac{u_0^2}{\|B\|^2 \|A_0\|^2 \|S(\tau)\|\beta_2} = \bar{c}(\tau)$$

where $\beta_2$ is such that

$$\left\| \left( R + B'S(\tau)B \right)^{-1} \right\| \le \beta_2$$

It is important to note that the above approximation of the ellipsoid of maximum size, where the control does not saturate, explicitly depends on the matrices $A_0$ and $B$. With this explicit dependence, it is not easy to cope with uncertainties. The approximation proposed here is

$$x'S(\tau)x \le \frac{u_0^2}{R^{-1}B'(S(\tau)^{-1} + B'R^{-1}B)^{-1}BR^{-1}}$$

It does not depend explicitly on $A_0$.

*Remark 5.7* When the model is certain and if the eigenvalues of $A_r$ are in the unit disk, we have the following lemma.

**Lemma 5.8** *Suppose that the system without uncertainty is such that $A_r$ has all its eigenvalues in the unit disk. Then the state feedback $u(x)$ is a global stabilizer for the system.*

The proof is exactly the same as for Theorem 3.1 in [4]. It has to be noticed that the above lemma is true for a continuous-time as well as for a discrete-time system.

*Remark 5.8* Suppose we have found a d-quadratic stabilizing control $K$. It can be interesting to find $R, Q, S$ such that the Riccati equation (5.30) is satisfied. This problem can be seen as an inverse problem and has been solved for the LQ design in [45]. First remark that the Riccati equation can be written as

$$(A_r + B_r K)'(S^{-1} - \varepsilon_* D_r D_r')^{-1}(A_r + B_r K) - S + \varepsilon_*^{-1}E_r'E_r + K'RK + Q = 0$$

$K$ being given. Then we have

$$\begin{bmatrix} -(S^{-1} - \varepsilon_* D_r D_r') & (A_r + B_r K) \\ (A_r + B_r K)' & -S + \varepsilon_*^{-1} E_r' E_r + K' R K \end{bmatrix} < 0$$

Letting $P = (S^{-1} - \varepsilon_* D_r D_r')^{-1}$ the previous inequality becomes

$$\begin{bmatrix} -P^{-1} & (A_r + B_r K) \\ (A_r + B_r K)' & -(P^{-1} + \varepsilon_* D_r D_r')^{-1} + \varepsilon_*^{-1} E_r' E_r + K' R K \end{bmatrix} < 0$$

Applying the inversion lemma and Schur's complement, we get

$$\begin{bmatrix} -P^{-1} & (A_r + B_r K) & 0 \\ (A_r + B_r K)' & -P + \varepsilon_*^{-1} E_r' E_r + K' R K & P D_r \\ 0 & D_r' P & -(\varepsilon_*^{-1} I + D_r' P D_r) \end{bmatrix} < 0$$

Right- and left-multiplying by

$$\begin{bmatrix} P & 0 & 0 \\ 0 & I & 0 \\ 0 & 0 & I \end{bmatrix}$$

leads to

$$\begin{bmatrix} -P & P(A_r + B_r K) & 0 \\ (A_r + B_r K)' P & -P + \varepsilon_*^{-1} E_r' E_r + K' R K & P D_r \\ 0 & D_r' P & -(\varepsilon_*^{-1} I + D_r' P D_r) \end{bmatrix} < 0 \tag{5.46}$$

The problem can be stated as follows: Find $R = R' > 0$ and $P = P' > 0$ satisfying

$$\begin{cases} \begin{bmatrix} -P & P(A_r + B_r K) & 0 \\ (A_r + B_r K)' P & -P + \varepsilon_*^{-1} E_r' E_r + K' R K & P D_r \\ 0 & D_r' P & -(\varepsilon_*^{-1} I + D_r' P D_r) \end{bmatrix} < 0 \\ (R + B_r' P B_r) K + B_r' P A_r = 0 \end{cases}$$

If this problem is feasible, $Q$ is given by

$$Q = -\left[ (A_r + B_r K)' P (A_r + B_r K) - \left( P^{-1} + \varepsilon_* D_r D_r' \right)^{-1} + \varepsilon_*^{-1} E_r' E_r + K' R K \right]$$

### 5.4.3 The Multi-variable Case

In this section, we consider the multi-input case. The matrix $B_r$ is written as $B_r = [B_{r1} \ \ldots \ B_{rm}]$, $B_{ri} \in \Re^n$ and we take

$$R = \text{Diag}(r_1, \ldots, r_m) > 0$$

5.4 Multi-objective Control: Eigenvalues Placement and Guaranteed Cost 239

Define also

$$c_i(\tau) = \frac{u_{0(i)}^2 r_i^2}{B'_{ri}(S(\tau)^{-1} + \Delta)^{-1}B_{ri}}, \quad i = 1, \ldots, m$$

and

$$\mathcal{C}(\tau) = \min_i c_i(\tau)$$

Since $c_i(\tau)$, $i = 1, \ldots, m$ are increasing functions of $\tau$, $\mathcal{C}(\tau)$ is an increasing function of $\tau$. However, note that in the continuous-time case, $\mathcal{C}(\tau)$ is not necessarily differentiable for any $\tau > 0$, but its right-side derivative is defined as

$$DC = \lim_{\xi \to 0^+} \frac{\mathcal{C}(\tau + \xi) - \mathcal{C}(\tau)}{\xi}$$

With the previous notation, it is possible to extend the results of the previous sections.

**Proposition 5.11** *Let $S(\tau)$ be the positive definite symmetric solution to* (5.33). *Define the function $\tau(x)$ in the following way:*

- *For $x \in \Xi(1) = \{x \in \mathfrak{R}^n; \; x'S(1)x \leq \mathcal{C}(1)\}$, $\tau(x) = 1$.*
- *For $x \in \Xi \setminus \Xi(1)$, $\tau(x)$ is the positive solution of the equation:*

$$x'S(\tau)x - \mathcal{C}(\tau) = 0$$

*Then the control $u(x)$ defined by*

$$u(x) = \begin{cases} -R^{-1}B'_r(S(\tau)^{-1} + B_r R^{-1}B'_r - \varepsilon_* D_r D'_r)^{-1}A_r x & \text{if } x \in \Xi \setminus \Xi(1) \\ -R^{-1}B'_r(S(1)^{-1} + B_r R^{-1}B'_r - \varepsilon_* D_r D'_r)^{-1}A_r x & \text{if } x \in \Xi(1) \end{cases}$$

(5.47)

*satisfies the constraints*

$$-u_{0(i)} \leq u_{(i)}(x) \leq u_{0(i)}, \quad i = 1, \ldots, m$$

*and drives any point of $\Xi$ to the origin.*

*Proof* The proof of the asymptotic stability in $\Xi$, $\Xi$ being included in the region of attraction of the origin, mimics the one of Proposition 5.10 in the case of single-input case. Thus in the continuous-time case, we must consider the right-side derivative of $\mathcal{C}(\tau)$ defined above. Hence, following the same lines as in the single-input case, one gets the following case.

*Continuous-Time System* The time variation of $\tau$ is described by (5.45) in which $\frac{dc(\tau)}{d\tau}$ is replaced by $DC$ where

$$DC(\tau) = \lim_{\xi \to 0^+} \xi^{-1} \left[ \min_i \frac{u_{0(i)} r_i{}^2}{B'_{ri}(S(\tau+\xi)^{-1} + \Delta)^{-1} B_{ri}} \right.$$

$$\left. - \min_i \frac{u_{0(i)} r_i{}^2}{B'_{ri}(S(\tau)^{-1} + \Delta)^{-1} B_{ri}} \right]$$

Note that $S(\tau + \xi) < S(\tau)$, $\xi > 0$. Then we have $DC(\tau) \geq 0$ and therefore $\tau$ is a decreasing function of time.

*Discrete-Time System* The proof is exactly the same as for a single-input system. $\qquad\square$

Suppose now that $\tau \longrightarrow \infty$, the Riccati equation (5.33) becomes

$$A'_r H(\infty) A_r - S(\infty) + \varepsilon_*^{-1} E'_r E_r = 0 \tag{5.48}$$

with

$$H(\infty) = S(\infty) - S(\infty)[B_r \, D_r]$$

$$\times \begin{bmatrix} R + B'_r S(\infty) B_r & B'_r S(\infty) D_r \\ D'_r S(\infty) B_r & D'_r S(\infty) D_r - \varepsilon_*^{-1} I \end{bmatrix}^{-1} \begin{bmatrix} B'_r \\ D'_r \end{bmatrix} S(\infty)$$

It is well known that

- if $(E_r, A_r)$ is observable and if $(A_r, B_r)$ is stabilizable, $S(\infty)$ is positive definite;
- if $(E_r, A_r)$ is detectable and if $(A_r, B_r)$ is stabilizable, $S(\infty)$ is positive semi-definite.

The notions of stabilizability and detectability invoked above are those of discrete-time systems. Hence the set $\varXi$ is characterized in the following proposition.

**Proposition 5.12** *Suppose that the pair $(A_r, B_r)$ is stabilizable. Then $\varXi$ is defined by*

1. *If $(A_r, E_r)$ is observable*

$$\varXi = \left\{ x \in \mathfrak{R}^n; \ \max_i \left[ \frac{B'_i H(\infty) B_i}{u_{0(i)}^2 r_i^2} \right] x' S(\infty) x < 1 \right\}$$

2. *If $(A_r, E_r)$ is detectable, $\varXi$ is equal to the Cartesian product of $\mathfrak{R}^{\dim(\mathrm{Ker}\, S(\infty))}$ and a $k$-dimensional open ellipsoid $(k = n - \dim(\mathrm{Ker}\, S(\infty))$. This ellipsoid is a linear transformation of the set*

$$\varXi = \left\{ y \in \mathfrak{R}^k; \ \max_i \left[ \frac{B'_i H(\infty) B_i}{u_{0(i)}^2 r_i^2} \right] y' S_k(\infty) y < 1 \right\}$$

*where*

$$M^{-1} S(\infty) M = \begin{bmatrix} 0 & 0 \\ 0 & S_k(\infty) \end{bmatrix}$$

*with $M$ an appropriate change of coordinates.*

## 5.4 Multi-objective Control: Eigenvalues Placement and Guaranteed Cost

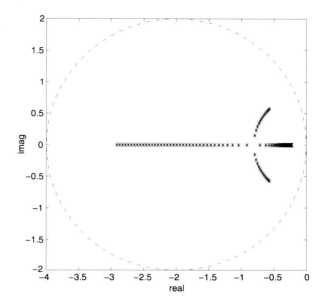

Fig. 5.6 Example 5.4—closed-loop system poles for $-1 \leq F \leq 1$

*Proof* It follows from the above arguments. □

Algorithms 5.1, 5.2 and 5.3 can be directly transposed to deal with the problem handled in this section. The main changes are to replace continuous ARE and the associated gain by their discrete counterparts. All the remarks associated with Algorithms 5.1, 5.2 and 5.3 remain valid. In the same way, it is also possible to develop a parameterized LMI approach. In that case, it is easy to derive LMI optimization problems similar to problems (5.17), (5.18) and (5.19). All these transpositions can easily be done by the reader.

*Example 5.4* Let the uncertain system (5.25) be defined by

$$A = \begin{bmatrix} 0 & 1 \\ 0 & 0 \end{bmatrix}; \quad B = \begin{bmatrix} 0 \\ 1 \end{bmatrix}; \quad D = \begin{bmatrix} 0 \\ 1 \end{bmatrix}; \quad E = \begin{bmatrix} 0 & 1 \end{bmatrix}$$

$$Q = \begin{bmatrix} 1 & 0 \\ 0 & 1 \end{bmatrix}; \quad R = 1; \quad u_0 = 0.5$$

The guaranteed cost problem with pole placement in a disk centered at $(-2, 0)$ and with radius 2 is solved. We obtain $\varepsilon_* = 0.219$ and a control gain given by

$$K = \begin{bmatrix} -0.65 & -2.13 \end{bmatrix}$$

Figure 5.6 shows the closed-loop system poles for the uncertain parameter $F$ varying between $-1$ and 1. The curve of the guaranteed cost versus $\varepsilon$ is represented in Fig. 5.7. We see that the criterion is convex with respect to $\varepsilon$. Algorithm 5.3 is executed with $\tau_N = 10000$, $N = 9$ and a sequence generated following

**Fig. 5.7** Example 5.4—guaranteed cost versus $\varepsilon$

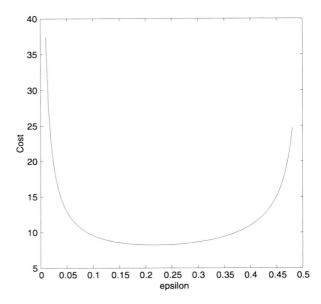

**Fig. 5.8** Example 5.4—ellipsoids for ten values of $\tau$ between 1 and 10000

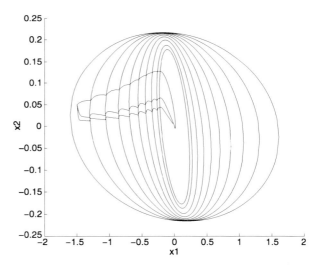

Remark 5.1. Algorithm 5.3 returns 10 switching surfaces shown in Fig. 5.8. In the same figure, system trajectories for $F = -1, 0$ and 1, issued from initial condition $x_0 = [-1.5 \ 0.05]'$, are also depicted. Figures 5.9 and 5.10 show, respectively, the guaranteed cost control designed without taking into account the limits of actuator and our piecewise-linear control. We see that in the first case, the control limitation ($u_0 = 0.5$) is not respected for $F = 1$. With the approach proposed in this section, the control respects the limit of the actuator but does not exploit its total capability. In our ellipsoids, the control never reaches the limit of actuators. In fact our

**Fig. 5.9** Example 5.4—linear controls for $F = -1$ and $F = 1$

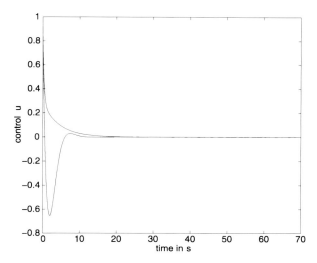

**Fig. 5.10** Example 5.4—nonlinear control for $F = -1$ and $F = 1$

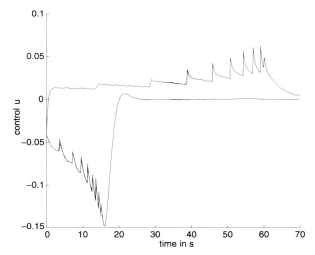

ellipsoids are included in the ellipsoids in which the control attains its limit. This is illustrated in Fig. 5.11 representing the ellipsoid obtained with $\widetilde{c}(1)$ and the one obtained with $c(1)$ (see proof of Proposition 5.10).

## 5.5 Disturbance Tolerance

When a perturbation affects the system, the state can be translated far from its original trajectory under perturbation influence. The closed-loop control depending of the system state and being limited, it can be out of the limits of the actuators, inducing saturations. In this section, the objective is to take into account in the control

**Fig. 5.11** Example 5.4—ellipsoids for $c(1)$ and $\widetilde{c}(1)$

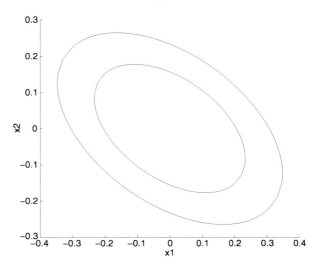

design that the system is affected by perturbations whose amplitude is bounded with known bounds. Consider the following uncertain system:

$$\dot{x}(t) = (A_0 + DFE_1)x(t) + Bu(t) + B_w w(t) \tag{5.49}$$

where $x \in \Re^n$ is the state vector, $u \in \Re^m$ is the control vector and $w \in \Re^r$ is an exogenous disturbance. Matrices $A_0$, $B$, $B_w$, $D$ and $E_1$ are real constant matrices of appropriate dimensions. $F$ is the uncertain matrix allowed to be time-varying. The disturbance $w(t)$ is amplitude limited, that is, belongs to the set[2]

$$\mathcal{W} = \left\{ w \in \Re^r;\ w(t)'w(t) \leq w_0,\ \forall t \geq 0 \right\} \tag{5.50}$$

In the context of the approach developed in the previous sections, the problem we want to solve is formulated as follows.

**Problem 5.2** Find a control function $u(x)$ and a set $\varXi$ such that:

1. $u(x)$ belongs to $\mathcal{U}$, for each trajectory of system (5.49) initialized in $\varXi$ (i.e., $x(0) \in \varXi$) and any disturbance $w \in \mathcal{W}$, that is, the control constraints are respected.
2. Regarding the closed-loop system obtained from $u(x)$ defined as

$$\dot{x}(t) = (A + DFE_1)x(t) + Bu(x(t)) + B_w w(t) \tag{5.51}$$

one has:

---

[2]Note that in this section, we consider a notation $w_0$ instead of $\delta^{-1}$ (used in the previous chapters) for the bound of the disturbance to simplify the presentation of the results.

5.5 Disturbance Tolerance                                                                245

(a) When $w = 0$ (disturbance-free case) one gets $x(t) \to 0$ as $t \to \infty$, $\forall x(0) \in \Xi$. That corresponds to the asymptotic stability of the disturbance-free closed-loop system.
(b) When $w \neq 0$, $w \in \mathcal{W}$, one gets $x(t) \in \Xi$, $\forall x(0) \in \Xi$. That corresponds to verify the invariance of the set $\Xi$.

## 5.5.1 Preliminaries

We present some preliminaries used to derive our main results.

**Lemma 5.9** *Consider system (5.49). If there exist symmetric positive definite matrices $P \in \Re^{n \times n}$, $R \in \Re^{m \times m}$, $Q \in \Re^{n \times n}$ and positive scalars $\beta$ and $\alpha$ satisfying the following relations:*

$$A_0' P + P A_0 + \alpha P - P B R B' P + E_1' E_1 + P D D' P + \frac{1}{\beta} P B_w B_w' P + Q = 0$$
$$(5.52)$$

$$\beta w_0 - \alpha c \leq 0 \tag{5.53}$$

*then the state feedback*

$$u(x) = -R B' P x \tag{5.54}$$

*and the set $\mathcal{E}(P, c^{-1})$ defined by*

$$\mathcal{E}(P, c^{-1}) = \left\{ x \in \Re^n; \ x' P x \leq c = \min_i \frac{u_{0(i)}^2}{R_{(i)} B' P B R_{(i)}'}, \ i = 1, \ldots, m \right\} \tag{5.55}$$

*solve Problem 5.2.*

*Proof* By considering the disturbance-free case ($w = 0$), one obtains with respect to the quadratic Lyapunov function $V(x) = x' P x$:

$$\dot{V}(x) = x' \left( A_0' P + P A_0 - 2 P B R B' P + P D F E_1 + E_1' F' D' P \right) x$$

By majoring the terms depending of $F$ [288], one gets

$$\dot{V}(x) \leq x' \left( A_0' P + P A_0 - 2 P B R B' P + P D D' P + E_1' E_1 \right) x$$

Thus, from (5.52)

$$\dot{V}(x) \leq -\alpha x' P x < 0$$

and therefore point 2(a) of Problem 5.2 is verified.

When $w \neq 0$, we want to prove that the trajectories of the closed-loop system (5.51) remain confined in $\mathcal{E}(P, c^{-1})$, $\forall x(0) \in \mathcal{E}(P, c^{-1})$ and $\forall w(t) \in \mathcal{W}$. Thus, we can prove that if relation (5.52) is satisfied then:

$$\dot{V}(x) \leq \beta w' w - \alpha x' P x < 0 \tag{5.56}$$

Indeed, we have

$$\dot{V}(x) \leq x' \left( A_0' P + P A_0 - 2 P B R B' P + P D D' P + E_1' E_1 \right) x + 2 x' P B_w w$$

and from the Riccati equation (5.52), we have

$$\begin{bmatrix} x \\ w \end{bmatrix}' \begin{bmatrix} A_0' P + P A_0 - 2 P B R B' P + \alpha P + P D D' P + E_1' E_1 & P B_w \\ B_w' P & -\beta I \end{bmatrix} \begin{bmatrix} x \\ w \end{bmatrix} < 0$$

and inequality (5.56) follows. Hence, to prove the invariance of the considered ellipsoid let $x$ belong to the boundary of the ellipsoid, that is, verify $x' P x = c$. From (5.50) and (5.53), it follows that

$$\dot{V}(x) \leq \beta w' w - \alpha x' P x \leq \beta w_0 - \alpha c \leq 0$$

That means that when $w \neq 0$ one gets $\dot{V}(x) \leq 0$ for any $x$ belonging to the boundary of $\mathcal{E}(P, c^{-1})$. Therefore point 2(b) of Problem 5.2 is satisfied.

Furthermore since the invariance of $\mathcal{E}(P, c^{-1})$ is ensured along the closed-loop trajectories, we can guarantee that $u(x) = -R B' P x$ satisfies amplitude limitations, $\forall x \in \mathcal{E}(P, c^{-1})$ and for any $w(t) \in \mathcal{W}$. Thus, point 1 of Problem 5.2 is also verified. $\qquad \square$

*Remark 5.9* Concerning the Riccati equation (5.52), the following remarks can be made.

- The presence of the positive scalar $\alpha$ in the Riccati equation (5.52) can be interpreted as a pole placement specification in the sense that the existence of a solution $P = P' > 0$ to (5.52) guarantees that the poles of the matrix $A_0 - B R B' P + D F E_1$ are located in the complex plane at the left of the line $-\frac{\alpha}{2}$.
- Consider the controlled output $z(t) = Q^{\frac{1}{2}} x(t)$ and the initial condition $x(0) = 0$. The positive scalar $\beta$ can be interpreted as a $\mathcal{L}_2$-attenuation bound in the sense that, provided that the pair $(Q^{\frac{1}{2}}, A_0)$ is detectable, the existence of a solution $P = P' > 0$ to (5.52) guarantees that

$$\int_0^\infty z(t)' z(t) \, dt \leq \sqrt{\beta} \int_0^\infty w(t)' w(t) \, dt \tag{5.57}$$

provided $w \in \mathcal{L}_2$ ($w$ is energy bounded).

5.5 Disturbance Tolerance                                                                 247

### 5.5.2 τ-Parameterized Control

Our objective is to solve Problem 5.2 by considering a controller depending on the location of the state in $\Re^n$. To express the main result, we first parameterize (5.52) similarly to the case in Sect. 5.2:

$$A_0'P(\tau) + P(\tau)A_0 + \alpha P(\tau) - P(\tau)BRB'P(\tau) + E_1'E_1 + P(\tau)DD'P(\tau)$$
$$+ \frac{1}{\beta}P(\tau)B_w B_w'P(\tau) + \frac{Q}{\tau} = 0 \tag{5.58}$$

with $1 \leq \tau < \infty$. For $\tau = 1$, we recover equation (5.52). In the sequel, we suppose that for $\tau = 1$, solutions $P(1) = P(1)' > 0$, $R = R' > 0$, $Q = Q'$, $\alpha > 0$ and $\beta > 0$ satisfying conditions of Lemma 5.9 exist. In addition, we fix matrices $R$, $Q$ and scalars $\alpha$ and $\beta$ in (5.58) as the solutions obtained for $\tau = 1$ (i.e., $R$ and $Q$ are constant matrices and $\alpha$ and $\beta$ are constant scalars in the remainder of the section). The computation of these matrices when they exist, will be presented later. Now, we follow the steps of the approach presented in Sect. 5.2. We begin by the following lemma obtained from classical arguments of monotonicity [296, 353].

**Lemma 5.10** *If there exists $P(1) = P(1)' > 0$ solution to (5.58) then for all $\tau > 1$, (5.58) has a solution $P(\tau) = P(\tau)'$ which satisfies $P(1) \geq P(\tau) > 0$.*

Define now the ellipsoid:

$$\mathcal{E}\big(P(\tau), c(\tau)^{-1}\big) = \big\{x \in \Re^n; \ x'P(\tau)x \leq c(\tau)\big\} \tag{5.59}$$

$$c(\tau) = \min_i \frac{u_{0(i)}^2}{R_{(i)}B'P(\tau)BR_{(i)}'}, \quad i = 1, \dots, m \tag{5.60}$$

$\mathcal{E}(P(\tau), c(\tau)^{-1})$ reduces to $\mathcal{E}(P, c^{-1})$ as defined in (5.55) for $\tau = 1$. Considering the normalized matrix

$$\mathcal{X}(\tau) = \frac{P(\tau)}{c(\tau)} \tag{5.61}$$

and associated normalized ellipsoid

$$\mathcal{E}\big(\mathcal{X}(\tau), 1\big) = \big\{x \in \Re^n; \ x'\mathcal{X}(\tau)x \leq 1\big\} \tag{5.62}$$

the normalized τ-parameterized family of ellipsoidal sets is defined as

$$\Xi(\tau) = \big\{\mathcal{E}\big(\mathcal{X}(\tau_i), 1\big), \ 1 \leq \tau_i \leq \tau\big\} \tag{5.63}$$

Following the steps of Sect. 5.2, we can prove that $\frac{d\mathcal{X}(\tau)}{d\tau}$ is negative definite.

**Lemma 5.11** *Consider $\mathcal{X}(\tau)$ defined in (5.61) with $P(\tau)$ solution to (5.58) and $c(\tau)$ defined in (5.60), then matrix $\frac{d\mathcal{X}(\tau)}{d\tau}$ is negative definite.*

248                    5  Synthesis via a Parameterized ARE Approach or a Parameterized LMI

Application of Lemmas 5.10 and 5.11 allows to deduce the following lemma.

**Lemma 5.12** *The $\tau$-parameterized family of ellipsoids $\Xi(\tau)$ is a nested family set. In other words, $\Xi(\tau_1) \subset \Xi(\tau_2)$ whenever $\tau_1 < \tau_2$ with the maximal element $\Xi = \bigcup_\tau \Xi(\tau)$.*

We are in a position to propose a solution to Problem 5.2.

**Proposition 5.13** *Let $P(\tau)$, $c(\tau)$ solutions to (5.58), (5.53) and $\Xi(\tau)$ the normalized family of ellipsoids defined from (5.63). Define the function $\tau(x)$ in the following way:*

*– for $x \in \Xi(1)$ one gets $\tau(x) = 1$;*
*– for $x \in \Xi \setminus \Xi(1)$ one determines $\tau(x)$ as the positive solution to the equation*

$$x'P(\tau)x = c(\tau)$$

*Then the controller*

$$u(x) = \begin{cases} -RB'P(\tau)x & \text{if } x \in \Xi \setminus \Xi(1) \\ -RB'P(1)x & \text{if } x \in \Xi(1) \end{cases} \tag{5.64}$$

*and $\Xi = \bigcup_\tau \Xi(\tau)$ solve Problem 5.2.*

*Proof* From the previous results, it is clear that $u(x)$ defined in (5.64) satisfies point 1 of Problem 5.2. In order to verify that the point 2(a) of Problem 5.2 holds, consider that $w = 0$. Then by applying Lemma 5.9 we prove that the trajectories of the disturbance-free closed-loop system are asymptotically stable $\forall x \in \Xi(1)$. We have now to prove that the compact set $\Xi(1)$ is a finite attractor in $\Xi$ for the disturbance-free closed-loop system. For this, it suffices to prove that for all $x$ belonging to the closure of $\Xi \setminus \Xi(1)$ one has $\frac{d\tau}{dt} = \dot{\tau} < 0$. Thus, the time derivative of $\tau$ is computed from the equality $x'P(\tau)x = c(\tau)$ in which $\tau$ implicitly appears. Therefore $\dot{\tau}$ is defined by

$$\dot{\tau} = \frac{-2x'P(\tau)\dot{x}}{x'\frac{dP(\tau)}{d\tau}x - Dc(\tau)} \tag{5.65}$$

where $Dc(\tau)$ is the right-side derivative of $c(\tau)$ with respect to $\tau$ and $\dot{x} = (A_0 + DFE_1)x + Bu(x)$. Then one obtains

$$2x'P(\tau)\dot{x} = x'\left[\left(A_0 + DFE_1 - BRB'\right)'P(\tau) + P(\tau)\left(A_0 + DFE_1 - BRB'\right)\right]x$$
$$\leq -\alpha x'P(\tau)x < 0$$

similarly as in the proof of Lemma 5.9 and therefore $-2x'P(\tau)\dot{x} > 0$. Furthermore,

$$\frac{dP(\tau)}{d\tau} < 0 \quad \text{and} \quad Dc(\tau) \geq 0$$

## 5.5 Disturbance Tolerance

then it follows that $\dot{\tau} < 0$. Thus, $u(x)$ drives any $x \in \Xi \setminus \Xi(1)$ to $\Xi(1)$ and therefore to the origin. Then, from Lemma 5.12, we can conclude that $\Xi(1)$ is an attractor for the disturbance-free closed-loop system ($w = 0$) with $u(x)$ defined in (5.64). Moreover, to verify that point 2(b) of Problem 5.2 holds, we prove that the trajectories of the closed-loop system with $w \neq 0$ and $u(x)$ remain confined in $\Xi$. With $V(x) = x'P(\tau)x$, one gets

$$\dot{V}(x) = 2x'P(\tau)\dot{x} + x'\frac{dP(\tau)}{d\tau}x\dot{\tau}$$

Hence, from (5.65) one can verify that $\dot{V}(x) = \dot{\tau}Dc(\tau)$. From relations (5.53) and (5.58) it follows that

$$2x'P(\tau)\dot{x} + \alpha x'P(\tau)x - \beta w'w \leq 0$$

From Lemma 5.11 one gets

$$\frac{d\mathcal{X}(\tau)}{d\tau} = \frac{\frac{dP(\tau)}{d\tau}c(\tau) - P(\tau)Dc(\tau)}{c(\tau)^2}$$

and then

$$\frac{dP(\tau)}{d\tau}c(\tau) - P(\tau)Dc(\tau) < 0$$

and because $c(\tau) > 0$, we can deduce that

$$-x'\frac{dP(\tau)}{d\tau}x + Dc(\tau) > 0$$

it follows that

$$\dot{\tau} \leq \frac{-\alpha x'P(\tau)x + \beta w'w}{-x'\frac{dP(\tau)}{d\tau}x + Dc(\tau)}$$

Since $c(\tau)$ is an increasing function, one gets $c(1) = c \leq c(\tau)$ and therefore $0 \geq \beta w_0 - \alpha c(1) \geq \beta w_0 - \alpha c(\tau)$. Hence, to prove the invariance of the ellipsoid $\mathcal{E}(P(\tau), c(\tau)^{-1})$, one considers $x'P(\tau)x = c(\tau)$ and therefore from

$$0 \geq \beta w_0 - \alpha c(1) \geq \beta w_0 - \alpha c(\tau)$$

we have

$$\dot{\tau} \leq \frac{-\alpha c(\tau) + \beta w'w}{-x'\frac{dP(\tau)}{d\tau}x + Dc(\tau)} \leq \frac{-\alpha c(1) + \beta w_0}{-x'\frac{dP(\tau)}{d\tau}x + Dc(\tau)} \leq 0$$

That means that when $w \neq 0$, $w \in \mathcal{W}$, $\dot{V}(x) = \dot{\tau}Dc(\tau) \leq 0$ for any $x$ belonging to the boundary of the ellipsoid $\mathcal{E}(P(\tau), c(\tau)^{-1})$. $\quad\square$

*Remark 5.10* Concerning the function $\tau(x) > 0$, the following holds:

250         5 Synthesis via a Parameterized ARE Approach or a Parameterized LMI

- When $w = 0$ one verifies that $\dot{\tau}(x) < 0$ along the closed-loop trajectories.
- When $w \neq 0$ one verifies $\dot{\tau}(x) \leq 0$ for any $x$ belonging to the boundary of the ellipsoid $\mathcal{E}(P(\tau), c(\tau)^{-1})$.

Hence, it means that the function $\tau(x)$ could be considered as a Lyapunov function for the closed-loop system.

Consider now that $\tau \to \infty$. Then in this case, (5.58) reads

$$\left[A_0 + \frac{\alpha}{2} I_n\right]' P(\infty) + P(\infty)\left[A_0 + \frac{\alpha}{2} I_n\right]$$

$$+ E_1' E_1 + P(\infty)\left[-BRB' + \frac{1}{\beta} B_w B_w' + DD'\right] P(\infty) = 0 \qquad (5.66)$$

The following proposition gives conditions for existence of $P(\infty) = P(\infty)'$ solution to (5.66) and for characterization of $\varXi$.

**Proposition 5.14** *If the pair $((A_0 + \frac{\alpha}{2} I_n), B)$ is stabilizable then*

1. *if $(E_1, (A_0 + \frac{\alpha}{2} I_n))$ is observable then there exists a solution $P(\infty) = P(\infty)' > 0$ to (5.66) and, in this case,*

$$\varXi = \left\{x \in \mathfrak{R}^n; \ x' P(\infty) x < \min_i \left[\frac{u_{0(i)}^2}{R_{(i)} B' P(\infty) B R_{(i)}'}\right]\right\}$$

2. *if $(E_1, (A_0 + \frac{\alpha}{2} I_n))$ is detectable then there exists a solution $P(\infty) = P(\infty)' \geq 0$ and, in this case, $\varXi$ is equal to the Cartesian product of $\mathfrak{R}^{\dim(\mathrm{Ker}(P(\infty)))}$ and a $k$-dimensional open ellipsoid $(k = n - \dim(\mathrm{Ker}(P(\infty))))$. This ellipsoid is a linear transformation of the set*

$$\varXi = \left\{y \in \mathfrak{R}^k; \ \max_i \left[\frac{R_{(i)} B' P(\infty) B R_{(i)}'}{u_{0(i)}^2}\right] y' P_k(\infty) y < 1\right\}$$

*where $M' P(\infty) M = \begin{bmatrix} 0 & 0 \\ 0 & P_k(\infty) \end{bmatrix}$, $M$ being an appropriate change of coordinates.*

*Remark 5.11* In the case where matrix $B$ is also uncertain with $\Delta B = DF(t)E_2$, one can check that Lemma 5.12 is not necessarily verified. In fact, in this case, the denominator of $c(\tau)$ depends both on $P(\tau)$ and $P(\tau)^{-1}$. Thus, there exist some counter-examples in which the inequality $c(\tau) \geq c(1)$ is not satisfied for some $\tau > 1$. Hence, without additional conditions on $E_2$, the current approach does not seem to be directly applicable and some modifications would have to be done to deal with uncertainties on $B$ [173].

### 5.5 Disturbance Tolerance

### 5.5.3 *Control Law Computation and Implementation*

From the previous developments, an algorithm for designing the control law may be expressed as follows.

**Algorithm 5.6**

1. Solve the optimization program by a one-line search on parameter $\alpha > 0$:

$$\alpha^\star = \max_{W,\alpha,\beta,\eta,R} \alpha$$

$$\text{subject to} \quad \begin{bmatrix} WA_0' + A_0 W - BRB' \\ + \alpha W + DD' & B_w & WE_1' & W \\ B_w' & -\beta I & 0 & 0 \\ E_1 W & 0 & -I & 0 \\ W & 0 & 0 & -\varepsilon I \end{bmatrix} < 0 \tag{5.67}$$

$$\begin{bmatrix} W & BR_{(i)}' \\ R_{(i)} B' & \eta u_{0(i)}^2 \end{bmatrix} \geq 0, \quad i = 1, \ldots, m$$

$$W - \eta P_0 \geq 0$$

$$W = W' > 0, \quad R = R' > 0, \quad \alpha > 0, \quad \beta > 0, \quad \eta > 0$$

Set

$$\bar{Q} = -\Bigl[ A_0' W^{-1} + W^{-1} A_0 - W^{-1} BRB' W^{-1} + E_1' E_1 + W^{-1} DD' W^{-1} \\ + \frac{1}{\beta} W^{-1} B_w B_w' W^{-1} \Bigr]$$

2. Keep the previous values of $\alpha = \alpha^\star$, $R$, $\bar{Q}$ and $\beta$ found in Step 1. Choose scalars $\xi_j$, $j = 1, \ldots, N$, such that $0 < \xi_j \leq 1$. Set $\xi_1 = 1$. For each $\xi_j$ compute $P(\xi_j) = P(\xi_j)' > 0$ solution to

$$A_0' P(\xi_j) + P(\xi_j) A_0 + \alpha P(\xi_j) - P(\xi_j) BRB' P(\xi_j) + E_1' E_1$$

$$+ P(\xi_j) DD' P(\xi_j) + \frac{1}{\beta} P(\xi_j) B_w B_w' P(\xi_j) + \frac{\bar{Q}}{\xi_j} = 0$$

For each solution $P(\xi_j)$ compute

$$K(\xi_j) = -RB' P(\xi_j)$$

$$c(\xi_j) = \min_i \frac{u_{0(i)}^2}{R_{(i)} B' P(\xi_j) BR_{(i)}'}, \quad i = 1, \ldots, m$$

$$w_0(\xi_j) = \frac{\alpha c(\xi_j)}{\beta}$$

$w_0(\xi_j)$ satisfies $\beta w_0(\xi_j) - \alpha c(\xi_j) = 0$. It constitutes the maximum bound for the disturbance associated to the ellipsoid $\mathcal{E}(P(\xi_j), c(\xi_j)^{-1})$.

3. The objective of this third step is to enlarge the disturbance tolerance bound. For each $P(\xi_j)$ solve the following optimization program (see Remark 5.9):

$$\beta^\star = \min_{\beta > 0} \beta$$

$$\text{subject to} \quad A_0' P(\xi_j) + P(\xi_j) A_0 + \alpha P(\xi_j) - P(\xi_j) B R B' P(\xi_j)$$

$$+ E_1' E_1 + P(\xi_j) D D' P(\xi_j) + \frac{1}{\beta} P(\xi_j) B_w B_w' P(\xi_j) < 0 \tag{5.68}$$

Then with the optimal value $\beta^\star$, compute the maximal upper bound $w_{\text{up}}(\xi_j)$ as $w_{\text{up}}(\xi_j) = \frac{\alpha c(\xi_j)}{\beta^\star}$.

4. Let us denote $\mathcal{E}(P(\xi_j), c(\xi_j)^{-1})$ by $\Xi(\xi_j)$, $j = 1, \ldots, N$. The piecewise-linear control law can be computed as follows:

$$u(x) = \begin{cases} K(\xi_1) x(t) & \text{if } x(t) \in \Xi(\xi_1), \ x(t) \notin \{\Xi(\xi_2), \Xi(\xi_3), \ldots, \Xi(\xi_N)\} \\ \vdots & \vdots \\ K(\xi_N) x(t) & \text{if } x(t) \in \Xi(\xi_N) \end{cases}$$

*Remark 5.12* Some comments concerning the previous algorithm are formulated. In Step 1, the idea is to compute the parameters $R$, $Q$, $\alpha$ and $\beta$ used in the sequel in the Riccati equation (5.58). These parameters are determined in a way allowing one to define a minimum size, if possible, for $\Xi$. The last inequality of problem (5.67) ensures that

$$\mathcal{E}(P_0^{-1}, 1) \subseteq \mathcal{E}(W^{-1}, \rho^{-1})$$

and it imposes a lower bound in the sense of inclusion of sets on the size of $\mathcal{E}(W^{-1}, \rho^{-1})$ where $\rho$ is defined, taking into account the second inequality of problem (5.67), by

$$c(\infty) > \min_i \frac{u_{0(i)}^2}{R_{(i)} B' W^{-1} B R_{(i)}'} = \rho \geq \eta^{-1}, \quad i = 1, \ldots, m$$

Moreover, we have

$$\mathcal{E}(W^{-1}, \rho^{-1}) \subset \Xi \quad (\text{see Proposition 5.14})$$

The second inequality also imposes that

$$\mathcal{E}(W^{-1}, \rho^{-1}) \subset S(|R B' W^{-1}|, u_0)$$

## 5.5 Disturbance Tolerance 253

where

$$S\left(\left|RB'W^{-1}\right|, u_0\right) = \left\{x \in \Re^m; \ -u_{0(i)} \leq -R_{(i)} B' W^{-1} x \leq u_{0(i)}, \ i = 1, \ldots, m\right\}$$

The first inequality is equivalent to:

$$A_0' W^{-1} + W^{-1} A_0 - W^{-1} B R B' W^{-1} + E_1' E_1 + W^{-1} D D' W^{-1}$$

$$+ \frac{1}{\beta} W^{-1} B_w B_w' W^{-1} + \bar{Q} = 0$$

with

$$\bar{Q} \geq \frac{1}{\varepsilon} I_n$$

At the end of Step 1, we have computed the set of initial conditions $\mathcal{E}(W^{-1}, \rho^{-1})$, the controller can drive to the origin, taking into account all the constraints. This set corresponds to the outermost ellipsoid and is included in the maximal set of initial conditions $\Xi$. In Step 2, the nested family of ellipsoids is determined with the corresponding controls and maximal bounds for disturbance associated with each of them. For this, we take $0 < \xi_{min} < \xi_j \leq \xi_{max} = 1$. In terms of values of $\tau$, this corresponds to the following. In the Riccati equation (5.58), take the values of $\beta$, $R$, $\alpha$ found in Step 1 and $Q = \frac{\bar{Q}}{\xi_{min}}$ with

$$\tau = \frac{\xi_j}{\xi_{min}}$$

For $\tau_{max} = 1/\xi_{min}$, the corresponding ellipsoid is $\mathcal{E}(W^{-1}, \rho^{-1})$ (outermost ellipsoid) in which all the others are nested. The step 3 tries to improve the bound computation of the admissible disturbances and the last step computes the control law.

*Remark 5.13* It can be shown, from the results presented in [246], that in the case without uncertainty (i.e., $F = 0$) and when $B = B_w$ (additive disturbance in the input), it is always possible to find a solution to (5.67) by considering an arbitrarily large but ball $\mathcal{E}(P_0^{-1}, 1)$, provided that matrix $A_0$ is not strictly unstable. Otherwise, we cannot ensure that a solution exists for any arbitrarily large ball.

*Example 5.5* In order to illustrate the method proposed in this section, consider the uncertain system (5.49) described by the following matrices:

$$A_0 = \begin{bmatrix} 0 & 1 \\ 0 & 0 \end{bmatrix}; \quad B = \begin{bmatrix} 0 \\ 1 \end{bmatrix}; \quad B_w = \begin{bmatrix} -0.1 \\ 0.2 \end{bmatrix}$$

$$E_1 = \begin{bmatrix} 0 & 1 \end{bmatrix}; \quad D = \begin{bmatrix} 0 \\ 1 \end{bmatrix}; \quad u_0 = 1$$

Our first objective is to ensure the stability of a region of initial states as large as possible. For this, we solve the optimization problem (5.67). The larger ball, for

**Table 5.1** Example 5.5—evolution of $w_{up}$ in function of $\beta^*$

| $j$ | $\xi_j$ | $\beta^*$ | $\rho$ | $w_{up}$ | $\sigma(A + BK(\xi_j))$ |
|---|---|---|---|---|---|
| 1 | $10^0$ | 37.403 | 0.203 | 0.0004 | $-0.08; -2.15$ |
| 2 | $10^{-1}$ | 3.915 | 0.193 | 0.0035 | $-0.13; -2.20$ |
| 3 | $10^{-2}$ | 0.460 | 0.165 | 0.0257 | $-0.30; -2.43$ |
| 4 | $10^{-3}$ | 0.073 | 0.111 | 0.1099 | $-0.63; -3.50$ |
| 5 | $10^{-4}$ | 0.077 | 0.054 | 0.0500 | $-0.89; -7.46$ |

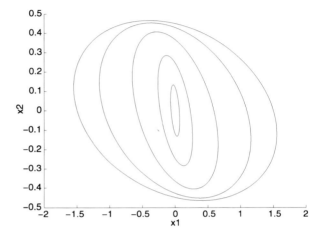

**Fig. 5.12** Example 5.5—nested ellipsoids. The *larger one* corresponds to $\xi_j = 1$, the *smaller one* to $\xi_j = 5$

which it is possible to find a solution is given by $P_0 = 5I_2$. In this case, considering $\frac{1}{\varepsilon} = 10^{-3}$ the following matrices are obtained:

$$Q = \begin{bmatrix} 0.001 & 0 \\ 0 & 0.001 \end{bmatrix}; \quad W = \begin{bmatrix} 0.0903 & 0.0784 \\ 0.0784 & 1.0060 \end{bmatrix}$$

$$R = 2.2155; \quad \alpha = 0.0717; \quad \beta = 1.3855 \times 10^5$$

Results obtained by applying the steps described in Sect. 5.5 are summarized in Table 5.1, for which the following comments can be made.

- The control laws for the inner ellipsoids are "stronger" which implies that the time-domain performance is improved as the state approaches to the origin. This fact can be seen by regarding the eigenvalues of $A + BK(\xi_j)$, denoted by $\sigma(A + BK(\xi_j))$. Recall that small values of $\xi$ correspond to large values of $\tau$ and that the small value of $\xi$ corresponds to $\tau = 1$, see Remark 5.12.
- Concerning $j = 1, 2, 3, 4$, the larger is the domain of stability, the smaller is the bound on the disturbance tolerance ($w_{up}$). This can be justified by the fact that by applying "stronger" control laws the disturbance rejection is more effective. On the other hand, from a certain $j$ ($j = 4$ in this example), this bound cannot be improved any more. Actually, the bound starts to decrease if we consider outer ellipsoids. This can be justified by Remark 5.9.

5.6 Nonlinear Bounded Control for Time-Delay Systems 255

Figure 5.12 depicts the ellipsoidal stability regions associated to each $\xi_j$, $j = 1, 2, 3, 4, 5$.

## 5.6 Nonlinear Bounded Control for Time-Delay Systems

This section extends the method proposed in the previous sections to the case of time-delay systems. As before, the controllers are based on an $\tau$-parameterized family of algebraic Riccati equations or on an $\tau$-parameterized family of LMI optimization problems. Hence, nested ellipsoidal neighborhoods of the origin are determined. Thus, from the Lyapunov–Krasovskii theorem, the uniform asymptotic stability of the closed-loop system is guaranteed and a certain performance level is attained through a quadratic cost function. Since almost all the proofs are quite similar to the ones of the previous sections, the reader can find all the details of the proofs of the main results in [113].

### 5.6.1 Problem Statement

Consider the time-delay linear system described by

$$\dot{x}(t) = Ax(t) + A_d x(t - d) + Bu(t) \tag{5.69}$$

with the initial condition

$$x(t_0 + \theta) = \phi(\theta), \quad \forall \theta \in [-d, 0], \ (t_0, \phi) \in \Re^+ \times \mathcal{C}_\tau^v x(t_0) = x_0 \tag{5.70}$$

where $x(t) \in \Re^n$ is the state, $u(t) \in \Re^m$ is the control input, $d$ is the time-independent delay of the system. $A$, $A_d$ and $B$ are constant matrices of appropriate dimensions and pair $(A, B)$ is supposed to be stabilizable. $\mathcal{C}_j = \mathcal{C}([-d, 0], \Re^n)$ denotes the Banach space of continuous vector functions mapping the interval $[-d, 0]$ into $\Re^n$ with the topology of uniform convergence. We assume now that the control takes values in a compact set:

$$\mathcal{U} = \left\{ u \in \Re^m; \ -u_{0(i)} \leq u_{(i)} \leq u_{0(i)}, \ i = 1, \ldots, m \right\} \tag{5.71}$$

Associated with system (5.69)–(5.70), let us define the following quadratic cost function which defines a performance criterion:

$$J = \int_0^{+\infty} \left( x(t)' Q x(t) + u(t)' R u(t) \right) dt, \quad Q = Q' > 0, \ R = R' > 0 \tag{5.72}$$

The problem addressed in this section is to find a control $u(x)$ such that, for all $t$, $u(x) \in \mathcal{U}$, and such that system (5.69) is asymptotically stable. Moreover, among all possible controls satisfying these properties, we want to select a control which minimizes $J$. In order to solve this problem, some preliminaries now are introduced.

## 5.6.2 Preliminaries

In [342], the problem of designing a linear state feedback which stabilizes system (5.69) is addressed. An important result stated in this paper is presented in the following lemma.

**Lemma 5.13** *Given symmetric positive definite matrices* $Q \in \Re^{n \times n}$ *and* $R \in \Re^{m \times m}$, *if there exist two symmetric positive definite matrices* $P \in \Re^{n \times n}$ *and* $S \in \Re^{n \times n}$ *solutions to*

$$A'P + PA + PA_d S^{-1} A_d' P - PBR^{-1}B'P + S + Q = 0 \qquad (5.73)$$

*then system* (5.69) *closed by the state feedback*

$$u = Kx = -R^{-1}B'Px \qquad (5.74)$$

*is asymptotically stable for all initial conditions* $\phi \in \mathcal{B}(\sigma)$ *where* $\mathcal{B}(\sigma)$ *is defined by*

$$\mathcal{B}(\sigma) = \left\{ \phi \in \mathcal{C}_j^v;\ \|\phi\|_c^2 \leq \sigma \right\} \quad \text{with } \sigma = \frac{c}{\lambda_{\max}(P) + \tau \lambda_{\max}(S)} \qquad (5.75)$$

$c > 0$ *corresponds to the largest ellipsoid*

$$\mathcal{E}\left(P, c^{-1}\right) = \left\{ x \in \Re^n;\ x'Px \leq c \right\} \qquad (5.76)$$

*contained in* $\mathcal{U}$.

The proof is obtained by showing that

$$V(x_t) = x(t)'Px(t) + \int_{t-d}^{t} x(\theta)'Sx(\theta)\,d\theta, \quad P = P' > 0,\ S = S' > 0 \qquad (5.77)$$

where $x_t$, $\forall t \geq t_0$, denotes the restriction of $x$ to the interval $[t - d, t]$ translated to $[-d, 0]$, that is,

$$x_t(\theta) = x(t + \theta), \quad \forall \theta \in [-d, 0]$$

is a Lyapunov functional for the closed-loop system. If the system satisfies Lemma 5.13, it is asymptotically stable and as described above, $V(x_t)$ defined by (5.77) is a Lyapunov functional for the closed-loop system. We can write

$$
\begin{aligned}
\frac{dV(x_t)}{dt} &= x(t)'\left[(A + BK)'P + P(A + BK)\right]x(t) + 2x(t)'PA_d x(t - d) \\
&\quad + x(t)'Sx(t) - x(t - d)'Sx(t - \tau) \\
&\leq x(t)'\left[(A + BK)'P + P(A + BK)\right]x(t) + x(t)'Sx(t) \\
&\quad + x(t)'PA_d S^{-1} A_d' Px(t) \\
&\leq -x(t)'\left[Q + K'RK\right]x(t) \quad \text{by Lemma 5.13}
\end{aligned}
$$

5.6  Nonlinear Bounded Control for Time-Delay Systems

Then

$$J = \int_0^{+\infty} x(t)' \big[ Q + K'RK \big] x(t)\, dt \leq -\int_0^{+\infty} dV(x_t) = V(x(0))$$

because the system is stable. We have

$$J \leq x_0' P x_0 + \int_{-d}^{0} x(\theta)' S x(\theta)\, d\theta$$

This inequality suggests the following optimization problem in order to minimize $J$:

$$\begin{cases} \min \left\{ \operatorname{trace}\big(P x_0' x_0\big) + \operatorname{trace}\left( S \int_\tau^0 x(\theta) x(\theta)'\, d\theta \right) \right\} \\ \text{under } P = P' > 0, \quad S = S' > 0, \quad \text{and} \quad (5.73) \end{cases} \tag{5.78}$$

We may note that the criterion is linear with respect to $P$ and $S$. But $P$ and $S$ appears nonlinearly in (5.73). A possibility to solve (5.73) by standard algorithms consists in fixing $S$. In fact a compromise has to be found between the value of $J$ and the size of the initial conditions domains (5.75) and (5.76).

We can have considered model uncertainties, for example, on matrix $A$:

$$A = A_0 + DFE_1 \tag{5.79}$$

where $A_0$, $D$ and $E_1$ are constant matrices of appropriate dimensions and $F'F \leq I$. In this case the Riccati equation (5.73) would have been replaced by (see Appendix B):

$$A_0' P + P A_0 + P A_d S^{-1} A_d' P + \varepsilon P D D' P - P B R^{-1} B' B P$$

$$+ S + \varepsilon^{-1} E_1' E_1 + Q = 0 \tag{5.80}$$

where the unknowns are $P$, $S$ and $\varepsilon > 0$. However, for the sake of simplicity, only the case without uncertainty is presented in the sequel.

### 5.6.3 Riccati Equation Approach

The idea is to fully use the capabilities of actuators. If that is possible, it is hoped that the performance of the system in terms of time response will be better. Suppose that optimization problem (5.78) has been solved obtaining a good compromise between performances and size of initial conditions domains. Then there exist symmetric positive definite matrices $P$ and $S$ solutions to the Riccati equation (5.73), with given symmetric positive definite matrices $R$ and $Q$. Now the idea is to parameterize the Riccati equation (5.73) in the following way:

$$A'P(\tau) + P(\tau)A + P(\tau)A_d S^{-1} A_d' P(\tau) - P(\tau)BR^{-1}B'P(\tau) + S + \frac{Q}{\tau} = 0 \tag{5.81}$$

with $1 \leq \tau < \infty$. It has to be noted that $S$ is maintained as constant, only $P$ varies with $\tau$. For $\tau = 1$, we recover the Riccati equation (5.73). Matrix $B \in \Re^{n \times m}$ is written as $B = [B_1 \ \ldots \ B_m]$ with $B_i \in \Re^n$, $i = 1, \ldots, m$, and we take for simplicity $R = rI > 0$.

Define also

$$c_i(\tau) = \frac{u_{0(i)}^2 r^2}{B_i' P(\tau) B_i}, \quad \forall i = 1, \ldots, m$$

$$\mathcal{C}(\tau) = \min_i c_i(\tau)$$

and

$$\mathcal{B}(\tau) = \left\{ \phi \in \mathcal{C}_d^v; \ \|\phi\|_c^2 \leq \tilde{c}(\tau) \right\} \quad \text{with} \ \tilde{c}(\tau) = \frac{c(\tau)}{\lambda_{\max}(P(\tau)) + d\lambda_{\max}(S)} \tag{5.82}$$

Note that $c_i(\tau)$, $i = 1, \ldots, m$, are increasing functions of $\tau$. Hence $\mathcal{C}(\tau)$ is also an increasing function of $\tau$. Nevertheless, $\mathcal{C}(\tau)$ is not necessarily differentiable for any $\tau > 0$, but its right-hand side derivative is well-defined. Using this definition and the previous notation, it is possible to extend the results of the previous sections, as follows.

**Proposition 5.15** *Let $P(\tau)$ and $S$ be the positive definite symmetric solutions to the Riccati equation (5.81). Define the function $\tau(x)$ in the following way:*

- *For $x \in \Xi(1)$ and $\phi \in \mathcal{B}(1)$ as the positive solution $\tau(x) = 1$.*
- *For $x \in \Xi \setminus \Xi(1)$ and $\phi \in \mathcal{B}(1)$ as the positive solution to the equation*

$$x'P(\tau)x + \int_{t-d}^{t} x(\theta)'Sx(\theta)\,d\theta - \mathcal{C}(\tau) = 0 \tag{5.83}$$

*Then the control $u(x)$ defined by*

$$u(x) = \begin{cases} -R^{-1}B'P(\tau(x))x & \text{if } x \in \Xi \setminus \Xi(1) \\ -R^{-1}B'P(1)x & \text{if } x \in \Xi(1) \end{cases} \tag{5.84}$$

*satisfies the constraints $-u_{0(i)} \leq u_{(i)}(x) \leq u_{0(i)}$, $i = 1, \ldots, m$, and drives any point of $\Xi$ to the origin.*

Suppose now that $\tau \to +\infty$, and the Riccati equation (5.81) becomes

$$A'P(\infty) + P(\infty)A + P(\infty)A_d S^{-1} A_d' P(\infty) - P(\infty)BR^{-1}B'P(\infty) + S = 0$$

## 5.6 Nonlinear Bounded Control for Time-Delay Systems

If the pair $(A, S^{\frac{1}{2}})$ is observable and if $(A, B)$ is stabilizable, then $P(\infty)$ is positive definite. Hence the set $\varXi$ is characterized by the following proposition.

**Proposition 5.16** *Suppose that pair $(A, B)$ is stabilizable. Then*

$$\varXi = \left\{ x \in \mathfrak{R}^n ; \ \max_i \left[ \frac{B_i' P(\infty) B_i}{u_{0(i)}^2 r^2} \right] x' P(\infty) x < 1 \right\}, \quad i = 1, \dots, m$$

From a practical point of view, it is not possible to solve equation (5.83). In practice, to implement the control it is possible to use the following algorithm.

**Algorithm 5.7**

0. Choose $N$ values of $\tau$ such that $\tau_0 = 1 < \tau_1 < \tau_2 < \cdots < \tau_N < \infty$. For $\tau = \tau_N$, solve Riccati equation (5.73). We obtain the corresponding ellipsoid $\mathcal{E}(P(\tau_N), C(\tau_N)^{-1})$, the set $\mathcal{B}(\tau_N)$ and the control $K(\tau_N)$. Set

$$\varXi = \left\{ \varXi(\tau_N) \right\}, \qquad \mathcal{B} = \left\{ \mathcal{B}(\tau_N) \right\}, \qquad K = \left\{ K(\tau_N) \right\}$$

i. Take $\tau = \tau_{N-i}$. Solve the Riccati equation (5.73) for $\tau = \tau_{N-i}$. We obtain $\mathcal{E}(P(\tau_{N-i}), C(\tau_{N-i})^{-1})$, $\mathcal{B}(\tau_{N-i})$ and $K(\tau_{N-i})$. Set

$$\varXi = \left\{ \varXi, \mathcal{E}\left( P(\tau_{N-i}), C(\tau_{N-i})^{-1} \right) \right\},$$
$$\mathcal{B} = \left\{ \mathcal{B}, \mathcal{B}(\tau_{N-i}) \right\}, \qquad K = \left\{ K, K(\tau_{N-i}) \right\}$$

Go to step $i + 1$.

At the end of the algorithm, we obtain a nested family of ellipsoids $\varXi$, sets $\mathcal{B}$ and corresponding control gains. To apply the control, we measure $x(t)$ and identify the outer ellipsoid in $\varXi$, which contains $x(t)$, and the corresponding control is applied. With this method, a piecewise control is obtained.

*Example 5.6* Let us consider the system described by the following matrices

$$A = \begin{bmatrix} 0 & 1 \\ 0 & 0 \end{bmatrix}; \qquad A_d = \begin{bmatrix} 0 & 0 \\ 0 & 1 \end{bmatrix}; \qquad B = \begin{bmatrix} 0 \\ 1 \end{bmatrix}$$

with $d = 0.5s$ and $u_0 = 2$. By selecting $S = I$, $R = 1$ and $Q = I$, the solution to the Riccati equation can be written

$$P(\tau) = \begin{bmatrix} \frac{1}{2}\sqrt{4 + \frac{2}{\tau}}\sqrt{4 + \frac{2}{\tau} + 4\sqrt{4 + \frac{2}{\tau}}} & \sqrt{4 + \frac{2}{\tau}} \\[2ex] \sqrt{4 + \frac{2}{\tau}} & \sqrt{4 + \frac{2}{\tau} + 4\sqrt{4 + \frac{2}{\tau}}} \end{bmatrix}$$

and

$$c(\tau) = \frac{4}{\sqrt{4 + \frac{2}{\tau} + 4\sqrt{4 + \frac{2}{\tau}}}}$$

When $\tau \to \infty$, one gets

$$\lim P(\tau) = 2 \begin{bmatrix} \sqrt{3} & 1 \\ 1 & \sqrt{3} \end{bmatrix}$$

$$\lim c(\tau) = \frac{2}{\sqrt{3}}$$

The set of initial conditions is defined as

$$\mathcal{B}(1) = \left\{ \phi \in \mathcal{C}_{\frac{1}{2}}; \ \|\phi\|_c^2 \leq \frac{c(1)}{\lambda_{\max}(P(1)) + 0.5\lambda_{\max}(S)} = 0.136 \right\}$$

which implies that

$$\|\phi\|_c^2 = \left[ \sup_{0 \leq \theta \leq 0.5} \|\phi(\theta)\| \right]^2 \leq 0.136 \quad \Rightarrow \quad \sup_{\substack{0 \leq \theta \leq 0.5 \\ \phi \in \bar{\mathcal{C}}_{\frac{1}{2}}}} \|\phi(\theta)\| \leq 0.369$$

### 5.6.4 LMI Approach

Another parameterization of the control gain matrix is possible using an LMI formulation. From (5.73), it is easy to see that condition of Lemma 5.13 can be expressed as follows.

– Find matrices $K$, $P = P' > 0$ and $S = S' > 0$ such that

$$(A + BK)'P + P(A + BK) + PA_dS^{-1}A_d'P + S < 0$$

– Right- and left-multiplying by $P^{-1} = W$ and denoting $Y = KW$, we obtain after some manipulations

$$AW + WA' + BY + Y'B' + \begin{bmatrix} A_dS^{-1} & W \end{bmatrix} \begin{bmatrix} S & 0 \\ 0 & S \end{bmatrix} \begin{bmatrix} S^{-1}A_d' \\ W \end{bmatrix} < 0$$

– Introducing $U = S^{-1}$ we arrive at

$$\begin{bmatrix} AW + WA' + BY + Y'B' & A_dU & W \\ UA_d' & -U & 0 \\ W & 0 & -U \end{bmatrix} < 0$$

## 5.6 Nonlinear Bounded Control for Time-Delay Systems

We can deduce the following lemma which is similar to Lemma 5.13 in the context of the LMI formulation.

**Lemma 5.14** *If there exist positive definite symmetric matrices* $W \in \mathfrak{R}^{n \times n}$, $U \in \mathfrak{R}^{n \times n}$ *and a matrix* $Y \in \mathfrak{R}^{m \times n}$ *such that*

$$\begin{bmatrix} AW + WA' + BY + Y'B' & A_d U & W \\ U A'_d & -U & 0 \\ W & 0 & -U \end{bmatrix} < 0 \tag{5.85}$$

$$\begin{bmatrix} W & Y'_{(i)} \\ Y_{(i)} & u^2_{0(i)} \end{bmatrix} \le 0, \quad i = 1, \ldots, m \tag{5.86}$$

*then system* (5.69) *closed by the state feedback*

$$u = Kx = YW^{-1}x \tag{5.87}$$

*is asymptotically stable for all initial conditions* $\phi \in \mathcal{B}(\sigma)$ *where* $\mathcal{B}(\sigma)$ *is defined by*

$$\mathcal{B}(\sigma) = \left\{ \phi \in \mathcal{C}^v_d; \ \|\phi\|^2_c \le \sigma \right\} \quad \text{with } \sigma = \frac{1}{\lambda_{\max}(W^{-1}) + d\lambda_{\max}(U^{-1})} \tag{5.88}$$

*and*

$$\mathcal{E}\left(W^{-1}, 1\right) = \left\{ x \in \mathfrak{R}^n; \ x'W^{-1}x \le 1 \right\} \tag{5.89}$$

*is contained in* $\mathcal{U}$.

The proof follows from the previous manipulations and from Lemma 5.13. $\mathcal{E}(W^{-1}, 1)$ is contained in $\mathcal{U}$ from inequalities (5.86).

Now to deal with a quadratic cost as defined in (5.72), a similar development as previously leads to the inequality

$$\begin{bmatrix} AW + WA' + BY + Y'B' & A_d U & W & Y' & W \\ U A'_d & -U & 0 & 0 & 0 \\ W & 0 & -U & 0 & 0 \\ Y & 0 & 0 & -R^{-1} & 0 \\ W & 0 & 0 & 0 & -Q^{-1} \end{bmatrix} < 0 \tag{5.90}$$

with $K = YW^{-1}$ and

$$J \le \text{trace}\left(W^{-1}x_0 x'_0\right) + \text{trace}\left(U^{-1} \int_{-d}^0 x(\theta)x(\theta)' \, d\theta\right) \tag{5.91}$$

Although the inequality (5.90) is linear with respect to the unknowns, it is not easy to minimize $J$ because it is nonlinear in the unknowns. A way to obtain a linear

262    5  Synthesis via a Parameterized ARE Approach or a Parameterized LMI

problem consists in minimizing the following problem:

$$(P2) \begin{cases} \min_{W,U,Y,\gamma,\delta} \left\{ \gamma \, \operatorname{trace}(x_0 x_0') + \delta \, \operatorname{trace}\left( \int_{-d}^{0} x(\theta) x(\theta)' d\theta \right) \right\} \\ \text{under relations (5.90), (5.86)} \\ \begin{bmatrix} \gamma I & I \\ I & W \end{bmatrix} \geq 0, \quad \begin{bmatrix} \delta I & I \\ I & U \end{bmatrix} \geq 0 \end{cases} \tag{5.92}$$

The main advantage is that now the problem (5.92) is linear. Conditions $\begin{bmatrix} \gamma I & I \\ I & W \end{bmatrix} \geq 0$ and $\begin{bmatrix} \delta I & I \\ I & U \end{bmatrix} \geq 0$ ensure that $\operatorname{trace}(W^{-1}) \leq n\gamma$ and $\operatorname{trace}(U^{-1}) \leq n\delta$, respectively. Problem (5.92) is solvable by an LMI solver when a solution exists. We have the following lemma.

**Lemma 5.15** *If Problem* (5.92) *is solvable, then for all initial conditions belonging to* $\mathcal{B}(\sigma)$ *defined in* (5.88)*, the system is asymptotically stable by the control* $u = Y W^{-1} x$*, which belongs to* $\mathcal{U}$*.*

The idea is to parameterize problem (5.92) in order to obtain a nested family of ellipsoids and sets $\mathcal{B}(\sigma)$ as in the case of the Riccati equation approach. For that, introduce the following optimization problem:

$$\begin{cases} \min_{W,U,Y,\gamma} \left\{ \tau \gamma + (1 - \tau) \log\left(\det(W^{-1})\right) \right\} \\ \text{under} \quad \begin{bmatrix} W & Y'_{(i)} \\ Y_{(i)} & u^2_{0(i)} \end{bmatrix} \geq 0, \quad i = 1, \ldots, m \\ \begin{bmatrix} AW + WA' + BY + Y'B' & A_d U & W & \tau Y' & \tau W \\ UA'_d & -U & 0 & 0 & 0 \\ W & 0 & -U & 0 & 0 \\ \tau Y & 0 & 0 & -R^{-1} & 0 \\ \tau W & 0 & 0 & 0 & -Q^{-1} \end{bmatrix} < 0 \\ \begin{bmatrix} \gamma I & I \\ I & W \end{bmatrix} \geq 0, \quad 0 \leq \tau \leq 1 \end{cases} \tag{5.93}$$

## 5.6  Nonlinear Bounded Control for Time-Delay Systems

When $\tau = 0$, problem (5.93) reduces to the following problem:

$$
\begin{cases}
\displaystyle\min_{W,U,Y} \left\{ \log\left(\det\left(W^{-1}\right)\right) \right\} \\[2ex]
\text{under} \quad \begin{bmatrix} W & Y'_{(i)} \\ Y_{(i)} & u_{0(i)}^2 \end{bmatrix} \geq 0, \quad i = 1, \dots, m \\[3ex]
\begin{bmatrix} AW + WA' + BY + Y'B' & A_d U & W \\ & U A'_d & -U & 0 \\ & W & 0 & -U \end{bmatrix} < 0
\end{cases}
$$

This problem, when a solution exists, solves the stabilization problem by a control belonging to $\mathcal{U}$ and maximizes the size of $\mathcal{E}(W^{-1}, 1)$ and $\mathcal{B}(\sigma)$.

When $\tau = 1$, we obtain Problem (5.93) in which performances are taken into account by minimizing $\gamma$, the size of $\mathcal{E}(W^{-1}, 1)$ being not a priori maximized. The control gain depends on the parameter $\tau$ and is written

$$
K(\tau) = Y(\tau)W(\tau)^{-1} \tag{5.94}
$$

where $Y(\tau)$ and $W(\tau)$ are the solutions obtained by solving (5.93). Now the idea is to let $\tau$ vary from 0 to 1. When $\tau_1 < \tau_2$, since constraints (5.86) are satisfied, the size of $K(\tau_1)$ is lower than the one of $K(\tau_2)$. To be sure that the domains $\mathcal{E}(W^{-1}, 1)$ and $\mathcal{B}(\sigma)$ are nested, for two values of $\tau$, $\tau_1$ and $\tau_2$ such that

$$
0 \leq \tau_1 < \tau_2 \leq 1
$$

we have to impose that

$$
\begin{aligned}
W(\tau_1) &< W(\tau_2) \\
U(\tau_1) &\leq U(\tau_2)
\end{aligned} \tag{5.95}
$$

All these remarks suggest the following algorithm to build a piecewise-linear control law.

### Algorithm 5.8

0. Choose $N$ values of $\tau$ such that $\tau_0 = 0 < \tau_1 < \tau_2 < \cdots < \tau_N = 1$. Solve LMI problem (5.93) for $\tau = \tau_N$. We obtain the corresponding ellipsoid $\mathcal{E}(W(\tau_N)^{-1}, 1)$, the set $\mathcal{B}(\tau_N)$ and the control $K(\tau_N) = Y(\tau_N)W(\tau_N)^{-1}$. Set

$$
\mathcal{B} = \left\{ \mathcal{B}(\tau_N) \right\}, \qquad K = \left\{ K(\tau_N) \right\}
$$

i. Take $\tau = \tau_{N-i}$. Solve the LMI problem (5.93) for $\tau = \tau_{N-i}$ by adding the constraints

$$
\begin{aligned}
W(\tau_{N-i-1}) &< W(\tau_{N-i}) \\
U(\tau_{N-i-1}) &\leq U(\tau_{N-i})
\end{aligned}
$$

We obtain $\mathcal{E}(W(\tau_{N-i})^{-1}, 1)$, $\mathcal{B}(\tau_{N-i})$ and $K(\tau_{N-i})$. Set

$$\Xi = \left\{ \Xi, \mathcal{E}\left(W(\tau_{N-i})^{-1}, 1\right) \right\}, \qquad \mathcal{B} = \left\{ \mathcal{B}, \mathcal{B}(\tau_{N-i}) \right\},$$
$$K = \left\{ K, K(\tau_{N-i}) \right\}$$

Go to step $i + 1$.

At the end of the algorithm we obtain a nested family of ellipsoids $\Xi$, sets $\mathcal{B}$, with the corresponding control gains. To implement this control, we proceed as for the Riccati equation approach.

## 5.7 Conclusion

In this chapter, the synthesis of control laws taking into account the limits in amplitude of actuators has been developed. The idea is to exploit the well-known results of optimal control for linear systems where a quadratic criterion is considered. Parameterizing adequately the algebraic Riccati equation involved in the optimal control of linear systems, it is possible to derive a nonlinear bounded control which takes into account the limits of actuators. The LMI counterpart is also presented and in some cases, its superiority has been demonstrated. Following the same lines, some extensions are proposed. In particular, the case of a multi-objective control or disturbance rejection problems are studied. The chapter ends by an extension to the case of time-delay systems. All the methods are illustrated by numerical examples which allow us to point out the potentialities and the drawbacks of the approach.

# Part III
# Anti-windup

# Chapter 6
# An Overview of Anti-windup Techniques

## 6.1 Introduction—Philosophy

Most control engineers are acutely aware of the simultaneous blessing and curse of linearity. Actually, linear techniques allow powerful mathematical results to be stated while combining a convenient and tractable framework for controller design. It is thus common practice for control engineers to design controllers, for a local linear model of the process, which attempt to keep signals small so that deviation from the linear operating point is also small, in particular to avoid undesirable effects of saturation elements.

In Part II of the book, various techniques have been presented, in the stabilization context, which refer to the one step approach. The controller was designed simultaneously taking into account performance specifications and a safe domain of operation. It has been shown that the size of the region of stability is strongly related to the control design specifications, and that large domains of stability are often associated to poor performance. Low and high gains strategies have been discussed as a successful alternative to cope with this compromise. While these approaches are satisfactory in principle, and have a significant portion of the literature devoted to them, they have also often been criticized because of their conservatism, lack of intuition (in terms of tuning rules etc.) and lack of applicability to some practical problems.

On the other hand, anti-windup compensation is a fruitful alternative which allows to perform some separation in the controller such that one part is devoted to achieving nominal performance and the other one is devoted to constraint handling. Such a two-steps approach is considered attractive in practice because no restriction is placed upon the nominal linear controller design and, assuming no saturation is encountered, this controller alone dictates the behavior of the linear closed-loop. It is only when saturation occurs that the anti-windup compensator becomes active and acts to modify the closed-loop behavior such that it is more resilient to saturation. The implication of the above is that anti-windup techniques can be retro-fitted to existing controllers which may function very well except during saturation, making them a popular choice with practicing engineers. It must be kept in mind, as

S. Tarbouriech et al., *Stability and Stabilization of Linear Systems with Saturating Actuators*, DOI 10.1007/978-0-85729-941-3_6, © Springer-Verlag London Limited 2011

discussed in Part II of the book, that there are several other methods which can be used to handle saturation nonlinearities in control system design and, furthermore, that in some situations these techniques maybe more suitable than the anti-windup strategy. A vast literature on this more general topic has been cited in the previous Parts I and II: let us just cite some books or special issues in journals as, for example, [25, 188, 208, 331, 361] and references therein. In the current part, we concentrate on anti-windup techniques, and especially "modern" anti-windup techniques by following the same lines as in the recent survey [352] and tutorial [108].

One of the problems with the emergence of anti-windup techniques is that they were developed from many sources (some unpublished) and they began to evolve in an almost organic way. Although now and then there have been attempts to unify the results, since 2000 no real effort to continue this unification has been made. Despite the existence of several book chapters and technical papers devoted to the presentation of new theoretical advances, before the publication of the two surveys [352] and [108], no concise overview of the existing literature allowing the reader to connect and assess the various techniques was really available. Thus, largely inspired by these references [108, 352], we will try to summarize the more modern anti-windup compensation techniques which have emerged, with a bias toward those which give rigorous stability guarantees. For more details regarding the anti-windup techniques the reader can also consult the recent book fully dedicated to anti-windup synthesis [406].

The chapter is structured as follows. Section 6.2 is devoted to the description of general anti-windup architecture, and some historical elements are given in Sect. 6.3. The problem on which Part III focuses is stated in Sect. 6.4. Then, a quick discussion on some solutions in the literature in regional (local) and global contexts is presented in Sect. 6.5 The development of conditions with their proofs will be detailed in Chap. 7 and illustrated in Chap. 8.

## 6.2 General Anti-windup Architecture

The basic idea inherent to anti-windup design for linear systems with saturating actuators is to introduce control modifications in order to recover, as much as possible, the performance induced by a previous design carried out on the basis of the unsaturated system. Thus, the general principle of the anti-windup scheme can be depicted in Fig. 6.1. In this figure, the (unconstrained) signal produced by the controller is compared to that which is actually fed into the plant (the constrained signal). This difference is then used to adjust the control strategy in a manner conducive to stability and performance preservation.

The general scheme of Fig. 6.1 can be simplified to depict the basic anti-windup architecture.

The established architecture for anti-windup is shown in Fig. 6.2, where the linear closed-loop is "disturbed" by the dead-zone signal corresponding to $u - y_c$ and where this same signal is exploited to activate the anti-windup filter action whenever necessary (namely whenever the saturation limits are exceeded or, equivalently, whenever the dead-zone function becomes different from zero).

## 6.2 General Anti-windup Architecture

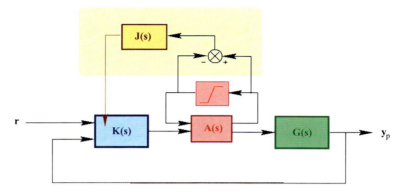

**Fig. 6.1** General principle of anti-windup. $G(s)$, $K(s)$, $A(s)$ and $J(s)$ are the plant, the controller, the actuator and the anti-windup controller, respectively

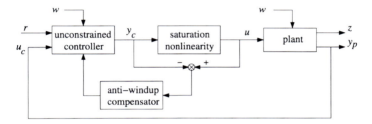

**Fig. 6.2** Basic anti-windup architecture

*Remark 6.1* It should be emphasized that some anti-windup schemes do not fall into the general structure of Fig. 6.2. In particular, MPC-based discrete-time schemes named "reference (or command, or error) governors" (see [7, 125] and references therein) follow a different design paradigm, which is not covered in this book.

Note that the depiction in Figs. 6.1 or 6.2 is very general and over the years has been refined into the form depicted in Fig. 6.3. In this figure, there is a separation between the so-called "unconstrained controller" and the "anti-windup" compensator. As with Fig. 6.1, the anti-windup compensator is driven by the difference between the constrained and unconstrained control signal. To do this, the knowledge of both signals (i.e., the output produced by the linear controller and the saturated version of this) is assumed. In some case, it is not realistic because only the saturated version of the signal is known, especially in the case of complex nonlinear actuators or output saturations (see Chap. 8). This may be problematic for the two-stage (anti-windup) approach, and hence an observer can be used to overcome this difficulty [361].

The anti-windup compensator itself emits two signals, one which is fed directly into the constrained control signal and one which may be used to drive the controller state equation directly. Virtually all anti-windup compensators which are present in the literature, in the paradigm covered by this book, can be represented in this form

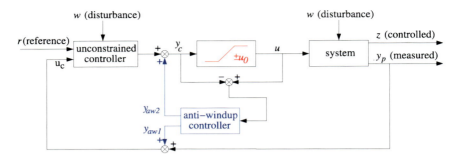

**Fig. 6.3** Principle of anti-windup

and thus the anti-windup compensators discussed herein will be assumed to have this form.

*Remark 6.2* Many anti-windup strategies inject the signal $y_{aw2}$ directly into the controller state equation, rather than additively with the measured outputs. While the former strategy may give more freedom in the anti-windup design, the stability and optimality conditions remain roughly similar with both strategies. Thus, not much attention is devoted to this issue.

*Remark 6.3* Practical implementation of the scheme in Fig. 6.3 considers that the anti-windup signal is actually injected to the input of the controller. Then the anti-windup compensator may be easily appended to an existing controller.

*Remark 6.4* Sometimes, a filter $F(s)$ can be added such that $\bar{y}_{aw2} = F(s)y_{aw2}$ and the signal $\bar{y}_{aw2}$ is therefore added to the controller output instead of $y_{aw2}$. In the sequel, we suppose that such a filter is omitted. However, in practice, a low-pass filter can be used to avoid some chattering effects on the control signals as well as algebraic loops (inducing possibly nested saturations) [27, 151].

In Figs. 6.2 or 6.3, each block (controller, plant, anti-windup compensator) can be either linear or nonlinear, thus leading to different anti-windup problem formulations. However, most of the anti-windup literature deals with linear plants (before saturation) which highly simplifies the design problem. As for the unconstrained controller, it is typically assumed that it is linear too, but we will emphasize later that several of the approaches summarized here allow for nonlinear controllers as well, with very mild assumptions on their properties (see Sect. 7.6 in Chap. 7). Finally, the main distinction that will be made here is whether the anti-windup compensator block is linear or not. In particular, we will talk about *direct linear* anti-windup compensators when the dynamics driven by the dead-zone function are linear and we will talk about *fully nonlinear* or simply *nonlinear* anti-windup in all other cases.

By considering as in [108] that anti-windup can be seen as a bounded stabilization problem with extra constraints on the small- and medium-signal behavior, it is relevant to summarize what the intrinsic limitations of anti-windup scheme are in order to clarify what are reasonable anti-windup design goals. As observed in Chap. 1 (see Sect. 1.6.2), the intrinsic limitations for saturated systems are directly related to the stability properties of the open-loop system. Then, regarding anti-windup compensators, it should be recognized that not all solutions will work for all types of plants because the stability limitations provide the following constraints:

1. global exponential stability will only be achievable with exponentially stable plants;
2. global asymptotic stability will only be achievable with non-exponentially unstable plants;
3. regional asymptotic stability is achievable with any type of plant but then large operating regions become most desirable.

These three items represent a very natural classification among the several constructive anti-windup recipes, in particular those illustrated in Part III. This classification, which is in terms of achievable closed-loop properties, could be similarly stated in terms of type of plant under consideration, namely:

1. with exponentially stable plants all the provided algorithms will be applicable;
2. with marginally stable/unstable plants, only the algorithms providing global asymptotic stability or regional exponential stability will be applicable; here it will be in general necessary that the anti-windup strategy be nonlinear to achieve global results;
3. with exponentially unstable plants only algorithms yielding regional (local) guarantees will be applicable.

## 6.3 A Bit of History

The study of anti-windup probably began in industry where practitioners noticed performance degradation in systems where saturation occurred. In the book [127], the authors cite boiler regulators as sources of saturation problems, although similar problems were probably found elsewhere too. Many different authors consider that the term "windup" is a phenomenon associated with saturation in systems with integral controllers and alluded to the build-up of charge on the integrator's capacitor during saturation. The subsequent dissipation of this charge would then cause long settling times and excessive overshoot, thereby degrading the performances of the system. Actually, for the more complicated modern multivariable controllers, this *winding up* phenomenon is not as simple, but the word "windup" still represents the fact that a controller designed ignoring saturation can often become confused about the unexpected saturated plant response and induce closed-loop performance (and possibly stability) loss. Windup has been always regarded from the industrial point of view as a problem associated with performance deterioration after saturation has occurred. This hiddenly conveys the fact that whenever saturation does not occur, the underlying controller (which we will call the "unconstrained controller" in this

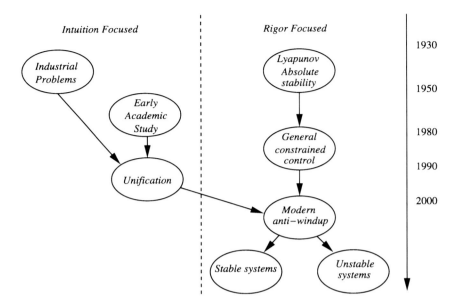

**Fig. 6.4** The development of anti-windup compensation

part) induces the most desirable (often linear) performance on the closed-loop. Modifications to the controller which avoided this charge build-up were often termed "anti-windup" modifications and hence the term anti-windup was born. Since then, however, the term "anti-windup" has evolved and it now means the generic two-step procedure for controller design which was described earlier.

It is then hard to pin-point exactly the origins of anti-windup compensation due to lack of published works on the subject in the early years of control. Teel and co-authors, in their ACC03 Workshop T-1: Modern Anti-Windup Synthesis, point out the discovery back to the 1930s and in particular cite the paper of Lozier as being one of the key early academic papers in identifying the windup problem. Teel and co-authors then describe the evolution of anti-windup compensation on a time-line spanning the last 80 years or so. In a similar manner, in [352] the authors view the development of anti-windup as that depicted in Fig. 6.4. From this figure, we can partition the history of the studies on anti-windup technique into four main stages, namely:

- the first stages in anti-windup development (around 1940–1950);
- early academic study;
- constrained input control;
- modern anti-windup.

Let us now give some precision about each stage above mentioned. Regarding **The First Stage in Anti-windup Development**, it seems certain that practitioners were aware of the problems which saturation caused from the early days of control and that they adopted ad hoc solutions to the problem, without a formal treatment.

## 6.3 A Bit of History

This awareness also manifests itself in some of the early work on absolute stability which was published from the 1940s onwards, and where formal treatment of the saturation problem seems to have begun. Note, however, there is a separation from the anti-windup problem and the work on absolute stability, which was more general than just anti-windup, but arguably less practical.

**Early Academic Study** Some years after this, academics began to study the problem of saturation in control systems. Teel traces this back to Lozier [250] who was able to explain saturation problems in the integrator portion of PI controllers. Somewhat later, Fertik and Ross [90] proposed perhaps the first properly documented anti-windup methodology. This was also accompanied by a variety of work on intelligent integrators [228] and later the celebrated conditioning technique of Hanus et al. [163, 164, 390]. This early literature heralded the beginnings of more formal studies of the saturation problem but, at this point, most papers still focused on developing ad hoc techniques which seemed to overcome certain practical problems, but provided no formal guarantees. Several years after Hanus' conditioning technique was proposed, several researchers [10, 53] provided a unification of many anti-windup schemes and in [226] related this to modern robust control ideas. This unification was later continued in [87] where a state space interpretation was given in "generic" form for many compensators.

**Constrained Input Control** While anti-windup was being developed from a very applied perspective, new results were beginning to be developed in the more general constrained input control field, particularly from the mid-1980s onwards. Such work was focused on using Lyapunov's second method to develop *one step* controllers which could guarantee stability of the *nonlinear* closed-loop systems. Although this work was not directly related to anti-windup, this line of work provided important theoretical results which anti-windup would later benefit from. Some useful references on this subject are [245, 309, 335, 378] and the survey [25] contains an excellent overview.

**Modern Anti-windup** It is difficult to define exactly what constitutes a modern anti-windup technique but in this book we view it as a *systematic* method which can be used to design an anti-windup compensator which provides rigorous guarantees of stability and or performance. Most of these techniques were developed from the late 1990s [138, 150, 261, 269, 320, 369] onwards and were developed in part, thanks to the constrained input control research which preceded them. At about this time there was also a split in the development of anti-windup controllers. Some researchers chose to investigate the problem of enforcing global stability and performance properties for anti-windup compensators [150, 152, 394] while others began to look at regional (local) stability and performance properties, which was necessary for exponentially unstable plants [56, 138, 139].

The above description in terms of four main stages of the development of anti-windup is only a broad approximation, which does not include every possible nuance in the history of this topic. Nevertheless, such a classification provides a useful overview of how anti-windup developed into its present form. It is important to note that it has developed from the problem-specific ad hoc solutions [10, 90] aimed particularly at PID controllers, to sophisticated synthesis methods which populate actual anti-windup literature. In particular, major improvements in this field have been achieved in the late 1990s, as can be observed in [14, 51, 211, 224, 225, 261, 367, 369] among others.

Many LMI-based approaches now exist to design the anti-windup gains in a systematic way (see, for example, [375] for a quick overview). Most often, these are based on the optimization of either a stability domain [56, 139], or a nonlinear $\mathcal{L}_2$-induced performance level [195, 251, 373]. Recently, based on the LFT/LPV framework, extended anti-windup schemes were proposed (see [251, 302, 312, 398]). In these contributions, the saturations are viewed as sector nonlinearities and anti-windup controller design is cast into a convex optimization problem under LMI constraints. Following a similar path, alternative techniques using less conservative representations of the saturation nonlinearities, yet with sector nonlinearities, are proposed in [28, 138, 139, 195, 360]. Moreover, during the last phases previously evoked, we can point out several papers dealing with practical experiments with application of anti-windup strategies in various fields like aeronautical or spatial domains [28, 46, 121, 295, 368], mechanical domains [357, 371], open water channels [409], nuclear fusion [317], telecommunication networks [286, 391] and hard disk drive control [178].

## 6.4 Formulation of Problems

We consider the continuous-time linear plant, using the input–output notation of Fig. 6.3:[1]

$$\dot{x}_p = A_p x_p + B_{pu} u + B_{pw} w$$
$$y_p = C_p x_p + D_{pu} u + D_{pw} w \qquad (6.1)$$
$$z = C_z x_p + D_{zu} u + D_{zw} w$$

where $x_p \in \Re^n$, $u \in \Re^m$, $w \in \Re^q$ and $y_p \in \Re^p$ are the state, the input, the exogenous input and the measured output vectors of the plant, respectively. $z \in \Re^l$ is the regulated output vector used for performance purposes. Matrices $A_p$, $B_{pu}$, $B_{pw}$, $C_p$, $C_z$, $D_{pu}$, $D_{pw}$, $D_{pu}$ and $D_{zw}$ are real constant matrices of appropriate dimensions. Pairs $(A_p, B_{pu})$ and $(C_p, A_p)$ are assumed to be controllable and observable, respectively.

*Remark 6.5* This assumption could be relaxed to stabilizability and detectability of pairs $(A_p, B_{pu})$ and $(C_p, A_p)$. Controllability and observability is considered here only to simplify the hypotheses and the presentation of results in the sequel.

---

[1]For simplicity, the time dependence in the vector will be omitted.

6.4 Formulation of Problems 275

Furthermore, with respect to system (6.1), we assume that an $n_c$th-order dynamic output stabilizing compensator

$$\dot{x}_c = A_c x_c + B_c u_c + B_{cw} w$$
$$y_c = C_c x_c + D_c u_c + D_{cw} w \tag{6.2}$$

where $x_c \in \mathfrak{R}^{n_c}$ is the controller state, $u_c \in \mathfrak{R}^p$ is the controller input and $y_c \in \mathfrak{R}^m$ is the controller output, has been designed in order to guarantee some performance requirements and the stability of the closed-loop system in the absence of control saturation.

It is important to note that the interconnection considered to compute the stabilizing controller (6.2) is the linear interconnection defined by

$$u = y_c; \qquad u_c = y_p \tag{6.3}$$

*Remark 6.6* System (6.1)–(6.2) is assumed to be well-posed. Hence, the interconnection (6.3) is defined from

$$y_c = \Delta^{-1} C_c x_c + \Delta^{-1} D_c C_p x_p + \Delta^{-1} (D_c D_{pw} + D_{cw}) w \tag{6.4}$$

with $\Delta = I_m - D_c D_{pu}$.

*Remark 6.7* By assumption the closed-loop without saturation (with connection (6.3)) is supposed internally stable and well-posed. Therefore, it is evidently necessary that the matrix $\Delta = I_m - D_c D_{pu}$ and the matrix $I_p - D_{pu} D_c$ are both nonsingular.[2]

Moreover, the closed-loop matrix $\mathbb{A}$ defined as

$$\mathbb{A} = \begin{bmatrix} A_p + B_{pu} \Delta^{-1} D_c C_p & B_{pu} \Delta^{-1} C_c \\ B_c (I + D_{pu} \Delta^{-1} D_c) C_p & A_c + B_c D_{pu} \Delta^{-1} C_c \end{bmatrix} \tag{6.5}$$

is supposed to be Hurwitz, i.e., in absence of control bounds, the closed-loop system would be globally exponentially stable.

Suppose now that the input vector $u$ is subject to magnitude limitations as follows:

$$-u_{0(i)} \le u_{(i)} \le u_{0(i)}, \qquad u_{0(i)} > 0, \ i = 1, \dots, m \tag{6.6}$$

*Remark 6.8* To keep the discussion simple we use here symmetric saturation levels, but all the approaches illustrated in this part (potentially conservatively) work as well with nonsymmetric saturations as long as $u_0$ is a lower bound on the two saturation levels. Other symmetrizing steps may, however, be necessary in some particular cases [41, 42].

---

[2] The matrix $I_p - D_{pu} D_c$ is, however, never explicitly used in what follows.

As a consequence of the control bounds (6.6), the actual control signal to be injected in the system is a saturated one, that is, the real interconnection between the plant (6.1) and the controller (6.2) is a nonlinear one described by

$$u = \text{sat}(y_c) = \text{sat}\left(\Delta^{-1}C_c x_c + \Delta^{-1}D_c C_p x_p + \Delta^{-1}(D_c D_{pw} + D_{cw})w\right) \quad (6.7)$$

with each component of the saturation term $\text{sat}(y_c)$ classically defined $\forall i = 1, \ldots, m$ by

$$\text{sat}(y_{c(i)}) = \text{sign}(y_{c(i)}) \min\left(|y_{c(i)}|, u_{0(i)}\right) \quad (6.8)$$

In order to alleviate the undesirable effects of windup caused by input saturation [184], one can consider a dynamic anti-windup controller, of dimension $n_{aw}$, whose input is related to the dead-zone nonlinearity $\phi(y_c) = \text{sat}(y_c) - y_c$, and its output is a vector $\begin{bmatrix} v_x \\ v_y \end{bmatrix} \in \Re^{n_c + m}$.

Such an anti-windup controller can be connected to the controller by acting both on the dynamics of the controller (through $v_x$) and on its output (through $v_y$) [140, 195, 358] as follows:

$$\begin{aligned} \dot{x}_c &= A_c x_c + B_c y_p + B_{cw} w + v_x \\ y_c &= C_c x_c + D_c y_p + D_{cw} w + v_y \end{aligned} \quad (6.9)$$

It allows one to define a closed-loop system related the augmented system state,

$$x = \begin{bmatrix} x_p \\ x_c \\ x_{aw} \end{bmatrix} \in \Re^{n + n_c + n_{aw}}$$

It is now possible to identify two issues: the stability and performance problems. When $w = 0$, it is of interest to estimate the basin of attraction of the closed-loop system, denoted $R_a$ which is defined as the set of all $x \in \Re^{n + n_c + n_{aw}}$ such that for any $x(0)$ belonging to $R_a$, the corresponding trajectory converges asymptotically to the origin. In particular, when, global stability of the system is ensured the basin of attraction corresponds to the whole state space. However, as mentioned in Chap. 1, the exact characterization of the basin of attraction is, in general, not possible. In this case, it is important to obtain estimates of the basin of attraction. In this sense, regions of asymptotic stability can be used to estimate the basin of attraction [215]. On the other hand, in some practical applications one can be interested in ensuring the stability for a given set of admissible initial conditions. This set can be seen as a practical operating region for the system, or a region where the system states can be brought about by the action of temporary disturbances ($w \neq 0$).

In the case where $w = 0$, one of the problems of interest with respect to the closed-loop system (6.9) modified by the addition of the two static anti-windup loops $v_x$ and $v_y$ then consists of computing the anti-windup gains in order to enlarge the basin of attraction of the resulting closed-loop system. In the case where $w \neq 0$, the problem of interest is to ensure a certain level of performance which can be measured, for example, from the finite $\mathcal{L}_2$-gain from the exogenous input $w$ to the performance output $z$. In this case, the problem can then be formulated as follows.

6.4 Formulation of Problems

**Problem 6.1** Determine the anti-windup compensator and a region of asymptotic stability, denoted $\mathcal{E}$, as large as possible, such that

1. the closed-loop system with $w = 0$ is asymptotically stable for any initial condition belonging to the set $\mathcal{E}$;
2. the map from $w$ to $z$ is finite $\mathcal{L}_2$-gain stable with gain $\gamma > 0$.

Note that the implicit objective in the first item of Problem 6.1 is to optimize the size of the basin of attraction for the closed-loop system (6.9) (with $w = 0$) over the choice of the anti-windup compensator. This can be accomplished indirectly by searching for an anti-windup compensator that leads to a region of stability for the closed-loop system as large as possible. Considering quadratic Lyapunov functions and ellipsoidal regions of stability, the maximization of the region of stability can be carried out by using some well-known size optimization criteria for ellipsoidal sets, such as: minor axis maximization, volume maximization, or even the maximization of the ellipsoid in certain given directions (see Sect. 2.2.5 in Chap. 2). On the other hand, when the open-loop system is asymptotically stable, it can be possible to search for an anti-windup controller which guarantees the global asymptotic stability of the origin of the closed-loop system. Different strategies to build such anti-windup compensators, both in the local and the global context, will be presented in Chap. 7.

*Remark 6.9* Points 1 and 2 of Problem 6.1 can be studied in an analysis context by considering the saturated closed-loop system without anti-windup loop, i.e., $n_{\mathrm{aw}} = 0$, $v_x = 0$ and $v_y = 0$. This allows us in particular to *measure* the improvement due to the anti-windup loop.

Different classes of disturbance can be considered, such as been done in the previous chapters of Part II. To ease the presentation, in this Part III, the results are presented only for the class of disturbance vectors limited in energy, that is, $w \in \mathcal{L}_2$ and, for some scalar $\delta$, $0 < \frac{1}{\delta} < \infty$, one gets

$$\|w\|_2^2 = \int_0^\infty w(s)'w(s)\,ds \leq \delta^{-1} \tag{6.10}$$

*Remark 6.10* A nonlinear $\mathcal{L}_2$-induced performance level does not unfortunately provide a direct answer to the most standard anti-windup control problem which consists in minimizing the saturation effects for a restricted class of reference signals. The class of $\mathcal{L}_2$-bounded signals which is typically considered is indeed very large, while in practice, it is most often sufficient to consider step-like reference inputs with bounded magnitudes. Hence, a particular case consists in considering that the exogenous signal (which can represent reference signals or disturbance) is generated by a stable autonomous linear system as used, for example, in [27, 304]

$$\tau \dot{w} + w = 0, \qquad w(0) = w_0 \in \Re^q \tag{6.11}$$

Thus, it is easily verified that, for small values of $\tau$, the exogenous signals $w$ can be interpreted as bounded step inputs. This bound is clearly fixed by $\|w_0\|$. Furthermore, $w$ is also $\mathcal{L}_2$-bounded since, according to (6.10), one gets

$$\|w\|_2^2 = \int_0^\infty w(s)'w(s)\,ds = \frac{w_0^2}{2\tau} \tag{6.12}$$

## 6.5 Regional (Local) Versus Global Strategies

### 6.5.1 A Quick Overview in the Regional (Local) Context

Only a few papers have been dedicated to anti-windup strategy for exponentially unstable systems, that is, in a regional (local) context, until around 2000. The majority of those papers presented algorithms for computing anti-windup compensators but without characterization of the region of stability: see, for example [401] and [369]. However, a key element in the local case is the ability to guarantee the stability and therefore to characterize the region of stability for the closed-loop system (6.9). One of the first paper addressing clearly the local case with a guarantee of stability was Teel's paper [367]. In [367], an algorithm, which requires measurement of the exponentially unstable modes, was proposed. The results provided an anti-windup compensator extending those presented in [369] by removing some restrictions on the transient behavior of the unsaturated feedback loop. In [367], the conditions were not, however, in LMI form. Indeed, one of the first applications of LMI's to the anti-windup synthesis problem in the local framework was given in [138, 139] by considering only static anti-windup loop. [139] came later than the papers such as [56, 143] in which the conditions proposed were not into LMI form but in BMI (bilinear matrix inequalities) form due mainly to the way chosen to model the saturation terms based on linear differential inclusions or classical sector conditions (see Chap. 1).

Furthermore, in [397], an extension of [151] was proposed allowing for the computation of dynamic anti-windup compensators. Nevertheless, contrary to [139], the region in which the stability of the closed-system was guaranteed was not clearly described. Recently, several papers dealing with performance, like $\mathcal{L}_2$-performance, have been published mainly in the context of dynamic anti-windup compensator design: see, for example, [361] in which the six first chapters are dedicated to anti-windup strategies and their applications. See also [29, 195, 197, 406].

*Remark 6.11* It should be pointed out that the generalized (or modified) sector condition (see Sect. 1.7.2.2), proposed initially in [138, 139, 359], played a key role in the development of "true" LMI conditions for the synthesis of anti-windup compensators (static as well as dynamic). Actually, the classical sector condition (see Sect. 1.7.2.1) leads to LMI conditions only in the global stabilization case (see for instance [269] and [151]). On the other hand, the use of polytopic modeling (see Sect. 1.7.1) will always lead to BMI conditions for the anti-windup synthesis in the local context (see for instance [56]).

## 6.5.2 A Quick Overview in the Global Context

Solutions to the anti-windup problem in the local case are typically of somewhat higher complexity than the global anti-windup problem, because a key element of their solution requires some sort of description of an associated region of asymptotic stability for the saturated system. The geometry of this region is generally not easy to describe exactly and thus, as mentioned previously, normally the region is estimated using an ellipsoid or a polyhedral set (see also Chap. 1). For stable linear plants (i.e. $\Re e\lambda_i(A_p) < 0, \ \forall i$) it is possible to design anti-windup compensators which go beyond regional (local) guarantees and provide stability for all $\xi \in \Re^{n+n_c+n_{aw}}$. This simplifies the anti-windup problem somewhat as now a description of the region of attraction is not required, as it is the whole state space.

Although in principle the design of anti-windup compensators for stable linear systems subject to input saturation could be achieved using absolute stability tools such as the circle and Popov criteria [215] (see also Appendix A), these were most useful for single-loop systems and analysis. Systematic design was somewhat more complicated to derive until LMIs began to emerge, although useful classical tools are reported in [212, 396]. One of the first applications of LMIs to the anti-windup synthesis problem was given by [256] which considered the anti-windup synthesis problem as an application of absolute stability theory involving common Lyapunov functions. [256] followed [226] but considering only static anti-windup compensators. Similar ideas to the above were also exploited in [301] where the observer-based structure of anti-windup compensators was used (i.e. again the anti-windup compensator was static). Even if the papers [256] and [301] were important steps in anti-windup design and they both used the LMI framework as part of the anti-windup synthesis procedure, the design methods were not wholly LMI-based, as the inequalities were bilinear matrix inequalities (BMIs) which were linearized by fixing one of the free variables. The late part of the 20th and early part of the 21st century saw the development of two anti-windup synthesis methods which were wholly LMI-based. In [269], a method continues the static anti-windup theme but effectively uses the circle criterion with an $\mathcal{L}_2$-gain constraint to devise a purely LMI-based synthesis method. The work in [269] is similar to that in [312] where a small gain approach is used to synthesize an anti-windup compensator. The results in [312] are generally stated in terms of bilinear matrix inequalities which transpire to be linear in the special case of single-input-single-output systems with fixed static multiplier. Note that the advantage of the results of [312] is that they can also be applied, in the regional (local) context, to unstable systems, although no estimate of the region of attraction is provided. A similar result was reported in [372] except that the results were improved in two ways. Firstly provision for low order anti-windup synthesis was made, which significantly enlarges the class of compensators which can be designed and also allows compensators with superior performance (in terms of their $\mathcal{L}_2$-gains) to be obtained. Secondly, [372] proposed a performance map which allowed a linear performance optimization explicitly via LMIs.

The main problem with the results of [269] was, because the anti-windup compensator was restricted to a static structure, that it was not always possible to construct an anti-windup compensator for any given plant-controller combination. Mulder and colleagues tried to relax this by using piecewise linear Lyapunov functions [268] to decrease the conservatism introduced by the circle criterion, but this did not lead to linear matrix inequalities and resulted in some rather complicated constructions for the anti-windup compensator. On the other hand, in [150] and [151] conditions were given which allowed a general $n_{aw}$th order anti-windup compensator to be constructed using "almost" LMI conditions. In the general case they were nonconvex but under certain conditions (tricks to guarantee the rank condition), they could be relaxed to be linear.

## 6.6 Mismatch-Based Anti-windup Synthesis

The anti-windup problem is often interpreted as one of keeping the behavior of the system during saturation as close as possible to the behavior of the system, had saturation not been present. This idea had been around for many years in the anti-windup community but was not formalized until the work of [261, 370, 393], although preliminary work in this spirit can be found in [185, 253]. In this case, it is often useful to define the nominal linear system as

$$\mathcal{P}_{\text{lin}} = \begin{cases} \dot{\bar{\xi}}_{\text{lin}} &= \mathbb{A}\bar{\xi}_{\text{lin}} + \mathbb{B}_2 w \\ y_{c,\text{lin}} &= \mathbb{C}_2\bar{\xi}_{\text{lin}} + \mathbb{D}_{22} w \\ z_{\text{lin}} &= \mathbb{C}_1\bar{\xi}_{\text{lin}} + \mathbb{D}_{12} w \end{cases} \tag{6.13}$$

This system represents the behavior of the system when no saturation is encountered and hence does not include the dynamics of the anti-windup compensator. In this representation $z_{\text{lin}}$ denotes the performance output when no saturation is present, $y_{c,\text{lin}}$ the linear control signal and $\bar{\xi}_{\text{lin}} \in \mathfrak{R}^{n+n_c}$ the state when no saturation is present. With this in mind, it is possible (see [212, 372, 377, 386, 393]) to represent the anti-windup compensated system as depicted in Fig. 6.5.

In Fig. 6.5, the nominal linear system $\mathcal{P}_{\text{lin}}$ is clearly decoupled from the nonlinear behavior of the system during saturation. Note that $M(s) - I$ and $G_2(s)M(s)$ represent transfer function matrices derived from the dynamics of the linear plant, linear controller and anti-windup compensator and also have attractive "robust control" interpretations [212, 372, 386]. This figure gives a clear interpretation of the stability and performance problems when trying to keep linear performance deviation minimal: the stability problem is to keep the nonlinear loop stable (as by assumption $\mathbb{A}$ is Hurwitz) and the performance problem is to keep $z_d$ as small as possible.

In [370], a quite similar anti-windup problem represented by the block diagram in Fig. 6.5 was defined in order to ensure the following.

1. The performance of the "real" saturated system and the mismatch system are identical unless saturation occurs. i.e. $z - z_{\text{lin}} = 0$, $\forall t \geq 0$ if $y_{c,\text{lin}} \preceq u_0$, $\forall t \geq 0$.
2. If $y_{c,\text{lin}}$ eventually falls below the saturation threshold, which can be captured as ensuring $\phi(y_{c,\text{lin}}) \in \mathcal{L}_2$, then the real output $z$ will converge to the linear output, $z_{\text{lin}}$ asymptotically.

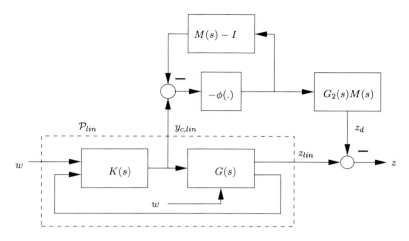

**Fig. 6.5** Structure of mismatch system for anti-windup analysis

This is a convenient way in which to look at the anti-windup problem and many authors, [76, 153, 179, 261, 372, 386, 393, 394, 396] have adopted this approach for studying the anti-windup problem. One can also strengthen the anti-windup problem to ensure that the "gain" from the linear control signal, $y_{c,\text{lin}}$, to the performance output $z_d$ is less than a certain bound, i.e.

$$\|z_d\|_2 \leq \gamma \|y_{c,\text{lin}}\|_2$$

The papers [181, 372, 376] and others have given constructive design procedures for minimizing such a cost function. The reader may also consult the related other references proposing interesting work in this context [18, 404, 405].

## 6.7 Conclusion

This chapter has presented a quick tour of the different anti-windup techniques developed in the literature and in particular those we can classify as modern anti-windup techniques. More precisely, a quick overview regarding both regional (local) and global contexts was provided by focusing in formal anti-windup strategies in opposition to ad hoc methods. A large part of this chapter has followed [352] and [108]. The techniques evoked are presented in detail in Chap. 7 of this book or else can be found in [406].

# Chapter 7
# Anti-windup Compensator Synthesis

## 7.1 Introduction

According to the context described in Chap. 6, the current chapter considers the design problem of anti-windup compensators. Several theoretical conditions both in regional (local) and global context are developed and proven. Associated algorithms allowing one to guarantee stability and performance are described and commented. Two complementary ways are chosen in the framework of anti-windup techniques based on LMI conditions: direct linear anti-windup (DLAW) and model recovery anti-windup (MRAW).

The chapter is structured as follows. Section 7.2 is devoted to the description of the system under consideration and problems we intend to solve. Hence, inspired two main architectures, which are of course closely related, are discussed: In Sect. 7.2.1, the named direct linear anti-windup is discussed, whereas the second architecture named model recovery anti-windup is considered in Sect. 7.2.2. The conditions associated to the first architecture (DLAW) are then presented in Sect. 7.3, both in the global and regional (local) contexts. Several algorithms are then described and commented in Sect. 7.3.4. Section 7.4 is similarly dedicated to the second architecture (MRAW) and algorithms mainly associated to LMI-based conditions are presented in Sect. 7.4.2. Algorithms described in Sects. 7.3 and 7.4 corresponding to different alternatives are summarized and compared in Sect. 7.5. Section 7.6 is dedicated to pointing out several possible extensions. Finally, a short conclusion ends the chapter.

## 7.2 Problems Setup

Let us reproduce Fig. 6.2 to ease the presentation. As explained in the previous chapter, Fig. 7.1 basically depicts the anti-windup architecture, with the notation used in the sequel.

However, a further classification should be made between two families of anti-windup compensation which most of the modern constructive anti-windup tech-

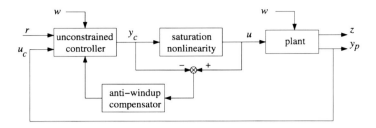

**Fig. 7.1** Basic anti-windup architecture

niques fall into. To do this, it is useful to introduce some notation for the block diagrams in Fig. 7.1 in order to introduce these two families of solutions. Consider the following state space representation for the continuous-time linear plant:[1]

$$\begin{aligned} \dot{x}_p &= A_p x_p + B_{pu} u + B_{pw} w \\ y_p &= C_p x_p + D_{pu} u + D_{pw} w \\ z &= C_z x_p + D_{zu} u + D_{zw} w \end{aligned} \quad (7.1)$$

where $x_p \in \mathfrak{R}^{n_p}$, $u \in \mathfrak{R}^m$, $w \in \mathfrak{R}^q$ and $y_p \in \mathfrak{R}^p$ are the state, the input, the exogenous input and the measured output vectors of the plant, respectively. $z \in \mathfrak{R}^l$ is the regulated output vector used for performance purposes. Matrices $A_p$, $B_{pu}$, $B_{pw}$, $C_p$, $C_z$, $D_{pu}$, $D_{pw}$, $D_{zu}$ and $D_{zw}$ are real constant matrices of appropriate dimensions. Pairs $(A_p, B_{pu})$ and $(C_p, A_p)$ are assumed to be controllable and observable, respectively.

Considering the plant (7.1), we assume that an $n_c$th-order dynamic output stabilizing compensator

$$\begin{aligned} \dot{x}_c &= A_c x_c + B_c u_c + B_{cw} w + v_x \\ y_c &= C_c x_c + D_c u_c + D_{cw} w + v_y \end{aligned} \quad (7.2)$$

has been designed, where $x_c \in \mathfrak{R}^{n_c}$ is the controller state, $u_c \in \mathfrak{R}^p$ is the controller input, $y_c \in \mathfrak{R}^m$ is the controller output and $v_x$, $v_y$ are extra inputs used for anti-windup purposes, specified later. Controller (7.2) has been designed in order to guarantee some performance requirements and the stability of the closed-loop system in absence of control saturation, that is, when the following unconstrained interconnection is used:

$$u = y_c, \quad u_c = y_p, \quad v_x = 0, \quad v_y = 0 \quad (7.3)$$

The closed-loop system (7.1), (7.2), (7.3) is assumed to be well-posed (see Remark 6.6). The matrix $\Delta$ is defined as

$$\Delta = I_m - D_c D_{pu}$$

---

[1] For simplicity, the time dependence in the vector will be omitted.

## 7.2 Problems Setup

and the unconstrained closed-loop dynamic matrix

$$\mathbb{A} = \begin{bmatrix} A_p + B_{pu}\Delta^{-1}D_cC_p & B_{pu}\Delta^{-1}C_c \\ B_c(I_p + D_{pu}\Delta^{-1}D_c)C_p & A_c + B_cD_{pu}\Delta^{-1}C_c \end{bmatrix} \tag{7.4}$$

is necessarily Hurwitz, i.e., in the absence of control bounds, the closed-loop system is globally exponentially stable.

Suppose now that the input vector $u$ is subject to symmetric magnitude limitations:

$$-u_{0(i)} \le u_{(i)} \le u_{0(i)}, \quad u_{0(i)} > 0, \ i = 1, \dots, m \tag{7.5}$$

As a consequence of the control bounds, the actual control signal to be injected into the system is a saturated one, that is, there is an input nonlinearity before the input variable $u$ which is defined as

$$u = \text{sat}(y_c) \tag{7.6}$$

where each component of the saturation function $\text{sat}_{u_0}(\cdot)$ is classically defined $\forall i = 1, \dots, m$ by

$$\text{sat}(y_{c(i)}) = \text{sign}(y_{c(i)}) \min\left(|y_{c(i)}|, u_{0(i)}\right) \tag{7.7}$$

It is now possible to introduce the two main architectures for anti-windup design. The first one, somewhat natural, is called *direct linear anti-windup* (DLAW) and will then be addressed in detail in Sect. 7.3. The second one, called *model recovery anti-windup* (MRAW) will be addressed in Sect. 7.4.

### 7.2.1 Direct Linear Anti-windup

The direct linear anti-windup (DLAW) strategy consists in selecting the anti-windup filter in Fig. 7.1 as a linear filter which produces the signals $v_x$ and $v_y$ as an output.

$$\dot{x}_{\text{aw}} = A_{\text{aw}}x_{\text{aw}} + B_{\text{aw}}\left(\text{sat}(y_c) - y_c\right)$$
$$\begin{bmatrix} v_x \\ v_y \end{bmatrix} = C_{\text{aw}}x_{\text{aw}} + D_{\text{aw}}\left(\text{sat}(y_c) - y_c\right) \tag{7.8}$$

where $x_{\text{aw}} \in \mathfrak{R}^{n_{\text{aw}}}$ is the anti-windup state, $u_{\text{aw}} = \text{sat}(y_c) - y_c = \phi(y_c)$ is the anti-windup input and $[\, v_x' \ v_y' \,]' \in \mathfrak{R}^{n_c+m}$ is the anti-windup output. The goal of DLAW design is to compute suitable matrices $A_{\text{aw}}, B_{\text{aw}}, C_{\text{aw}}, D_{\text{aw}}$ in (7.8) so that the so-called anti-windup closed-loop system (7.1), (7.2), (7.8) satisfies desirable stability and performance properties. Representative algorithms solving this problem with different guaranteed properties will be presented in Sect. 7.3 and are summarized here.

A tentative history of DLAW is given in Chap. 6. We recall just some quick elements below. Historically, DLAW belongs to the large family of anti-windup controllers summarized in the early survey [226], but the most recent literature, addressing the corresponding design problem from the point of view of guaranteed stability and performance properties is strongly coupled with the use of linear matrix inequalities (see, e.g., [45]). Perhaps the first important paper in this direction is [269], which addresses static anti-windup design, namely only $D_{aw}$ is to be designed in (7.8) and the remaining matrices are empty. In this paper, the design procedure is based on global sector conditions on the saturation. Later works using global sector conditions and providing globally stabilizing results are given in [102, 151, 153, 372]. Since the use of LMIs, typically, can only show exponential stability of the arising closed-loop, these results are, however, only applicable to control systems with exponentially stable plants. Later works, using generalized sector conditions for the saturation (this condition allows for regional stability estimates) turn out to algorithms applicable also with non exponentially stable plants [28, 139, 197, 303, 360, 361, 397]. Discrete-time counterparts are reported in [140, 181, 257, 339].

## 7.2.2 Model Recovery Anti-windup

This second architecture follows a different paradigm from anti-windup design and is based on selecting the anti-windup compensator in Fig. 7.1 as a dynamical system incorporating a model of the plant transfer function from $u$ to $y_p$. In particular, the compensator is selected as follows:

$$
\begin{aligned}
\dot{x}_{aw} &= A_p x_{aw} + B_{pu}\big(\text{sat}(y_c) - y_c + v_y\big) \\
v_x &= B_c\big(C_p x_{aw} + D_{pu}\big(\text{sat}(y_c) - y_c + v_y\big)\big)
\end{aligned}
\tag{7.9}
$$

The signal $v_y$ is purposely left unspecified because it corresponds to a degree of freedom to be exploited in the anti-windup design. The advantage of the model recovery anti-windup (MRAW) architecture lies in the fact that the anti-windup filter (7.9) keeps track (via $x_{aw}$) of the amount of plant state response that is missing in the saturated closed-loop due to the undesired effects of saturation. A different way of interpreting this is that, regardless of what the selection of $v_y$ is, the closed-loop system (7.1), (7.2), (7.9) is such that $x_p + x_{aw}$ always coincides with the unconstrained plant state response (see Sect. 7.4.1 for details) so that driving $x_{aw}$ to zero forces the plant state $x_p$ to recover the unconstrained response. This fact allows one to prove closed-loop stability and performance by mean of a cascaded systems analysis and provides two useful advantages:

(1) the controller (7.2) does not need to be linear (the MRAW scheme is controller independent) and robust closed-loop properties hold under the assumption that it satisfies mild Lipschitz conditions;
(2) the stabilizing signal $v_y$ can be selected nonlinearly, thereby allowing for the design of nonlinear anti-windup laws, suitable for solving the trickiest anti-windup design problems (where nonlinear compensation is necessary to get global results).

7.3 Direct Linear Anti-windup Design

The key ingredients of model recovery anti-windup designs were laid down in the companion papers [369, 370] (the approach has been called $\mathcal{L}_2$ anti-windup for a long time). Later on, the architecture has been further illustrated for exponentially stable linear plants in [18, 404, 405]. Nontrivial extensions to exponentially unstable linear plants have been given in [105, 367] while approaches guaranteeing global asymptotic stability for non exponentially stable nor exponentially unstable plants (those globally asymptotically but not globally exponentially stabilizable by bounded inputs) are given in [104, 409].

Discrete-time counterparts can be found in [154]. A further property of the MRAW structure is that it can be applied (in a nontrivial way) to nonlinear plants too. An example where this has been done is [266].

*Remark 7.1* The mismatch based anti-windup synthesis discussed in Sect. 6.6 may be explained as a particular case of MRAW.

## 7.3 Direct Linear Anti-windup Design

In this section, the system defined through (7.1), (7.2) and (7.8) can also be written as

$$
\begin{aligned}
\dot{x}_p &= A_p x_p + B_{pu} \operatorname{sat}(y_c) + B_{pw} w \\
y_p &= C_p x_p + D_{pu} \operatorname{sat}(y_c) + D_{pw} w \\
z &= C_z x_p + D_{zu} \operatorname{sat}(y_c) + D_{zw} w \\
\dot{x}_c &= A_c x_c + B_c y_p + B_{cw} w + v_x \\
y_c &= C_c x_c + D_c y_p + D_{cw} w + v_y \\
\dot{x}_{\mathrm{aw}} &= A_{\mathrm{aw}} x_{\mathrm{aw}} + B_{\mathrm{aw}} \big(\operatorname{sat}(y_c) - y_c\big) \\
v_x &= \begin{bmatrix} I_{n_c} & 0 \end{bmatrix} \big(C_{\mathrm{aw}} x_{\mathrm{aw}} + D_{\mathrm{aw}} (\operatorname{sat}(y_c) - y_c)\big) \\
v_y &= \begin{bmatrix} 0 & I_m \end{bmatrix} \big(C_{\mathrm{aw}} x_{\mathrm{aw}} + D_{\mathrm{aw}} (\operatorname{sat}(y_c) - y_c)\big)
\end{aligned}
\tag{7.10}
$$

*Remark 7.2* The presence of the implicit loop in the closed-loop system due to $v_y$ can be removed by considering a simplified anti-windup controller: the anti-windup output is only injected in the dynamics of $x_c$ ($v_y = 0$). In this case, the anti-windup output reduces to $y_{\mathrm{aw}} = v_x \in \Re^{n_c}$. Another way consists of filtering the signal $v_y$ and therefore of injecting $\bar{v}_y = F(s)v_y$. Thus, a low-pass filter $F(s)$ can be used to avoid some chattering effects on the control signals as well as algebraic or implicit loops [27].

As underlined in Chap. 6, we can distinguish two issues of interest with respect to system (7.10), namely stability ($w = 0$) and performance problems ($w \neq 0$). It is important to point out that in the case where $w = 0$, one of the problems of interest with respect to the closed-loop system (7.10), consists of computing the anti-windup compensator in order to enlarge the basin of attraction of the resulting closed-loop system. In the case where $w \neq 0$, the problem of interest is to ensure a certain level of performance which can be measured, for example, by the finite $\mathcal{L}_2$-gain from the exogenous input $w$ to the performance output $z$. For this, the disturbance vector $w$ is assumed to be limited in energy, that is, $w \in \mathcal{L}_2$ and for some scalar $\delta$, $0 \leq \frac{1}{\delta} < \infty$, it follows that

$$\|w\|_2^2 = \int_0^\infty w(t)'w(t)\,dt \leq \delta^{-1} \tag{7.11}$$

In this case, the problem can be summarized through Problem 6.1 with respect to the closed-loop system (7.10).

### 7.3.1 Preliminary Elements

A great part of the results developed in the DLAW context are based upon the use of dead-zone nonlinearities and associated modified sector conditions. Indeed, it is important to underline that every system, which involves saturation-type nonlinearities, may be easily rewritten with dead-zone nonlinearities (see Chap. 3). Considering the saturation function $\mathrm{sat}(y_c)$, the resulting dead-zone nonlinearity $\phi(y_c)$ is obtained from $\phi(y_c) = \mathrm{sat}(y_c) - y_c$. Thus, by considering the extended state vector $x$ defined by

$$x = \begin{bmatrix} x_p \\ x_c \\ x_{\mathrm{aw}} \end{bmatrix} \in \mathfrak{R}^{n_p + n_c + n_{\mathrm{aw}}} \tag{7.12}$$

the closed-loop system is described by

$$\dot{x} = \mathcal{A}x + \mathcal{B}_1\phi(y_c) + \mathcal{B}_2 w$$
$$y_c = \mathcal{C}_1 x + \mathcal{D}_{11}\phi(y_c) + \mathcal{D}_{12} w \tag{7.13}$$
$$z = \mathcal{C}_2 x + \mathcal{D}_{21}\phi(y_c) + \mathcal{D}_{22} w$$

where

$$\mathcal{A} = \begin{bmatrix} \mathbb{A} & B_v C_{\mathrm{aw}} \\ 0 & A_{\mathrm{aw}} \end{bmatrix}; \qquad \mathcal{B}_1 = \begin{bmatrix} B_\phi + B_v D_{\mathrm{aw}} \\ B_{\mathrm{aw}} \end{bmatrix}$$

$$\mathcal{B}_2 = \begin{bmatrix} B_2 \\ 0 \end{bmatrix}; \qquad \mathcal{C}_1 = \begin{bmatrix} C_1 & C_{v1} C_{\mathrm{aw}} \end{bmatrix} \tag{7.14}$$

$$\mathcal{D}_{11} = D_1 + C_{v1} D_{\mathrm{aw}}$$

$$\mathcal{C}_2 = \begin{bmatrix} C_2 & C_{v2} C_{\mathrm{aw}} \end{bmatrix}; \qquad \mathcal{D}_{21} = D_2 + C_{v2} D_{\mathrm{aw}}$$

7.3 Direct Linear Anti-windup Design

with $\mathbb{A}$ defined in (7.4) and

$$B_v = \begin{bmatrix} B_{pu}\Delta^{-1}\begin{bmatrix} 0 & I_m \end{bmatrix} \\ B_c D_{pu}\Delta^{-1}\begin{bmatrix} 0 & I_m \end{bmatrix} + \begin{bmatrix} I_{n_c} & 0 \end{bmatrix} \end{bmatrix}$$

$$B_\phi = \begin{bmatrix} B_{pu}(I_m + \Delta^{-1}D_c D_{pu}) \\ B_c D_{pu}(I_m + \Delta^{-1}D_c D_{pu}) \end{bmatrix}$$

$$C_1 = \begin{bmatrix} \Delta^{-1}D_c C_p & \Delta^{-1}C_c \end{bmatrix}$$

$$C_{v1} = \Delta^{-1}\begin{bmatrix} 0 & I_m \end{bmatrix}; \qquad D_1 = \Delta^{-1}D_c D_{pu}$$

$$C_2 = \begin{bmatrix} C_z + D_{zu}\Delta^{-1}D_c C_p & D_{zu}\Delta^{-1}C_c \end{bmatrix} \tag{7.15}$$

$$C_{v2} = D_{zu}\Delta^{-1}\begin{bmatrix} 0 & I_m \end{bmatrix}$$

$$D_2 = D_{zu}(I_m + \Delta^{-1}D_c D_{pu})$$

$$B_2 = \begin{bmatrix} B_{pu}\Delta^{-1}(D_{cw} + D_c D_{pw}) + B_{pw} \\ B_c D_{pu}\Delta^{-1}(D_{cw} + D_c D_{pw}) + B_{cw} + B_c D_{pw} \end{bmatrix}$$

$$\mathcal{D}_{12} = \Delta^{-1}(D_{cw} + D_c D_{pw})$$

$$\mathcal{D}_{22} = D_{zw} + D_{zu}\Delta^{-1}(D_{cw} + D_c D_{pw})$$

The results developed in the sequel rest on the use of the modified sector condition, which is applied specifically to dead-zone nonlinearities, as detailed in Sect. 1.7.2.2 and described in Lemma 1.6. Hence, in this context, let us define the following set:

$$S(u_0) = \left\{ y_c \in \Re^m, \ \omega \in \Re^m; \ -u_0 \preceq y_c - \omega \preceq u_0 \right\} \tag{7.16}$$

and let us reproduce the statement of Lemma 1.6 in our symmetric saturation case.

**Lemma 7.1** [359] *If $y_c$ and $\omega$ are elements of $S(u_0)$ then the nonlinearity $\phi(y_c)$ satisfies the following inequality*:

$$\phi(y_c)'S^{-1}\big(\phi(y_c) + \omega\big) \leq 0 \tag{7.17}$$

*for any diagonal positive definite matrix $S \in \Re^{m \times m}$.*

*Remark 7.3* Particular formulations of Lemma 7.1 can be found in [139] (concerning the case of systems with a single saturation function) and in [295, 360] (concerning systems presenting both magnitude and dynamics restricted actuators). Moreover, it should be pointed out that $\omega = \Lambda y_c$, where $\Lambda$ is a diagonal matrix such that $0 < \Lambda \leq I_m$ (see Lemma 1.5 of Chap. 1), is a particular case of the generic formulation (7.17) and corresponds to the classical sector condition (see for instance [183, 215]). Moreover, Lemma 7.1 caters easily for nested saturations.

*Remark 7.4* Particular formulations of Lemma 7.1 can be stated by considering

$$S_\omega(u_0) = \left\{ \omega \in \Re^m; \ -u_0 \preceq \omega \preceq u_0 \right\} \tag{7.18}$$

$$\phi(y_c)' S^{-1} \left( \phi(y_c) + y_c + \omega \right) \le 0 \tag{7.19}$$

or equivalently

$$\phi(y_c)' S^{-1} \left( \mathrm{sat}_{u_0}(y_c) + \omega \right) \le 0 \tag{7.20}$$

instead of (7.16) and (7.17), respectively (see Remark 3.1 in Chap. 3). Such a formulation is also used in [195] and [197] and gives simplified conditions in the case when $y_c$ depends on $\phi(y_c)$, that is, in the presence of an implicit function and nested conditions.

Lemma 7.1, as written, is rather dedicated to the regional case, but it can be considered in a global context. For this, it suffices to consider $\omega = y_c$ and therefore the set $S(u_0)$ is the whole state space (see also Lemma 1.4 in Chap. 1). In that case, the sector condition (7.17) is globally satisfied. Note, however, that the global stability of the closed-loop system subject to such a nonlinearity will be obtained only if some assumptions on the stability of the open-loop system are verified (see Sect. 1.6.2.1 in Chap. 1).

### 7.3.2 DLAW Schemes with Global Stability Guarantees

Let us first suppose that the anti-windup compensator is given (that is matrices $A_{\mathrm{aw}}$, $B_{\mathrm{aw}}$, $C_{\mathrm{aw}}$, $D_{\mathrm{aw}}$ are known) and let us denote $n = n_p + n_c + n_{\mathrm{aw}}$. The following general result can be stated by using the framework developed in [27] (see also [151]).

**Proposition 7.1** *If there exist a symmetric positive definite matrix $Q \in \Re^{n \times n}$, a diagonal matrix $S \in \Re^{m \times m}$ and a positive scalar $\gamma$ such that the following conditions hold*:

$$\begin{bmatrix} Q\mathcal{A}' + \mathcal{A}Q & \mathcal{B}_1 S - Q\mathcal{C}_1' & \mathcal{B}_2 & Q\mathcal{C}_2' \\ \star & -2S - \mathcal{D}_{11}S - S\mathcal{D}_{11}' & -\mathcal{D}_{12} & S\mathcal{D}_{21}' \\ \star & \star & -I & \mathcal{D}_{22}' \\ \star & \star & \star & -\gamma I \end{bmatrix} < 0 \tag{7.21}$$

*then*

1. *when $w = 0$, the global asymptotic stability (of the origin) for the saturated system (7.13) is guaranteed;*
2. *when $w \neq 0$,*

   - *the closed-loop trajectories remain bounded for any $w(t) \in \mathcal{L}_2$ and any initial conditions;*
   - *the map from $w$ to $z$ is finite $\mathcal{L}_2$-gain stable with:*

## 7.3 Direct Linear Anti-windup Design

$$\int_0^T z(t)'z(t)\,dt \le \gamma \int_0^T w(t)'w(t)\,dt + \gamma x(0)'Q^{-1}x(0), \quad \forall T \ge 0 \qquad (7.22)$$

*Proof* Consider a quadratic Lyapunov function $V(x) = x'Px$, with $P = P' = Q^{-1} > 0$. Then, the nonlinear closed-loop plant defined by (7.13) is stable for any initial condition $x(0) \in \Re^n$ if one verifies: $\dot{V}(x) + \frac{1}{\gamma}z'z - w'w < 0$. Note further that Lemma 7.1 is globally satisfied provided that $\omega = y_c$. Then, one has $\phi(y_c)'S^{-1}(\phi(y_c) + y_c) \le 0$, where $S$ is any diagonal positive definite matrix in $\Re^{m \times m}$. From the Schur complement, the satisfaction of relation (7.21) means that

$$\dot{V}(x) + \frac{1}{\gamma}z'z - w'w - 2\phi(y_c)'S^{-1}\big(\phi(y_c) + y_c\big) < 0$$

The stability immediately follows from the above inequality which obviously implies that $\dot{V}(x) < 0$ along any closed-loop trajectory when $w = 0$. The condition (7.22) is obtained by integration of $\dot{V}(x) + \frac{1}{\gamma}z'z - w'w < 0$ and, since $V(x(T)) > 0$, $\forall T > 0$,

$$\int_0^T z(t)'z(t)\,dt \le \gamma\big(V\big(x(0)\big) - V\big(x(T)\big)\big) \le \gamma V\big(x(0)\big) + \gamma \int_0^T w(t)'w(t)\,dt$$

$$\le \gamma x(0)'Q^{-1}x(0) + \gamma \int_0^T w(t)'w(t)\,dt \qquad \square$$

Let us now focus on the synthesis issue. Then, the analysis variable $Q$ which is introduced in Proposition 7.1 and the matrices $A_{aw}$, $B_{aw}$, $C_{aw}$, $D_{aw}$ of the anti-windup compensator have to be computed simultaneously. As a result, the inequality (7.21) which is a priori a BMI, is no longer convex. However, in the full-order case (i.e. $n_{aw} = n_p + n_c$), some particular congruence transformations and the Projection Lemma (see Appendix C) can be exploited to derive a convex characterization.

**Proposition 7.2** *There exists an anti-windup controller $(A_{aw}, B_{aw}, C_{aw}, D_{aw})$ such that the conditions of Proposition 7.1 are satisfied if there exist two symmetric positive definite matrices $X \in \Re^{(n_p + n_c) \times (n_p + n_c)}$, $Y \in \Re^{(n_p + n_c) \times (n_p + n_c)}$ and a positive scalar $\gamma$ such that the following conditions hold:*

$$\begin{bmatrix} \mathbb{A}'X + X\mathbb{A} & XB_2 & C_2' \\ \star & -I & \mathcal{D}_{22}' \\ \star & \star & -\gamma I \end{bmatrix} < 0 \qquad (7.23)$$

$$\begin{bmatrix} Y_1 A_p' + A_p Y_1 & B_{pw} & Y_1 C_z' \\ \star & -I & D_{zw}' \\ \star & \star & -\gamma I \end{bmatrix} < 0 \qquad (7.24)$$

$$\begin{bmatrix} X & I \\ I & Y \end{bmatrix} > 0 \qquad (7.25)$$

*where $Y_1 \in \Re^{n_p \times n_p}$ is the block $(1,1)$ of $Y$.*

*Proof* Let us partition the $Q$-matrix used in Proposition 7.1 as follows:

$$Q = \begin{bmatrix} Y & N' \\ N & W \end{bmatrix} = \begin{bmatrix} Y & I \\ N & 0 \end{bmatrix} \begin{bmatrix} I & X \\ 0 & M \end{bmatrix}^{-1} \quad \text{with } M'N = I - XY \qquad (7.26)$$

and

$$Q^{-1} = \begin{bmatrix} X & M^{-1} \\ M & \hat{W} \end{bmatrix} \qquad (7.27)$$

Let us define the following four matrices:

$$\Psi_1 = \begin{bmatrix} Y\mathbb{A}' + \mathbb{A}Y & \mathbb{A}N' & B_\phi S - YC_1' & B_2 & YC_2' \\ \star & 0 & -NC_1' & 0 & NC_2' \\ \star & \star & -2S - D_1 S - SD_1' & -\mathcal{D}_{12} & SD_2' \\ \star & \star & \star & -I & \mathcal{D}_{22}' \\ \star & \star & \star & \star & -\gamma I \end{bmatrix} \qquad (7.28)$$

$$F = \begin{bmatrix} 0 & I & 0 & 0 & 0 \\ B_v' & 0 & -C_{v1}' & 0 & C_{v2}' \end{bmatrix} \qquad (7.29)$$

$$G = \begin{bmatrix} N & W & 0 & 0 & 0 \\ 0 & 0 & S & 0 & 0 \end{bmatrix} \qquad (7.30)$$

$$\Omega = \begin{bmatrix} A_{\text{aw}} & B_{\text{aw}} \\ C_{\text{aw}} & D_{\text{aw}} \end{bmatrix} \qquad (7.31)$$

It can be checked that the inequality (7.21) can be written as

$$\Psi_1 + F'\Omega G + G'\Omega'F < 0 \qquad (7.32)$$

Thus, by using the Projection Lemma, relation (7.32) is equivalent to:

$$\mathcal{N}_F' \Psi_1 \mathcal{N}_F < 0 \qquad (7.33)$$

$$\mathcal{N}_G' \Psi_1 \mathcal{N}_G < 0 \qquad (7.34)$$

where $\mathcal{N}_F$ and $\mathcal{N}_G$ denote any basis of the null spaces of $F$ and $G$, respectively. The basis of $\text{Ker}(G)$ being defined by

$$\mathcal{N}_G = \begin{bmatrix} X & 0 & 0 \\ M & 0 & 0 \\ 0 & 0 & 0 \\ 0 & I & 0 \\ 0 & 0 & I \end{bmatrix}$$

it follows that (7.34) is equivalent to relation (7.23). By the same way, the basis of $\text{Ker}(F)$ can be defined by

$$\mathcal{N}_F' = \begin{bmatrix} \begin{bmatrix} I_{n_p} & 0 \end{bmatrix} & 0 & B_{pu} & 0 & 0 \\ 0 & 0 & 0 & I & 0 \\ 0 & 0 & D_{zu} & 0 & I \end{bmatrix}$$

## 7.3 Direct Linear Anti-windup Design

On the other hand, by noting also that the following properties are satisfied:

$$\begin{bmatrix} I_{n_p} & 0 \end{bmatrix} B_v - B_{pu} C_{v1} = 0$$

$$-D_{zu} C_{v1} + C_{v2} = 0$$

$$B_{pu} S B_\phi' \begin{bmatrix} I_{n_p} \\ 0 \end{bmatrix} + \begin{bmatrix} I_{n_p} & 0 \end{bmatrix} B_\phi S B_{pu}' = 0$$

$$D_{zu}\left(-2S - D_1 S - S D_1'\right) D_{zu}' + D_2 S D_{zu}' + D_{zu} S D_2' = 0$$

and by considering $Y_1 \in \mathfrak{R}^{n_p \times n_p}$ the block $(1, 1)$ of $Y$, it follows that (7.33) is equivalent to relation (7.24). Finally, relation (7.25) allows one to verify the existence of a positive definite matrix $Q$, and therefore positive definite matrices $Y$ and $X$ satisfying (7.26). □

It is important to underline that the notion of full-order anti-windup has a different meaning depending on the authors in the literature. For example, in [151, 195, 370, 398], the authors use full order to mean plant order (i.e., $n_{aw} = n_p$). On the contrary, in [29, 140, 358] or in the current chapter, the full order means $n_{aw} = n_p + n_c$. In [151], a theorem involving two relations similar to (7.23) and (7.24) together with a rank condition is provided. This result is an existence condition, which is in general difficult to satisfy due to the non-convex rank constraint. Such a result is indeed similar to the LMI $\mathcal{H}_\infty$ synthesis problem which is generally a number of LMIs coupled with a rank constraint [100]. However, it transpires that, for the special cases of $n_{aw} = 0$ and $n_{aw} \geq n_p$, the rank constraint vanishes, leaving simply a set of linear matrix inequalities. Although the above theorem is an existence condition, an anti-windup compensator yielding the $\mathcal{L}_2$-gain $\gamma > 0$ can then be constructed by solving an extra LMI [151], as will be seen in Sect. 7.3.4. Furthermore, mirroring techniques used in low-order robust controller design, a trace minimization can be performed to "remove" the rank constraint as advocated in [102]. Although this procedure is not guaranteed to work, experience in other areas of the control field has shown this to be sometimes successful.

Let us specify the solution when we consider the simple case of static anti-windup. In this case, we just have to compute $D_{aw}$ ($n_{aw} = 0$, $A_{aw} = 0$, $B_{aw} = 0$ $C_{aw} = 0$) and the augmented vector $x$ is constituted from $x_p$ and $x_c$. Hence, Proposition 7.1 can be modified as follows.

**Proposition 7.3** *If there exist a symmetric positive definite matrix* $Q \in \mathfrak{R}^{(n_p+n_c)\times(n_p+n_c)}$, *a diagonal matrix* $S \in \mathfrak{R}^{m \times m}$, *a matrix* $E \in \mathfrak{R}^{(n_c+m)\times m}$ *and a positive scalar* $\gamma$ *such that the following conditions hold:*

$$\begin{bmatrix} Q\mathbb{A}' + \mathbb{A}Q & B_\phi S + B_v E - Q C_1' & B_2 & Q C_2' \\ \star & -2S - D_1 S - S D_1' - C_{v1} E - E' C_{v1}' & -\mathcal{D}_{12} & S D_2' + E' C_{v2}' \\ \star & \star & -I & \mathcal{D}_{22}' \\ \star & \star & \star & -\gamma I \end{bmatrix} < 0 \qquad (7.35)$$

*then the static anti-windup gain* $D_{aw} = E S^{-1}$ *is such that*

1. when $w = 0$, the global asymptotic stability (of the origin) for the saturated system (7.13) is guaranteed;
2. when $w \neq 0$,

- the closed-loop trajectories remain bounded for any $w(t) \in \mathcal{L}_2$ and any initial conditions;
- the map from $w$ to $z$ is finite $\mathcal{L}_2$-gain stable with

$$\int_0^T z(t)'z(t)\,dt \leq \gamma \int_0^T w(t)'w(t)\,dt + \gamma x(0)'Q^{-1}x(0), \quad \forall T \geq 0 \quad (7.36)$$

*Proof* The proof follows the same lines as the one of Proposition 7.1. More details can be found in [139]. $\qquad\square$

### 7.3.3 DLAW Schemes with Regional Stability Guarantees

Regarding the regional (local) context, the following general result can be stated, by using the same framework as in Proposition 7.1.

**Proposition 7.4** *If there exist a symmetric positive definite matrix $Q \in \mathfrak{R}^{n \times n}$, a matrix $Z \in \mathfrak{R}^{m \times n}$, a positive diagonal matrix $S \in \mathfrak{R}^{m \times m}$ and a positive scalar $\gamma$ such that the following conditions hold:*

$$\begin{bmatrix} Q\mathcal{A}' + \mathcal{A}Q & \mathcal{B}_1 S - Q\mathcal{C}_1' - Z' & \mathcal{B}_2 & Q\mathcal{C}_2' \\ \star & -2S - \mathcal{D}_{11}S - S\mathcal{D}_{11}' & -\mathcal{D}_{12} & S\mathcal{D}_{21}' \\ \star & \star & -I & \mathcal{D}_{22}' \\ \star & \star & \star & -\gamma I \end{bmatrix} < 0 \quad (7.37)$$

$$\begin{bmatrix} Q & Z_{(i)}' \\ \star & \delta u_{0(i)}^2 \end{bmatrix} \geq 0, \quad i = 1, \dots, m \quad (7.38)$$

*then*

1. when $w = 0$, the ellipsoid $\mathcal{E}(Q^{-1}, \delta) = \{x \in \mathfrak{R}^n; \ x'Q^{-1}x \leq \delta^{-1}\}$ is a region of asymptotic stability (RAS) for the saturated system (7.13);
2. when $w \neq 0$, satisfying condition (7.11), for $x(0) = 0$,

- the closed-loop trajectories remain bounded in the set $\mathcal{E}(Q^{-1}, \delta)$;
- the map from $w$ to $z$ is finite $\mathcal{L}_2$-gain stable with:

$$\int_0^T z(t)'z(t)\,dt \leq \gamma \int_0^T w(t)'w(t)\,dt, \quad \forall T \geq 0 \quad (7.39)$$

7.3 Direct Linear Anti-windup Design

*Proof* Consider a quadratic Lyapunov function $V(x) = x'Px$, with $P = P' = Q^{-1} > 0$. Note that Lemma 7.1 applies by choosing $\omega = y_c + Gx$, that is, the relation (7.17) is satisfied and reads

$$\phi(y_c)'S^{-1}\big(\phi(y_c) + y_c + Gx\big) \leq 0$$

for any diagonal positive definite matrix $S \in \Re^{m \times m}$ and any $x \in S(u_0)$ with

$$S(u_0) = \big\{y_c \in \Re^m, \ \omega \in \Re^m; \ -u_0 \preceq y_c - \omega \preceq u_0\big\} = \big\{x \in \Re^n; \ -u_0 \preceq Gx \preceq u_0\big\}$$

Hence, by setting $Z = GQ$, the satisfaction of relation (7.38) means that the ellipsoid $\mathcal{E}(Q^{-1}, \delta) = \{x \in \Re^n; \ x'Q^{-1}x \leq \delta^{-1}\}$ is included in $S(u_0)$. Then, the satisfaction of relation (7.37) means that

$$\dot{V}(x) + \frac{1}{\gamma}z'z - w'w \leq \dot{V}(x) + \frac{1}{\gamma}z'z - w'w - 2\phi(y_c)'S^{-1}\big(\phi(y_c) + y_c + Gx\big) < 0$$

for any $x \in \mathcal{E}(Q^{-1}, \delta)$. The stability immediately follows from the above inequality which obviously implies that $\dot{V}(x) < 0$ along any closed-loop trajectory when $w = 0$. The condition (7.39) is obtained by integration of $\dot{V}(x) + \frac{1}{\gamma}z'z - w'w < 0$, by taking $x(0) = 0$ and noticing that $V(x(T)) > 0$. □

*Remark 7.5* In the case of a non-null initial condition $x(0)$, a positive scalar $\beta$ has to be considered in order to ensure that the closed-loop trajectories remain bounded in $\mathcal{E}(Q^{-1}, (\beta^{-1} + \delta^{-1})^{-1})$, $\forall x(0) \in \mathcal{E}(Q^{-1}, \beta)$. Hence, $\mathcal{E}(Q^{-1}, \beta)$ corresponds to a set of admissible initial conditions. From this, there clearly appears a trade-off between the size of the set of admissible conditions (given basically by $\beta$), the size of the region of stability (depending on $\beta^{-1} + \delta^{-1}$) and the bound on the admissible disturbance (given by $\delta$). Furthermore, in this case, the finite $\mathcal{L}_2$-gain from $w$ to $z$ presents a bias term and reads

$$\int_0^T z(t)'z(t)\,dt \leq \gamma \int_0^T w(t)'w(t)\,dt + \gamma x(0)'Q^{-1}x(0)$$

A detailed discussion about this issue can be found in [62], where the stabilization via state feedback of systems presenting actuator saturation is considered (see also Chaps. 2 and 3).

Similarly to the global context studied in Sect. 7.3.2, we can derive a convex condition in order to design the anti-windup compensator satisfying Proposition 7.4.

**Proposition 7.5** *There exists an anti-windup controller* $(A_{\mathrm{aw}}, B_{\mathrm{aw}}, C_{\mathrm{aw}}, D_{\mathrm{aw}})$ *such that the conditions of Proposition 7.4 are satisfied if there exist two symmetric positive definite matrices* $X \in \Re^{(n_p+n_c) \times (n_p+n_c)}$, $Y \in \Re^{(n_p+n_c) \times (n_p+n_c)}$, *a matrix* $Z_1 \in \Re^{m \times (n_p+n_c)}$, *a matrix* $U \in \Re^{m \times (n_p+n_c)}$ *and a positive scalar* $\gamma$ *such that the following conditions hold*:

$$\begin{bmatrix} \mathbb{A}'X + X\mathbb{A} & XB_2 & C_2' \\ \star & -I & \mathcal{D}_{22}' \\ \star & \star & -\gamma I \end{bmatrix} < 0 \tag{7.40}$$

$$\begin{bmatrix} A_p Y_1 - B_{pu} Z_1 \begin{bmatrix} I_{n_p} \\ 0 \end{bmatrix} + Y_1 A'_p - [I_{n_p} \ 0] Z'_1 B'_{pu} & \star & \star \\ B'_{pw} & -I & \star \\ C_z Y'_1 - D_{zu} Z_1 \begin{bmatrix} I_{n_p} \\ 0 \end{bmatrix} & D_{zw} & -\gamma I \end{bmatrix} < 0 \quad (7.41)$$

$$\begin{bmatrix} X & \star & \star \\ I & Y & \star \\ U_{(i)} & Z_{1(i)} & \delta u_{0(i)}^2 \end{bmatrix} \geq 0, \quad i = 1, \ldots, m \quad (7.42)$$

with $Y_1 \in \Re^{n_p \times n_p}$ is the block $(1, 1)$ of $Y$.

*Proof* The proof mimics that one of Proposition 7.2. Consider matrix $Q$ partitioned as in (7.26). By pre- and post-multiplying relation (7.38) by

$$\begin{bmatrix} X & M' & 0 \\ I & 0 & 0 \\ 0 & 0 & 1 \end{bmatrix} \quad \text{and} \quad \begin{bmatrix} X & I & 0 \\ M & 0 & 0 \\ 0 & 0 & 1 \end{bmatrix}$$

respectively, it is written:

$$\begin{bmatrix} X & \star & \star \\ I & Y & \star \\ U_{(i)} & Z_{1(i)} & \delta u_{0(i)}^2 \end{bmatrix} \geq 0, \quad i = 1, \ldots, m \quad (7.43)$$

where $Z = [Z_1 \ Z_2]$, $U_{(i)} = Z_{1(i)} X + Z_{2(i)} M$.

Let us define the following matrix:

$$\Psi_2 = \begin{bmatrix} Y\mathbb{A}' + \mathbb{A}Y & \mathbb{A}N' & B_\phi S - YC'_1 - Z'_1 & B_2 & YC'_2 \\ \star & 0 & -NC'_1 - Z'_2 & 0 & NC'_2 \\ \star & \star & -2S - D_1 S - SD'_1 & -\mathcal{D}_{12} & SD'_2 \\ \star & \star & \star & -I & \mathcal{D}'_{22} \\ \star & \star & \star & \star & -\gamma I \end{bmatrix} \quad (7.44)$$

The end of the proof is the same as in Proposition 7.2. Using the Projection Lemma and the basis of $\text{Ker}(G)$ and $\text{Ker}(F)$, few tedious calculations allow us to recover conditions (7.40) and (7.41). $\qquad\square$

*Remark 7.6* In order to optimize the size of the stability domain $\mathcal{E}(Q^{-1}, \delta)$, one considers the system without disturbance ($w = 0$). In that case, inequality (7.37) is modified by removing the two last lines and columns, and therefore the LMI constraints (7.40) and (7.41) become respectively,

$$\mathbb{A}' X + X \mathbb{A} < 0 \quad (7.45)$$

$$A_p Y_1 - B_{pu} Z_1 \begin{bmatrix} I_{n_p} \\ 0 \end{bmatrix} + Y_1 A'_p - [I_{n_p} \quad 0] Z'_1 B'_{pu} < 0 \quad (7.46)$$

*Remark 7.7* In the case where the open-loop matrix $A_p$ is Hurwitz, the global asymptotic stabilization problem can be addressed. Thus, considering $U = 0$ and $Z_1 = 0$ in Proposition 7.5 leads to the conditions of Proposition 7.2.

## 7.3 Direct Linear Anti-windup Design

*Remark 7.8* It is important to underline that conditions of Proposition 7.4 (and Proposition 7.1) imply the satisfaction $Q\mathbb{A}' + \mathbb{A}Q < 0$. This guarantees the asymptotic stability of the matrix $A_{\text{aw}}$ of the anti-windup compensator, contrarily to the approach suggested in [124]. Moreover, in the spirit of [303], it would also be possible to modify conditions of Propositions 7.1, 7.2, 7.4 or 7.5 in order to force only the poles of the anti-windup controller and not the whole closed-loop plant dynamics.

Proposition 7.4 may be particularized to the simple case of static anti-windup ($n_{\text{aw}} = 0$, $A_{\text{aw}} = 0$, $B_{\text{aw}} = 0$, $C_{\text{aw}} = 0$) as follows.

**Proposition 7.6** *If there exist a symmetric positive definite matrix $Q \in \mathfrak{R}^{(n_p+n_c)\times(n_p+n_c)}$, a matrix $Z \in \mathfrak{R}^{m\times n}$, a positive diagonal matrix $S \in \mathfrak{R}^{m\times m}$, a matrix $E \in \mathfrak{R}^{(n_c+m)\times m}$ and a positive scalar $\gamma$ such that the following conditions hold:*

$$\begin{bmatrix} Q\mathbb{A}' + \mathbb{A}Q & B_\phi S + B_\upsilon E - QC_1' - Z' & B_2 & QC_2' \\ \star & -2S - D_1 S - SD_1' - C_{\upsilon 1}E - E'C_{\upsilon 1}' & -\mathcal{D}_{12} & SD_2' + E'C_{\upsilon 2}' \\ \star & \star & -I & \mathcal{D}_{22}' \\ \star & \star & \star & -\gamma I \end{bmatrix} < 0 \tag{7.47}$$

$$\begin{bmatrix} Q & Z_{(i)}' \\ \star & \delta u_{0(i)}^2 \end{bmatrix} \geq 0, \quad i = 1, \ldots, m \tag{7.48}$$

*then the static anti-windup gain $D_{\text{aw}} = ES^{-1}$ is such that*

1. *when $w = 0$, the ellipsoid $\mathcal{E}(Q^{-1}, \delta) = \{x \in \mathfrak{R}^n;\ x'Q^{-1}x \leq \delta^{-1}\}$ is a region of asymptotic stability (RAS) for the saturated system (7.13);*
2. *when $w \neq 0$, satisfying condition (7.11), for $x(0) = 0$,*

   - *the closed-loop trajectories remain bounded in the set $\mathcal{E}(Q^{-1}, \delta)$;*
   - *the map from $w$ to $z$ is finite $\mathcal{L}_2$-gain stable with:*

$$\int_0^T z(t)'z(t)\, dt \leq \gamma \int_0^T w(t)'w(t)\, dt, \quad \forall T \geq 0 \tag{7.49}$$

*Proof* The proof follows the same lines as that one of Proposition 7.4. More details in the disturbance free case can be found in [139]. □

### 7.3.4 Some Algorithms

Based on DLAW Propositions 7.1–7.2 and 7.4–7.5, several algorithms are now proposed in order to compute full and fixed-order anti-windup compensators. Actually, several LMI-based optimization problems can be suggested to compute the anti-windup controller which optimizes one of the following criteria:

- maximization of the $\mathcal{L}_2$-bound on the admissible disturbances (disturbance tolerance maximization which corresponds to minimize $\delta$);
- minimization of the induced $\mathcal{L}_2$-gain between the disturbance $w$ and the regulated output $z$ (disturbance rejection maximization which corresponds to minimize $\gamma$ for a given $\delta$);
- maximization of the region where the asymptotic stability of the closed-loop system is ensured (maximization of the region of attraction).

Using Propositions 7.2 (in the global context) or 7.5 (in the regional (local) context), the existence of a full-order anti-windup compensator is easily checked by solving a finite set of LMIs. It also allows us to build the matrix $Q$ obtained from (7.26), that is,

$$Q = \begin{bmatrix} Y & I \\ N & 0 \end{bmatrix} \begin{bmatrix} I & X \\ 0 & M \end{bmatrix}^{-1} \quad \text{with } M'N = I - XY$$

The conditions of Propositions 7.1 and 7.4 then become convex as soon as $Q$ is fixed, and the decision variables $A_{aw}$, $B_{aw}$, $C_{aw}$ and $D_{aw}$ can finally be calculated, using a change of variables $\bar{B}_{aw} = B_{aw}S$, $\bar{D}_{aw} = D_{aw}S$. Moreover, it can be observed that the matrices $S$ and $Z$ do not have to be fixed. This offers some additional degrees of freedom that can be used for example to add constraints on the anti-windup dynamics through the pole placement of matrix $A_{aw}$ [303].

In the global context (Propositions 7.1 and 7.2), the main optimization problem of interest is a performance problem, i.e., minimization of $\delta$ (disturbance tolerance problem) or minimization of $\gamma$ (disturbance rejection problem) when $\delta$ is a priori given. This latter case is used in the following algorithm.

**Algorithm 7.1** Full-order synthesis (Global case)—performance

1. Minimize $\gamma$ under the LMI constraints (7.23), (7.24), (7.25) with respect to the variables $\gamma$, $X$, $Y$.
2. Compute $Q$ as the solution of (7.26).
3. Fix $Q$ in inequality (7.21) and solve the convex feasibility problem with respect to the variables $A_{aw}$, $\bar{B}_{aw}$, $C_{aw}$ and $\bar{D}_{aw}$. Then, compute $B_{aw} = \bar{B}_{aw}S^{-1}$ and $D_{aw} = \bar{D}_{aw}S^{-1}$.

Its local counterpart, using Propositions 7.4 and 7.5 reads

**Algorithm 7.2** Full-order synthesis (Regional case)—performance

1. Given $\delta$, minimize $\gamma$ under the LMI constraints (7.40), (7.41), (7.42) with respect to the variables $\gamma$, $X$, $Y$, $U$ and $Z_1$.
2. Compute $Q$ as the solution of (7.26).
3. Fix $Q$ in inequality (7.37) and solve the convex feasibility problem with respect to the variables $A_{aw}$, $\bar{B}_{aw}$, $C_{aw}$ and $\bar{D}_{aw}$. Then, compute $B_{aw} = \bar{B}_{aw}S^{-1}$ and $D_{aw} = \bar{D}_{aw}S^{-1}$.

# 7.3 Direct Linear Anti-windup Design

Another optimization problem of interest is to compute the largest admissible stability region without a performance constraint. When the open-loop matrix $A_p$ is asymptotically stable, if the conditions shown in Proposition 7.1 or Proposition 7.2 are feasible then the region of stability is the whole state space. Otherwise, in the regional (local) case, the problem of maximizing the region of stability consists of maximizing the size of $\mathcal{E}(Q^{-1}, \delta)$. Different linear optimization criteria $J(\cdot)$, associated to the size of $\mathcal{E}(Q^{-1}, \delta)$, can be considered, like the volume: $J = -\log(\det(\delta^{-1}Q))$, or the size of the minor axis: $J = -\lambda$, with $Q \geq \lambda I_n$. A given *shape set* $\mathcal{X}_0 \in \mathfrak{R}^n$ and a scaling factor $\beta$, where $\mathcal{X}_0 = \mathrm{Co}\{v_r \in \mathfrak{R}^n; \ r = 1, \ldots, n_r\}$ can also be considered and the associated criterion may then be to maximize the scaling factor $\beta$ such that $\beta \mathcal{X}_0 \subset \mathcal{E}(Q^{-1}, \delta)$. In particular it is interesting to address this problem in the plant space. These size criteria have been more precisely described in Sect. 2.2.5.1. Using Remark 7.6, this can be done with the following algorithm.

**Algorithm 7.3** Full-order synthesis—region of stability

1. Choose a set of interesting directions $v_i \in \mathfrak{R}^{n_p}$, $i = 1, \ldots, q$. Define a shape set $\mathcal{X}_0$ through $\bar{v}_i = [v_i' \ 0]' \in \mathfrak{R}^{n_p + n_c}$, $i = 1, \ldots, q$.
2. Given $\delta$, minimize a positive scalar $\mu$ under the LMI constraints (7.45), (7.46), (7.42) and $\mu - \bar{v}_i' X \bar{v}_i \geq 0$, $i = 1, \ldots, q$, with respect to the variables $\mu$, $X$, $Y$, $U$ and $Z_1$.
3. Fix $Q$ in the modified inequality (7.37) and solve the convex feasibility problem with respect to the variables $A_{\mathrm{aw}}$, $\bar{B}_{\mathrm{aw}}$, $C_{\mathrm{aw}}$ and $\bar{D}_{\mathrm{aw}}$. Then, compute $B_{\mathrm{aw}} = \bar{B}_{\mathrm{aw}} S^{-1}$ and $D_{\mathrm{aw}} = \bar{D}_{\mathrm{aw}} S^{-1}$.

*Remark 7.9* Recall that

$$Q^{-1} = \begin{bmatrix} X & M^{-1} \\ M & \hat{W} \end{bmatrix}$$

Hence, in step 2, the minimization of $\mu$ signifies that we are actually minimizing $\mathcal{E}(Q^{-1}, \delta)$ in the directions of the plant state, with the scaling factor of $\mathcal{X}_0$ being $\beta = \frac{1}{\sqrt{\mu}}$.

Many other options for the criteria could be investigated, in particular to find a "good" compromise between the size of the domain and the level of disturbance rejection. This can be done, for example, by considering multi-objective criteria. Another option can be, in Algorithm 7.3, to transform the feasibility problem of step 3 in a minimization problem of $\gamma$, $\delta$ being given.

To go further, it may be of strong interest to consider that the matrices $A_{\mathrm{aw}}$ and $C_{\mathrm{aw}}$ of the anti-windup compensator are a priori fixed. Indeed, the BMI constraints (7.21) and (7.37) of Propositions 7.1 and 7.4, respectively, are convex as soon as the matrices $A_{\mathrm{aw}}$ and $C_{\mathrm{aw}}$ of the anti-windup controller are fixed. This also allows the order of the anti-windup compensator to be reduced, and also simplifies the computational effort. Based on this remark, simple algorithms can be derived, both in the

300                                                          7  Anti-windup Compensator Synthesis

context of minimization of the induced $\mathcal{L}_2$-gain between $w$ and $z$ (Algorithm 7.4) and in the context of maximization of the stability region (Algorithm 7.5). They are given here in the regional (local) case, but the global case obviously follows.

**Algorithm 7.4**  Fixed-dynamics synthesis—performance

1. Choose appropriate $A_{aw}$ and $C_{aw}$.
2. Given $\delta$, minimize $\gamma$ under the LMI problem constraints (7.37) and (7.38) with respect to the variables $Q$, $S$, $Z$, $\bar{B}_{aw}$, $\bar{D}_{aw}$.
3. Compute $B_{aw} = \bar{B}_{aw} S^{-1}$ and $D_{aw} = \bar{D}_{aw} S^{-1}$.

**Algorithm 7.5**  Fixed-dynamics synthesis—region of stability

1. Choose appropriate $A_{aw}$ and $C_{aw}$.
2. Choose a set of interesting directions $v_i \in \Re^{n_p}$, $i = 1, \ldots, q$. Define $\bar{v}_i = [\, v_i'\ 0 \,]' \in \Re^{n_p + n_c + n_{aw}}$, $i = 1, \ldots, q$.
3. Given $\delta$, minimize a positive scalar $\mu$ under the LMI constraints (7.37) and (7.38) with respect to the variables $Q$, $S$, $Z$, $\bar{B}_{aw}$, $\bar{D}_{aw}$.
4. Compute $B_{aw} = \bar{B}_{aw} S^{-1}$ and $D_{aw} = \bar{D}_{aw} S^{-1}$.

It is important to point out that the main difficulty in the above algorithms resides in the first step, i.e., in choosing the matrices $A_{aw}$ and $C_{aw}$ adequately. However, according to the approach developed in [27] (see also [178, 372]) this choice may be carried out by considering the poles of the anti-windup controller. These poles can be chosen by selecting a part of those obtained in the full order design case. Typically, the slow and fast dynamics are eliminated. Alternatively, an iterative procedure starting from the static case can be used. The list of poles is then progressively enriched until the gap between the full and reduced order cases becomes small enough. Note then that the order of the controller is now given by $n_{aw}$, chosen small enough, so that $n_{aw} < n_p + n_c$. Exploiting a similar idea, the algorithm proposed in [213] is based on a different decomposition of the anti-windup controller using dyadic forms. Even if such forms are slightly more general, the associated algorithm requires the user to specify output pole directions which may not be trivial.

*Remark 7.10*  Static anti-windup compensators are easily computed as particular solutions of previous algorithms since relation (7.21) in the global case or relation (7.37) in the regional case are simplified and allow one to obtain directly $\bar{D}_{aw} = D_{aw} S^{-1}$ [56, 139, 352], or directly by using the conditions (7.35) (global case, see Proposition 7.3) or (7.47)–(7.48) (local case, see Proposition 7.6).

## 7.4  Model Recovery Anti-windup Design

### 7.4.1  Preliminary Elements on the Architecture

Model recovery anti-windup corresponds to select the anti-windup compensator as a dynamical filter containing a (possibly approximated) model of the plant dynamics.

## 7.4 Model Recovery Anti-windup Design

The aim of this filter is to keep track of what the closed-loop response would be in the absence of saturation. In particular, call $y_{c,\ell in}$ the controller output response that one would get without saturation and $z_{\ell in}$ the plant performance output obtained without saturation. Then, under the assumption of perfect knowledge of the plant model, inserting the filter (7.9) in the closed-loop corresponds to introduce in the closed-loop the following dynamics:

$$\begin{aligned}
\dot{x}_{aw} &= A_p x_{aw} + B_{pu}\big(\text{sat}(y_{c,\ell in} + v_y) - y_{c,\ell in}\big) \\
z - z_{\ell in} &= C_z x_{aw} + D_{zu}\big(\text{sat}(y_{c,\ell in} + v_y) - y_{c,\ell in}\big)
\end{aligned} \tag{7.50}$$

which clarifies the relation between anti-windup dynamics (whose state is $x_{aw}$) and performance output deviation from the desirable unconstrained response $z_{\ell in}$.

*Remark 7.11* It should be emphasized that the notation adopted in this book is different from that we used in the previous works on model recovery anti-windup (i.e., $\mathcal{L}_2$ anti-windup). We adopt this notation here for consistency with the DLAW architecture, however, for clarification purposes, we should point out that often in the MRAW literature the controller dynamics (7.2) is given without the signals $v_x$ and $v_y$. Moreover, the MRAW compensator (7.9) is formulated as follows:

$$\begin{aligned}
\dot{x}_{aw} &= A_p x_{aw} + B_{pu}\big(\text{sat}(y_c + v_1) - y_c\big) \\
v_2 &= C_p x_{aw} + D_{pu}\big(\text{sat}(y_c + v_1) - y_c\big) \\
u &= \text{sat}(y_c + v_1), \qquad u_c = y_p + v_2
\end{aligned} \tag{7.51}$$

where $v_1$ coincides with the signal $v_y$ that we use here and $v_2$ is often also denoted as $y_{aw}$ (for obvious reasons arising from its parallel definition to $y$). All the results reported here are equivalent to those found in the literature, however, stated with this different notation motivated by the fact that the discussion in this book groups DLAW and MRAW schemes.

Based on the interpretation of (7.50), it appears that the anti-windup goal of forcing, as much as possible, $z$ to recover the unconstrained response $z_{\ell in}$ amounts to the goal of driving to zero the anti-windup output $z_{aw} = z - z_{\ell in}$ in (7.50) as effectively as possible. In other words, the anti-windup goal is transformed into a bounded stabilization problem with an input matched disturbance $y_{c,\ell in}$. Furthermore, the way the "disturbance" $y_{c,\ell in}$ acts on this bounded stabilization problem is quite peculiar, indeed it shifts up and down the saturation levels of an equivalent time varying saturation nonlinearity affecting the stabilizing signal $v_y$. Due to this reason, the MRAW architecture branches out into many solutions, different from each other depending on how the signal $v_y$ is designed to guarantee that, eventually, the anti-windup compensator state $x_{aw}$ converges to zero thereby causing the output $z_{aw} = z - z_{\ell in}$ to converge to zero and, consequently, the plant output $z$ to converge to the desirable unconstrained performance output response $z_{\ell in}$.

Note that the MRAW architecture is independent of the controller dynamics. By this fact, any stabilizing controller can be used within the MRAW scheme and closed-loop stability will be guaranteed by the scheme. However, extra mild assumptions on the controller dynamics are highly desirable because they are necessary for robustness. These properties are some suitable incremental stability properties (see [369, 404] for details) which allow us to carry out a small gain type argument on the closed-loop in the presence of uncertainties.

Furthermore, the most important drawback of the MRAW scheme as compared to the DLAW solutions is maybe that the compensation scheme is of the same order as the plant, which is often too high. However, the robustness property commented above becomes a key fact when implementing MRAW in real-life problems where one would like to use a rough model of the plant dynamics. This fact is certainly possible within the scheme and typically allows for a strong reduction of the order of the arising anti-windup compensator (see, e.g., [278] for an application study where several anti-windup order reduction techniques were implemented in the discrete-time MRAW setting).

### 7.4.2 Some Algorithms

In this section, similarly to Sect. 7.3.4, we present several algorithms based on the MRAW schemes and addressing global and regional (local) stability. For the sake of simplicity, only the case with $w = 0$ is discussed in this section, but it can be extended, without difficulty, to the case with $w \neq 0$. More details can be found in [108, 406].

First, in the context of exponentially stable plants, the simplest possible compensation scheme is given by the so-called IMC-based anti-windup, which corresponds to blindly use the controller as if no saturation was in place and deliver to the plant (after trimming it by way of saturation) the same signal that the controller would have been produced without any saturation. This scheme is described in [267] (see also [411]) and was also discussed independently in [162].

**Algorithm 7.6**  IMC-based MRAW for exponentially stable plants

1. Select $v_y = 0$.

A more sophisticated strategy for selecting the compensation signal $v_y$ is given by the Lyapunov-based procedure issued from [369], where the degrees of freedom available in the design process can often lead to better performance, even though this should be tuned by means of trial and error approaches. Furthermore, from [404], the anti-windup gains can be selected with the goal of guaranteeing closed-loop global exponential stability (by global sector conditions for the saturation function) while minimizing the LQ index:

$$
J = \int_0^\infty \left( x'_{\mathrm{aw}} Q_P x_{\mathrm{aw}} + v'_y R_P v_y \right) dt
$$

where $Q_P$ and $R_P$ are suitable positive definite matrices corresponding to design parameters.

## 7.4 Model Recovery Anti-windup Design

**Algorithm 7.7** LQ-based MRAW for exponentially stable plants

1. Select positive definite matrices $Q_P$ and $R_P$.
2. Solve the following LMI:

$$\min \; \beta$$

$$\text{subject to} \quad \begin{bmatrix} QA'_p + A_p Q & B_{pu}U + X'_1 \\ UB'_{pu} + X_1 & X'_2 + X_2 - 2U \end{bmatrix} < 0$$

$$\begin{bmatrix} A_p Q + QA'_p + B_{pu}X_1 + X'_1 B'_{pu} & Q & X'_1 \\ Q & -Q_P^{-1} & 0 \\ X_1 & 0 & -R_P^{-1} \end{bmatrix} < 0$$

$$\begin{bmatrix} \beta I & I \\ I & Q \end{bmatrix} > 0$$

in the unknowns $Q > 0$, $X_1$, $X_2$ and $U > 0$ diagonal.
3. Select the compensation signal $v_y$ as

$$v_y = K x_{\text{aw}} + L\big(\text{sat}(y_c + v_y) - y_c\big)$$

where $K = X_1 Q^{-1}$ and $L = X_2 U^{-1}$.

The obtained gain $K$ is such that $A_p + B_{pu}K$ is Hurwitz.

The following algorithm, taken from [104], is based on the use of the generalized sector condition for the design of $v_y$ in the case of exponentially unstable plants (regional (local) case). Note that in this algorithm, first the stability region is maximized under a certain constraint on a desired performance level $\bar{\gamma}$. The larger $\bar{\gamma}$, the larger the stability region will be. Then, among all the compensator gains that induce that performance level in that region, the selected one is the one that maximizes the speed of convergence to zero.

**Algorithm 7.8** LMI-based MRAW for exponentially unstable plants

1. Choose a certain desired $\mathcal{L}_2$ performance level $\bar{\gamma}$.
2. Solve the following LMI optimization problem:

$$\min \; \eta$$

$$\text{subject to} \quad \begin{bmatrix} A_p Q + B_{pu}Y + QA'_p + Y'B'_{pu} & B_{pu} & QC'_z - Y'D'_z \\ B'_{pu} & -I & D'_{zu} \\ C_z Q - D_z Y & D_{zu} & -\bar{\gamma}I \end{bmatrix} < 0$$

$$\begin{bmatrix} \eta u_{0(k)}^2 & Y_{(k)} \\ Y'_{(k)} & Q \end{bmatrix}, \quad k = 1, \dots, m$$

in the unknowns $Q > 0$, $Y$ and $\eta$. Minimizing $\eta$ corresponds to maximizing the arising stability region.

3. Based on the solution $Q, Y$ issued from the previous step, solve the following generalized eigenvalue problem:

$$\min \lambda$$

subject to

$$(A_p + B_{pu}\tilde{K})Q + Q(A_p + B_{pu}\tilde{K})' - 2\lambda Q < 0$$

$$\begin{bmatrix} A_p Q + Q A'_p + B_{pu}Y + Y'B'_{pu} & \star & \star & \star \\ B'_{pu} & -I & \star & \star \\ C_z Q - D_{zu}Y & D_z & -\bar{\gamma}I & \star \\ \tilde{K}Q - Y - \tilde{L}B' + B'_{pu} & -\tilde{L} & D'_{zu} - \tilde{L}D'_{zu} & \tilde{L} + \tilde{L}' - 2I \end{bmatrix} < 0$$

in the unknowns $\tilde{K}$ and $\tilde{L}$.

4. Select the compensation signal $v_y$ as

$$v_y = K x_{\text{aw}} + L\big(\text{sat}(y_c + v_y) - y_c\big)$$

where $L = -(I - \tilde{L})^{-1}\tilde{L}$ and $K = (I - \tilde{L})^{-1}\tilde{K}$.

In [108], several other algorithms are expressed, in particular using Lyapunov-based procedure, when dealing with marginally unstable plants (namely plants without poles with positive real part but with poles on the imaginary axis). Moreover, it is also possible to develop an algorithm providing a fully nonlinear compensation scheme for the case when it is crucial to keep the exponentially unstable plant dynamics within a certain safety region (of course this is only possible if the disturbances do not push the plant state outside of the null controllability region). This algorithm has been originally presented in [106] and is written in a simplified form in [108].

## 7.5 Anti-windup Algorithms Summary

The algorithms described in Sects. 7.3 and 7.4 correspond to different alternatives to address and solve suitable anti-windup problems. It is important to underline that we have chosen to focus mainly on LMI-based algorithms. For extensions and clarifications, the reader is referred to [108].

Table 7.1 contains a summary of all the algorithms reported previously, emphasizing for each of them the following aspects.

Applicability: Characterizes what type of windup problems can be addressed with that algorithm, based on the properties of the plant to be controlled. Thus, ES stands for exponentially stable (namely $A_p$ Hurwitz), and ANY stands for any plant.

Architecture: Characterizes whether the anti-windup architecture is Direct Linear (Sect. 7.3) or Model Recovery (Sect. 7.4).

Guarantees: What type of guarantees are given on the closed-loop. These come either from theorems reported in this paper or from results in other references. The

## 7.6 Some Extensions

**Table 7.1** Overview of the anti-windup algorithms

| Algorithm | Applic | Arch | Guarant | Proofs | Order | Method | Notes |
|---|---|---|---|---|---|---|---|
| 7.1 | ES | DLAW | GAS | Propositions 7.1, 7.2 | $n_p + n_c$ | LMI | $\mathcal{L}_2$-gain optimization |
| 7.2 | ANY | DLAW | RAS | Propositions 7.4, 7.5 | $n_p + n_c$ | LMI | $\mathcal{L}_2$-gain optimization |
| 7.3 | ANY | DLAW | RAS | Propositions 7.4, 7.5 | $n_p + n_c$ | LMI | Region of stability maximization |
| 7.4 | ANY | DLAW | RAS | Proposition 7.4 | any | LMI | A priori fixed AW dynamics |
| 7.5 | ANY | DLAW | RAS | Proposition 7.4 | any | LMI | A priori fixed AW dynamics |
| 7.6 | ES | MRAW | GES | [369, 411] | $n_p$ | N/A | Simplest possible MRAW scheme |
| 7.7 | ES | MRAW | GES | [404] | $n_p$ | LMI | Constrained LQ |
| 7.8 | ANY | MRAW | RES | [104] | $n_p$ | LMI | Uses generalized sector |

guarantees can be GES for global exponential stability, GAS for global asymptotic stability, RAS for regional asymptotic stability and RES for regional exponential stability, meaning that there is a guaranteed nontrivial basin of attraction.

Proofs: Indicates where to find the proofs of the properties of the previous column when no sketches are provided.

Order: Indicates the order of the anti-windup compensator. Note that by construction all the MRAW schemes are of order $n_p$. This may make the static/low-order DLAW a better candidate for simple anti-windup solutions.

Method: Characterizes the synthesis method.

Notes: Brief notes on the peculiarities of the specific algorithm.

Note that the results developed in Propositions 7.1, 7.2 and Propositions 7.4, 7.5 also ensure exponential stability in global and local contexts. The conditions of these propositions are based on the strict negativity of the time-derivative of the quadratic Lyapunov function $V(x) = x'Px$. Then, we can always find a small enough positive scalar allowing us to ensure $\dot{V} \le -\alpha V(x)$ along the closed-loop trajectories when $w = 0$ and $\dot{V} + \frac{1}{\gamma}z'z - w'w \le -\alpha V(x)$ along the closed-loop trajectories when $w \ne 0$.

## 7.6 Some Extensions

Several extensions can be considered: the presence of rate and magnitude saturation, the presence of saturation in the output, or more generally the presence of nested saturations, the presence of delays in the systems... Furthermore, the nonlinear nature of the system is also a very important feature to tackle. Some of the extensions which are discussed in this section will be illustrated in Chap. 8.

## 7.6.1 Rate Saturation

Quite frequently limits on the magnitude of the control signal which an actuator can handle is less important than limits on the rate of the control signal. Rate limits are of particular importance in mechanical systems where the inertia in various components of the actuator prevents it from moving very fast, thereby limiting the rate of the control signal which it can pass to the plant.

Note that modeling of rate saturation nonlinearities is not unique within the constrained control literature [239, 273, 308] and is even disparate within the anti-windup literature [14, 307, 325]. Some complications are added when the rate limit is combined with the magnitude limit which appears to be related to the physics of the actuator (contrast, for example, the electro-hydraulic actuator discussed in the first chapter of [361] with the aircraft actuators used in [121]). However, a useful model of the rate limit [144, 307, 308, 325] may be obtained by cascading the standard saturation function with an integrator and a gain, and enclosing this within a feedback loop. In this representation, the limits of the saturation function, namely $u_1$, now represent the rate limits of the system and the gain determines the actuator's linear bandwidth when no rate limits are reached. The state space equations are thus given by

$$\dot{u} = \text{sat}\big(T_0(y_c - u)\big) \tag{7.52}$$

Let us point out that even if such a representation of a rate limit is attractive, we have to take care of the presence of the integrator (which is not asymptotically stable). See [325] for example. Furthermore, a good paper which discusses the merits of different ways to model actuator position and rate limits is [308]. Actuator rate limits have recently attracted a lot of interest due to their role in pilot-induced-oscillations (PIOs) and the subsequent untimely demise of several aircrafts due to rate-limited actuators (see [8, 46, 86, 295] for further details). The reader may also consult [103] and [107] in which a different model of rate saturation is proposed. An anti-windup scheme is developed in a similar way to that used when dealing with only magnitude saturation.

## 7.6.2 Sensor Saturations

Sensor saturation is normally found in applications where cost prohibits the use of sensors with adequate range, leading to sensor saturation for large reference/disturbance inputs. Alternatively sensor saturation can model the situation where only the sign of the output is known. In this case, the sign function can be modeled by a saturation function with a steep gradient.

Naively, the sensor saturation problem suggests that it is similar to the actuator saturation problem with the plant and the controller interchanged. In fact this is not the case [241, 347]; one of the crucial differences between the two problems is the availability of the "un-saturated" signal. In the actuator saturation case, knowledge

7.6 Some Extensions

of both the output produced by the linear controller and the saturated version of this (i.e. the signals at both sides of the saturation block) can be soundly assumed. In the case of sensor saturation, it is not realistic to assume that the actual plant output is known; only the saturated version of this is known (otherwise there would be no problem!). This is problematic for the anti-windup approach, and hence an observer may be used to overcome this difficulty.

Comparatively to the actuator saturation case, systems subject to sensor saturation have been less studied, with only a few papers devoted to this topic [57, 206, 222, 227, 241, 347]. Moreover, if the study of systems subject to sensor saturation is under-developed, the study of anti-windup compensation for this class of systems is still less developed. See, for example, the papers [347, 373, 374] that establish conditions for both local stability and global stability with $\mathcal{L}_2$ performance, respectively. These three papers are also related to the paper of Park [284] in which state-constrained systems are considered and an anti-windup type approach is proposed.

### 7.6.3 Nested Saturations

Based on the discussion of the two sections above, we can consider that an important class of systems of interest consists of systems presenting nested saturations. In particular, such structures appear when we deal with nonlinear actuators and sensors. One may recall in that context the problem of rate or dynamics and amplitude limitations representation as mentioned in Sect. 7.6.1. For instance, it is common in aerospace control systems (e.g., launcher and aircraft control) that actuators are both limited in amplitude and rate, even in dynamics: see, for example, [208, 375, 388, 398]. However, only a few results relative to anti-windup strategies can be found for such systems.

Different models are used in the literature [208, 361] to describe the simultaneous presence of rate/dynamics and amplitude saturations. The most common one is directly obtained from system (7.52) as follows:

$$\dot{u} = \mathrm{sat}_{u_1}\big(T_0\big(\mathrm{sat}_{u_0}(y_c) - u\big)\big) \tag{7.53}$$

where $\mathrm{sat}_{u_0}(\cdot)$ and $\mathrm{sat}_{u_1}(\cdot)$ denote the saturation functions with $u_0$ and $u_1$ as levels of saturation in magnitude and rate, respectively.

Actually, in this context both conditions in local and global cases can be derived to design static or dynamic anti-windup compensators. Non-constructive conditions are proposed in [14] to characterize a plant-order anti-windup controller. Constructive conditions to exhibit anti-windup schemes are proposed in [361] for magnitude and rate saturation and in [360] for magnitude and dynamics saturation. Some applications of such studies are given, for example, in [260, 295, 368]. This case will also be illustrated in Chap. 8.

Another case of nested saturation resides in the presence of both sensor and actuator amplitude limitations, which extends the class of systems discussed in Sect. 7.6.2. The saturation of the sensor output induces an incorrect action of the controller, since the actual state or output of the plant is no longer precisely measured. This is the case, for instance, in linear systems controlled by dynamic output feedback controllers in the presence of saturating sensors and actuators [119, 347]. Some application oriented studies can be found for example in [357, 361].

## 7.6.4 Time-Delay Systems

In the last few years, the study of systems presenting time-delays has received special attention in the control systems literature; see for example [67, 155, 274, 298]. This interest comes from the fact that time-delays appear in many kinds of control systems (e.g. chemical, mechanical and communication systems) and their presence can be source of performance degradation and instability. In this sense, we can find in the literature many works giving conditions for ensuring stability as well as performance and robustness requirements, considering or not the delay dependence. Concerning the delay independent results, the stability is ensured no matter the size of the delay, whereas in the delay dependent results, the size of the delay is directly taken into account and this fact can lead, especially when the time-delays are small, to less conservative results.

Due to the fact that many practical systems present both time-delays and saturating inputs, from the considerations above, it becomes important to study the stability issues regarding this kind of systems. With this aim, different techniques can be investigated: in particular the characterization of admissible regions of stability is often based on the use of Razumikhin or Lyapunov–Krasovskii functionals. In parallel, another popular technique consists of approximating the delay through a Padé approximation, which implies an increase in the order of the closed-loop system. It may be used in order to prove some robustness properties with respect to the presence of delays. Techniques based on Lyapunov functionals or Padé approximations can be used to analyze the stability and the performance of saturated systems.

In the global anti-windup context, [285] and [407] have considered dynamic anti-windup strategies for systems with input and output delays. It should be highlighted that the results in [285] can be applied only to open-loop stable systems and that in [407], the main focus is the formal definition and characterization of the $\mathcal{L}_2$ gain based anti-windup. Differently from [285, 407], in [145, 355, 356], the design of anti-windup gains was studied with the aim at enlarging the region of attraction of the closed-loop system. In [355] a method for computing anti-windup gains for systems presenting only input delays is proposed leading to BMI conditions (see also [145, 356]). In contrast to [285], the proposed techniques can be applied both to stable and to unstable open-loop systems. Finally, considering systems with state delay, in [21, 124, 148], the design of a rational or a non-rational (in the sense that the anti-windup compensator also presents delayed terms) dynamic anti-windup controller is done in order to guarantee both that the trajectories of the system are bounded and that a certain $\mathcal{L}_2$ performance level is achieved by the regulated outputs.

Some application oriented studies can be found in [409] in the context of open water channels and in [27] in the context of a fighter aircraft, where the delays are replaced by first-order Padé approximations.

### 7.6.5 Anti-windup for Nonlinear Systems

Let us stress that most of the anti-windup literature has concentrated on the development of anti-windup techniques for systems which are largely linear, or at least represent linear approximations of nonlinear systems. This has been because even for linear systems, the anti-windup problem has only just begun to be understood in a rigorous technical way and again, only recently, have modern control techniques been harnessed to address the problem. Furthermore, when dealing with saturated linear systems, the control engineer can draw upon the large body of knowledge on, for example, absolute stability theory which is now well-developed.

Systems which are naturally nonlinear have problems with saturation as well, and this has not been lost on the research community. Several anti-windup techniques for nonlinear systems have recently been proposed and the intention of many of these is to mimic the anti-windup techniques for linear systems in some way. The literature base is too sparse to warrant a full discussion here but it suffices to say that techniques based on particular structure of nonlinear systems like quadratic systems [382] and rational systems [276], feedback linearization (nonlinear dynamic inversion) [180, 186, 210], adaptive control [191, 203, 204] and neural network control [182] are beginning to emerge and promise to be exciting and fruitful areas of research.

## 7.7 Conclusion

In this chapter, inspired by [108], we have introduced and summarized two approaches to anti-windup design. The DLAW approach, a category into which most modern AW compensators belong, has been described and sample algorithms which one may use to synthesize this type of compensator have been given. Next, the MRAW approach to anti-windup designs and several specific algorithms have been described. The main goal of this chapter was to propose a useful summary of some of the various approaches for the design of anti-windup compensators. Furthermore, several extensions have also been discussed to show the related problems and the potential forthcoming issues. For more details, the reader is referred to [108, 352, 406].

# Chapter 8
# Applications of Anti-windup Techniques

## 8.1 Introduction

The objective of this chapter is to present several examples to illustrate the theoretical conditions developed in Chap. 7. Regional (local) and global cases are considered. Beyond the illustration of the results of Chap. 7, systems with more complex actuators, as those with nested saturations, are handled. Actually, the treatment of such more complex closed-loop systems can be achieved from slight modifications of the framework proposed in Chap. 7. Academic examples as well as examples closer to practical cases and borrowed from the literature are treated in-depth in order to illustrate the effectiveness and the drawbacks of the anti-windup strategies. With these objectives, static and (full-order and reduced-order) dynamic anti-windup compensators are computed to guarantee the asymptotic stability (in a local or global sense) of the closed-loop system and to achieve performance improvement under saturation.

The chapter is organized as follows. Section 8.2 is concerned with the static anti-windup case. The motivation example used in several chapters of the manuscript and an academic example borrowed from the literature are considered. Dynamic anti-windup compensators are computed and tested in Sect. 8.3. The cart–spring–pendulum system and the F-8 aircraft MIMO system are studied to illustrate the improvements in terms of the size of the region of asymptotic stability and in terms of performance. Section 8.4 is devoted to more complex actuators in order to underline how the machinery developed in Chap. 7 is adaptable. In Sect. 8.5, the case when the difference between the saturated and the linear signal (i.e. the dead-zone nonlinearity) is not accessible is discussed. Finally, some concluding remarks end the chapter.

## 8.2 Static Anti-windup Examples

Static anti-windup strategies represent the simplest way to evaluate the potential benefits of the anti-windup philosophy for systems presenting actuator saturation.

S. Tarbouriech et al., *Stability and Stabilization of Linear Systems with Saturating Actuators*, DOI 10.1007/978-0-85729-941-3_8, © Springer-Verlag London Limited 2011

Associated optimization procedures are convex, and when the anti-windup action is injected only in the dynamics of the controller ($v_y = 0$), there is no implicit (algebraic) loop in the closed-loop system.

*Example 8.1* (Balancing Pointer)   Let us consider again the motivation example used throughout the manuscript, i.e., the balancing pointer [215] introduced in Chap. 1. With the notation of system (7.10), one has[1]

$$A_p = \begin{bmatrix} 0 & 1 \\ 1 & 0 \end{bmatrix}; \qquad B_{pu} = \begin{bmatrix} 0 \\ -1 \end{bmatrix}; \qquad u_0 = 5$$

To evaluate the effect of various anti-windup strategies in the presence of saturation in the control loop, it is more interesting to consider closed-loop systems involving a dynamic controller. Instead of the state-feedback gain used in the previous manipulations of this motivation example, we now apply a dynamic output feedback built from a PI controller ($y_c = k_p y_p + \frac{k_p}{T_i} \int_0^t y_p$), with $k_p = 2$ and $T_i = 0.1$ and using the sum of the state components as the plant output, that is,

$$C_p = [1 \quad 1]; \qquad D_{pu} = 0$$

The dynamic feedback controller data are

$$A_c = 0; \qquad B_c = \frac{k_p}{T_i}; \qquad C_c = 1; \qquad D_c = k_p$$

Let us consider the stability problem ($w = 0$), such as defined in Chap. 7. The open-loop system being unstable, DLAW schemes with regional guarantees have to be applied (see Sect. 7.3.3). Let us consider the simplest case where a static anti-windup scheme is proposed ($A_{\text{aw}} = B_{\text{aw}} = C_{\text{aw}} = 0$) as suggested in Remark 7.10, and considering that the anti-windup output is only injected in the dynamics of $x_c$ (see Remark 7.2). Proposition 7.6 is applied to find an anti-windup gain $D_{\text{aw}} = \begin{bmatrix} E_c \\ 0 \end{bmatrix}$ which maximizes the estimate of the region of attraction of the closed-loop saturated system with anti-windup action. As suggested in Algorithm 7.3, a set of interesting directions is fixed as a square box of $x_p$ components:

$$v_1 = \begin{bmatrix} 1 \\ 1 \end{bmatrix}; \qquad v_2 = \begin{bmatrix} 1 \\ -1 \end{bmatrix}; \qquad v_3 = \begin{bmatrix} -1 \\ 1 \end{bmatrix}; \qquad v_4 = \begin{bmatrix} -1 \\ -1 \end{bmatrix}$$

The solution of the optimization problem is given by

$$P = Q^{-1} = \begin{bmatrix} 0.0400 & 0.0400 & 0.0000 \\ 0.0400 & 0.0400 & 0.0000 \\ 0.0000 & 0.0000 & 0.0040 \end{bmatrix}$$

$$\mu = 0.1600 \quad (\beta = 2.5000); \qquad E_c = 10.0098 \quad \left( D_{\text{aw}} = \begin{bmatrix} 10.0098 \\ 0 \end{bmatrix} \right)$$

---

[1]For the sake of simplicity, all along this chapter, system matrices not considered or equal to 0 (of appropriate dimension) are not explicitly given. Similarly, for simulation, initial components of the state vector which are not explicitly given are actually equal to 0. In particular, initial states of the controller and dynamic anti-windup are systematically equal to 0.

## 8.2 Static Anti-windup Examples

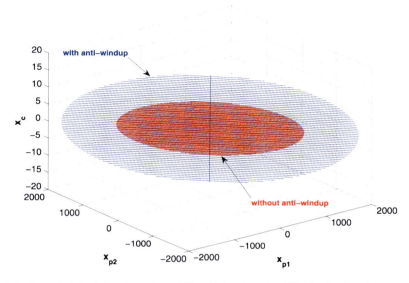

**Fig. 8.1** Example 8.1—balancing pointer. Ellipsoidal region of stability for the closed-loop system with and without anti-windup action

where $\mu$ is the optimization criterion, directly related to the scaling factor $\beta = \frac{1}{\sqrt{\mu}}$ of the shape set $\mathcal{X}_0$ formed by the directions $v_1$ to $v_4$ (see also Remark 7.9). Figure 8.1 depicts the obtained ellipsoid of closed-loop stability in the presence of an anti-windup action.

In this figure, the ellipsoid of closed-loop stability for the saturated system without anti-windup, obtained from the same optimization criterion (min $\mu$), is also depicted. In this case, one obtains[2] $\mu_{\text{noaw}} = 0.9898$ ($\beta_{\text{noaw}} = 1.0052$). It can be noticed the improvement in terms of the size of the region of asymptotic stability obtained with the use of an anti-windup strategy.

The influence of the anti-windup action may also be appreciated on the time evolution of the system state ($x_p$) and input ($u = \text{sat}(y_c)$) for various initial conditions. Figure 8.2 illustrates the case with the initial condition ($x_p = [-4\ 1]'$, $x_c = 0$). This initial condition belongs to the region of asymptotic stability with anti-windup but does not belong to the region of asymptotic stability of the closed-loop system without anti-windup. It can be checked that the trajectory of the closed-loop system with anti-windup converges toward 0 when the trajectory without anti-windup goes toward infinity.

A full static anti-windup (anti-windup action both on the dynamics and the output of the controller) can also be employed but it does not bring about any significant improvements in this example.

---

[2] To avoid ambiguity, the solution of the analysis problem without anti-windup action is denoted with subscript noaw.

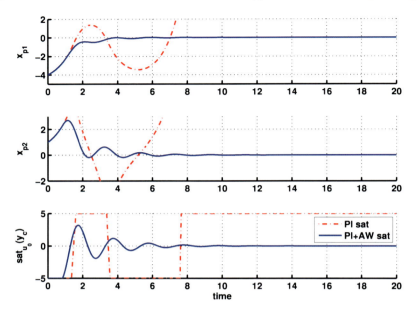

**Fig. 8.2** Example 8.1—balancing pointer. Time evolution of the closed-loop system with anti-windup (*solid line*) and without anti-windup (*dashed line*) initialized at $x_p = [-4\ 1]'$

*Example 8.2* (One-Dimension Example)  Consider the linear open-loop unstable system (local schemes are then addressed) given by [139]

$$A_p = 0.1; \qquad B_{pu} = 1; \qquad C_p = 1; \qquad D_{pu} = 0$$

with saturation level $u_0 = 1$. A stabilizing PI control is given by

$$A_c = 0; \qquad B_c = -0.2; \qquad C_c = 1; \qquad D_c = -2$$

As in the previous example, a static anti-windup action may be examined ($A_{\mathrm{aw}} = B_{\mathrm{aw}} = C_{\mathrm{aw}} = 0$, anti-windup injected only in the dynamics of $x_c$ (see Remark 7.2)). No disturbance is considered, and the optimization procedure is related to the region of asymptotic stability. Let the shape set $\mathcal{X}_0$ of directions $v_i$ be defined as a unit square region in the space $\mathfrak{R}^2$. A solution to the static anti-windup synthesis problem is

$$Q_{\mathrm{aw}} = \begin{bmatrix} 99.8738 & 32.3026 \\ 32.3026 & 144.9908 \end{bmatrix}$$

$$D_{\mathrm{aw}} = \begin{bmatrix} 0.1269 \\ 0 \end{bmatrix}$$

with associated region of asymptotic stability $\mathcal{E}(Q^{-1}, 1)$ which includes $\beta\,\mathcal{X}_0$, $\beta = 6.5888$. Considering the case without anti-windup, one obtains $\beta_{\mathrm{noaw}} = 4.3423$.

The ellipsoidal regions of stability for the closed-loop system with (solid ellipsoid) and without (dashed ellipsoid) anti-windup action are plotted in Fig. 8.3. It

8.3 Dynamic Anti-windup Examples

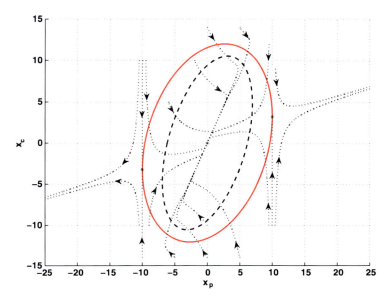

**Fig. 8.3** Example 8.2—one-dimension example. Ellipsoidal regions of stability for the closed-loop system with (*solid-line ellipsoid*) and without (*dashed-line ellipsoid*) anti-windup action

is worth noticing that, with the computed anti-windup gain, the closed-loop system presents unstable equilibrium points in $\pm[10 \ 3.2348]'$. These points (stars in the figure) are very close to the boundary of the stability domain obtained. State phase closed-loop trajectories are also plotted to illustrate the approximation of the basin of attraction.

## 8.3 Dynamic Anti-windup Examples

Although less directly computable than static anti-windup strategies, dynamic anti-windup may be more beneficial than static anti-windup, both for enlarging the domain of safe behavior and for increasing disturbance rejection performance. Moreover, reduced-order strategies generally present a good compromise between the attainable performance and the increase of size of the control law. These aspects are illustrated in the two examples which follow.

*Example 8.3* (Cart–Spring–Pendulum System) Let us consider the cart–spring–pendulum system shown in Fig. 8.4 which has been fully described in [151].

By considering the plant state $x_p = [p \ \dot{p} \ \theta \ \dot{\theta}]'$, a stable linearized model around the origin is given by

**Fig. 8.4** Example 8.3—scheme of the cart–spring–pendulum system (Courtesy L. Zaccarian)

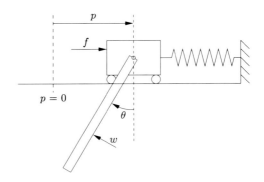

$$A_p = \begin{bmatrix} 0 & 1 & 0 & 0 \\ -330.46 & -12.15 & -2.44 & 0 \\ 0 & 0 & 0 & 1 \\ -812.61 & -29.87 & -30.10 & 0 \end{bmatrix}$$

$$B_{pu} = \begin{bmatrix} 0 \\ 2.71762 \\ 0 \\ 6.68268 \end{bmatrix}; \quad B_{pw} = \begin{bmatrix} 0 \\ 0 \\ 0 \\ 15.61 \end{bmatrix}$$

$$C_p = \begin{bmatrix} 1 & 0 & 0 & 0 \\ 0 & 0 & 1 & 0 \end{bmatrix}; \quad D_{pu} = D_{pw} = \begin{bmatrix} 0 \\ 0 \end{bmatrix}$$

In this system, the input $u$ is the voltage applied to the armature of the DC motor, constrained to belong to the range $[-5, 5]$, and the disturbance $w$ is a force in the plane of motion orthogonal to the pendulum. As suggested in [151], an observer-based controller issued from a LQG construction is designed with

$$K = [64.81 \quad 213.12 \quad 1242.27 \quad 85.82]$$

$$L = \begin{bmatrix} 64 & 2054 & -8 & -1432 \\ -8 & -280 & 142 & 10169 \end{bmatrix}'$$

which results in $A_c = A_p - B_p K - LC_p$, $B_c = L$, $C_c = -K$. The disturbance does not affect this dynamic controller. For analysis and anti-windup synthesis purposes, the bound on the disturbance $\delta^{-1}$ is set to 0.1261 (i.e. $\delta = 7.931$), which corresponds to the larger test tap modeled as a force of 1.588 Newton with duration 0.05 s. The performance output $z$ is related to the angle $\theta$, that is, with

$$C_z = [0 \quad 0 \quad 1 \quad 0]; \quad D_z = D_{zw} = 0$$

As already commented in [151], the stability analysis (without anti-windup), the static anti-windup synthesis in the global case, associated to Proposition 7.3, as well as in the local case, associated to Proposition 7.6, are unfeasible. Note, however, that these are only sufficient conditions and that we can only comment on the fact that it has not been possible to find feasible static anti-windup compensators with the proposed strategy.

## 8.3 Dynamic Anti-windup Examples

Algorithm 7.1 may then be applied, in the global case, to derive an admissible dynamic anti-windup controller which minimizes the $\mathcal{L}_2$-gain $\gamma$. The solution is, however, very high ($\gamma = 32812.9789$) and local anti-windup strategies may be preferably investigated to find a good compromise between the $\mathcal{L}_2$-gain $\gamma$ and the set of stabilizable initial conditions associated to $Q$. In this context, Algorithms 7.2 and 7.4 are applied to compute an admissible dynamic anti-windup controller. As discussed in Chap. 7, a reduced-order version of the full-order anti-windup controller may be used as an appropriate choice for $A_{aw}$ and $C_{aw}$ in Algorithm 7.4. Issued from Algorithm 7.2, a full-order dynamic anti-windup controller of order 8 ($n_p + n_c$) is obtained, which eigenvalues are $\{-5.3350 \pm 15.7535j; -277.9061; -309.3775; -732.6719; -1217.269; -7154.618; -8500.701\}$, and which guarantees a performance level $\gamma = 272.8731$. Then, for the fixed-dynamics algorithm, only a reduced version of $A_{aw}$ and $C_{aw}$ is considered with only two middle range frequency modes $-5.3350 \pm 15.7535j$. Algorithm 7.4 is then applied, which results in the reduced-order anti-windup compensator:

$$A_{aw} = \begin{bmatrix} -0.0838 & 26.2559 \\ -10.5023 & -10.5862 \end{bmatrix}; \qquad B_{aw} = \begin{bmatrix} -1.2744 \\ -0.3587 \end{bmatrix}$$

$$C_{aw} = \begin{bmatrix} 0.0324 & 0.0615 \\ 0.9658 & 1.9085 \\ 0.2993 & 0.4978 \\ 20.7082 & 34.4974 \\ -0.0189 & -0.0313 \end{bmatrix}; \qquad D_{aw} = \begin{bmatrix} -0.0044 \\ 2.4053 \\ 0.0059 \\ 5.4987 \\ -0.1828 \end{bmatrix}$$

with associate performance index $\gamma = 271.7159$.

Simulated responses of the closed-loop cart–spring–pendulum system without (dashed line) and with (solid line) anti-windup action are presented in Figs. 8.5, 8.6 and 8.7.

The time responses of the closed-loop system, initially at rest, with a perturbation force of 1.588 Newton with duration 0.05 seconds, are plotted in Fig. 8.5, for the case with anti-windup (solid line) and without anti-windup (dashed line). It illustrates that the anti-windup action prevents the system from being pushed in a slowly damping oscillation when facing a slap. Similarly, if the simulation is initialized from a non-zero initial state (a small angle $\theta(0)$ of 0.3 radians is considered in Fig. 8.6 and a larger angle $\theta(0) = 1$ radians is considered in Fig. 8.7 at $t = 0$), the anti-windup action allows us to drastically reduce the settling time. For the larger initial angle, the closed-loop response without anti-windup converges to a limit cycle where the control input oscillates between $-5$ and $5$.

*Example 8.4* (F-8 Aircraft MIMO Example) This unstable F-8 aircraft MIMO example is borrowed from [399], who considered an anti-windup strategy to deal with actuator saturation. Note that a stable version of this F-8 aircraft had been initially proposed in [207] to illustrate the use of an "Error Governor" to manage input saturations. It has also been used by [57] to evaluate the effect of sensor saturation, and by [120] to examine the synthesis of controllers in the presence of both actuator and sensor saturations. In the current case, this example is first used to illustrate how

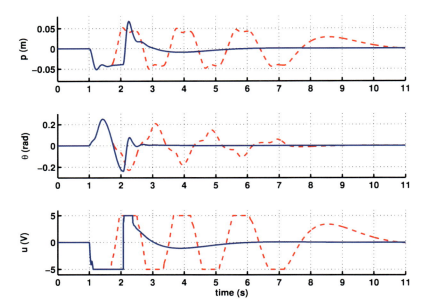

**Fig. 8.5** Example 8.3—cart–spring–pendulum system. Time evolution of the closed-loop system with anti-windup (*solid line*) and without anti-windup (*dashed line*), initially at rest, with a perturbation force of 1.588 Newton with duration 0.05 seconds

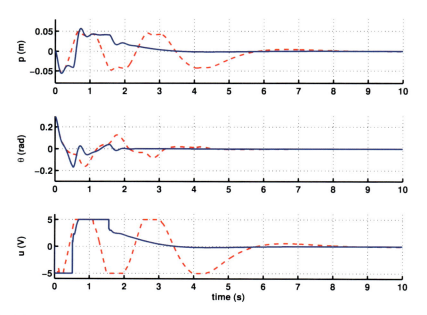

**Fig. 8.6** Example 8.3—cart–spring–pendulum system. Time evolution of the closed-loop system with anti-windup (*solid line*) and without anti-windup (*dashed line*), initialized with a small angle of $\theta(0) = 0.3$ radians

8.3 Dynamic Anti-windup Examples

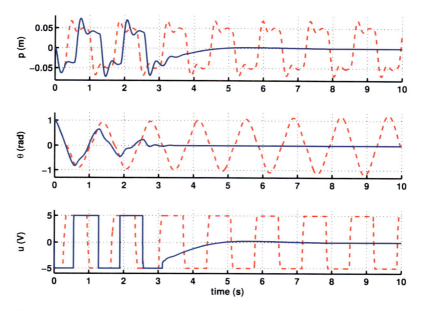

**Fig. 8.7** Example 8.3—cart–spring–pendulum system. Time evolution of the closed-loop system with anti-windup (*solid line*) and without anti-windup (*dashed line*) initialized with an angle $\theta(0) = 1$ radians

anti-windup strategies may be used to enlarge the region of asymptotic stability (in absence of additive disturbance) of the closed-loop system.

The longitudinal dynamics of an F-8 aircraft is described by the matrices

$$A_p = \begin{bmatrix} -0.8 & -0.006 & -12 & 0 \\ 0 & -0.014 & -16.64 & -32.2 \\ 1 & -0.0001 & -1.5 & 0 \\ 1 & 0 & 0 & 0 \end{bmatrix}; \quad B_{pu} = \begin{bmatrix} -19 & -3 \\ -0.66 & -0.5 \\ -0.16 & -0.5 \\ 0 & 0 \end{bmatrix}$$

$$C_p = \begin{bmatrix} 0 & 0 & 0 & 1 \\ 0 & 0 & -1 & 1 \end{bmatrix}$$

with saturation limits of the actuator given by $u_0 = [15\ 15]'$. This open-loop unstable system is controlled by the observer-based controller

$$A_c = \begin{bmatrix} -4.2676 & 0.0362 & -11.7964 & -31.7599 \\ -1.7022 & -0.0182 & 2703.8 & 470.5213 \\ -0.9265 & -0.0066 & -7.0109 & 6.2734 \\ 1 & 0 & 0.0801 & -3.2993 \end{bmatrix}$$

$$B_c = \begin{bmatrix} 0.3711 & 1.9263 \\ -3217.1 & 2717.5 \\ 7.6163 & -9.1867 \\ 3.2192 & 0.0801 \end{bmatrix}$$

**Table 8.1** Example 8.4—F-8 aircraft. Comparison of the enlargement of the domain of asymptotic stability for various anti-windup schemes. $\beta$ represents the maximal scaling factor of the shape set $\mathcal{X}_0$ formed by vertices $v_r$, such that $\beta \mathcal{X}_0 \subset \mathcal{E}(Q^{-1}, 1)$

| Strategy | $\beta$ |
|---|---|
| Analysis (without anti-windup) | 0.1888 |
| Partial static anti-windup synthesis (acts only on the controller dynamics) | 1.4851 |
| Static anti-windup synthesis ($D_{aw}$) | 2.0379 |
| Full-order dynamic anti-windup synthesis (order 8, Algorithm 7.3) | 2.7270 |
| Full-order dynamic anti-windup, fixed-dynamics synthesis (order 8, Algorithm 7.5) | 5.1483 |
| Reduced-order dynamic anti-windup, fixed-dynamics synthesis (order 2) | 4.2614 |

$$
C_c = \begin{bmatrix} -0.4485 & -0.0045 & 1.3181 & 3.1974 \\ 3.9966 & 0.0144 & -7.7734 & -10.4292 \end{bmatrix}; \qquad D_c = \begin{bmatrix} 0 & 0 \\ 0 & 0 \end{bmatrix}
$$

This controller presents rather poor performance regarding the stabilization of non-zero initial states and disturbance rejection, and would definitively not be selected to control the F-8 aircraft. Its objective is here to help us to illustrate various aspects of anti-windup strategies.

**Enlargement of the Domain of Admissible Initial States**  Anti-windup strategies may be applied to enlarge the local domain of admissible initial conditions for this system. Typically, a shape set $\mathcal{X}_0$ of interesting directions $v_r$ may be described as a unit square box for the components of the system state $x_p$,[3] or, to limit the number of additional LMI constraints, and taking into account the symmetry of the ellipsoidal domains built up, as a partial box of the components of the system state $x_p$.[4] Algorithm 7.3 for the full-order AW synthesis or Algorithm 7.5 for the fixed-dynamics AW synthesis may then be used to provide anti-windup controllers. Table 8.1 summarizes the results obtained by the different strategies, in terms of the maximal scaling factor $\beta$ of the shape set $\mathcal{X}_0$ formed by vertices $v_r$, such that $\beta \mathcal{X}_0 \subset \mathcal{E}(Q^{-1}, 1)$. It can be noticed that any anti-windup strategy significantly increases the set of admissible initial states with respect to the analysis of the closed-loop system without anti-windup scheme. Dynamic anti-windup strategies outclass static ones, but to the detriment of an additional dynamic system. It has to be noted that the full-order dynamic anti-windup issued from Algorithm 7.3 may be improved by an additional optimization step with fixed $A_{aw}$, $C_{aw}$ issued from the first step, using Algorithm 7.5. Finally, the reduced dynamic anti-windup con-

---

[3]Such a shape set may of course also account for non-zero initial states of the controller if requested.

[4]Only 8 of the 16 vertices which compose the unit square box are considered.

8.3 Dynamic Anti-windup Examples

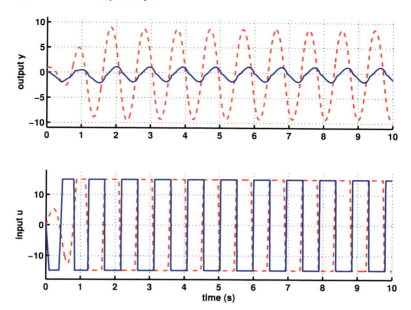

**Fig. 8.8** Example 8.4—F-8 aircraft. Time evolution of the output $y_p$ and the input $u$ of the closed-loop system without anti-windup, initialized at $x_p(0) = [0\ 0\ 1\ 1]'$. Case without disturbance

troller gives almost the same scaling factor with much less computational burden (in particular for the implementation).

Figures 8.8 and 8.9 illustrate the behavior of the closed-loop system initialized at $x_p(0) = [0\ 0\ 1\ 1]'$, in the cases without and with anti-windup, respectively. The reduced-order anti-windup compensator is given by

$$A_{\text{aw}} = \begin{bmatrix} -1.8720 & 4.0979 \\ -1.6392 & -3.5111 \end{bmatrix}; \quad B_{\text{aw}} = \begin{bmatrix} -0.0001 & 0.0032 \\ 0.0002 & 0.0175 \end{bmatrix}$$

$$C_{\text{aw}} = \begin{bmatrix} 0.1250 & 0.7584 \\ 65.6183 & -311.4135 \\ -0.4377 & 0.4340 \\ 0.1608 & 0.4048 \\ -0.0594 & -0.0313 \\ 0.4916 & 0.1248 \end{bmatrix}; \quad D_{\text{aw}} = \begin{bmatrix} -0.0003 & -0.0143 \\ -0.0026 & 1.0326 \\ 0.0001 & -0.0044 \\ 0.0000 & 0.0024 \\ -1.0000 & -0.0006 \\ -0.0000 & -0.9941 \end{bmatrix} \quad (8.1)$$

The selected initial state belongs to the region of asymptotic stability for all the computed anti-windup compensators, but does not belong to the region of asymptotic stability issued from the analysis of the saturated closed-loop system without anti-windup. This is illustrated on Figs. 8.8 and 8.9 which exhibit a limit cycle behavior in the case without anti-windup while, using an additional anti-windup loop, the closed-loop system converges toward 0 after a transient period where saturations are active.

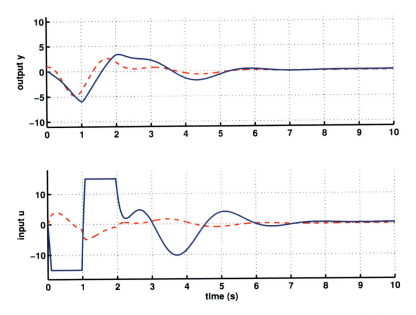

**Fig. 8.9** Example 8.4—F-8 aircraft. Time evolution of the output $y_p$ and the input $u$ of the closed-loop system with the second-order anti-windup compensator, initialized at $x_p(0) = [0\ 0\ 1\ 1]'$. Case without disturbance

**Maximization of Disturbance Rejection** To illustrate the ability of the anti-windup action to counteract the effect of additive disturbance, let us turn now to the case where a disturbance of limited energy defined as $\|w\|_2^2 \leq 20$ ($\delta = 0.05$) acts on the system, through the matrix

$$B_{pw} = \begin{bmatrix} 1 & 0 \\ 0 & 1 \\ 0 & 0 \\ 0 & 0 \end{bmatrix}$$

The LMI conditions for stability analysis of the closed-loop system facing disturbance $\delta$ are not feasible, which suggests that the system, without anti-windup compensation, is not able to reject the disturbance. The simulation of the system, facing disturbance $\delta$ at $t = 1$ second, confirms the analysis study (see Fig. 8.10).

Moreover, the static anti-windup synthesis acting only on the dynamics of the controller (through the gain $E_c$) is also unfeasible. On the other hand, using Algorithms 7.2 and 7.4, it is shown that the other anti-windup strategies allow one to guarantee the external stability of the system in the presence of the disturbance $\delta$. The analysis and anti-windup synthesis results are summarized in Table 8.2. It has to be noticed that, in the full-order synthesis (Algorithm 7.2), the first step related to

## 8.3 Dynamic Anti-windup Examples

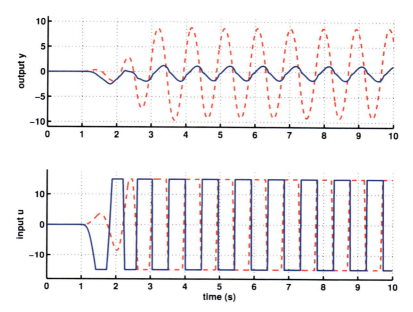

**Fig. 8.10** Example 8.4—F-8 aircraft. Time evolution of the output $y_p$ and the input $u$ of the closed-loop system without anti-windup, initially at rest ($x = 0$), in the presence of a disturbance $\delta = 0.05$ at $t = 1$ second on the first disturbance input

**Table 8.2** Example 8.4—F-8 aircraft. Comparison of disturbance rejection $\gamma$ and scaling $\beta$ of the shape set $\mathcal{X}_0$

| Strategy | $\gamma$ | $\beta$ |
|---|---|---|
| Analysis (without anti-windup) | – | – |
| Partial static anti-windup synthesis (acts only on the controller dynamics) | – | – |
| Static anti-windup synthesis ($D_{aw}$) | 2755.2 | 0.1925 |
| Full-order dynamic anti-windup synthesis (order 8, Algorithm 7.2) | 1117.7 | 0.2808 |
| Full-order dynamic anti-windup, fixed-dynamics synthesis (order 8, Algorithm 7.4) | 1117.5 | 0.3166 |
| Reduced-order dynamic anti-windup, fixed-dynamics synthesis (order 2) | 1117.5 | 0.3213 |

Proposition 7.4 is solved by considering $\gamma$ as the optimization criterion. In the third step of this algorithm and in Algorithm 7.4 related to the fixed-dynamics synthesis, the optimization criterion is actually the sum of $\gamma$ and $\mu = \frac{1}{\beta^2}$, i.e. the optimization step evaluates an anti-windup compensator which gives a good compromise between performance and size of the domain of asymptotic stability.

A simulation of the closed-loop system with the reduced-order anti-windup compensator given by matrices

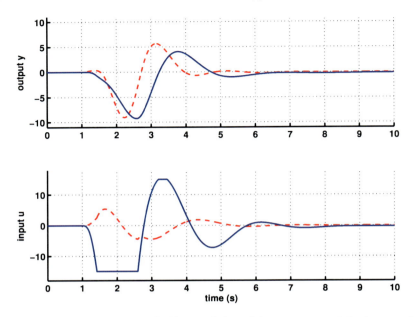

**Fig. 8.11** Example 8.4—F-8 aircraft. Time evolution of the output $y_p$ and the input $u$ of the closed-loop system with the second-order anti-windup compensator, initially at rest ($x = 0$), in the presence of a disturbance $\delta = 0.05$ at $t = 1$ second on the first disturbance input

$$A_{\text{aw}} = \begin{bmatrix} -0.2849 & 4.7104 \\ -1.8842 & -2.1691 \end{bmatrix}; \quad B_{\text{aw}} = \begin{bmatrix} -0.0068 & -0.0018 \\ -0.0044 & -0.0013 \end{bmatrix}$$

$$C_{\text{aw}} = \begin{bmatrix} 0.0023 & -0.0125 \\ 5.4812 & 13.7443 \\ -0.0160 & -0.0425 \\ -0.0041 & -0.0047 \\ -0.0024 & -0.0036 \\ 0.0131 & 0.0191 \end{bmatrix}; \quad D_{\text{aw}} = \begin{bmatrix} -0.0004 & -0.0001 \\ 0.0602 & 0.0150 \\ -0.0002 & -0.0001 \\ 0.0001 & 0.0000 \\ -1.0000 & 0.0000 \\ 0.0000 & -1.0000 \end{bmatrix} \quad (8.2)$$

is plotted in Fig. 8.11. The disturbance exogenous signal is defined as $w(t) = [10\ 0]'$ during 0.2 seconds. It occurs at $t = 1$ second.

## 8.4 Toward More Complex Nonlinear Actuators

As suggested in Sect. 7.6, the machinery which has been described in Chap. 7 for a position saturation to represent the actuator may be extended, following exactly the same developments, to the cases of more complex models of dynamic actuators involving position, rate or any saturation elements encapsulated in an actuator block.

## 8.4 Toward More Complex Nonlinear Actuators

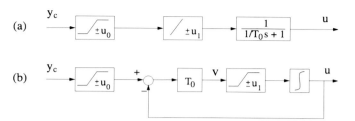

**Fig. 8.12** (a) Actuator with position and rate saturations. (b) Surrogate structure for the rate saturation. Scalar case

The three case studies which follow intend to illustrate how anti-windup strategies may be employed to deal with such complex nonlinear actuators.

### 8.4.1 Actuator with Position and Rate Saturations

Let us first consider a dynamic actuator involving position and rate saturations, such as described in Fig. 8.12(a). Roughly speaking, the rate limiter may be placed before or after the position saturation element. The case considered here corresponds to the situation proposed in the working group AG15 of GARTEUR (Group for Aeronautical Research and Technology in Europe, http://www.garteur.org) and illustrated in [295]. The case with the position saturation positioned after the rate limiter has been investigated, for example, in [107].

The rate limiter may be replaced by an equivalent saturation model, as shown in Fig. 8.12(b), in order to obtain a mathematical representation involving only position saturation elements, for which one can directly extend the previous anti-windup strategies. Note that the scalar case presented in Fig. 8.12 may be directly extended to the vector case by considering a diagonal matrix representation with gains $T_{0i}$ corresponding to the $i$th input/output relationship between the output $y_c$ of the controller and the input $u$ of the system.

In that case, the actuator which interconnects the plant (7.1) and the controller (7.2) is no more represented by a single saturation function such as considered in (7.6) but is described by a dynamic saturated system as follows:

$$\dot{x}_a = \mathrm{sat}_{u_1}(v)$$
$$v = T_0 \,\mathrm{sat}_{u_0}(y_c) - T_0 x_a \qquad (8.3)$$
$$u = x_a$$

where $\mathrm{sat}_{u_0}(\cdot)$ and $\mathrm{sat}_{u_1}(\cdot)$ denote the saturation functions with $u_0$ and $u_1$ as levels of saturation in magnitude and rate, respectively.

The direct linear anti-windup (DLAW) architecture (7.8) is then modified to

$$\dot{x}_{\mathrm{aw}} = A_{\mathrm{aw}} x_{\mathrm{aw}} + B_{\mathrm{aw}}^0 \left(\mathrm{sat}_{u_0}(y_c) - y_c\right) + B_{\mathrm{aw}}^1 \left(\mathrm{sat}_{u_1}(v) - v\right)$$
$$\begin{bmatrix} v_x \\ v_y \end{bmatrix} = C_{\mathrm{aw}} x_{\mathrm{aw}} + D_{\mathrm{aw}}^0 \left(\mathrm{sat}_{u_0}(y_c) - y_c\right) + D_{\mathrm{aw}}^1 \left(\mathrm{sat}_{u_1}(v) - v\right) \qquad (8.4)$$

Two dead-zone nonlinearities are considered:

$$\phi_0(y_c) = \mathrm{sat}_{u_0}(y_c) - y_c \quad \text{and} \quad \phi_1(v) = \mathrm{sat}_{u_1}(v) - v$$

With the extended state vector $x = [\,x_p'\ x_a'\ x_c'\ x_{\mathrm{aw}}'\,]' \in \mathfrak{R}^n$, $n = n_p + m + n_c + n_{\mathrm{aw}}$, the closed-loop system (7.13)–(7.15) is updated under a similar standard form:[5]

$$
\begin{aligned}
\dot{x} &= \mathcal{A}x + \mathcal{B}_0\phi_0(y_c) + \mathcal{B}_1\phi_1(v) + \mathcal{B}_2 w \\
y_c &= \mathcal{C}_0 x + \mathcal{D}_{00}\phi_0(y_c) + \mathcal{D}_{01}\phi_1(v) + \mathcal{D}_{0w} w \\
v &= \mathcal{C}_1 x + \mathcal{D}_{10}\phi_0(y_c) + \mathcal{D}_{11}\phi_1(v) + \mathcal{D}_{1w} w \\
z &= \mathcal{C}_2 x + \mathcal{D}_{20}\phi_0(y_c) + \mathcal{D}_{21}\phi_1(v) + \mathcal{D}_{2w} w
\end{aligned}
\tag{8.5}
$$

where the matrices are updated as follows:

$$
\begin{aligned}
\mathcal{A} &= \begin{bmatrix} \mathbb{A} & B_v C_{\mathrm{aw}} \\ 0 & A_{\mathrm{aw}} \end{bmatrix}; &
\mathcal{B}_0 &= \begin{bmatrix} B_{\phi 0} + B_v D_{\mathrm{aw}}^0 \\ B_{\mathrm{aw}}^0 \end{bmatrix} \\
\mathcal{B}_1 &= \begin{bmatrix} B_{\phi 1} + B_v D_{\mathrm{aw}}^1 \\ B_{\mathrm{aw}}^1 \end{bmatrix}; &
\mathcal{B}_2 &= \begin{bmatrix} B_2 \\ 0 \end{bmatrix} \\
\mathcal{C}_0 &= [\,C_0 \quad C_{v0} C_{\mathrm{aw}}\,]; &
\mathcal{C}_1 &= [\,C_1 \quad C_{v1} C_{\mathrm{aw}}\,] \\
\mathcal{D}_{00} &= C_{v0} D_{\mathrm{aw}}^0; &
\mathcal{D}_{01} &= C_{v0} D_{\mathrm{aw}}^1 \\
\mathcal{D}_{10} &= D_1 + C_{v1} D_{\mathrm{aw}}^0; &
\mathcal{D}_{11} &= C_{v1} D_{\mathrm{aw}}^1
\end{aligned}
\tag{8.6}
$$

with

$$
\begin{aligned}
\mathbb{A} &= \begin{bmatrix} A_p & B_{pu} & 0 \\ T_0 D_c C_p & T_0 D_c D_{pu} - T_0 & T_0 C_c \\ B_c C_p & B_c D_{pu} & A_c \end{bmatrix}; &
B_v &= \begin{bmatrix} 0 \\ T_0[\,0\ I_m\,] \\ [\,I_{n_c}\ 0\,] \end{bmatrix} \\
B_{\phi 0} &= \begin{bmatrix} 0 \\ T_0 \\ 0 \end{bmatrix}; \quad
B_{\phi 1} = \begin{bmatrix} 0 \\ I_m \\ 0 \end{bmatrix}; &
B_2 &= \begin{bmatrix} B_{pw} \\ T_0 D_c D_{pw} + T_0 D_{cw} \\ B_{cw} + B_c D_{pw} \end{bmatrix} \\
C_0 &= [\,D_c C_p \quad D_c D_{pu} \quad C_c\,] \\
C_1 &= [\,T_0 D_c C_p \quad T_0 D_c D_{pu} - T_0 \quad T_0 C_c\,] \\
C_{v0} &= [\,0 \quad I_m\,]; \quad C_{v1} = T_0[\,0 \quad I_m\,] \\
D_1 &= T_0; \quad C_2 = [\,C_z \quad D_{zu} \quad 0\,] \\
\mathcal{D}_{2w} &= D_{zw}; \quad \mathcal{D}_{0w} = D_{cw} + D_c D_{pw}; \quad \mathcal{D}_{1w} = T_0(D_{cw} + D_c D_{pw}) \\
\mathcal{C}_2 &= [\,C_2 \quad 0\,]; \quad \mathcal{D}_{20} = 0; \quad \mathcal{D}_{21} = 0
\end{aligned}
\tag{8.7}
$$

---

[5] $\mathcal{D}_{20}$ and $\mathcal{D}_{21}$ are actually equal to 0, but they are written here since this almost standard formulation of system (8.5) will be also used later in the chapter.

## 8.4 Toward More Complex Nonlinear Actuators

Regarding the regional context, an updated version of Proposition 7.4 may be written which considers a symmetric positive definite matrix $Q \in \mathfrak{R}^{n \times n}$, matrices $Z_0 \in \mathfrak{R}^{m \times n}$ and $Z_1 \in \mathfrak{R}^{m \times n}$, positive diagonal matrices $S_0 \in \mathfrak{R}^{m \times m}$ and $S_1 \in \mathfrak{R}^{m \times m}$, a positive scalar $\gamma$ and the following conditions:

$$
\begin{bmatrix}
Q\mathcal{A}' + \mathcal{A}Q & \mathcal{B}_0 S_0 - Q C_0' - Z_0' & \mathcal{B}_1 S_1 - Q C_1' - Z_1' & \mathcal{B}_2 & Q C_2' \\
\star & -2S_0 - \mathcal{D}_{00} S_0 - S_0 \mathcal{D}_{00}' & -\mathcal{D}_{01} S_1 - S_0 \mathcal{D}_{10}' & -\mathcal{D}_{0w} & S_0 \mathcal{D}_{20}' \\
\star & \star & -2S_1 - \mathcal{D}_{11} S_1 - S_1 \mathcal{D}_{11}' & -\mathcal{D}_{1w} & S_1 \mathcal{D}_{21}' \\
\star & \star & \star & -I & \mathcal{D}_{2w}' \\
\star & \star & \star & \star & -\gamma I
\end{bmatrix}
< 0
\tag{8.8}
$$

$$
\begin{bmatrix} Q & Z_{0(i)}' \\ \star & \delta u_{0(i)}^2 \end{bmatrix} \geq 0, \quad i = 1, \dots, m
\tag{8.9}
$$

$$
\begin{bmatrix} Q & Z_{1(i)}' \\ \star & \delta u_{1(i)}^2 \end{bmatrix} \geq 0, \quad i = 1, \dots, m
\tag{8.10}
$$

*Remark 8.1* This is the simplest case where it is assumed that the input and output signals saturation elements are known. It is considered only to explain the underlying principles when considering complex models of actuators. More realistic strategies have been considered, either by considering an anti-windup scheme which compares the nonlinear dynamic actuator with a fictitious linear actuator (same dynamics, but without saturations) [295] or by using a state observer for the non accessible bounds of the saturations [363]. Such aspects will be discussed in Sect. 8.5.

*Example 8.5* (F-8 Aircraft MIMO Example with Position and Rate Saturation) Let us consider again the unstable F-8 aircraft MIMO example presented in Example 8.4. The same model and controller are considered, while the actuator is now changed by a magnitude saturation + rate-limiter actuator, as shown in Fig. 8.12, with:

$$
u_0 = \begin{bmatrix} 15 \\ 15 \end{bmatrix}; \qquad u_1 = \begin{bmatrix} 25 \\ 25 \end{bmatrix}; \qquad T_0 = \begin{bmatrix} 100 & 0 \\ 0 & 100 \end{bmatrix}
$$

To illustrate how anti-windup strategies may be used to enlarge the domain of asymptotic stability (in absence of additive disturbance) of the closed-loop system, an initial shape set $\mathcal{X}_0$ is defined, as previously, by considering a partial unit square box for the components of the system state $x_p$. An optimization procedure is applied which mimics that one described in Algorithms 7.3 or 7.5, but considering conditions (8.8)–(8.10). A reduced-order anti-windup is evaluated, based on a fixed-dynamics strategy where matrices $A_{\mathrm{aw}}$ and $C_{\mathrm{aw}}$ of order 2 are those that had been considered in the case of a single saturation actuator (8.1). The other matrices are updated as follows:

$$
B_{\mathrm{aw}}^0 = \begin{bmatrix} -0.0000 & -0.0001 \\ -0.0000 & -0.0000 \end{bmatrix}; \qquad B_{\mathrm{aw}}^1 = \begin{bmatrix} -1.1580 & -0.0982 \\ -0.0892 & -0.0097 \end{bmatrix}
$$

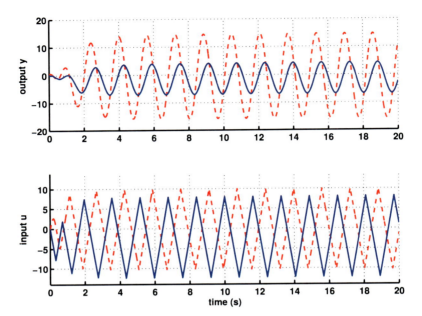

**Fig. 8.13** Example 8.5—F-8 aircraft with position and rate limiter. Time evolution of the output $y_p$ and the input $u$ of the closed-loop system without anti-windup, initialized at position $x_p(0) = [0\ 0\ 0.69\ 0.69]'$

$$D_{aw}^0 = \begin{bmatrix} 0.0000 & 0.0001 \\ -0.0017 & -0.0053 \\ 0.0000 & 0.0000 \\ -0.0000 & -0.0000 \\ -1.0000 & -0.0000 \\ -0.0000 & -1.0000 \end{bmatrix} ; \quad D_{aw}^1 = \begin{bmatrix} 0.5414 & 0.0473 \\ -49.1045 & -4.2698 \\ 0.2060 & 0.0181 \\ -0.0844 & -0.0077 \\ -0.0099 & -0.0000 \\ -0.0002 & -0.0099 \end{bmatrix}$$

This dynamic anti-windup corresponds to an optimal scaling factor for $\mathcal{X}_0$ given by $\beta = 0.6315$. Such value may be compared to $\beta = 0.1388$ obtained in the analysis case without anti-windup and $\beta = 0.2643$ solution to the optimization problem with a static anti-windup. It means that the initial condition considered in Example 8.4, i.e., $x_p(0) = [0\ 0\ 1\ 1]'$, does not belong to the set $\mathcal{E}(Q^{-1}, 1)$ associated to this dynamic anti-windup compensator. Then, the behavior of the closed-loop system initialized in $x_p(0) = [0\ 0\ 0.69\ 0.69]'$, (which belongs to the region of asymptotic stability with anti-windup) is plotted in Figs. 8.13 and 8.14 considering the control architecture without and with anti-windup, respectively.

It may be noticed on that figures that this is the rate saturation which is mainly active. Moreover, the saturated signals $\text{sat}_{u_1}(v)$ and $\text{sat}_{u_0}(y_c)$ are plotted in Fig. 8.15. The input $u$ of the aircraft system presents a reduced activity with respect to the case of a single position saturation, as shown in Fig. 8.9, but to the detriment of the time response.

8.4 Toward More Complex Nonlinear Actuators

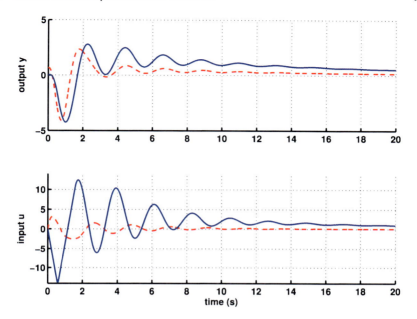

**Fig. 8.14** Example 8.5—F-8 aircraft with position and rate limiter. Time evolution of the output $y_p$ and the input $u$ of the closed-loop system with the second-order anti-windup compensator, initialized at position $x_p(0) = [0\ 0\ 0.69\ 0.69]'$

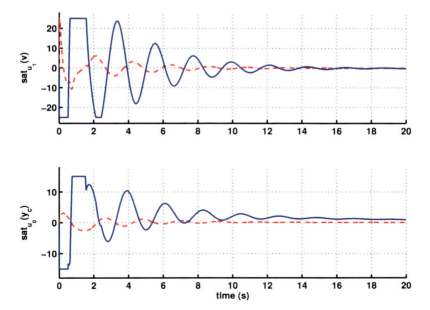

**Fig. 8.15** Example 8.5—F-8 aircraft with position and rate limiter. Time evolution of the saturated signals $\text{sat}_{u_1}(v)$ and $\text{sat}_{u_0}(y_c)$ of the closed-loop system with the second-order anti-windup compensator, initialized at position $x_p(0) = [0\ 0\ 0.69\ 0.69]'$

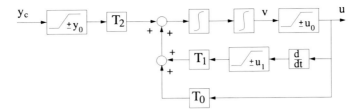

**Fig. 8.16** Model of a launcher actuator in the case $q = 2$ (one derivation of the system input introduced in the actuator element)

### 8.4.2 Dynamics Restricted Actuator

To go further, not only magnitude and rate of actuator dynamics may saturate, but higher order dynamics may also be subjugated to limitations. As an illustration of this, let us consider now a model of actuator such as those encountered in the control of launchers [233, 234]. The actuator involving limitations on the thruster angle of deflection and its time-derivative, present in some phases of the flight path of launchers and in particular during the atmospheric phase, is presented in Fig. 8.16. It corresponds to the case of an actuator dynamics involving the first-order time-derivative of the system input ($q = 2$).

By extension, the generic case involving $(q - 1)$-order derivatives of the input signal, $q > 2$, can be described from the following actuator model [360]:

$$\dot{x}_a = A_a x_a + B_{a0} \text{sat}_{u_0}(C_a x_a) + \sum_{j=1}^{q-1} B_{aj} \text{sat}_{u_j}\left(\text{sat}_{u_0}(C_a x_a)^{(j)}\right) + B_{aq} \text{sat}_{y_0}(y_c)$$

$$v = C_a x_a \qquad (8.11)$$

$$u = \text{sat}_{u_0}(v)$$

where $x_a = [v' \; \dot{v}' \; \ldots \; v^{(q-1)'}]' \in \mathfrak{R}^{m.q}$ is the state of the actuator and the superscript $(j)$ denotes the $j$th derivative of the corresponding signal. Matrices $A_a \in \mathfrak{R}^{m.q \times m.q}$, $B_{aj} \in \mathfrak{R}^{m.q \times m}$, $j = 0, \ldots, q - 1$, $B_{aq} \in \mathfrak{R}^{m.q \times m}$ and $C_a \in \mathfrak{R}^{m \times m.q}$ are defined by

$$A_a = \begin{bmatrix} 0 & I & 0 & 0 & \ldots & 0 \\ 0 & 0 & I & 0 & \ldots & 0 \\ \vdots & & & \ddots & & \vdots \\ \vdots & & & \ldots & I & 0 \\ 0 & \ldots & & \ldots & 0 & I \\ 0 & \ldots & & \ldots & 0 & 0 \end{bmatrix} \; ; \quad B_{aj} = \begin{bmatrix} 0 \\ 0 \\ \vdots \\ 0 \\ T_j \end{bmatrix}$$

## 8.4 Toward More Complex Nonlinear Actuators

$$B_{aq} = \begin{bmatrix} 0 \\ 0 \\ \vdots \\ 0 \\ T_q \end{bmatrix}; \qquad C_a = [\, I \quad 0 \quad \ldots \quad 0 \quad 0 \,]$$

The system then involves $q + 1$ dead-zone nonlinearities $\phi_0(v)$, $\phi_j(v)$, $j = 1, \ldots, q - 1$, and $\phi_c(y_c)$

$$\phi_0(v) = \mathrm{sat}_{u_0}(C_a x_a) - C_a x_a$$

$$\phi_j(v) = \mathrm{sat}_{u_j}\left(u^{(j)}\right) - u^{(j)} = \mathrm{sat}_{u_j}\left(\mathrm{sat}_{u_0}(C_a x_a)^{(j)}\right) - \mathrm{sat}_{u_0}(C_a x_a)^{(j)} \qquad (8.12)$$

$$\phi_c(y_c) = \mathrm{sat}_{y_0}(y_c) - y_c$$

Moreover, from the definition of $\phi_0$, one gets

$$\phi_0^{(j)}(v) = \mathrm{sat}_{u_0}(C_a x_a)^{(j)} - C_a x_a^{(j)} = \left(\mathrm{sat}_{u_0}(C_a x_a) - C_a x_a\right)^{(j)}$$

Let us consider an anti-windup scheme which may consider for its input any of the dead-zone terms (8.12) and/or the time-derivatives of $\phi_0(v)$. Typically, the following anti-windup architecture may be considered:

$$\dot{x}_{\mathrm{aw}} = A_{\mathrm{aw}} x_{\mathrm{aw}} + B_{\mathrm{aw}}^0 \phi_0(v) + \sum_{j=1}^{q-1} B_{\mathrm{aw}}^j \phi_0^{(j)}(v) + B_{\mathrm{aw}}^c \phi_c(y_c)$$

$$\begin{bmatrix} v_x \\ v_y \end{bmatrix} = C_{\mathrm{aw}} x_{\mathrm{aw}} + D_{\mathrm{aw}}^0 \phi_0(v) + \sum_{j=1}^{q-1} D_{\mathrm{aw}}^j \phi_0^{(j)}(v) + D_{\mathrm{aw}}^c \phi_c(y_c) \qquad (8.13)$$

The closed-loop system may then be written in a standard form,

$$\dot{x} = \mathcal{A}x + \mathcal{B}_0 \phi_0(v) + \sum_{j=1}^{q-1}\left(\mathcal{B}_j \phi_j(v) + \mathcal{B}_{0j}\phi_0^{(j)}(v)\right)$$
$$\quad + \mathcal{B}_y \phi_c(y_c) + \mathcal{B}_2 w$$

$$y_c = \mathcal{C}_0 x + \mathcal{D}_{00}\phi_0(v) + \sum_{j=1}^{q-1}\left(\mathcal{D}_{0j}\phi_j(v) + \mathcal{D}_{00j}\phi_0^{(j)}(v)\right) \qquad (8.14)$$
$$\quad + \mathcal{D}_{0y}\phi_c(y_c) + \mathcal{D}_{0w} w$$

$$v = \mathcal{C}_1 x$$

$$z = \mathcal{C}_2 x + \mathcal{D}_{20}\phi_0(v)$$

with the extended state vector $x = [\, x_p' \ x_a' \ x_c' \ x_{\mathrm{aw}}' \,]' \in \mathfrak{R}^n$, $n = n_p + m.q + n_c + n_{\mathrm{aw}}$. The construction of the matrices involved in this system (8.14) does not present any difficulty although it may represent, in the general case, a tedious task. Then an updated version of Proposition 7.4 directly follows, using exactly the same developments as previously. Such conditions are described in [360] for the particular case of a static anti-windup action (only $D_{\mathrm{aw}}$ gain).

332                                       8   Applications of Anti-windup Techniques

*Example 8.6* (F/A-18 HARV Aircraft Example) This example intends to illustrate how an anti-windup scheme may be actually employed to deal with such an actuator. The model of the linearized F/A-18 HARV aircraft lateral dynamics is borrowed from [322]. It describes the time evolution of the state composed with the roll rate, yaw rate, sideslip and bank angle at a flight condition of Mach 0.7 and altitude 20000 ft. The control governs are the aileron, stabilor and rudder. The sideslip and bank angle are assumed to be measured. In this case the open-loop dynamics is described by

$$
A_p = \begin{bmatrix} -2.3142 & 0.5305 & -15.5763 & 0 \\ -0.0160 & -0.1287 & 3.0081 & 0 \\ 0.0490 & -0.9980 & -0.1703 & 0.0440 \\ 1.0000 & 0.0491 & 0 & 0 \end{bmatrix}
$$

$$
B_{pu} = \begin{bmatrix} 23.3987 & 21.4333 & 3.2993 \\ -0.1644 & 0.3313 & -1.9836 \\ -0.0069 & -0.0153 & 0.0380 \\ 0 & 0 & 0 \end{bmatrix} \tag{8.15}
$$

$$
C_p = \begin{bmatrix} 0 & 0 & 1 & 0 \\ 0 & 0 & 0 & 1 \end{bmatrix}
$$

The controller is the one built in [290], taking into account a position saturation for the actuator. We have

$$
A_c = \begin{bmatrix} -0.98 & 0.05 & -0.03 & -1.84 \\ 32.55 & -4.09 & 0.42 & -16.22 \\ 65.56 & -2.90 & -6.85 & -9.77 \\ 10.91 & 0.20 & -0.05 & -9.92 \end{bmatrix}
$$

$$
B_c = \begin{bmatrix} 0.24 & -0.03 \\ 0.205 & -0.2897 \\ -46.23 & 0.89 \\ 1.59 & -0.14 \end{bmatrix} \tag{8.16}
$$

$$
C_c = \begin{bmatrix} 32.55 & -0.00 & -0.63 & -10.57 \\ 20.11 & 0.18 & -0.26 & -7.73 \\ -1.61 & -0.73 & -0.47 & 5.40 \end{bmatrix}
$$

$$
D_c = \begin{bmatrix} -2.77 & -0.1 \\ -0.64 & -0.11 \\ -4.22 & 0.19 \end{bmatrix}
$$

It is assumed that the control governs are limited in magnitude, rate and acceleration, corresponding to the case $q = 2$ illustrated in Fig. 8.16, with the dynamics

$$
T_0 = \begin{bmatrix} -25 & 0 & 0 \\ 0 & -25 & 0 \\ 0 & 0 & -25 \end{bmatrix} ; \quad T_1 = \begin{bmatrix} -20 & 0 & 0 \\ 0 & -20 & 0 \\ 0 & 0 & -20 \end{bmatrix}
$$

$$
T_2 = \begin{bmatrix} 25 & 0 & 0 \\ 0 & 25 & 0 \\ 0 & 0 & 25 \end{bmatrix}
$$

## 8.4 Toward More Complex Nonlinear Actuators

and the bounds

$$u_0 = \begin{bmatrix} 25 \\ 10.5 \\ 30 \end{bmatrix}; \qquad u_1 = \begin{bmatrix} 100 \\ 40 \\ 82 \end{bmatrix}; \qquad y_0 = \begin{bmatrix} 30 \\ 20 \\ 35 \end{bmatrix}$$

A static anti-windup strategy is evaluated as follows:

$$\begin{bmatrix} v_x \\ v_y \end{bmatrix} = D_{aw}^0 \phi_0(v) + D_{aw}^1 \phi_0^{(1)}(v) + D_{aw}^c \phi_c(y_c)$$

where $\phi_1(v)$ is not used since it is not easily accessible. This corresponds to the case already implemented in [360]. Additional constraints are also considered such that $\phi_c(y_c)$ only acts on the dynamics of the controller and $\phi_0^{(1)}(v)$ only acts on the output of the controller.

It has to be noticed that the open-loop system is asymptotically stable, but with a spectrum containing a very slow eigenvalue $(-0.0003)$ and a pair of poorly damped eigenvalues $(-0.2016 \pm 1.8855j)$. Associated to the fact that the open-loop system is only a linearized model of the F/A-18 HARV aircraft lateral dynamics, it does not make much sense to consider global anti-windup. The optimization problem is set as the optimization of the region of asymptotic stability, i.e. we aim at maximizing $\beta$, the scaling factor of the shape set $\mathcal{X}_0$. Such a set is defined by values 1 or $-1$ for the components of the state vector associated to the states of the aircraft. The solution issued from the execution of Algorithm 7.5 is

$$D_{aw}^0 = \begin{bmatrix} -2.0647 & -2.0247 & -1.3470 \\ -45.2868 & -43.8489 & -68.0036 \\ -5.3265 & -5.2806 & 11.1035 \\ -6.2804 & -6.0695 & -5.7257 \\ 0.8896 & 0.0824 & -0.5242 \\ 0.0206 & 1.1768 & -0.1399 \\ -0.3258 & -0.3929 & 0.0070 \end{bmatrix}; \qquad D_{aw}^1 = \begin{bmatrix} 0 & 0 & 0 \\ 0 & 0 & 0 \\ 0 & 0 & 0 \\ 0 & 0 & 0 \\ 0.8 & 0 & 0 \\ 0 & 0.8 & 0 \\ 0 & 0 & 0.8 \end{bmatrix}$$

$$D_{aw}^c = \begin{bmatrix} -5.2436 & -6.8113 & -6.5449 \\ -101.0102 & -124.1467 & -78.5259 \\ -35.4659 & -32.4052 & -146.2122 \\ -15.0439 & -19.7528 & -14.3537 \\ 0 & 0 & 0 \\ 0 & 0 & 0 \\ 0 & 0 & 0 \end{bmatrix}$$

which corresponds to the optimal value $\beta = 41.7292$. The time simulations of the closed-loop system without and with the anti-windup scheme are plotted on Figs. 8.17 and 8.18, respectively. They are initialized at $x_p(0) = [1500\,0\,0\,0]'$. This initial condition belongs to the region of asymptotic stability $\mathcal{E}(Q^{-1}, 1)$ of the closed-loop system involving the static anti-windup. It may be checked on the figure that the anti-windup action prevents the system from entering in a limit cycle, contrarily to the case without anti-windup.

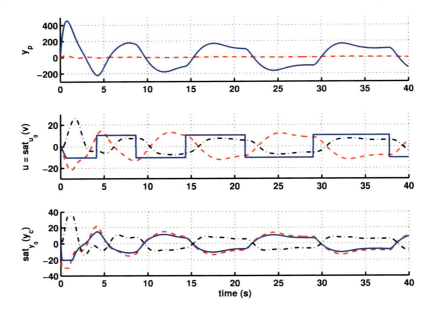

**Fig. 8.17** Example 8.6—F/A-18 HARV aircraft. Time evolution of the system output $y_p$, the system input $u$ and the vector $\text{sat}_{y_0}(y_c)$ for the closed-loop system initialized at $x_p(0) = [1500\ 0\ 0\ 0]'$. Case without anti-windup

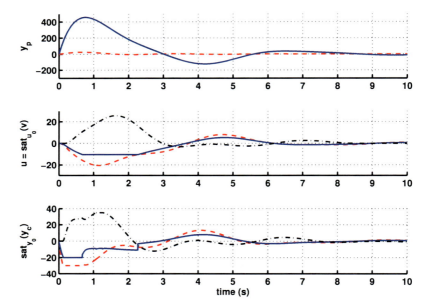

**Fig. 8.18** Example 8.6—F/A-18 HARV aircraft. Time evolution of the system output $y_p$, the system input $u$ and the vector $\text{sat}_{y_0}(y_c)$ for the closed-loop system initialized at $x_p(0) = [1500\ 0\ 0\ 0]'$. Case with anti-windup

## 8.4 Toward More Complex Nonlinear Actuators

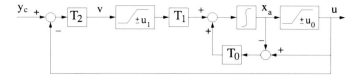

**Fig. 8.19** Model of an electro-hydraulic actuator

### 8.4.3 Electro-Hydraulic Actuator

Yet another case of nonlinear actuator consisted of several interconnected saturation elements is the electro-hydraulic actuator described in Fig. 8.19, which has been used as a benchmark actuator in [361]. The mechanical stops of the servomotor piston are at $u_0$ (stroke constraints), $u_1$ represents the flow constraints from the pilot servovalve (slew constraints), and $T_2$ is the finite, moderate gain of the P-controller. $T_0$ and $T_1$ are interconnection gains.

Such an actuator is described by

$$\begin{aligned}\dot{x}_a &= T_1 \operatorname{sat}_{u_1}(v) + T_0\big(\operatorname{sat}_{u_0}(x_a) - x_a\big) \\ v &= T_2\big(y_c - \operatorname{sat}_{u_0}(x_a)\big) \\ u &= \operatorname{sat}_{u_0}(x_a)\end{aligned} \qquad (8.17)$$

Taking into account the dead-zone elements $\phi_0(x_a)$ and $\phi_1(v)$, an anti-windup controller may be implemented following the same manipulations as presented or suggested in the previous sections. Thus, considering the extended state vector $x = [x_p'\ x_a'\ x_c'\ x_{\text{aw}}']' \in \Re^{n_p+m+n_c+n_{\text{aw}}}$, the closed-loop system is expressed by

$$\begin{aligned}\dot{x} &= \mathcal{A}x + \mathcal{B}_0\phi_0(x_a) + \mathcal{B}_1\phi_1(v) + \mathcal{B}_2 w \\ x_a &= \mathcal{C}_0 x \\ v &= \mathcal{C}_1 x + \mathcal{D}_{10}\phi_0(x_a) + \mathcal{D}_{11}\phi_1(v) + \mathcal{D}_{1w} w \\ z &= \mathcal{C}_2 x + \mathcal{D}_{20}\phi_0(x_a) + \mathcal{D}_{2w} w\end{aligned} \qquad (8.18)$$

with an anti-windup architecture as follows:

$$\begin{aligned}\dot{x}_{\text{aw}} &= A_{\text{aw}} x_{\text{aw}} + B_{\text{aw}}^0\big(\operatorname{sat}_{u_0}(x_a) - x_a\big) + B_{\text{aw}}^1\big(\operatorname{sat}_{u_1}(v) - v\big) \\ \begin{bmatrix}v_x \\ v_y\end{bmatrix} &= C_{\text{aw}} x_{\text{aw}} + D_{\text{aw}}^0\big(\operatorname{sat}_{u_0}(x_a) - x_a\big) + D_{\text{aw}}^1\big(\operatorname{sat}_{u_1}(v) - v\big)\end{aligned} \qquad (8.19)$$

Matrices which appear in (8.18) have the standard form already used in (8.6), but updated with

$$\mathbb{A} = \begin{bmatrix} A_p & B_{pu} & 0 \\ T_1 T_2 D_c C_p & T_1 T_2 (D_c D_{pu} - I_m) & T_1 T_2 C_c \\ B_c C_p & B_c D_{pu} & A_c \end{bmatrix}$$

$$B_v = \begin{bmatrix} 0 \\ T_1 T_2 [0 \; I_m] \\ [I_{n_c} \; 0] \end{bmatrix}$$

$$B_{\phi 0} = \begin{bmatrix} B_{pu} \\ T_1 T_2 (D_c D_{pu} - I_m) + T_0 \\ B_c D_{pu} \end{bmatrix}; \qquad B_{\phi 1} = \begin{bmatrix} 0 \\ T_1 \\ 0 \end{bmatrix} \qquad (8.20)$$

$$B_2 = \begin{bmatrix} B_{pw} \\ T_1 T_2 (D_c D_{pw} + D_{cw}) \\ B_{cw} + B_c D_{pw} \end{bmatrix}$$

$$C_0 = [0 \quad I_m \quad 0]; \qquad C_1 = [T_1 T_2 D_c C_p \quad T_2 D_c D_{pu} - T_2 \quad T_2 C_c]$$

$$C_{v0} = 0; \qquad C_{v1} = T_1 [0 \quad I_m]; \qquad D_1 = T_2 D_c D_{pu} - T_2$$

$$C_2 = [C_z \quad D_{zu} \quad 0]; \qquad \mathcal{D}_{2w} = D_{zw}$$

$$\mathcal{D}_{1w} = T_2 (D_{cw} + D_c D_{pw}); \qquad \mathcal{D}_{20} = D_{zu}$$

One can therefore derive updated versions of the propositions presented in Chap. 7. They are very close to those given for the case with position and rate saturations, except that in (8.8), the terms $\mathcal{D}_{00}$, $\mathcal{D}_{01}$, $\mathcal{D}_{0w}$ and $\mathcal{D}_{21}$ are equal to 0. These conditions are also partially given in [362] for a static anti-windup scheme (and with in addition sensor saturations).

*Example 8.7* (F-8 Aircraft MIMO Example (Continued)) The F-8 aircraft MIMO example presented in Example 8.4 is used another time to illustrate the role of this actuator regarding the stability of the closed-loop system, and the role of an anti-windup action to mitigate such saturation effects. System and controller dynamics are unchanged. The actuator dynamics is given by

$$T_0 = \begin{bmatrix} 10 & 0 \\ 0 & 10 \end{bmatrix}; \qquad T_1 = \begin{bmatrix} 4 & 0 \\ 0 & 4 \end{bmatrix}; \qquad T_2 = \begin{bmatrix} 25 & 0 \\ 0 & 25 \end{bmatrix}$$

This actuator dynamics is faster than the system dynamics so as to have a limited effect on the linear closed-loop dynamics of the system, with the bounds on the saturation elements:

$$u_0 = \begin{bmatrix} 15 \\ 15 \end{bmatrix}; \qquad u_1 = \begin{bmatrix} 20 \\ 20 \end{bmatrix}$$

The control problems are formulated as in Example 8.4, but now with the actuator described in (8.17). In a first stage, the control problem is to find the largest ellipsoidal set of admissible initial conditions, with the objective to stabilize the

## 8.4 Toward More Complex Nonlinear Actuators

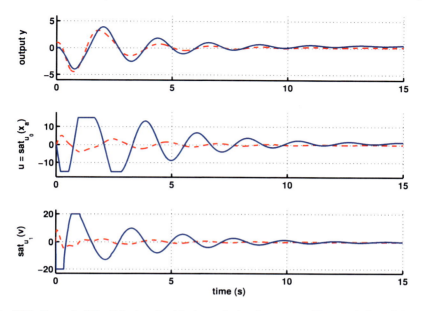

**Fig. 8.20** Example 8.7—F-8 aircraft with electro-hydraulic actuator. Time evolution of the output $y_p$, the input $u = \text{sat}_{u_0}(x_a)$ and $\text{sat}_{u_1}(v)$ of the closed-loop system with the second-order anti-windup compensator, initialized at $x_p(0) = [0\ 0\ 1\ 1]'$. Case without disturbance

system initialized, as previously, at $x_p(0) = [0\ 0\ 1\ 1]'$. Without anti-windup, the initial state cannot be stabilized. Although not belonging to the ellipsoidal set issued from the optimization procedure, the initial state is transferred to 0 when a static anti-windup is employed. Finally, a reduced-order anti-windup may be evaluated, based on a fixed-dynamics strategy where the matrices $A_{\text{aw}}$ and $C_{\text{aw}}$ are those that had been considered in the case of only magnitude saturation actuator (8.1). The solution of the fixed-dynamics synthesis in the regional stability context is

$$B_{\text{aw}}^0 = \begin{bmatrix} -356.2287 & -53.2496 \\ -6.9050 & 2.3313 \end{bmatrix}; \quad B_{\text{aw}}^1 = \begin{bmatrix} -0.0001 & -0.0002 \\ -0.0000 & -0.0000 \end{bmatrix}$$

$$D_{\text{aw}}^0 = \begin{bmatrix} 133.3896 & 16.6304 \\ -12501.0979 & -1138.4453 \\ 55.5195 & 5.4674 \\ -22.5448 & -2.7121 \\ 1.0184 & -0.0081 \\ -0.1103 & 1.0802 \end{bmatrix}; \quad D_{\text{aw}}^1 = \begin{bmatrix} 0.0000 & 0.0001 \\ -0.0035 & -0.0078 \\ 0.0000 & 0.0000 \\ -0.0000 & -0.0000 \\ -0.0400 & -0.0000 \\ 0.0000 & -0.0400 \end{bmatrix}$$

with $\beta = 1.9027$, to be compared to $\beta = 0.2042$ in the case without anti-windup. A simulation of the closed-loop system with the reduced-order anti-windup compensator, initialized at $x_p(0) = [0\ 0\ 1\ 1]'$, is plotted in Fig. 8.20. It has to be noticed that only the fifteen first seconds are plotted to show the initial transitory behav-

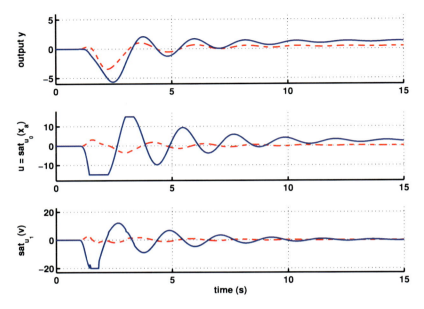

**Fig. 8.21** Example 8.7—F-8 aircraft. Time evolution of the output $y_p$ and the input $u$ of the closed-loop system with a second-order anti-windup compensator, initially at rest ($x(0) = 0$), in the presence of a disturbance $\delta = 0.05$ at $t = 1$ second on the first disturbance input

ior. The output and input signals then (slowly) converge toward 0 in about thirty seconds.

The problem of disturbance attenuation is now investigated, considering the case where a disturbance of limited energy $\|w\|_2^2 \leq 20$ ($\delta = 0.05$) acts on the system. The only feasible case is that one where a dynamic anti-windup is considered. Based on the matrices $A_{aw}$ and $C_{aw}$ issued from the case with saturation (8.2), the solution of the fixed-dynamics synthesis in the performance context is

$$B_{aw}^0 = \begin{bmatrix} 3919.1259 & 137.8012 \\ -1728.9321 & -211.8088 \end{bmatrix}; \quad B_{aw}^1 = \begin{bmatrix} 0.0000 & 0.0000 \\ -0.0000 & -0.0000 \end{bmatrix}$$

$$D_{aw}^0 = \begin{bmatrix} 23.8233 & -4.8793 \\ -3912.7866 & 847.6312 \\ 15.3991 & -2.9896 \\ -3.4997 & 0.7040 \\ 0.9938 & 0.0166 \\ -0.6859 & 1.2515 \end{bmatrix}; \quad D_{aw}^1 = \begin{bmatrix} 0.0000 & -0.0000 \\ -0.0000 & 0.0001 \\ 0.0000 & -0.0000 \\ -0.0000 & 0.0000 \\ -0.0400 & 0.0000 \\ -0.0000 & -0.0400 \end{bmatrix}$$

The simulation of the system, facing a disturbance given by $\delta$ defined with $w(t) = [10 \ 0]'$ during 0.2 seconds, occurring at time $t = 1$ second, is plotted in Fig. 8.21.

## 8.5 Pseudo Anti-windup Strategies when the Dead-Zone Element is not Accessible

Another important feature to be taken into account when dealing with an anti-windup strategy is the accessibility of the signals on both sides (input and output) of the saturation elements. In the case of an actuator involving a single magnitude saturation block, it is legitimate to consider that both signals are accessible (controlled output of the controller, from one side, and actual input of the plant, on the other side). In the case of more complex actuator, this hypothesis has to be reconsidered. As suggested in Remark 8.1, typical cases of this situation may appear as for example:

- When the internal state of the nonlinear actuator cannot be measured: such a case is briefly presented in Sect. 8.5.1.
- In the case of sensor saturation, for which it is not realistic to assume that the actual plant output is known; only its saturated version is known. This issue is discussed in Sect. 8.5.2.

### *8.5.1 Anti-windup Scheme Using a Fictitious Linear Element*

Until now, it has been considered that decentralized dead-zone nonlinearities could be used to directly manipulate the saturation terms. Implicitly it has been assumed that both the input and the output signals of each saturation elements are available. On the other hand, it has been suggested that, when dealing with rate saturation, for example, an equivalent saturation model could be used to manipulate only magnitude saturation elements. In such a case, it may be considered that the rate-limiter block input and output are known, but not the signal inside this block. To overcome this problem, a fictitious linear actuator can be used to evaluate the occurrence as well as the level of saturations in the true actuator. The basic idea in this case consists in injecting the signal issued from the difference between the nonlinear actuator output and the fictitious linear one at the output of the controller. This is the case which has been considered in [295].

Let us consider again the case of position and rate saturation actuator, as described in Fig. 8.12. To compare the output of the nonlinear dynamic actuator with a fictitious linear actuator (same dynamics, but without saturations), the state vector is now augmented with the fictitious linear dynamics, or, equivalently, with the additional state $e(t) = x_a(t) - x_{al}(t)$, where $x_{al}$ is the state of the fictitious linear actuator. The error dynamics is given by

$$\dot{e} = \phi_1(v) + T_0\phi_0(y_c) - T_0 e \tag{8.21}$$

and the direct linear anti-windup (DLAW) architecture (7.8), which cannot anymore directly manipulate the dead-zone elements, is then related to this error $e$ by

$$\dot{x}_{\text{aw}} = A_{\text{aw}} x_{\text{aw}} + B^e_{\text{aw}} e$$

$$\begin{bmatrix} v_x \\ v_y \end{bmatrix} = C_{\text{aw}} x_{\text{aw}} + D^e_{\text{aw}} e \tag{8.22}$$

Then, denoting the augmented state vector $x = [\, x'_p \; x'_a \; x'_c \; e' \; x'_{\text{aw}} \,]' \in \Re^n$, $n = n_p + m + n_c + m + n_{\text{aw}}$, the closed-loop system (7.13)–(7.15) can be written in an almost standard form:

$$\dot{x} = \mathcal{A}x + \mathcal{B}_0 \phi_0(y_c) + \mathcal{B}_1 \phi_1(v) + \mathcal{B}_2 w$$

$$y_c = \mathcal{C}_0 x + \mathcal{D}_{0w} w$$

$$v = \mathcal{C}_1 x + \mathcal{D}_{10} \phi_0(y_c) + \mathcal{D}_{1w} w \tag{8.23}$$

$$z = \mathcal{C}_2 x + \mathcal{D}_{2w} w$$

where the matrices are updated as follows:

$$\mathcal{A} = \begin{bmatrix} \mathbb{A} & B_v D^e_{\text{aw}} & B_v C_{\text{aw}} \\ 0 & -T_0 & 0 \\ 0 & B^e_{\text{aw}} & A_{\text{aw}} \end{bmatrix}$$

$$\mathcal{B}_0 = \begin{bmatrix} 0 \\ T_0 \\ 0 \\ T_0 \\ 0 \end{bmatrix}; \qquad \mathcal{B}_1 = \begin{bmatrix} 0 \\ I_m \\ 0 \\ I_m \\ 0 \end{bmatrix}; \qquad \mathcal{B}_2 = \begin{bmatrix} B_{pw} \\ T_0 D_c D_{pw} + T_0 D_{cw} \\ B_{cw} + B_c D_{pw} \\ 0 \\ 0 \end{bmatrix} \tag{8.24}$$

$$\mathcal{C}_0 = [\, C_0 \;\; C_{v0} D^e_{\text{aw}} \;\; C_{v0} C_{\text{aw}} \,]; \qquad \mathcal{C}_1 = [\, C_1 \;\; C_{v1} D^e_{\text{aw}} \;\; C_{v1} C_{\text{aw}} \,]$$

$$\mathcal{D}_{10} = T_0; \qquad \mathcal{C}_2 = [\, C_2 \;\; 0 \;\; 0 \,]$$

with matrices $\mathbb{A}$, $B_v$, $C_0$, $C_1$, $C_{v0}$, $C_{v1}$, $C_2$, $\mathcal{D}_{0w}$, $\mathcal{D}_{1w}$ and $\mathcal{D}_{2w}$ unchanged with respect to (8.7).

At this point, it should be observed that all the elements of the anti-windup scheme now appear in matrices $\mathcal{A}$, $\mathcal{C}_0$ and $\mathcal{C}_1$. Then the conditions extracted from Proposition 7.1 in the global context, or in Proposition 7.4 in the local context are now nonlinear for the dynamic anti-windup scheme (as discussed in Chap. 7) as well as for the static anti-windup scheme, due to the non-direct products of anti-windup matrices with $Q$. Any relaxation procedure may be used to get over this problem. For example, the following algorithm, similar to the one proposed in Sect. 2.2.5, can be considered.

**Algorithm 8.1** Fixed-dynamics synthesis—region of stability (case of a fictitious linear actuator)

## 8.5 Pseudo Anti-windup Strategies when the Dead-Zone Element 341

1. Initialize the anti-windup matrices $A_{aw}$, $C_{aw}$, $B_{aw}$ and $D_{aw}$. Typically, we consider that $A_{aw}$ and $C_{aw}$ are a priori given, while $B_{aw}$ and $D_{aw}$ may be either a priori given or set to $0^6$.
2. Solve the following optimization problem for $Q$ and $\delta$:

$$\min \; f\left(\mathcal{E}(Q^{-1}, \delta)\right)$$

subject to inequalities (8.8)–(8.10) with updated matrices (8.24)

3. Keep the previous value of $Q$, solve the following problem for $\delta$, $B_{aw}$ and $D_{aw}$

$$\min \; \delta$$

subject to inequalities (8.8)–(8.10) with updated matrices (8.24)

4. Go to step 2 until no significant change on the size of the ellipsoid $\mathcal{E}(Q^{-1}, \delta)$ is obtained. When no significant change arises then stop.

*Example 8.8* (F-8 Aircraft MIMO Example (Last Time)) To illustrate the potential interest of such an anti-windup-inspired strategy, let us consider again the example of the F-8 aircraft used in Example 8.4 with a position saturation only, but also in Example 8.5 in the case of a position and rate saturation actuator. Let us assume now that the saturation elements are no more accessible. Then considering the pseudo anti-windup strategy proposed in this section, a reduced-order dynamic anti-windup is evaluated to feed back the error between the nonlinear actuator and its fictitious linear version. An iterative procedure is employed to evaluate $B_{aw}$ and $D_{aw}$, with, on the other hand, $A_{aw}$ and $C_{aw}$ fixed to their values given by (8.1). Using Algorithm 8.1, one obtains, after 39 iterations:

$$B_{aw}^e = \begin{bmatrix} -72.9480 & -14.6804 \\ 8.9834 & 8.7547 \end{bmatrix}; \quad D_{aw}^e = \begin{bmatrix} 4.7664 & 1.8561 \\ -66.0585 & -20.4844 \\ 1.5848 & 0.6052 \\ -1.1264 & -0.8093 \\ -0.7132 & -0.1320 \\ 0.0291 & 0.0504 \end{bmatrix}$$

with the scaling factor for $\mathcal{X}_0$ given by $\beta = 0.2128$, to be compared with the values $\beta = 0.1388$ without anti-windup, and $\beta = 0.6305$ with the reduced-order anti-windup computed in Example 8.5. An admissible initial condition is, for example, $x_p(0) = [0\ 0\ 0.239\ 0.239]'$. It may be checked in Fig. 8.22 that, without anti-windup action, this initial state cannot be brought back to 0. At the contrary, the pseudo anti-windup strategy allows us to stabilize the system (see the output and input of the system in Fig. 8.23 and the saturated signals $\text{sat}_{u_1}(v)$ and $\text{sat}_{u_0}(y_c)$ in Fig. 8.24).

---

[6]It has to be noticed that $A_{aw}$ and $C_{aw}$ could also be computed in this iterative procedure, although the fact to keep them fixed a priori drastically reduces the computational burden without much degrading the optimal solution (in terms of the size of region of asymptotic stability).

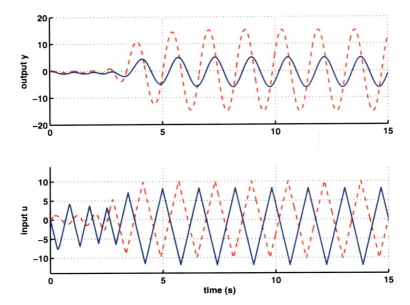

**Fig. 8.22** Example 8.8—F-8 aircraft with position and rate limiter. Time evolution of the output $y_p$ and the input $u$ of the closed-loop system without anti-windup, initialized at $x_p(0) = [0\ 0\ 0.239\ 0.239]'$

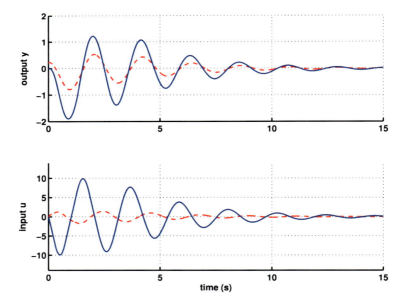

**Fig. 8.23** Example 8.8—F-8 aircraft with position and rate limiter. Time evolution of the output $y_p$ and the input $u$ of the closed-loop system with the pseudo anti-windup compensator based on the error between the nonlinear actuator and its fictitious linear version, initialized at $x_p(0) = [0\ 0\ 0.239\ 0.239]'$

8.5 Pseudo Anti-windup Strategies when the Dead-Zone Element 343

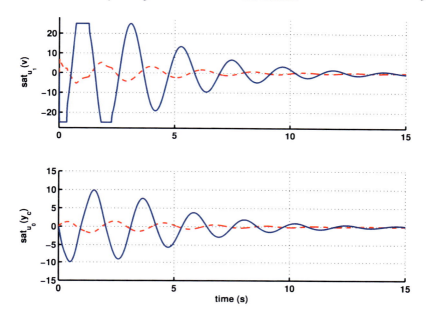

**Fig. 8.24** Example 8.8—F-8 aircraft with position and rate limiter. Time evolution of the saturated signals $\text{sat}_{u_1}(v)$ and $\text{sat}_{u_0}(y_c)$ of the closed-loop system with the pseudo anti-windup compensator based on the error between the nonlinear actuator and its fictitious linear version, initialized at $x_p(0) = [0\ 0\ 0.239\ 0.239]'$

Finally, the limits of this strategy with respect to a true position and rate anti-windup strategy is shown in Fig. 8.25 which presents the response of the system initialized at $x_p(0) = [0\ 0\ 0.69\ 0.69]'$, the other components of the augmented state being equal to 0. This initial state belonged to the region of asymptotic stability computed when using the position and rate anti-windup strategy (see Fig. 8.14), but it does not belong to the current region of asymptotic stability. Although not unstable, the time evolution of the system is now strongly influenced by the nonlinear actuator, and the use of a fictitious linear model is no more efficient enough to counteract the effect of these nonlinear elements. It shows, however, the potential interest of such pseudo anti-windup schemes when the nonlinear elements are hidden inside the actuator.

### 8.5.2 Observer-Based Anti-windup Scheme

Furthermore, when signals to be used by the anti-windup scheme are unaccessible, another alternative may be to consider the addition of an observer. This is particularly useful in the case of sensor saturation, where it is obviously inconceivable to contemplate the accessibility of the plant output (or sensor input), such as under-

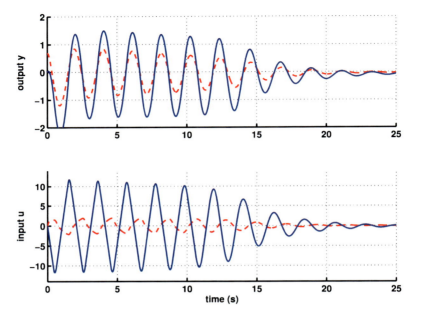

**Fig. 8.25** Example 8.8—F-8 aircraft with position and rate limiter. Time evolution of the output $y_p$ and the input $u$ of the closed-loop system with the pseudo anti-windup compensator based on the error between the nonlinear actuator and its fictitious linear version, initialized at $x_p(0) = [0\ 0\ 0.69\ 0.69]'$

lined in Sect. 7.6.2. The object of this section is then to show, via an illustrative example, the potential use of an observer-based anti-windup strategy.

Following the notation previously used in Chap. 7, the interconnection of systems (7.1) and (7.2) is now[7]

$$u = \mathrm{sat}_{u_0}(y_c), \qquad u_c = \mathrm{sat}_{y_0}(y_p) \tag{8.25}$$

considering also that $u$, $y_c$ and $u_c$ are accessible, On the other hand, $y_p$, the true output of the plan, is not accessible, which makes necessary to use a state observer. Various architectures can be employed, several of them being discussed in [374]. As an example, one considers here an architecture which manipulates the difference between the measured saturated output and the estimated unsaturated output:

$$\begin{aligned}\dot{\hat{x}}_p &= A_p\hat{x}_p + B_{pu}u + L\bigl(\mathrm{sat}_{y_0}(y_p) - \hat{y}_p\bigr) \\ \hat{y}_p &= C_p\hat{x}_p + D_{pu}u\end{aligned} \tag{8.26}$$

---

[7] Any other nonlinear actuators such as those presented in the sections above could also be considered within the same framework.

8.5 Pseudo Anti-windup Strategies when the Dead-Zone Element 345

This is not the most intuitive architecture but it seems more efficient than other options [374]. Equivalently, considering the error vector $\varepsilon = \hat{x}_p - x_p$, the error dynamics is described by

$$\dot{\varepsilon} = (A_p - LC_p)\varepsilon + L\phi_{y_0}(y_p) + LD_{pw}w - B_{pw}w$$

$$\hat{y}_p = C_p\varepsilon + C_p x_p + D_{pu}u \tag{8.27}$$

with the dead-zone elements

$$\phi_{y_0}(y_p) = \mathrm{sat}_{y_0}(y_p) - y_p$$

$$\phi_{u_0}(y_c) = \mathrm{sat}_{u_0}(y_c) - y_c \tag{8.28}$$

The direct linear anti-windup compensator (7.8) is then driven by $\mathrm{sat}_{u_0}(y_c) - y_c$ and by $\mathrm{sat}_{y_0}(y_p) - \hat{y}_p$:

$$\dot{x}_{\mathrm{aw}} = A_{\mathrm{aw}}x_{\mathrm{aw}} + B^u_{\mathrm{aw}}\big(\mathrm{sat}_{u_0}(y_c) - y_c\big) + B^y_{\mathrm{aw}}\big(\mathrm{sat}_{y_0}(y_p) - \hat{y}_p\big)$$

$$\begin{bmatrix} v_x \\ v_y \end{bmatrix} = C_{\mathrm{aw}}x_{\mathrm{aw}} + D^u_{\mathrm{aw}}\big(\mathrm{sat}_{u_0}(y_c) - y_c\big) + D^y_{\mathrm{aw}}\big(\mathrm{sat}_{y_0}(y_p) - \hat{y}_p\big) \tag{8.29}$$

or, equivalently, it is driven by $\phi_{u_0}(y_c)$ and, indirectly, by $\phi_{y_0}(y_p)$ as follows:

$$\dot{x}_{\mathrm{aw}} = A_{\mathrm{aw}}x_{\mathrm{aw}} + B^u_{\mathrm{aw}}\phi_{u_0}(y_c) + B^y_{\mathrm{aw}}\big(\phi_{y_0}(y_p) - C_p\varepsilon + D_{pw}w\big)$$

$$\begin{bmatrix} v_x \\ v_y \end{bmatrix} = C_{\mathrm{aw}}x_{\mathrm{aw}} + D^u_{\mathrm{aw}}\phi_{u_0}(y_c) + D^y_{\mathrm{aw}}\big(\phi_{y_0}(y_p) - C_p\varepsilon + D_{pw}w\big) \tag{8.30}$$

The same machinery may be employed to build the augmented closed-loop system with state vector $x = [x'_p \; x'_c \; \varepsilon' \; x'_{\mathrm{aw}}]' \in \Re^n$, $n = n_p + n_c + n_p + n_{\mathrm{aw}}$. It is then possible to extract a set of matrix inequalities conditions which give a solution to the anti-windup problem, expressed in terms of the region of stability (in the local context), admissible disturbance and/or $\mathcal{L}_2$-performance, anti-windup elements and observer $L$. The closed-loop system which is manipulated, extended from the standard expression stated in Sect. 7.3, is described by

$$\dot{x} = \mathcal{A}x + \mathcal{B}_0\phi_{u_0}(y_c) + \mathcal{B}_1\phi_{y_0}(y_p) + \mathcal{B}_2 w$$

$$y_c = \mathcal{C}_0 x + \mathcal{D}_{00}\phi_{u_0}(y_c) + \mathcal{D}_{01}\phi_{y_0}(y_p) + \mathcal{D}_{02}w$$

$$y_p = \mathcal{C}_1 x + \mathcal{D}_{10}\phi_{u_0}(y_c) + \mathcal{D}_{11}\phi_{y_0}(y_p) + \mathcal{D}_{12}w \tag{8.31}$$

$$z = \mathcal{C}_2 x + \mathcal{D}_{20}\phi_{u_0}(y_c) + \mathcal{D}_{21}\phi_{y_0}(y_p) + \mathcal{D}_{22}w$$

where the matrices are updated as follows:

$$\mathcal{A} = \begin{bmatrix} \mathbb{A} & -B_v D^y_{\mathrm{aw}} C_p & B_v C_{\mathrm{aw}} \\ 0 & A_p - LC_p & 0 \\ 0 & -B^y_{\mathrm{aw}} C_p & A_{\mathrm{aw}} \end{bmatrix}; \quad \mathcal{B}_0 = \begin{bmatrix} B_\phi + B_v D^u_{\mathrm{aw}} \\ 0 \\ B^u_{\mathrm{aw}} \end{bmatrix}$$

$$\mathcal{B}_1 = \begin{bmatrix} B_1 + B_v D_{\mathrm{aw}}^y \\ L \\ B_{\mathrm{aw}}^y \end{bmatrix}; \qquad \mathcal{B}_2 = \begin{bmatrix} B_2 \\ L D_{pw} - B_{pw} \\ B_{\mathrm{aw}}^y D_{pw} \end{bmatrix}$$

$$\mathcal{C}_0 = [\, C_0 \quad -C_{v0} D_{\mathrm{aw}}^y C_p \quad C_{v0} C_{\mathrm{aw}} \,]; \qquad \mathcal{D}_{00} = D_{00} + C_{v0} D_{\mathrm{aw}}^u \qquad (8.32)$$

$$\mathcal{D}_{01} = D_{01} + C_{v0} D_{\mathrm{aw}}^y; \qquad \mathcal{D}_{02} = D_{02} + C_{v0} D_{\mathrm{aw}}^y D_{pw}$$

$$\mathcal{C}_1 = [\, C_1 \quad -C_{v1} D_{\mathrm{aw}}^y C_p \quad C_{v1} C_{\mathrm{aw}} \,]; \qquad \mathcal{D}_{10} = D_{10} + C_{v1} D_{\mathrm{aw}}^u$$

$$\mathcal{D}_{11} = D_{11} + C_{v1} D_{\mathrm{aw}}^y; \qquad \mathcal{D}_{12} = D_{12} + C_{v1} D_{\mathrm{aw}}^y D_{pw}$$

$$\mathcal{C}_2 = [\, C_2 \quad -C_{v2} D_{\mathrm{aw}}^y C_p \quad C_{v2} C_{\mathrm{aw}} \,]; \qquad \mathcal{D}_{20} = D_{20} + C_{v2} D_{\mathrm{aw}}^u$$

$$\mathcal{D}_{21} = D_{21} + C_{v2} D_{\mathrm{aw}}^y; \qquad \mathcal{D}_{22} = D_{22} + C_{v2} D_{\mathrm{aw}}^y D_{pw}$$

with matrices $\mathbb{A}$, $B_\phi$, $B_v$, $B_2$, $C_2$ and $C_{v2}$ unchanged with respect to (7.15). The other matrices are given by

$$B_1 = \begin{bmatrix} B_{pu} \Delta^{-1} D_c \\ B_c + B_c D_{pu} \Delta^{-1} D_c \end{bmatrix}$$

$$C_0 = [\, \Delta^{-1} D_c C_p \quad \Delta^{-1} C_c \,]; \qquad C_{v0} = \Delta^{-1} [\, 0 \quad I_m \,]$$

$$C_1 = [\, C_p + D_{pu} \Delta^{-1} D_c C_p \quad D_{pu} \Delta^{-1} C_c \,]; \qquad C_{v1} = D_{pu} \Delta^{-1} [\, 0 \quad I_m \,]$$

$$D_{00} = \Delta^{-1} D_c D_{pu}; \qquad D_{01} = \Delta^{-1} D_c$$

$$D_{02} = \Delta^{-1} (D_c D_{pw} + D_{cw}) \qquad (8.33)$$

$$D_{10} = D_{pu} + D_{pu} \Delta^{-1} D_c D_{pu}; \qquad D_{11} = D_{pu} \Delta^{-1} D_c$$

$$D_{12} = D_{pu} \Delta^{-1} (D_{cw} + D_c D_{pw}) + D_{pw}$$

$$D_{20} = D_{zu} \Delta^{-1} D_c D_{pu} + D_{zu}; \qquad D_{21} = D_{zu} \Delta^{-1} D_c$$

$$D_{22} = D_{zu} \Delta^{-1} (D_{cw} + D_c D_{pw}) + D_{zw}$$

An updated version of Propositions 7.1 and 7.4 may then be directly written. Regarding the regional (local) context, conditions (8.8)–(8.10) directly apply, but with the updated versions of the matrices. In particular, it should be noted that the observer gain $L$, as well as the anti-windup matrices $A_{\mathrm{aw}}$, $B_{\mathrm{aw}}^y$, $C_{\mathrm{aw}}$ and $D_{\mathrm{aw}}^y$ are encapsulated in matrix $\mathcal{A}$. It signifies that these conditions are not LMIs and, as previously discussed, iterative procedures including, or not, Finsler's transformations, or a priori given values for these elements have to be considered. This approach is yet complicated to use, but it may be of interest to manipulate the conditions in an analysis context, i.e. considering gains matrices obtained a priori. This is illustrated in the example which follows.

## 8.5 Pseudo Anti-windup Strategies when the Dead-Zone Element 347

*Example 8.9* (F/A-18 HARV Aircraft Example (Continued)) To illustrate this section, let us consider again the F/A-18 HARV aircraft lateral dynamics described in Example 8.6. The open-loop system (8.15) and dynamic controller (8.16) are unchanged. On the other hand, the system is interconnected with a saturation actuator ($u = \mathrm{sat}_{u_0}(y_c)$) and a sensor saturation ($u_c = \mathrm{sat}_{y_0}(y_p)$) as above described, which bounds are given by

$$
u_0 = \begin{bmatrix} 25 \\ 10.5 \\ 30 \end{bmatrix}; \qquad y_0 = \begin{bmatrix} 30 \\ 20 \end{bmatrix}
$$

To simplify the computation of the anti-windup compensator and state observer $L$, we consider the following procedure.

- A state observer $L$ is constructed a priori, according to (8.26). The system being open-loop asymptotically stable, the observer may be just a copy of the plant ($L = 0$), as suggested by [284]. It is, however, not recommended in the current case with a very slow dominant mode (and poorly damped other ones), unless the initial state of the observer is accurately known. We then consider a pole-placement of the observer in $\{-1 \pm 0.5j; \ -1.5 \pm 0.5j\}$, close to the poles of the closed-loop system, to conceal this observer in the whole closed-loop system:

$$
L = \begin{bmatrix} -15.5809 & 0.8951 \\ 1.9944 & 0.0001 \\ 1.7022 & 0.0684 \\ -0.3482 & 0.6846 \end{bmatrix} \tag{8.34}
$$

- A static anti-windup is constructed under the hypothesis that both input and output signals of the sensor actuator are accessible. It signifies that $C_p \varepsilon - D_{pw} w = 0$ in (8.32) and that the block matrices related to $\varepsilon$ disappear in (8.33). This allows one to solve a convex optimization problem (min $f(\mathcal{E}(Q^{-1}, \delta))$) where the matrix $Q$ and the anti-windup gains $D_{\mathrm{aw}}^u$ and $D_{\mathrm{aw}}^y$ linearly appear. One obtains

$$
D_{\mathrm{aw}}^u = \begin{bmatrix} -0.0900 & -0.0024 & -0.2632 \\ -1.8561 & -0.2319 & -6.3646 \\ 2.6630 & 1.4390 & 4.7221 \\ -0.3759 & -0.0491 & -1.0669 \\ -0.0587 & -0.7369 & 0.0850 \\ -0.7641 & -0.1200 & 0.1657 \\ 0.2085 & 0.3164 & -0.0643 \end{bmatrix}; \qquad D_{\mathrm{aw}}^y = \begin{bmatrix} 0.1643 & 0.0510 \\ 5.2206 & 0.0009 \\ 22.4190 & -0.0643 \\ 0.2487 & 0.1640 \\ 1.5327 & 0.0930 \\ 0.4369 & 0.1345 \\ 1.6762 & -0.1880 \end{bmatrix}
$$

- Considering the static anti-windup and observer gains above constructed, it is now possible to solve the observer-based anti-windup analysis problem associated to the closed-loop system (8.31), involving the static part of the anti-windup scheme (8.30). The optimization problem related to the size of the region of

348                                                    8  Applications of Anti-windup Techniques

**Table 8.3** Example 8.9—F/A-18 HARV aircraft. Comparison of the enlargement of the region of asymptotic stability for various anti-windup schemes. $\beta$ represents the maximal scaling factor of the shape set $\mathcal{X}_0$ formed by vertices $v_r$, such that $\beta \mathcal{X}_0 \subset \mathcal{E}(Q^{-1}, 1)$

| Case | Strategy | $\beta$ |
|------|----------|---------|
| No. 1 | Analysis (without anti-windup) | 33.3443 |
| No. 2 | Static anti-windup synthesis of $D_{aw}^u$ and $D_{aw}^y$ (with $y_p$ available) | 1423.7367 |
| No. 3 | Static anti-windup analysis ($D_{aw}^u$ only, $D_{aw}^y = 0$) | 63.4284 |
| No. 4 | Static anti-windup analysis (including the observer $L$) | 71.5649 |
| No. 5 | Static anti-windup analysis (including the observer $L = 0$ (copy of the plant)) | 176.6561 |

asymptotic stability with this anti-windup scheme is linear in the decision variables.

Table 8.3 summarizes the results obtained in different cases, compared in terms of the maximal scaling factor $\beta$ of the shape set $\mathcal{X}_0$ formed by vertices $v_r$, such that $\beta \mathcal{X}_0 \subset \mathcal{E}(Q^{-1}, 1)$. It can be noticed that the choice of the observer (cases no. 4 and no. 5) has a significant influence about the scaling factor $\beta$. On the other hand, one cannot expect to obtain regions of asymptotic stability in the case of an observer-based anti-windup scheme (cases no. 4 and no. 5) as large as in the case where all the signals used in the anti-windup are available (case no. 2). On the other hand, this example illustrates that the use of an observer-based anti-windup scheme may allow us to enlarge the region of asymptotic stability with respect to the region which would be obtained with an anti-windup scheme manipulating only the actuator dead zone (case no. 3).

The output and input time evolutions of the closed-loop system for several simulation cases are plotted in the figures which follow. The system is initially in $x_p(0) = [0\ 0\ 80\ 180]'$. Figure 8.26 corresponds to the case without anti-windup. Figure 8.27 corresponds to the ideal case where both the input and output signals of the saturation elements are measured. In Fig. 8.28, the information about the sensor saturation is not used in the anti-windup compensator (case no. 3).

Finally, Figs. 8.29 and 8.30 correspond to the case of the observer-based anti-windup scheme, with $L$ given in (8.34) for two different initializations of the observer state, $\hat{x}_p(0) = 0.9\ x_p(0)$ and $\hat{x}_p(0) = 0.5\ x_p(0)$, respectively. It may be noticed that the use of the observer allows one to improve the closed-loop behavior of the system with respect to the time response of the case where the anti-windup only manipulates the actuator dead zone (Fig. 8.28). On the other hand, the performance of the observer-based anti-windup is yet far from what can be expected if all the signals were accessible (Fig. 8.27). It may also be noticed that the closer to the true state $x_p$ is the initial state of the observer, the better is the performance (in terms of time response) of the observer-based anti-windup scheme (Fig. 8.29 versus Fig. 8.30).

8.5  Pseudo Anti-windup Strategies when the Dead-Zone Element         349

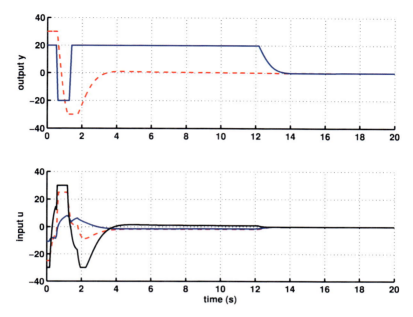

**Fig. 8.26** Example 8.9—F/A-18 HARV aircraft with actuator and sensor saturation. Time evolution of the output $\text{sat}_{y_0}(y_p)$ and the input $u$ of the closed-loop system without anti-windup, initialized at $x_p(0) = [0\ 0\ 80\ 180]'$

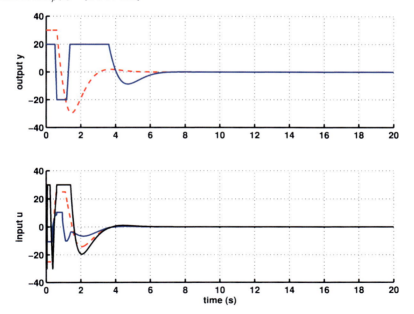

**Fig. 8.27** Example 8.9—F/A-18 HARV aircraft with actuator and sensor saturation. Time evolution of the output $\text{sat}_{y_0}(y_p)$ and the input $u$ of the closed-loop system with anti-windup, considering that $y_p$ is accessible, initialized at $x_p(0) = [0\ 0\ 80\ 180]'$

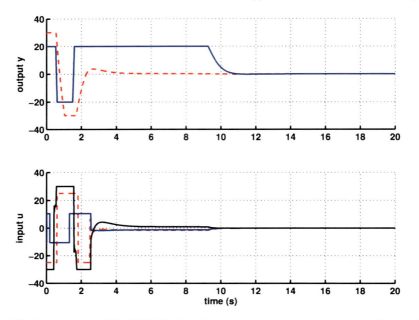

**Fig. 8.28** Example 8.9—F/A-18 HARV aircraft with actuator and sensor saturation. Time evolution of the output $\text{sat}_{y_0}(y_p)$ and the input $u$ of the closed-loop system with anti-windup only related to the actuator saturation, initialized at $x_p(0) = [0\ 0\ 80\ 180]'$

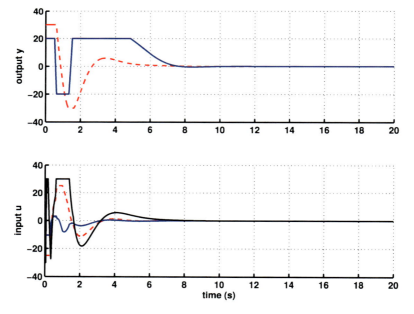

**Fig. 8.29** Example 8.9—F/A-18 HARV aircraft with actuator and sensor saturation. Time evolution of the output $\text{sat}_{y_0}(y_p)$ and the input $u$ of the closed-loop system with the observer-based anti-windup compensator, initialized at $x_p(0) = [0\ 0\ 80\ 180]'$, $\hat{x}_p(0) = 0.9 x_p(0)$

## 8.6 Conclusion

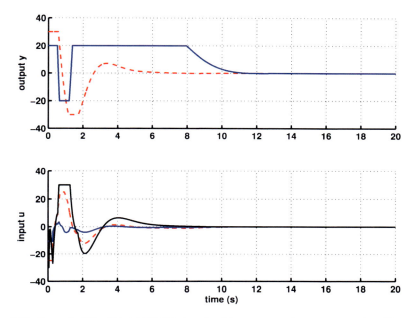

**Fig. 8.30** Example 8.9—F/A-18 HARV aircraft with actuator and sensor saturation. Time evolution of the output $\text{sat}_{y_0}(y_p)$ and the input $u$ of the closed-loop system with the observer-based anti-windup compensator, initialized at $x_p(0) = [0\ 0\ 80\ 180]'$, $\hat{x}_p(0) = 0.5x_p(0)$

## 8.6 Conclusion

This chapter has presented several examples with the common objective to illustrate the effectiveness but also the drawbacks of the theoretical conditions developed in Chap. 7. Academic examples but also examples closer to practical cases and borrowed from the literature (like for example the cart–spring–pendulum system and the F-8 aircraft MIMO system) were deeply treated and commented. Beyond the illustration of the results of Chap. 7, systems with more complex actuators, including nested saturations, were considered to show how the framework of Chap. 7 can be slightly modified to deal with such systems. Static and (full-order and fixed-order) dynamic anti-windup compensators were computed for stability or performance purposes. These examples have allowed us to highlight the potential gain in terms of enlargement of the region of stability or performance due to the anti-windup action. We have also briefly discussed some alternative uses of the anti-windup techniques, in particular when the input and/or output signals of a saturated element are not accessible.

Non-standard anti-windup strategies can be investigated, which employ the same ideas of additional loops manipulating dead-zone elements, but in different control contexts. For example, a non-standard use of anti-windup has been developed in order to improve the on-ground control system of a civilian aircraft thanks to an original approximation of the nonlinear ground forces by saturation-type nonlinearities [305].

We have not illustrated the MRAW approach considered in Sect. 7.4 of Chap. 7, since detailed examples can be found in [406]. An application in the aerospace context and more especially in the formation-flying missions problem can also be found in [380].

# Appendix A
# Some Concepts Related to Stability Theory

## A.1 Introduction

Stability theory plays a key role in the analysis of dynamical systems and in control design problems. In this sense, different kinds of stability definitions arise. Associated to these definitions, there are important concepts such as the notion of equilibrium point and region of attraction, as well as some important theoretical results that provide the basis for the analysis and synthesis methods in control system theory.

The objective of the appendix is then to recall some elements of the stability theory of dynamic systems, in particular, the Lyapunov theory. These elements constitute a necessary background for the results to be presented throughout this book concerning problems of robust stability and stabilization of linear systems subject to saturations. Moreover, as discussed in Chap. 1, we see that when the limits of actuators are taken into account, it is necessary to characterize the set of initial conditions for which some important properties, like stability for example, are guaranteed. These sets are often determined using Lyapunov functions [130, 170, 290]. In the context of systems described by linear state space models, the considered Lyapunov functions may quadratic or polyhedral leading to ellipsoidal or polyhedral sets [130, 170]. These sets play a fundamental role and some of their properties are also presented in Appendix C.

The material presented in this appendix is organized as follows. Section A.2 presents the characterization of the stability for a linear time-invariant system via the location of the eigenvalues of the dynamic matrix. Section A.3 deals with the stability of nonlinear systems by presenting Lyapunov's stability theorem and its extension due to LaSalle. A problem of major importance is to determine regions of the state space in which it is guaranteed that the solutions converge to a desired equilibrium point. Thus, the concepts of region of attraction (basin of attraction) and regions of stability are introduced. In Sect. A.4, some applications of Lyapunov stability are considered as the notion of absolute stability via circle and Popov criteria associated to Lure systems. Another important type of application of Lyapunov stability is the notion of quadratic stability widely used in the field of robust control theory.

S. Tarbouriech et al., *Stability and Stabilization of Linear Systems with Saturating Actuators*, DOI 10.1007/978-0-85729-941-3, © Springer-Verlag London Limited 2011

354                                                              A   Some Concepts Related to Stability Theory

It is opportune to say that the notions described in this chapter are carefully defined and proven in some books dealing with the stability of nonlinear systems or the robust control theory. The sections that follow are mainly inspired from the books of Khalil [215], Slotine and Li [324], Vidyasagar [385]. For this reason, we omit the proofs of the results. The reader interested in these proofs can consult these references.


## A.2 Stability of Linear Autonomous Systems

Consider the linear autonomous system defined by

$$\dot{x}(t) = A_0 x(t) \tag{A.1}$$

where $x \in \mathfrak{R}^n$, $A_0 \in \mathfrak{R}^{n \times n}$. $A_0$ is a constant known matrix.

The equilibrium points, $x_e$, of system (A.1) are solutions to

$$A_0 x_e = 0 \tag{A.2}$$

Hence, if matrix $A_0$ is nonsingular, there exists a unique equilibrium point $x_e = 0$ solution to (A.2). Otherwise, when $A_0$ is singular, there exist other points $x_e \neq 0$, in plus of $x_e = 0$, solutions to (A.2).

Let us consider the stability of the origin $x_e = 0$. The characterization of the stability of the origin with respect to system (A.1) can be done by examining the eigenvalues of matrix $A_0$. The following theorem can be recalled [230].

**Theorem A.1**  *With respect to system (A.1), the equilibrium point $x_e = 0$ is*

1. *Asymptotically stable if and only if all the eigenvalues of $A_0$ have strictly negative real parts: $\mathfrak{Re}(\lambda_i(A_0)) < 0$, $\forall \lambda_i(A_0) \in \sigma(A_0)$. Matrix $A_0$ is said to be Hurwitz.*
2. *Critically stable if and only if the two following conditions hold*
   (a) *All the eigenvalues of $A_0$ have non-positive real parts: $\mathfrak{Re}(\lambda_i(A_0)) \leq 0$, $\forall \lambda_i(A_0) \in \sigma(A_0)$.*
   (b) *Each eigenvalue $\lambda_j(A_0)$ with null real part is such that its algebraic multiplicity is equal to its geometrical multiplicity.*
3. *Critically unstable if and only if the previous condition 2(a) is satisfied but condition 2(b) is not verified.*
4. *Unstable if and only if there exists, at least, an eigenvalue of $A_0$ with strictly positive real part: $\exists \lambda_i(A_0) \in \sigma(A_0)$ such that $\mathfrak{Re}(\lambda_i(A_0)) > 0$.*

Let us propose a very simple example in order to focus on cases 2 and 3.

*Example A.1*  Matrix $A_0 = \begin{bmatrix} 0 & 0 \\ 0 & 0 \end{bmatrix}$ verifies case 2 whereas $A_0 = \begin{bmatrix} 0 & 1 \\ 0 & 0 \end{bmatrix}$ verifies case 3.

A.3 Stability for Nonlinear Systems

The very simple way to test the stability of the origin $x_e = 0$ provided by Theorem A.1 applies only for linear systems. Throughout this book, we focus our study on some classes of nonlinear systems, therefore Theorem A.1 is no more a sufficient tool.

## A.3 Stability for Nonlinear Systems

Consider the nonlinear system:

$$\dot{x}(t) = f\big(x(t)\big) \tag{A.3}$$

where $f$ is a locally Lipschitz map from a domain $\mathcal{D}$ into $\mathfrak{R}^n$.

The equilibrium points $x_e$ associated to system (A.3) are the points belonging to $\mathcal{D}$ and solutions to

$$f(x_e) = 0 \tag{A.4}$$

In the sequel, without loss of generality, we consider that the equilibrium point of interest is the origin $x_e = 0$. Indeed, since system (A.3) is an autonomous system, any equilibrium point can be shifted to the origin via a suitable change of variables. Thus, suppose that $x_e \neq 0$ and consider the change of variables $\bar{x} = x - x_e$. It follows that

$$\dot{\bar{x}}(t) = \dot{x}(t) = f\big(x(t)\big) = f\big(\bar{x}(t) + x_e\big) = g\big(\bar{x}(t)\big) \tag{A.5}$$

with $g(0) = 0$. Hence, in the new system (A.5) with unchanged dynamics with respect to system (A.3), the origin is an equilibrium point.

### A.3.1 Stability Definition (Lyapunov Stability)

Now, we study the stability of the following system:

$$\dot{x}(t) = f\big(x(t)\big), \qquad f(0) = 0 \tag{A.6}$$

in a neighborhood $\mathcal{W} \subseteq \mathcal{D}$ around the origin.

**Definition A.1** The equilibrium point $x_e = 0$ of system (A.6) is

1. Stable if for each $\varepsilon > 0$, there exists a positive scalar $\delta$ such that

$$\big\| x(t_0) \big\| < \delta \quad \Rightarrow \quad \big\| x(t) \big\| < \varepsilon, \quad \forall t \geq t_0$$

2. Asymptotically stable if it is stable and the scalar $\delta$ can be chosen such that

$$\big\| x(t_0) \big\| < \delta \quad \Rightarrow \quad \lim_{t \to \infty} x(t) = 0$$

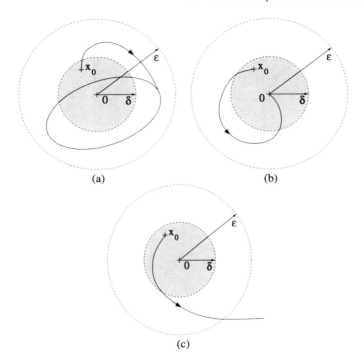

**Fig. A.1** Stability definitions

3. Unstable if it is not stable.

This definition is illustrated on Fig. A.1.

*Remark A.1* Note that from [319] if an equilibrium point $x_e$ is *attractive* then it has a region of attraction, namely a set $\Omega$ of initial set $x_0$ such that $x(t, x_0) \to x_e$ as $t \to \infty$ for all $x_e \in \Omega$ (see also Sect. 1.8). It is important to note that the notions of stability and attractivity are distinct properties. Indeed, for nonlinear systems, the exist some cases in which the origin $x_e = 0$ is attractive but unstable [161]. Cite, for example, the example due to Vinograd [387] in which the trajectories of an autonomous system, in a neighborhood around the origin, tend to the point $x_e = 0$ as $t$ tends to the infinity, but the point $x_e = 0$ is unstable because the trajectories initialized in $x_e = 0$ do not verify $x(t, 0) = 0$, $\forall t \geq 0$. However, for the systems that are linear around the origin, the attractivity implies the asymptotic stability.

The notion of stability as stated in Definition A.1 is based on the concept of neighborhood around an equilibrium point. In the application of Definition A.1, in order to conclude on the stability of the origin with respect to system (A.6), suppose that we explicitly know those trajectories. Nevertheless, the analytical determination of solutions to (A.6) may be difficult, or even impossible. Thus, having defined

A.3  Stability for Nonlinear Systems

stability and asymptotic stability of equilibrium points, our main task consists in finding ways to study stability. We can now state Lyapunov's stability theorem.

## A.3.2 Lyapunov's Stability Theorem

The objective of the Lyapunov's second method is to study the behavior of solutions (or trajectories) in a neighborhood around an equilibrium without explicit knowledge of those. This method is based on the use of some auxiliary functions having some properties (positivity, ...) and allows to conclude about the stability of an equilibrium point but studying the evolution of such functions along the trajectories of the considered system (without computing these trajectories).

Hence, let us consider $V : \mathcal{D} \to \mathfrak{R}$ be a continuously differentiable function defined in a domain $\mathcal{D} \subseteq \mathfrak{R}^n$ that contains the origin. The time-derivative of $V$ along the trajectories of system (A.6) is denoted $\dot{V}(x)$ and defined by [215]

$$\dot{V}(x) = \sum_{i=1}^{n} \frac{\partial V}{\partial x_{(i)}} \dot{x}_{(i)} = \sum_{i=1}^{n} \frac{\partial V}{\partial x_{(i)}} f_{(i)}(x)$$

$$= \begin{bmatrix} \frac{\partial V}{\partial x_{(1)}} & \cdots & \frac{\partial V}{\partial x_{(n)}} \end{bmatrix} \begin{bmatrix} f_{(1)}(x) \\ \vdots \\ f_{(n)}(x) \end{bmatrix}$$

$$= \frac{\partial V}{\partial x} f(x)$$

The derivative of $V$ along the trajectories of a system is dependent on the system equation.

With respect to system (A.6), the following theorem gives conditions making it possible to conclude about the stability of the origin.

**Theorem A.2**  *Let $x_e = 0$ be an equilibrium point for system (A.6). Let $V : \mathcal{D} \to \mathfrak{R}$ be a continuously differentiable function in a neighborhood $\mathcal{D}$ around $x = 0$, such that*

1. $V(0) = 0$ and $V(x) > 0$, $\forall x \in \mathcal{D}$, $x \neq 0$.
2. $\dot{V}(x) \leq 0$, $\forall x \in \mathcal{D}$.

*Then $x_e = 0$ is stable. Moreover, if*

3. $\dot{V}(x) < 0$, $\forall x \in \mathcal{D}$, $x \neq 0$.

*Then $x_e = 0$ is asymptotically stable.*

Note that Theorem A.2 guarantees the stability in a local sense since the stability is only ensured in a neighborhood around the origin. Furthermore, from Theorem A.2, we can exhibit the notions of positive definite and semi-definite functions.

**Definition A.2**

- The satisfaction of property 1 in Theorem A.2 means that the function $V(x)$ is a positive definite function in $\mathcal{D}$. Moreover, if $V(0) = 0$ and $V(x) \geq 0$, $\forall x \in \mathcal{D}$, $x \neq 0$, then $V(x)$ is a positive semi-definite function in $\mathcal{D}$.
- In the same way, a function $V(x)$ is said to be negative definite (resp. negative semi-definite) if $-V(x)$ is positive definite (resp. positive semi-definite).

**Definition A.3** A continuously differentiable function $V(x)$ satisfying the points 1 and 2 (or 1 and 3) of Theorem A.2 is called a Lyapunov function.

The main advantage of Theorem A.2 and the use of Lyapunov function reside in the fact that it can be applied without solving the differential equation (A.6). If Theorem A.2 applies in a certain neighborhood of the origin, it can be interesting to study under which conditions the stability of system (A.6) is obtained for any initial condition $x(t_0) = x(0) \in \mathfrak{R}^n$. In this case, system (A.6) is said globally stable. The idea is to use Theorem A.2 by considering $\mathcal{D} = \mathfrak{R}^n$. Nevertheless, this is not sufficient to obtain the global stability. An extra condition, which has to be considered with respect to function $V(x)$, is the notion of radially unbounded function.

**Theorem A.3** *Let $x_e = 0$ be an equilibrium point for system (A.6). Let $V : \mathfrak{R}^n \to \mathfrak{R}$ be a continuously differentiable function such that*:

1. $V(0) = 0$ and $V(x) > 0$, $\forall x \neq 0$.
2. $\|x\| \to \infty \Rightarrow V(x) \to \infty$.
3. $\dot{V}(x) < 0$, $\forall x \neq 0$.

*Then $x_e = 0$ is globally asymptotically stable.*

A geometrical interpretation of Theorems A.2 and A.3 can be given from the notion of Lyapunov surface defined as follows.

**Definition A.4** Let $V(x)$ a Lyapunov function. The surface defined in the state space by

$$L_s(V, c) = \left\{ x \in \mathfrak{R}^n; \ V(x) = c, \ c > 0 \right\} \tag{A.7}$$

is called a Lyapunov surface.

Hence, the condition $\dot{V}(x) \leq 0$ implies that when a trajectory of system (A.6) crosses a Lyapunov surface $L_s(V, c)$, it moves inside the set $S(V, c) = \{x \in \mathfrak{R}^n; \ V(x) \leq c\}$ and never comes out again. Note that as $c$ decreases, the Lyapunov surface $L_s(V, c)$ shrinks to the origin. Hence, when $\dot{V}(x) < 0$ the trajectory enters into inner Lyapunov surfaces that shrink to the origin as time increases. This shows that the trajectory converges to the origin. This reasoning is illustrated in Fig. A.2.

It is important to note that previous Theorems A.2 and A.3 provide a sufficient condition for the stability of system (A.6). Thus, the fact that a Lyapunov function

**Fig. A.2** Lyapunov surfaces

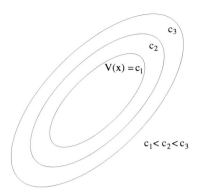

candidate does not satisfy conditions of Theorems A.2 or A.3 does not mean that the equilibrium point $x_e = 0$ is not stable (asymptotically stable). It only means that the stability property cannot be proven by using this Lyapunov function. Hence, the main difficulty in applying the previous theorems resides in the choice of the "good" Lyapunov function. Unfortunately, there does not exist a method allowing to choose a priori such a function. In general, the choice of the Lyapunov function requires the experience of the designer. In some cases, as in mechanical or electrical systems, based on the physical interpretation of the Lyapunov's theorem, one can choose for the Lyapunov function energy or storage functions. Thus, in this case, if the energy is dissipated along the trajectories of the system, then the energy function will be a potential Lyapunov function.

### A.3.3 The Invariance Principle

When the Lyapunov function $V(x)$ satisfies $\dot{V}(x) \leq 0$ along the trajectories of system (A.6), Theorems A.2 and A.3 do not allow to conclude on the asymptotic stability of the origin. Nevertheless, if we can guarantee that no system trajectory satisfies $\dot{V}(x) = 0$ except at $x = 0$, then the origin is asymptotically stable. This result is known as the LaSalle's invariance principle [235].

Before studying the main theorem, let us recall the notion of invariance.

**Definition A.5**

1. A set $\mathcal{D}$ is said to be invariant with respect to system (A.6) if $\forall x(0) \in \mathcal{D}$, one gets $x(t) \in \mathcal{D}$, $\forall t \in \Re$.
2. A set $\mathcal{D}$ is said to be positively invariant with respect to system (A.6) if $\forall x(0) \in \mathcal{D}$, one gets $x(t) \in \mathcal{D}$, $\forall t \geq 0$.

**Theorem A.4** *Let $\mathcal{D}$ be a positively invariant compact set with respect to system (A.6) and $V : \mathcal{D} \to \Re$ be a continuously differentiable function such that $\dot{V}(x) \leq 0$ in $\mathcal{D}$. Define $E$ as the set of all points in $\mathcal{D}$ where $\dot{V}(x) = 0$. Let $M$ be the largest invariant set in $E$. Then, every trajectory initialized in $\mathcal{D}$ converges to $M$ as $t \to \infty$.*

360                      A   Some Concepts Related to Stability Theory

It is important to remark that unlike Lyapunov's theorem, for Theorem A.4 the function $V(x)$ does not need to be positive definite. Moreover, the set $\mathcal{D}$ is not necessarily built from the function $V(x)$, even if in many cases the construction of $V(x)$ guarantees the existence of a set $\mathcal{D}$.

Since we are interested in the asymptotic stability of the origin, we need to prove that the largest invariant set in $E$ is the origin. This is stated in the following corollary.

**Corollary A.1** *Let $x_e = 0$ be an equilibrium for system (A.6). Let $V : \mathcal{D} \to \Re$ be a continuously differentiable function on a neighborhood $\mathcal{D}$ of the origin such that $\dot{V}(x) \leq 0, \forall x \in \mathcal{D}$. Let $S = \{x \in \mathcal{D}; \dot{V}(x) = 0\}$ and suppose that no solution can stay forever in $S$ other than the trivial solution $x = 0$. Then the origin is asymptotically stable.*

If in the above corollary we consider $\mathcal{D} = \Re^n$, we can conclude about the global asymptotic stability of the origin [215].

## A.3.4 Region of Attraction

In the previous sections, we have considered the property of stability of an equilibrium point, namely the origin. Nevertheless, it is not sufficient to verify whether the origin is asymptotically stable or not. It is also important to be able to characterize the region of attraction of $x_e = 0$ or, at least, to find an estimate of it.

Let us define the notion of attraction.

**Definition A.6** Denote by $x(t, x(0))$ the trajectory of system (A.6) initialized in $x(0)$. The region of attraction of the origin denoted by $R_A$ is defined by

$$R_A = \left\{ x \in \Re^n; \ x(t, x(0)) \to 0 \text{ as } t \to \infty \right\} \qquad (A.8)$$

The following properties of $R_A$ are proven in [215].

*Property A.1* If $x_e = 0$ is an asymptotically stable equilibrium point for system (A.6), then its region of attraction $R_A$, defined in (A.8), is an open invariant set. Moreover, the boundary of $R_A$ is formed by trajectories.

*Remark A.2* In the case where system (A.6) is globally asymptotically stable, the region of attraction of the origin is the whole state space, i.e. $R_A = \Re^n$.

From Property A.1, a way to determine the region of attraction $R_A$ consists in characterizing the trajectories of system (A.6) which allow to build the boundary of $R_A$. Hence, this fact means that the computation of $R_A$ is an hard task, even impossible in general [66, 215]. See the particular case of saturated systems discussed in Chap. 1 and [130, 300, 309, 334, 380, 381].

A.4  Application of Lyapunov Stability                                    361

However, an interesting way to overcome this difficulty is to search an estimate of this region with the objective to obtain good estimates. These estimates consist of regions in the state space in which the convergence of the trajectories to the origin is guaranteed. Such estimates, denoted, regions of stability, are defined as follows.

**Definition A.7** A region $R_S$ is called region of asymptotic stability with respect to the origin for system (A.6) if

$$\forall x(0) \in R_S, \quad x\big(t, x(0)\big) \to 0 \text{ as } t \to \infty \qquad (A.9)$$

Of course, according to Definition A.7, a region of asymptotic stability is always contained in the region of attraction, that is, $R_S \subseteq R_A$. Furthermore, the maximal region of asymptotic stability for the system is the region of attraction.

Different methods can be used in order to determine suitable regions $R_S$. The first one consists in determining a region of stability from a suitable Lyapunov function. In particular if $V(x)$ is a Lyapunov function for system (A.6), we can determine a suitable positive scalar $c$ such that the set

$$S(V, c) = \big\{ x \in \Re^n; \ V(x) \le c \big\}$$

is a region of stability for system (A.6). This method may, however, present some conservatism.

A second idea consists of using the LaSalle's invariance principle. Thus, according to Corollary A.1, we may consider more general sets as $S(V, c)$, but in this case we need to establish the positive invariance of the considered set.

Finally, a third method is based on the trajectory-reversing method. Nevertheless, such a method uses computer simulations and is quite efficient for systems for which the nonlinearity is of simple order.

### A.3.5  Set of Equilibrium Points

It is important to underline that a given nonlinear system may have many equilibrium points, including stable and unstable equilibrium points. In particular, system (A.6) may have stable or unstable equilibrium points $x_e$ satisfying $f(x_e) = 0$. The definition of stability of a given equilibrium point involves only initial states "close" to $x_e$, that is, concerns a local context around $x_e$. Thus, if an equilibrium point is attractive then it has a region of attraction $R_A^e = \{x \in \Re^n; \ x(t, x(0)) \to x_e \text{ as } t \to \infty\}$. See, for example, [319].

## A.4  Application of Lyapunov Stability

In this section, we consider the main applications of Lyapunov stability which are used in this book. At this aim, we consider only some particular classes of nonlinear systems (A.6).

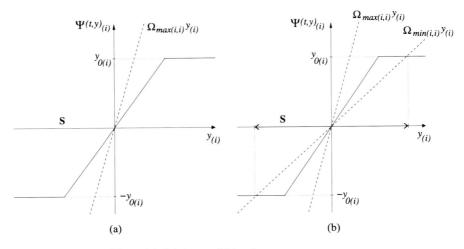

**Fig. A.3** Sector conditions: (**a**) global case; (**b**) local case

## A.4.1 Absolute Stability

There exist many practical systems that can be represented by the interconnection of a linear system and a nonlinear element [215, 319].

Such systems can be described, for example, as follows:

$$\begin{cases} \dot{x}(t) = Ax(t) + Bu(t) \\ y(t) = Cx(t) + Du(t) \\ u(t) = -\psi(t, y) \end{cases} \quad (A.10)$$

where $x \in \Re^n$, $u$ and $y$ belong to $\Re^m$, pairs $(A, B)$ and $(C, A)$ are, respectively, controllable and observable. The function $\psi(t, y) : [0, \infty) \times \Re^m \to \Re^m$ is a memoryless, possibly time-varying, nonlinearity. It is piecewise continuous in $t$ and locally Lipchitz in $y$. The nonlinearity $\psi(\cdot, \cdot)$ is required to satisfy a sector condition as defined as follows.

**Definition A.8** A memoryless nonlinearity $\psi : [0, \infty) \times \Re^m \to \Re^m$ is said to satisfy a sector condition if

$$\bigl(\psi(t, y) - \Omega_{\min} y\bigr)' \bigl(\psi(t, y) - \Omega_{\max} y\bigr) \leq 0, \quad \forall t \geq 0, \ \forall y \in \mathcal{S} \subseteq \Re^m \quad (A.11)$$

for some real matrices $\Omega_{\max}$ and $\Omega_{\min}$ where $\Omega = \Omega_{\max} - \Omega_{\min}$ is a symmetric positive definite matrix and $\mathcal{S}$ contains the origin in its interior. If $\mathcal{S} = \Re^m$, then $\psi(\cdot, \cdot)$ satisfies the sector condition globally.

When $\psi(t, y)$ satisfies (A.11) we say that $\psi(t, y)$ belongs to the sector $(\Omega_{\min}, \Omega_{\max})$. Figure A.3 illustrates this sector condition.

From this definition, we can study the stability of the origin with respect to system (A.10).

A.4 Application of Lyapunov Stability 363

**Definition A.9** Consider system (A.10), where $\psi(\cdot, \cdot)$ satisfies a sector condition as presented in Definition A.8. The system is absolutely stable if the origin is globally uniformly asymptotically stable for any nonlinearity in the given sector. It is absolutely stable with a finite domain if the origin is uniformly asymptotically stable.

The problem of the stability of system (A.10) is referred to in the literature as the Lure problem.

## A.4.2 Positive Real Function Concept

In order to be able to provide some links between the stability of system (A.10) and frequency-domain conditions, we define some concepts related to the notion of positive real transfer function. This notion is then related to the existence of Lyapunov functions.

**Definition A.10** A $m \times m$ proper rational function $Z(s)$ of functions of a complex variable $s$ is called positive real if

- $Z(s)$ has elements that are analytic for $\Re e(s) > 0$,
- $Z(s)^\star = Z(s^\star)$, for $\Re e(s) > 0$,
- $Z(s^\star)' + Z(s)$ is positive semi-definite for $\Re e(s) > 0$.

The star $\star$ denotes complex conjugation. The matrix is called strictly positive real if $Z(s - \varepsilon)$ is positive real for some $\varepsilon > 0$.

Let us now give the following lemma, known as the Kalman–Yakubovich–Popov lemma or the positive real lemma.

**Lemma A.1** *Let* $Z(s) = C(s I_n - A)^{-1} B + D$ *be a* $m \times m$ *transfer function matrix, where* $A$ *is Hurwitz,* $(A, B)$ *is controllable and* $(C, A)$ *is observable. Then* $Z(s)$ *is strictly positive real if and only if there exist a symmetric positive definite matrix* $P$, *matrices* $W$ *and* $L$, *and a positive scalar* $\varepsilon$ *such that*

$$A'P + PA = -L'L - \varepsilon P \tag{A.12}$$

$$PB = C' - L'W \tag{A.13}$$

$$W'W = D + D' \tag{A.14}$$

The proof of this lemma [215] concerning its sufficient part is based on the definition of a positive real function. Thus, the condition of stabilizability and detectability are sufficient. The necessary part is based on the spectral factorization. Thus, the condition of controllability and observability is necessary in order to prove the unicity of the matrix $P$.

The following lemma allows to characterize the strict positive realness.

364 A Some Concepts Related to Stability Theory

**Lemma A.2** *Let $Z(s) = C(sI_n - A)^{-1}B + D$ be a $m \times m$ transfer function matrix with $\det(D) \neq 0$. Then $Z(s)$ is strictly positive real if and only if*

- $\Re e(\lambda_i(A)) < 0, \forall i,$
- $Z(j\omega) + Z(-j\omega)' > 0, \forall \omega \in \Re.$

## *A.4.3 Circle Criterion*

Let us first give the preliminary result on which is based the proof of the circle criterion [215].

**Proposition A.1** *Let $\Omega_{\max}$ and $\Omega_{\min}$ be two diagonal matrices such that $\Omega = \Omega_{\max} - \Omega_{\min} > 0$. Consider system (A.10) where $A$ is Hurwitz, $(A, B)$ is controllable and $(C, A)$ is observable. Then system (A.10) is absolutely stable for $y \in \mathcal{S}$ if there exist a symmetric positive definite matrix $P \in \Re^{n \times n}$, a matrix $N \in \Re^{m \times n}$ and a positive scalar $\varepsilon$ such that*

$$A'P + PA = -N'N - \varepsilon P \tag{A.15}$$

$$PB = C'\Omega - \sqrt{2}N' \tag{A.16}$$

*If the set $\mathcal{S}$ is strictly included in $\Re^m$ then the stability is local. If $\mathcal{S} = \Re^m$ then the stability is global.*

The proof of this proposition is based on the contractivity property of the quadratic Lyapunov function $V(x) = x'Px$, with $P = P' > 0$ solution to relations (A.15) and (A.16). By using the condition (A.11), we can prove that the function $V(x)$ is strictly decreasing along the trajectories of system (A.10) satisfying the sector condition, i.e. for $y \in \mathcal{S}$.

By applying the sufficient condition of the Kalman–Yakubovich–Popov lemma and by setting $W'W = 2I_m$, one finally gets:

**Theorem A.5** (Circle Criterion, [215]) *Under the same assumptions as in Proposition A.1, system (A.10) is absolutely asymptotically stable for $y \in \mathcal{S}$ if*

$$Z(s) = I_m + \Omega G(s) = I_m + \Omega C(sI_n - A)^{-1}B \tag{A.17}$$

*is strictly positive real. Furthermore, if the domain $\mathcal{S}$ is strictly included in $\Re^m$ then the stability is local, and if $\mathcal{S} = \Re^m$ then the stability is global.*

Since only the sufficient condition of the Kalman–Yakubovich–Popov lemma is used, the condition of controllability and observability can be relaxed in a condition of stabilizability and detectability.

## A.4.4 Popov Criterion

As in the previous subsection, let us begin by a preliminary result on which is based the proof of the Popov criterion.

**Proposition A.2** *Let $\Omega_{max}$ and $\Omega_{min}$ be two diagonal matrices such that $\Omega = \Omega_{max} - \Omega_{min} > 0$. Consider system (A.10) where $A$ is Hurwitz, $(A, B)$ is controllable and $(C, A)$ is observable. Suppose that the nonlinearity $\psi(y)$ is time-invariant and belongs to the sector $(\Omega_{min}, \Omega_{max})$ for $y \in S \subseteq \Re^m$. Then system (A.10) is absolutely stable for $y \in S$ if there exist a symmetric positive definite matrix $P \in \Re^{n \times n}$, matrices $N \in \Re^{q \times n}$ and $W \in \Re^{q \times m}$, a positive scalar $\eta$ such that $-\frac{1}{\eta} \notin \sigma(A)$ and a positive scalar $\varepsilon$ such that*

$$A'P + PA = -N'N - \varepsilon P \tag{A.18}$$

$$PB = C'\Omega + \eta A'C'\Omega - N'W \tag{A.19}$$

$$W'W = 2I_m + \eta \Omega CB + \eta B'C'\Omega \tag{A.20}$$

*If the set $S$ is strictly included in $\Re^m$ then the stability is local. If $S = \Re^m$ then the stability is global.*

The proof of this proposition is based on the contractivity property of the Lure function:

$$V(x) = x'Px + 2\eta \sum_{i=1}^{m} \int_0^{y_{(i)}} \psi_{(i)}(\tau) \Omega_{(i,i)} d\tau$$

where $P = P' > 0$ is solution to relation (A.18) and (A.19). By using the condition (A.11), we can prove that the function $V(x)$ is strictly decreasing along the trajectories of system (A.10) satisfying the sector condition for $y \in S$.

By applying the sufficient condition of the Kalman–Yakubovich–Popov Lemma, one finally gets

**Theorem A.6** (Popov Criterion, [215]) *Under the same assumptions as in Proposition A.1, system (A.10) is absolutely asymptotically stable for $y \in S$ if*

$$Z(s) = I_m + (1 + \eta s)\Omega G(s) = I_m + (1 + \eta s)\Omega C(sI_n - A)^{-1}B \tag{A.21}$$

*is strictly positive real. Furthermore, if the domain $S$ is strictly included in $\Re^m$ then the stability is local, and if $S = \Re^m$ then the stability is global.*

Indeed, from the sufficient condition of the Kalman–Yakubovich–Popov lemma, the function $Z(s)$ is defined by

$$\begin{aligned} Z(s) &= I_m + \eta \Omega CB + (\Omega C + \eta \Omega CA)(sI_n - A)^{-1}B \\ &= I_m + \eta \Omega C(sI_n - A + A)(sI_n - A)^{-1}B + \Omega C(sI_n - A)^{-1}B \\ &= I_m + (1 + \eta s)\Omega C(sI_n - A)^{-1}B \end{aligned}$$

As in the case of the circle criterion we can relax the properties of controllability and observability.

Moreover, when $\eta = 0$, we retrieve the circle criterion, which can be considered, in a certain sense, as a particular case of the Popov criterion (if the nonlinearity is time-invariant). In this case, the Lure function reduces to the quadratic Lyapunov function.

There exist other ways to define both circle and Popov criteria, which are slightly different in the matrix relations. In particular, we can cite the following works:

- Haddad and Bernstein [159] replace the scalar $\eta$ in the Popov criterion by a non-negative matrix $H$.
- Park [283] replace the scalar $\eta$ in the Popov criterion by a real matrix $M$ without hypothesis of positivity. Furthermore, matrix $I_m + Ms$ is called a Popov multiplier.
- See also [9, 82, 165].

### A.4.5 Quadratic Stability

The notion of quadratic stability, originally published by Barmish [15], can be viewed as an extension of the Lyapunov stability when considering both uncertain systems and quadratic Lyapunov functions [77, 170, 287]. It will be discussed in Appendix B .

# Appendix B
# Quadratic Approach for Robust Control

## B.1 Introduction

This appendix describes some results concerning the control design for systems whose models are uncertain. There exists an immense literature dealing with this problem and numerous methods have been applied with success to practical control problems [74, 255, 414]. These methods can be classified following several ways. One of the most common, consists of a classification deduced from the uncertainty description or properties. For example, we distinguish structured and non structured uncertainty, parametric and non parametric uncertainty and so on. Generally the intersection between two classes is nonempty and it is possible to precisely characterize the context in which a method works, combining the corresponding terms. Another classification is related to the concepts and tools involved in the derived methods. For example, several methods consider systems described by uncertain models in the state space for which the stability is studied using Lyapunov theory [15, 74]. Other methods consider systems described by uncertain models expressed in the frequency domain for which stability is studied using the multivariable Nyquist criterion and so on [255, 414]. In fact, to give an exhaustive overview of robust control design methods is a difficult task. However, some good books giving a precise state of the art for the most interesting methods are available in the literature [74, 255, 414].

In this appendix, a summary of robust control design methods is presented, in which uncertain models are described in the state space, the uncertainty being parametric. Only methods used in the manuscript will be presented and in general no proofs will be provided. They can be found together with some extensions in the associated references. Section B.2 introduces the uncertain models considered in some parts of this book and the quadratic stabilizability concept. Section B.3 gives the main results concerning robust state feedback synthesis. Two approaches are presented: a Riccati equation approach and an LMI approach. These results can be extended to the output feedback synthesis case and can be found in the cited references. The problem of performances is addressed in Sect. B.4 through a minimization of a quadratic cost function and in Sect. B.5 through pole location in particular regions of complex plane.

S. Tarbouriech et al., *Stability and Stabilization of Linear Systems with Saturating Actuators*, DOI 10.1007/978-0-85729-941-3, © Springer-Verlag London Limited 2011

It has to be noticed that all the material presented in this appendix is associated to continuous-time systems since, although not exclusive, they correspond to the main class of systems studied in this book.

## B.2 Uncertain Models and Quadratic Stability

This section introduces uncertain models and the quadratic stability concept which is the extension of the classical stability concept to the case of uncertain models.

### *B.2.1 Uncertain Models*

The uncertain models considered in this chapter can be described by

$$\dot{x}(t) = A\big(F(t)\big)x(t) + B\big(G(t)\big)u(t)$$

where $F(\cdot)$ and $G(\cdot)$ belong, respectively, to $\mathcal{F}$ and $\mathcal{G}$ which are appropriately defined in the sequel and map $\mathfrak{R}_+$ to subsets, respectively, of $\mathfrak{R}^{f_1 \times f_2}$ and $\mathfrak{R}^{g_1 \times g_2}$. We suppose that matrix-valued functions $F(\cdot)$ and $G(\cdot)$ are at least Lebesgue measurable and in some cases depending on the context, satisfy some stronger assumptions. $A(\cdot)$ and $B(\cdot)$ are function matrices which associate at each element of $\mathcal{F}$ and $\mathcal{G}$, an element of $\mathfrak{R}^{n \times n}$ and $\mathfrak{R}^{n \times m}$, respectively. Now we particularize this general model to some cases often studied in the literature. For simplicity, only the dynamical matrix is considered as uncertain. We list below some examples which may be translated, after some elementary transformations, into a norm-bounded uncertainty. Their practical interest is discussed in some details in the robust control literature, particularly in [159] and [77]. Note that there also exist other ways to describe uncertainty.

#### B.2.1.1 Norm-Bounded Uncertainty (NB-Uncertainty) [114, 288]

The matrix $A(F(t))$ is expressed as

$$A\big(F(t)\big) = A_0 + DF(t)E \tag{B.1}$$

where $A_0 \in \mathfrak{R}^{n \times n}$, $D \in \mathfrak{R}^{n \times f_1}$, $E \in \mathfrak{R}^{f_2 \times n}$ define the structure of the uncertainty and the modeling uncertainty parameter $F(\cdot)$ belongs to the set

$$\mathcal{F} = \big\{ F(\cdot); \ \mathfrak{R}_+ \ \rightarrow \ \mathfrak{R}^{f_1 \times f_2} : \ F(\cdot)' F(\cdot) \leq I_{f_2} \big\} \tag{B.2}$$

It expresses that an ellipsoidal volume is defined as the uncertainty domain in the space of the entries of $A(\cdot)$, the nominal model $A_0$ being defined in the center of this ellipsoid.

## B.2 Uncertain Models and Quadratic Stability

### B.2.1.2 Bounded-Real Uncertainty (BR-Uncertainty) [159]

The uncertainty term is written as

$$A\big(F(t)\big) = A_0 + DF(t)\big(I_{f_2} - D_0F(t)\big)^{-1}E$$

where $A_0 \in \mathfrak{R}^{n \times n}$, $D \in \mathfrak{R}^{n \times f_1}$, $E \in \mathfrak{R}^{f_2 \times n}$ define the structure of the uncertainty and the modeling uncertainty parameter $F(\cdot)$ belongs to $\mathcal{F}$ previously defined. $D_0$ is a constant matrix satisfying $I_{f_1} - D_0'D_0 > 0$. We have after some simple manipulations:

$$A\big(F(t)\big) = \underbrace{A_0 + D\big(I_{f_1} - D_0'D_0\big)^{-1}D_0'E}_{\mathcal{A}_{0b}}$$

$$+ \underbrace{D\big(I_{f_1} - D_0'D_0\big)^{-1/2}}_{\mathcal{D}_b}\Delta(t)\underbrace{\big(I_{f_2} - D_0D_0'\big)^{-1/2}E}_{\mathcal{E}_b}$$

with [77]:

$$\Delta(t) = \Big[\big(I_{f_1} - D_0'D_0\big)^{1/2}F(t)\big(I_{f_2} - D_0F(t)\big)^{-1} - \big(I_{f_1} - D_0'D_0\big)^{-1/2}D_0'\Big]$$
$$\times \big(I_{f_1} - D_0'D_0\big)^{1/2}$$

We have $\Delta(t)'\Delta(t) \leq I_{f_2}$. This uncertainty can be considered as of norm-bounded type with a nominal matrix $\mathcal{A}_{0b}$.

### B.2.1.3 Positive Real Uncertainty (PR-Uncertainty) [117]

The uncertainty term is given by

$$A\big(F(t)\big) = A_0 - DF(t)\big(I_{f_1} + D_0F(t)\big)^{-1}E$$

where $A_0 \in \mathfrak{R}^{n \times n}$, $D \in \mathfrak{R}^{n \times f_1}$, $E \in \mathfrak{R}^{f_1 \times n}$ define the structure of the uncertainty and the modeling uncertainty parameter $F(\cdot)$ belongs to the set

$$\mathcal{F}_p = \big\{F(\cdot);\ \mathfrak{R}_+ \to \mathfrak{R}^{f_1 \times f_1} : F(\cdot)' + F(\cdot) \geq 0\big\} \qquad (B.3)$$

Moreover, $D_0$ is a constant matrix of appropriate dimension satisfying $D_0 + D_0' > 0$. This condition ensures that the matrix $I_{f_1} + D_0F(\cdot)$ is invertible for all $F(\cdot) \in \mathcal{F}_p$. We have

$$A\big(F(t)\big) = A_0 - D\big(D_0 + D_0'\big)^{-1}E + D\big(D_0 + D_0'\big)^{-1/2}\Delta(t)\big(D_0 + D_0'\big)^{-1/2}E$$

with:

$$\Delta(t) = I_{f_1} - \big(D_0 + D_0'\big)^{1/2}F(t)(I_{f_1} + D_0F)^{-1}\big(D_0 + D_0'\big)^{1/2}$$

We have $\Delta(t)'\Delta(t) \leq I_{f_1}$. As previously, this uncertainty can be considered as norm-bounded.

### B.2.1.4 Structured Uncertainty

The above defined uncertainties are called unstructured uncertainties in the sense that they are defined through a single uncertainty matrix $F(\cdot)$ which is defined in a very global and general setting. We can introduce some structural features on the uncertainty by defining multiblock uncertainty terms such as for instance

$$A\big(F(t)\big) = A\big(F_1(t), F_2(t), \dots, F_{n_F}(t)\big) = \text{diag}\big(A_1\big(F_1(t)\big), \dots, A_{n_F}\big(F_{n_F}(t)\big)\big)$$

where $A_i(F_i(t))$ can be expressed by one of the following expressions:

$$A_i\big(F_i(t)\big) = D_i F_i(t) E_i, \quad F_i \in \mathcal{F}_i$$

$$A_i\big(F_i(t)\big) = D_i F_i(t)\big(I_{f_{2i}} - D_{0i} F_i(t)\big)^{-1} E_i, \quad F_i \in \mathcal{F}_i, \ I_{f_{2i}} - D'_{0_i} D_{0_i} > 0$$

$$A_i\big(F_i(t)\big) = -D_i F_i(t)\big(I_{f_{2i}} + D_{0i} F_i(t)\big)^{-1} E_i, \quad F_i \in \mathcal{F}_{pi}, \ D_{0i} + D'_{0i} > 0$$

with $D_i$ and $E_i$ constant matrices of appropriate dimensions and the sets $\mathcal{F}_i$ and $\mathcal{F}_{pi}$ are, respectively, defined like $\mathcal{F}$ and $\mathcal{F}_p$ . In this way, one can account for more practical parametrical uncertainty, even if the derived methods are more involved and in some cases conservative than in the one block case.

### B.2.1.5 Polytopic Uncertainty (P-Uncertainty) [74, 116]

Another very studied uncertainty type is the well-known polytopic uncertainty defined as

$$A\big(F(t)\big) = \sum_{i=1}^{N_A} f_i(t)\, A_i$$

where $A_i$, $i = 1, \dots, N_A$, are $n \times n$ constant matrices (the vertices) and $F(t)$ belongs to the following set:

$$\mathcal{F}_{N_A} = \left\{ F(\cdot) : \Re_+ \to \Re^{1 \times N_A}, \ F(t) = \big[ f_1(t) \dots f_{N_A}(t) \big] \text{ and} \right.$$

$$\left. \sum_{i=1}^{N_A} f_i(t) = 1, \ f_i(t) \geq 0 \right\}$$

In the sequel, because BR and PR uncertainties can be translated into NB-uncertainty, only NB and polytopic uncertainties will be discussed.

## B.2.2 Quadratic Stability

The notion of quadratic stability can be viewed as an extension of the Lyapunov stability when considering both uncertain systems and quadratic Lyapunov functions.

B.2 Uncertain Models and Quadratic Stability

This concept, originally introduced by Barmish [15] has proven to be efficient to solve some robust analysis problems. Let us consider the following uncertain system:

$$\dot{x}(t) = A\big(F(t)\big)x(t) \tag{B.4}$$

where $x \in \Re^n$, $F(t) \in \mathcal{F}$.

**Definition B.1** [15] System (B.4) is quadratically stable if there exist a positive definite symmetric matrix $P \in \Re^{n \times n}$ and a positive scalar $\alpha$ satisfying:

$$x'\big[A\big(F(t)\big)'P + PA\big(F(t)\big)\big]x \le -\alpha\|x\|^2 \tag{B.5}$$

for any uncertainty $F(t) \in \mathcal{F}$.

If condition (B.5) is verified then the quadratic function $V(x) = x'Px$ is strictly decreasing along the trajectories of system (B.4). The quadratic stability concept is very restrictive because the same Lyapunov function allows to test the stability of all the possible system realizations and then it leads to conservative results. Recently, this conservatism has been reduced using parameter-dependent Lyapunov functions [287]. In general the obtained conditions are more involved and out of the scope of this book.

### B.2.2.1 P-Uncertainty

For polytopic uncertainty, we have

$$A\big(F(t)\big) = \sum_{i=1}^{N_A} f_i(t)\, A_i, \quad f_i \ge 0,\ \sum_{i=1}^{N_A} f_i(t) = 1$$

A necessary and sufficient condition to test the quadratic stability of a polytopic uncertain system (B.4) is proposed in [26] and recalled here.

**Theorem B.1** *System (B.4) with polytopic uncertainty is quadratically stable if and only if there exists a positive definite symmetric matrix $P \in \Re^{n \times n}$ satisfying:*

$$A_i'P + PA_i < 0 \quad for\ i = 1, \dots, N_A$$

The conditions of Theorem B.1 are LMIs and are easy to test with an LMI solver. We may note that:

$$\sum_{i=1}^{N_A} f_i(t)\big(A_i'P + PA_i\big) = A\big(F(t)\big)'P + PA\big(F(t)\big) < 0$$

and $x'Px$ is a Lyapunov function for system (B.4).

## B.2.2.2 NB-Uncertainty

The norm-bounded uncertainty is defined by

$$A\big(F(t)\big) = A_0 + DF(t)E, \quad F(t)'F(t) \le I_{f_2}$$

In this case, there exists a connection between quadratic stability and small gain theorem which allows to state the following results.

**Theorem B.2** [216]  *System (B.4) with NB-uncertainty is quadratically stable if and only if:*

1. $A_0$ *is a stable matrix (all its eigenvalues have negative real parts).*
2. *The following condition holds:*

$$\big\| E(s I_n - A_0)^{-1} D \big\|_\infty < 1 \tag{B.6}$$

It is possible to deduce from Theorem B.2, necessary and sufficient conditions expressed, respectively, through a Riccati equation or an LMI [45].

**Corollary B.1** [288]  *System (B.4) is quadratically stable if and only if given a positive definite symmetric matrix $Q \in \Re^{n \times n}$, there exist a positive definite symmetric matrix $P \in \Re^{n \times n}$ and a positive scalar $\varepsilon$ satisfying:*

$$A_0'P + PA_0 + \varepsilon PDD'P + \varepsilon^{-1} E'E + Q = 0 \tag{B.7}$$

*Proof* This result can be obtained remarking that if condition (B.6) is satisfied, the condition

$$\left\| \begin{bmatrix} E \\ \varepsilon Q^{1/2} \end{bmatrix} (s I_n - A_0)^{-1} D \right\|_\infty < 1$$

is also satisfied for all $Q = Q' > 0$ and $\varepsilon \in (0, \hat{\varepsilon}[$ where $\hat{\varepsilon}$ depends on $Q$. The use of standard results on $H_\infty$-norm optimization [414] allows one to come to the result. We have

$$x'\big[A\big(F(t)\big)'P + PA\big(F(t)\big)\big]x = x'\big[(A_0 + DFE)'P + P(A_0 + DFE)\big]x$$

$$\le x'\big[A_0'P + PA_0 + \varepsilon PDD'P + \varepsilon^{-1} E'E\big]x$$

$$= -x'Qx \le -\lambda_{\min}(Q)\|x\|^2$$

From the definition of quadratic stability, we can take $\alpha = \lambda_{\min}(Q)$ and $x'Px$ is a Lyapunov function for system (B.4). $\qquad\square$

To check the condition of Corollary B.1, remark that if for a given positive definite symmetric matrix $Q$, there exists an $\varepsilon_* > 0$ such that the Riccati equation (B.7) has a positive definite symmetric solution, then for all $\varepsilon \in ]0, \varepsilon_*]$, there exists a positive definite symmetric solution too. This property suggests the following algorithm to check the condition of Corollary B.1.

B.3 Quadratic Stabilizability                                                    373

**Algorithm B.1**

1. Select a positive definite symmetric matrix $Q$ and initialize $\varepsilon = \varepsilon_0 > 0$.
2. Solve the Riccati equation (B.7). If the solution is positive definite, stop. The system is quadratically stable. If not, go to Step 3.
3. Decrease the value of $\varepsilon$, for example $\varepsilon \leftarrow \frac{\varepsilon}{2}$. If $\varepsilon$ is lower than a define precision, stop. The system is not quadratically stable. If not go to Step 2.

It is clear that this algorithm converges in a finite number of steps. An alternative to Corollary B.1 is presented in the following corollary.

**Corollary B.2** *System (B.4) is quadratically stable if and only if there exists a positive definite symmetric matrix $W \in \mathfrak{R}^{n \times n}$ satisfying:*

$$
\begin{bmatrix} A_0 W + W A_0' + D D' & W E' \\ E W & -I \end{bmatrix} < 0 \tag{B.8}
$$

*The matrix $W$ and the matrix $P$ given in Corollary B.1 are related by*

$$
W = \varepsilon^{-1} P^{-1}
$$

The feasibility of the LMI of Corollary B.2 can be easily checked with a standard LMI solver. The linearity of this condition allows when necessary, to structure the unknowns, for example forcing some entries to be equal to zero. This is not possible solving the Riccati equation of Corollary B.1. From inequality (B.8), we deduce by Schur's complement that

$$
A_0 W + W A_0' + W E' E W + D D' < 0
$$

and it is possible to show that $x' W^{-1} x$ is a Lyapunov function for system (B.4).

## B.3 Quadratic Stabilizability

The notion of quadratic stability can be easily extended to quadratic stabilizability concept [15]. Let us consider the following uncertain system:

$$
\dot{x}(t) = A\big(F(t)\big) x(t) + B\big(G(t)\big) u(t) \tag{B.9}
$$

where $F(\cdot)$ and $G(\cdot)$ belong, respectively, to $\mathcal{F}$ and $\mathcal{G}$.

**Definition B.2** System (B.9) is said quadratically stabilizable if and only if there exist a control $u(t) = K x(t)$, $K \in \mathfrak{R}^{m \times n}$, a positive definite symmetric matrix $P \in \mathfrak{R}^{n \times n}$ and a positive scalar $\alpha$ such that:

$$
x'\big(A\big(F(t)\big)' P + P A\big(F(t)\big)\big) x + 2 x' P B\big(G(t)\big) K x \leq -\alpha \|x\|^2 \tag{B.10}
$$

for any $F(t) \in \mathcal{F}$ and $G(t) \in \mathcal{G}$.

If condition (B.10) is satisfied, then the quadratic function $V(x) = x'Px$ is strictly decreasing along the trajectories of the closed-loop system. As for the quadratic stability concept, the quadratic stabilizability one is very restrictive because a single Lyapunov function allows to test the stability of all possible controlled system realizations.

### B.3.1 P-Uncertainty

Consider the case of polytopic uncertainty described by

$$A\big(F(t)\big) = \sum_{i=1}^{N_A} f_i(t)\, A_i, \qquad B\big(G(t)\big) = \sum_{i=1}^{N_B} g_i(t)\, B_i,$$

$$\text{for all } t, \ f_i(t) \geq 0, \ g_i(t) \geq 0, \ \sum_{i=1}^{N_A} f_i(t) = 1 \text{ and } \sum_{i=1}^{N_B} g_i(t) = 1$$

The following theorem can be deduced [26].

**Theorem B.3** *System (B.9) with P-uncertainty is quadratically stabilizable if and only if there exist a positive definite symmetric matrix $W \in \mathfrak{R}^{n \times n}$ and a matrix $S \in \mathfrak{R}^{m \times n}$ satisfying*

$$A_i W + W A_i' + B_j S + S' B_j' < 0,$$

$$\textit{for all } i = 1, \ldots, N_A \textit{ and } j = 1, \ldots, N_B \tag{B.11}$$

*A quadratic stabilizing control is given by*

$$u(t) = SW^{-1}x(t)$$

*Proof* After some elementary manipulations, inequality (B.11) can written as

$$\big(A_i + B_j SW^{-1}\big)' W^{-1} + W^{-1}\big(A_i + B_j SW^{-1}\big) < 0$$

and

$$\sum_{i=1}^{N_A} f_i(t)\big[\big(A_i + B_j SW^{-1}\big)' W^{-1} + W^{-1}\big(A_i + B_j SW^{-1}\big)\big]$$

$$= \big(A\big(F(t)\big) + B_j RW^{-1}\big)' W^{-1} + W^{-1}\big(A\big(F(t)\big) + B_j RW^{-1}\big)$$

## B.3 Quadratic Stabilizability

$$\sum_{j=1}^{N_B} g_j(t)\left[\left(A\big(F(t)\big) + B_j S W^{-1}\right)' W^{-1} + W^{-1}\left(A\big(F(t)\big) + B_j S W^{-1}\right)\right]$$

$$= \left(A\big(F(t)\big) + B\big(G(t)\big) S W^{-1}\right)' W^{-1} + W^{-1}\left(A\big(F(t)\big) + B\big(G(t)\big) S W^{-1}\right) < 0$$

showing that $x' W^{-1} x$ is a Lyapunov function for the closed-loop system. $\qquad\square$

### B.3.2 NB-Uncertainty

For norm-bounded uncertainty, the system matrices are given by

$$A\big(F(t)\big) = A_0 + D F(t) E_1, \qquad B\big(F(t)\big) = B_0 + D F(t) E_2$$

where $E_1$ and $E_2$ are matrices of appropriate dimensions and $F(t) \in \mathcal{F}$. From Theorem B.2, the system is quadratically stabilizable if and only if

$$\left\| (E_1 + E_2 K)(s I_n - A_0 - B_0 K)^{-1} D \right\|_{\infty} < 1$$

which is equivalent to

$$\left\| \left( \begin{bmatrix} E_1 \\ 0 \\ (\varepsilon Q)^{1/2} \end{bmatrix} + \begin{bmatrix} E_2 \\ (\varepsilon R) \\ 0 \end{bmatrix} K \right) (s I_n - A_0 - B_0 K)^{-1} D \right\|_{\infty} < 1$$

for all $Q = Q' > 0$, $R = R' > 0$, $\varepsilon \in (0, \hat{\varepsilon}[$ where $\hat{\varepsilon}$ depends on $Q$ and $R$. The use of standard results on $H_\infty$ control synthesis [414] allows to design a control from the solution of a Riccati equation as stated in the next theorem.

**Theorem B.4** *System (B.9) with NB-uncertainty is quadratically stabilizable if and only if given positive definite symmetric matrices $Q \in \Re^{n \times n}$ and $R \in \Re^{m \times m}$, there exist a positive definite symmetric matrix $P \in \Re^{n \times n}$ and a positive scalar $\varepsilon$ satisfying:*

$$A'_\varepsilon P + P A_\varepsilon - P B R_\varepsilon^{-1} B' P + \varepsilon P D D' P + \varepsilon^{-1} E_1' \left(I - \varepsilon^{-1} E_2 R_\varepsilon^{-1} E_2'\right) E_1 + Q = 0 \tag{B.12}$$

*where:*

$$A_\varepsilon = A_0 - \varepsilon^{-1} B R_\varepsilon^{-1} E_2' E_1, \qquad R_\varepsilon = R + \varepsilon^{-1} E_2' E_2$$

*A quadratic stabilizing control is then given by*

$$u(t) = -R_\varepsilon^{-1}\left(B' P + \varepsilon^{-1} E_2' E_1\right) x(t)$$

By simple but tedious manipulations, it is possible to rewrite the Riccati equation as

$$(A_0 + B_0 K)' P + P(A_0 + B_0 K) + \varepsilon P D D' P$$

$$+ \varepsilon^{-1}(E_1 + E_2 K)'(E_1 + E_2 K) + K' R K + Q = 0$$

and then we have

$$x'\left[\left(A\left(F(t)\right)+B\left(G(t)\right)K\right)'P+P\left(A\left(F(t)\right)+B\left(G(t)\right)K\right)\right]x$$
$$=x'\left[\left(A_0+B_0K+DF(E_1+E_2K)\right)'P\right.$$
$$\left.+P\left(A_0+B_0K+DF(E_1+E_2K)\right)\right]x$$
$$\leq x'\left[(A_0+B_0K)'P+P(A_0+B_0K)+\varepsilon PDD'P\right.$$
$$\left.+\varepsilon^{-1}(E_1+E_2K)'(E_1+E_2K)\right]x$$
$$=-x'Qx\leq-\lambda_{\min}(Q)\|x\|^2$$

From the definition of quadratic stabilizability, we can take $\alpha=\lambda_{\min}(Q)$ and $x'Px$ is a Lyapunov function for the controlled system. An algorithm similar as for quadratic stability can be derived using the same arguments.

**Algorithm B.2**

1. Select positive definite symmetric matrices $Q$ and $R$, initialize $\varepsilon=\varepsilon_0>0$.
2. Solve the Riccati equation (B.12). If the solution is positive definite, stop. The system is quadratically stabilizable, compute the control gain matrix. If not, go to Step 3.
3. Decrease the value of $\varepsilon$, for example $\varepsilon\leftarrow\frac{\varepsilon}{2}$. If $\varepsilon$ is lower than a defined precision, stop. The system is not quadratically stabilizable. If not go to Step 2.

It is possible to obtain an alternative to the Riccati equation approach through an LMI formulation which is more adequate when the unknowns have a particular structure (as for example a block diagonal structure). The possibility to deal with structural constraints allow us to solve some specific control problems (decentralized, static output feedback, ...) as described in [45, 116].

**Theorem B.5** *System (B.9) with NB-uncertainty is quadratically stabilizable if and only if there exist a positive definite symmetric matrix $W\in\mathfrak{R}^{n\times n}$ and a matrix $S\in\mathfrak{R}^{m\times n}$ of appropriate dimensions such that*

$$\begin{bmatrix} A_0W+WA_0'+B_0S+S'B_0'+DD' & WE_1'+S'E_2' \\ E_1W+E_2S & -I \end{bmatrix}<0$$

*A quadratic stabilizing control is then given by*

$$u(t)=SW^{-1}x(t)$$

Using Schur's complement, the previous inequality can be transformed as

$$A_0W+WA_0'+B_0S+S'B_0+DD'+\left(WE_1'+S'E_2'\right)(E_1W+E_2S)<0$$

B.4 Guaranteed Cost Control

Right and left multiplying by $W^{-1}$ and denoting $K = SW^{-1}$, we have

$$(A_0 + B_0 K)'W^{-1} + W^{-1}(A_0 + B_0 K) + \varepsilon W^{-1}DD'W^{-1}$$
$$+ \varepsilon^{-1}(E_1 + E_2 K)'(E_1 + E_2 K) < 0$$

which signifies that $x'W^{-1}x$ is a Lyapunov function for the uncertain closed-loop system.

## B.4  Guaranteed Cost Control

The characterization of the set of all stabilizing controllers is central in practice because stability is a minimal requirement for a closed-loop system. Generally to satisfy some performance requirement, it is necessary to define a performance criterion and to pick up, among the stabilizing controllers, a controller which leads to a satisfactory closed-loop system behavior in the sense of this criterion. One way consists of associating to the system (B.9) the following quadratic cost function [230]:

$$J = \int_0^{+\infty} \left(x(t)'Qx(t) + u(t)'Ru(t)\right) dt \qquad (B.13)$$

where $Q = Q' > 0$ and $R = R' > 0$. For simplicity, suppose that uncertainty is time-invariant (that is $F(t) = F$, $G(t) = G$) and that system (B.9) is quadratically stabilizable, then it satisfies the conditions of the theorems stated in the previous sections. Let $V(x) = x'Px$, the associated Lyapunov function and $u(t) = Kx(t)$ the associated quadratic stabilizing controller. Replacing $u(t)$ in the quadratic criterion, we obtain

$$J(F, G) = \int_0^{+\infty} x(t)'\left(Q + K'RK\right)x(t) dt$$
$$= x(0)'\left[\int_0^{+\infty} e^{(A(F)+B(G)K)'t}\left(Q + K'RK\right)e^{(A(F)+B(G)K)t} dt\right]x(0)$$
$$= x(0)'\Lambda(F, G)x(0)$$

with $\Lambda(F, G) = \int_0^{+\infty} e^{(A(F)+B(G)K)'t}(Q + K'RK)e^{(A(F)+B(G)K)t} dt$. Because the system is quadratically stabilizable, for all $F \in \mathcal{F}$ and $G \in \mathcal{G}$, $\Lambda(F, G)$ is a positive semi-definite symmetric matrix solution of the following Lyapunov equation (see Appendix C):

$$\left[A(F) + B(G)K\right]'\Lambda(F, G) + \Lambda(F, G)\left[A(F) + B(G)K\right] + Q + K'RK = 0$$

We also have by quadratic stability of the closed-loop system for all $F \in \mathcal{F}, G \in \mathcal{G}$ and $x \in \mathfrak{R}^n$

$$\frac{dV(x)}{dt} = x'\left[\left(A(F) + B(G)K\right)'P + P\left(A(F) + B(G)K\right)\right]x < 0$$

378    B   Quadratic Approach for Robust Control

Thanks to some manipulations of the two equations above, one obtains

$$-\frac{dV(x)}{dt} = x'\big[\big(A(F) + B(G)K\big)'\big(\Lambda(F,G) - P\big)$$
$$+ \big(\Lambda(F,G) - P\big)\big(A(F) + B(G)K\big)\big]x + x'\big[Q + K'RK\big]x > 0$$

Inspecting this last inequality, we can deduce that if we impose

$$-\frac{dV(x)}{dt} \geq x'\big[Q + K'RK\big]x$$

we have for all $F \in \mathcal{F}$ and $G \in \mathcal{G}$:

$$\big(A(F) + B(G)K\big)'\big(P - \Lambda(F,G)\big) + \big(P - \Lambda(F,G)\big)\big(A(F) + B(G)K\big) < 0$$

and then for all $F \in \mathcal{F}$, $G \in \mathcal{G}$ and from the stability of $(A(F) + B(G)K)$:

$$\Lambda(F,G) \leq P$$

Consequently for all $F \in \mathcal{F}$ and $G \in \mathcal{G}$:

$$J(F,G) \leq V\big(x(0)\big)$$

The above discussion suggests the definition of a guaranteed cost control as follows [289].

**Definition B.3** Consider the system (B.9) and the quadratic cost function (B.13). A control $u(t) = Kx(t)$ is a guaranteed cost control associated to (B.13) if there exist a positive definite symmetric matrix $P \in \mathfrak{R}^{n \times n}$ and a positive scalar $\beta$ such that for all $F \in \mathcal{F}$, $G \in \mathcal{G}$ and $x \in \mathfrak{R}^n$:

$$x'\big[\big(A(F) + B(G)K\big)'P + P\big(A(F) + B(G)K\big) + Q + K'SK\big]x \leq -\beta\,\|x\|^2 \tag{B.14}$$

$P$ is called a guaranteed cost matrix and we have for all $F \in \mathcal{F}$ and $G \in \mathcal{G}$:

$$J(F,G) \leq x(0)'Px(0)$$

$x(0)'Px(0)$ is called a guaranteed cost.

In general, it is possible to obtain a guaranteed cost independent upon the initial condition $x(0)$ considering that the initial condition is a random vector with zero mean and with a covariance matrix equal to identity matrix $I_n$. In that case, we take the expectation of the quadratic cost function (B.13) and the last inequality in the previous definition becomes

$$J(F,G) \leq \text{trace}(P)$$

for all $F \in \mathcal{F}$ and $G \in \mathcal{G}$. The problem considered in this section, for system (B.9) involving polytopic or norm-bounded uncertainty, is the following.

**Guaranteed cost problem** "Find the optimal guaranteed cost control $u(t) = Kx(t)$ which minimizes the guaranteed cost trace$(P)$."

B.4 Guaranteed Cost Control 379

## B.4.1 P-Uncertainty

For polytopic uncertainty, we have the following theorem [74].

**Theorem B.6** *Consider system (B.9) with P-uncertainty and the quadratic cost function (B.13). A solution to the guaranteed cost control is obtained solving the following optimization problem:*

$$\min_{W=W'>0,\,S,\,\Gamma} \quad \text{trace}(\Gamma)$$

$$\begin{bmatrix} \Gamma & I_n \\ I_n & W \end{bmatrix} \geq 0$$

$$\begin{bmatrix} A_i W + W A_i' + B_j S + S' B_j' & S' & W \\ S & -R^{-1} & 0 \\ W & 0 & -Q^{-1} \end{bmatrix} < 0$$

$$\text{for all } i = 1, \dots, N_A \text{ and } j = 1, \dots, N_B$$

*Let $(W^*, S^*, \Gamma^*)$ the optimal solution. Then the optimal guaranteed cost control gain is given by $K = S^* W^{*-1}$, the optimal guaranteed cost matrix is $W^{*-1}$ and the optimal guaranteed cost is equal to $\Gamma^*$.*

*Proof* By Schur's complement, the last inequality is equivalent to

$$A_i W + W A_i' + B_j S + S' B_j' + S' R S + W Q W < 0$$

which, multiplied on the right and the left by $W^{-1}$, leads to

$$A_i' W^{-1} + W^{-1} A_i + W^{-1} B_j S W^{-1} + W^{-1} S' B_j' W^{-1} + W^{-1} S' R S W^{-1} + Q$$

$$= \left(A_i + B_j S W^{-1}\right)' W^{-1} + W^{-1}\left(A_i + B_j S W^{-1}\right) + W^{-1} S' R S W^{-1} + Q < 0$$

Then

$$\sum_{i=1}^{N_A} f_i \left[\left(A_i + B_j S W^{-1}\right)' W^{-1} + W^{-1}\left(A_i + B_j S W^{-1}\right) + W^{-1} S' R S W^{-1} + Q\right]$$

$$= \left(A(F) + B_j S W^{-1}\right)' W^{-1} + W^{-1}\left(A(F) + B_j S W^{-1}\right)$$

$$\quad + W^{-1} S' R S W^{-1} + Q < 0$$

$$\sum_{j=1}^{N_B} g_j \left[\left(A(F) + B_j S W^{-1}\right)' W^{-1} + W^{-1}\left(A(F) + B_j S W^{-1}\right) \right.$$

$$\quad \left. + W^{-1} S' R S W^{-1} + Q\right]$$

$$= \left(A(F) + B(G) S W^{-1}\right)' W^{-1} + W^{-1}\left(A(F) + B(G) S W^{-1}\right)$$

$$\quad + W^{-1} S' R S W^{-1} + Q < 0$$

The first inequality of Theorem B.6 is equivalent by Schur complement to

$$W^{-1} \leq \Gamma$$

At the optimum, the conclusions of the theorem follow. □

Find the solution of the optimization problem of Theorem B.6 can be done easily with a standard LMI solver because all the conditions and the criterion are linear [289].

## B.4.2 NB-Uncertainty

**Theorem B.7** *Consider system (B.9) with NB-uncertainty and the quadratic cost function (B.13). A solution to the guaranteed cost control is obtained solving the following optimization problem:*

$$\min_{P(\varepsilon)=P(\varepsilon)'>0,\ \varepsilon>0} \quad \text{trace } P(\varepsilon)$$

$$A'_\varepsilon P(\varepsilon) + P(\varepsilon)A_\varepsilon - P(\varepsilon)BR_\varepsilon^{-1}B'P(\varepsilon) + \varepsilon P(\varepsilon)DD'P(\varepsilon)$$

$$+ \varepsilon^{-1}E'_1\left(I - \varepsilon^{-1}E_2R_\varepsilon^{-1}E'_2\right)E_1 + Q = 0$$

*where:*

$$A_\varepsilon = A_0 - \varepsilon^{-1}BR_\varepsilon^{-1}E'_2E_1, \qquad R_\varepsilon = R + \varepsilon^{-1}E'_2E_2$$

*Let $(P(\varepsilon^*), \varepsilon^*)$ the optimal solution. Then the optimal guaranteed cost control gain is given by*

$$K = -R_\varepsilon^{-1}\left(B'P(\varepsilon^*) + \varepsilon^{*-1}E'_2E_1\right)$$

*The optimal guaranteed cost matrix is $P(\varepsilon^*)$ and the optimal guaranteed cost is equal to trace $P(\varepsilon^*)$. The above Riccati equation has a positive definite symmetric solution $P(\varepsilon)$ for $\varepsilon \in (0, \hat{\varepsilon}[$ where $\hat{\varepsilon}$ depends on $Q$ and $R$ and the function trace $P(\varepsilon)$ is convex with respect to $\varepsilon$.*

The results presented in this theorem follows from the results obtained for the quadratic stabilization problem and the fact that the Riccati equation can be written after some tedious manipulations as

$$(A_0 + B_0K)'P(\varepsilon) + P(\varepsilon)(A_0 + B_0K) + \varepsilon P(\varepsilon)DD'P(\varepsilon)$$

$$+ \varepsilon^{-1}(E_1 + E_2K)'(E_1 + E_2K) + K'RK + Q = 0$$

## B.4 Guaranteed Cost Control

Then we have

$$
\begin{aligned}
\big(A(F) &+ B(G)K\big)' P(\varepsilon) + P(\varepsilon)\big(A(F) + B(G)K\big) + K'RK + Q \\
&= \big(A_0 + B_0 K + DF(E_1 + E_2 K)\big)' P(\varepsilon) \\
&\quad + P(\varepsilon)\big(A_0 + B_0 K + DF(E_1 + E_2 K)\big) + K'RK + Q \\
&\leq (A_0 + B_0 K)' P(\varepsilon) + P(\varepsilon)(A_0 + B_0 K) + \varepsilon P(\varepsilon) DD' P(\varepsilon) \\
&\quad + \varepsilon^{-1}(E_1 + E_2 K)'(E_1 + E_2 K) + K'RK + Q = 0
\end{aligned}
$$

The proof of convexity of the function trace $P(\varepsilon)$ can be found in [109]. This last point is very important and guarantees that the optimum can be attained with an arbitrary precision through a one line search over $\varepsilon \in (0, \hat{\varepsilon}[$. $\hat{\varepsilon}$ can also be determined with an arbitrary precision by the algorithm presented in the previous section. An alternative to the Riccati equation approach is to derive conditions expressed in terms of LMIs [45]. This is done in the next theorem.

**Theorem B.8** *Consider system (B.9) with NB-uncertainty and the quadratic cost function (B.13). A solution to the guaranteed cost control is obtained solving the following optimization problem*:

$$
\min_{W = W' > 0, \ S, \ \Gamma} \quad \Gamma
$$

$$
\begin{bmatrix} \Gamma & I_n \\ I_n & W \end{bmatrix} \geq 0
$$

$$
\begin{bmatrix}
A_0 W + W A_0' + B_0 S + S' B_0' + DD' & W E_1' + S' E_2' & S' & W \\
E_1 W + E_2 S & -I_{f_1} & 0 & 0 \\
S & 0 & -R^{-1} & 0 \\
W & 0 & 0 & -Q^{-1}
\end{bmatrix} < 0
$$

*Let $(W^*, S^*, \Gamma^*)$ the optimal solution. Then the optimal guaranteed cost control gain is given by $K = S^* W^{*-1}$, the optimal guaranteed cost matrix is $W^{*-1}$ and the optimal guaranteed cost is equal to $\Gamma^*$.*

*Proof* By Schur's complement the latter inequality can be written as

$$
\begin{aligned}
A_0 W &+ W A_0' + B_0 S + S' B_0' + DD' + \big(W E_1' + S' E_2'\big)(E_1 W + E_2 S) \\
&+ S'RS + WQW < 0
\end{aligned}
$$

Multiplying on the right and the left by $W^{-1}$, we have

$$
\begin{aligned}
\big(A_0 &+ B_0 S W^{-1}\big)' W^{-1} + W^{-1}\big(A_0 + B_0 S W^{-1}\big) + W^{-1} DD' W^{-1} \\
&+ \big(E_1' + W^{-1} S' E_2'\big)(E_1 + E_2 S W^{-1}) + W^{-1} S' R S W^{-1} + Q < 0
\end{aligned}
$$

At the optimum, the result follows. $\qquad\square$

## B.5 Pole Placement

Another way to guarantee performances for the closed-loop system consists in pole placement for the closed-loop system in particular regions included in the left half complex plane. For that, we suppose as in the previous paragraph that uncertainty is time-invariant. Furthermore, it may be expected that for "slowly" time-varying uncertainty, good transient behavior for the controlled system will be ensured.

### B.5.1 Pole Placement in a Disk

The disk for pole location is an interesting region because it can be chosen in order to guarantee a good trade-off between overshoot and settling time [110]. For continuous-time systems, it suffices to include it in a sector located in the left half complex plane. The considered disk with center $\alpha + j0$ and radius $r$ is denoted by $d(\alpha, r)$ and defined by

$$d(\alpha, r) = \left\{ \lambda \in C : |\lambda - \alpha| < r \right\}$$

If $\lambda$ is a complex mode for the controlled system, $\omega_n = |\lambda|$, its undamped natural frequency, $\beta = \Re e[\lambda]$, its damping factor and $z = \beta \omega_n^{-1}$, its damping ratio, then one gets

$$\forall \lambda \in d(\alpha, r): \ \beta < r - \alpha \ ; \ z > \sqrt{1 - \frac{r^2}{\alpha^2}}$$

**Disk pole placement problem** "Find a control $u(t) = Kx(t)$ such that all the eigenvalues of $A(F) + B(G)K$ are located in $d(\alpha, r)$ for all $F \in \mathcal{F}$ and $G \in \mathcal{G}$."

#### B.5.1.1 $d$-Stabilizability

For solving the disk pole placement problem, we first introduce the concept of quadratic $d$-stabilizability [110] . The following notation will be used in the sequel.

$$A_r(F) = \frac{A(F) - \alpha I_n}{r}, \qquad B_r(G) = \frac{B(G)}{r}$$

**Definition B.4** [110]   System (B.9) is said to be quadratically $d$-stabilizable if there exist a linear state feedback $u(t) = Kx(t)$ and a positive definite symmetric matrix $P$ such that for all $F \in \mathcal{F}$ and $G \in \mathcal{G}$:

$$\left( A_r(F) + B_r(G)K \right)' P \left( A_r(F) + B_r(G)K \right) - P < 0 \qquad (B.15)$$

or equivalently

$$\begin{bmatrix} P^{-1} & (A_r(F) + B_r(G)K) \\ (A_r(F) + B_r(G)K)' & P \end{bmatrix} > 0$$

## B.5 Pole Placement

If $K$ and $P$ satisfy equation (B.15), the closed-loop poles of the system lie in $d(\alpha, r)$ for all $F \in \mathcal{F}$ and $G \in \mathcal{G}$ or equivalently the eigenvalues of the matrix $A_r(F) + B_r(G)K$ lie in $d(0, 1)$, see [110]. The relation between quadratic $d$-stabilizability and quadratic stabilizability is given in the following theorem.

**Theorem B.9** *If system (B.9) is quadratically d-stabilizable then it is quadratically stabilizable and P is a Lyapunov matrix for all the system realizations in the uncertainty domain.*

The proof of this theorem can be found in [110].

### B.5.1.2 P-Uncertainty

For polytopic uncertainty, we consider the following notation for the vertices:

$$A_{ri} = \frac{A_i - \alpha I_n}{r}, \qquad B_{rj} = \frac{B_j}{r}, \qquad i = 1, \ldots, N_A \text{ and } j = 1, \ldots, N_B$$

A solution to the disk pole placement problem is given in the following theorem [123]:

**Theorem B.10** *System (B.9) with P-uncertainty is quadratically d-stabilizable if and only if there exist a positive definite symmetric matrix $W \in \mathfrak{R}^{n \times n}$ and a matrix $S \in \mathfrak{R}^{m \times n}$ satisfying*

$$\begin{bmatrix} W & W A'_{ri} + S' B'_{rj} \\ A_{ri} W + B_{rj} S & W \end{bmatrix} > 0$$

$$\text{for all } i = 1, \ldots, N_A \text{ and } j = 1, \ldots, N_B$$

*A quadratic d-stabilizing gain is given by*

$$K = SW^{-1}$$

*and $x'W^{-1}x$ is a Lyapunov function for the closed-loop system.*

The proof of this theorem is obtained using Schur's complement and the property of polytopic uncertainty.

### B.5.1.3 NB-Uncertainty

For norm-bounded uncertainty, consider the following notation:

$$A_{0r} = \frac{A_0 - \alpha I_n}{r}, \qquad B_{0r} = \frac{B_0}{r}, \qquad D_r = \frac{D}{\sqrt{r}}, \qquad E_{1r} = \frac{E_1}{\sqrt{r}}, \qquad E_{2r} = \frac{E_2}{\sqrt{r}}$$

It is possible to connect quadratic $d$-stabilizability of system (B.9) to quadratic stabilizability of the system defined by [111]:

$$\dot{\eta} = A_r(F)\eta + B_r(G)v \qquad (B.16)$$

Define the Lebesgue space $\mathcal{L}_{d\infty}$ as the set of all measurable matrix-valued functions defined on the circle of center $\alpha + j0$ and radius $r$ denoted by $c(\alpha, r)$ and such that:

$$\|G\|_{d\infty} = \sup\{\bar{\sigma}(G(\omega)) : \omega \in c(\alpha, r)\} < \infty$$

When $c(\alpha, r) = c(0, 1)$, the norm $\|\cdot\|_{d\infty}$ is denoted as usually by $\|\cdot\|_\infty$. The subspace $\mathcal{H}_{d\infty}$ is the Hardy space consisting of all bounded functions with analytic continuation in

$$\bar{c}(\alpha, r) = \{\lambda \in \mathcal{C} : |\lambda - \alpha| > r\}$$

that is, the $d$-stable functions (with no singularities in $d(\alpha, r)$). We have the following important result for the sequel.

**Theorem B.11** [111] *The following statements are equivalent*:

1. *(B.9) with NB-uncertainty is quadratically $d$-stabilizable.*
2. *(B.16) is quadratically stabilizable.*
3. *There exists a matrix $K \in \mathfrak{R}^{m \times n}$ such that $A_{0r} + B_{0r}K$ is Schur stable and $\|(E_{1r} + E_{2r}K)(sI_n - A_{0r} - B_{0r}K)^{-1}D_r\|_\infty < 1$.*
4. *There exists a matrix $K \in \mathfrak{R}^{m \times n}$ such that $A_0 + B_0K$ is $d$-stable and $\|(E_1 + E_2K)(sI_n - A_0 - B_0K)^{-1}D\|_{d\infty} < 1$.*

From the points 1 and 3 of this theorem, to solve the disk pole placement problem, we can apply the results developed in [414] for standard discrete-time $H_\infty$ control synthesis. The Riccati equation approach leads to the following theorem [109].

**Corollary B.3** *System (B.9) with NB-uncertainty is quadratically $d$-stabilizable if and only if given positive definite symmetric matrices $Q \in \mathfrak{R}^{n \times n}$ and $R \in \mathfrak{R}^{m \times m}$, there exists a positive scalar $\varepsilon$ such that the Riccati equation*

$$A'_{\varepsilon r}\left(P^{-1} + B_{0r}R_\varepsilon^{-1}B'_{0r} - \varepsilon D_r D'_r\right)^{-1} A_{\varepsilon r} - P$$
$$+ \varepsilon^{-1} E'_{1r}\left(I_{f_2} - \varepsilon^{-1} E_{2r}R_{\varepsilon r}^{-1}E'_{2r}\right)E_{1r} + Q = 0$$

*has a positive definite symmetric solution $P \in \mathfrak{R}^{n \times n}$ such that*

$$\varepsilon^{-1}I_n - D'_r P D_r > 0$$

*where*

$$A_{\varepsilon r} = A_{0r} - \varepsilon^{-1}BR_\varepsilon^{-1}E'_{2r}E_{1r}, \qquad R_{\varepsilon r} = R + \varepsilon^{-1}E'_{2r}E_{2r}$$

B.5 Pole Placement

385

*A quadratic d-stabilizing control is then given by*

$$u(t) = -\left[ R_{\varepsilon r}^{-1} B_{0r}' \left( P^{-1} + B_{0r} R_\varepsilon^{-1} B_{0r}' - \varepsilon D_r D_r' \right)^{-1} A_{\varepsilon r} + \varepsilon^{-1} R_{\varepsilon r}^{-1} E_{2r}' E_{1r} \right] x(t)$$

*Proof* From Theorem B.11, the system is quadratically $d$-stabilizable if and only if

$$\left\| (E_{1r} + E_{2r} K)(s I_n - A_{0r} - B_{0r} K)^{-1} D_r \right\|_\infty < 1$$

which is equivalent to

$$\left\| \left( \begin{bmatrix} E_{1r} \\ 0 \\ (\varepsilon Q)^{1/2} \end{bmatrix} + \begin{bmatrix} E_{2r} \\ (\varepsilon R) \\ 0 \end{bmatrix} K \right) (s I_n - A_{0r} - B_{0r} K)^{-1} D \right\|_\infty < 1$$

for all $Q = Q' > 0$, $R = R' > 0$, $\varepsilon \in (0, \hat{\varepsilon}[$ where $\hat{\varepsilon}$ depends on $Q$ and $R$. Applying the standard results on discrete-time $H_\infty$ control synthesis, the result follows. $\quad\square$

Another formulation in terms of LMIs is presented below [115].

**Corollary B.4** *System (B.9) with NB-uncertainty is quadratically d-stabilizable if and only if there exist a positive definite symmetric matrix $W \in \mathfrak{R}^{n \times n}$ and a matrix $S \in \mathfrak{R}^{m \times n}$ such that*

$$\begin{bmatrix} -W & D_r & A_{0r} W + B_{0r} S & 0 \\ D_r' & -I_{f_1} & 0 & 0 \\ W A_{0r}' + S' B_{0r}' & 0 & -W & W E_{1r}' + S' E_{2r}' \\ 0 & 0 & E_{1r} W + E_{2r} S & -I_{f_2} \end{bmatrix} < 0$$

*A quadratic stabilizing control is then given by*

$$u(t) = S W^{-1} x(t)$$

This result can be deduced by simple manipulations using Schur's complement.

## *B.5.2 Pole Placement in LMI Regions*

It is also possible to locate the eigenvalues of the closed-loop system matrix in other regions included in the complex plane. Some particular regions, called LMI regions [69], present a great interest in practice. The disk is a particular case of an LMI region.

**Definition B.5** [69] An LMI region is any subset of the complex plane defined as

$$\mathcal{R} = \left\{ z \in \mathcal{C}; \ f(z) = L + z M + \bar{z} M' < 0 \right\}$$

where $L \in \mathfrak{R}^{k \times k}$ and $M \in \mathfrak{R}^{k \times k}$ are real matrices such that $L = L'$.

Intersections of LMI regions are LMI regions and any convex region symmetric with respect to the real axis can be approximated by an LMI region with an arbitrary accuracy. Examples of LMI regions are the following [69]:

- half-plane $\Re e(z) < \delta : f(z) = z + \bar{z} + 2\delta < 0$;
- disk centered at $(\alpha, 0)$ with radius $r : f(z) = \left[ \begin{smallmatrix} -r & -\alpha+z \\ -\alpha+\bar{z} & -r \end{smallmatrix} \right] < 0$;

A square matrix $A$ is called $\mathcal{R}$-stable if all its eigenvalues lie in $\mathcal{R}$. We have the following result useful for robust pole placement [68]:

**Theorem B.12** *The matrix $A$ is $\mathcal{R}$-stable if and only if there exists a positive definite symmetric matrix $X$ such that*:

$$M(A, X) = L \otimes X + M \otimes (XA) + M' \otimes (XA)' < 0$$

*where $\otimes$ denotes the Kronecker product of two matrices* [149].

For the disk and for $A \in \Re^{n \times n}$ we have

$$M(A, X) = \begin{bmatrix} -rX & (A - \alpha I_n)X \\ X(A - \alpha I_n)' & -rX \end{bmatrix}$$

and $M(A, X) < 0$ is equivalent to

$$\frac{(A - \alpha I_n)'}{r} X \frac{(A - \alpha I_n)}{r} - X < 0$$

### B.5.2.1  Quadratic $\mathcal{R}$-Stabilizability

The quadratic stabilizability concept can be easily extended to deal with pole placement control problem defined as follows:

$\mathcal{R}$ **pole placement problem** "Find a control $u(t) = Kx(t)$ such that all the eigenvalues of $A(F) + B(G)K$ are located in $\mathcal{R}$ for all $F \in \mathcal{F}$ and $G \in \mathcal{G}$."

**Definition B.6** [69] System (B.9) is said to be quadratically $\mathcal{R}$-stabilizable if there exist a linear state feedback $u(t) = Kx(t)$ and a positive definite symmetric matrix $X \in \Re^{n \times n}$ such that for all $F \in \mathcal{F}$ and $G \in \mathcal{G}$:

$$L \otimes X + M \otimes \left[ X\big(A(F) + B(G)K\big) \right] + M' \otimes \left[ X\big(A(F) + B(G)K\big) \right]' < 0$$

### B.5.2.2  P-Uncertainty

For polytopic uncertainty, we have the following theorem.

B.5 Pole Placement 387

**Theorem B.13** *System (B.9) with P-uncertainty is quadratically $\mathcal{R}$-stabilizable if and only if there exist a matrix $W \in \Re^{n \times n}$ and a matrix $S \in \Re^{m \times n}$ such that*

$$L \otimes W + M \otimes (A_i W + B_j S) + M' \otimes (W A_i' + S' B_j') < 0$$

$$\text{for all } i = 1, \ldots, N_A \text{ and } j = 1, \ldots, N_B \tag{B.17}$$

*A quadratic $\mathcal{R}$-stabilizing gain is given by*

$$K = SW^{-1}$$

*Proof* Multiplying (B.17) on the right and on the left by $I_k \otimes W^{-1}$ and using the following properties of the Kronecker product [149]:

$$(A + B) \otimes C = A \otimes C + B \otimes C$$

$$(A \otimes B)(C \otimes D) = AC \otimes BD$$

$$(A \otimes B)' = A' \otimes B'$$

$$(A \otimes B)^{-1} = A^{-1} \otimes B^{-1}$$

we obtain

$$L \otimes W^{-1} + M \otimes \left[ W^{-1} (A_i + BjSW^{-1}) \right] + M' \otimes \left[ (A_i' + W^{-1} S' B_j') W^{-1} \right] < 0$$

and then

$$\sum_{i=1}^{N_A} \sum_{j=1}^{N_B} f_i g_j \left\{ L \otimes W^{-1} + M' \otimes \left[ W^{-1} (A_i + B_j SW^{-1}) \right] \right.$$

$$\left. + M' \otimes \left[ W^{-1} (A_i' + W^{-1} S' B_j') \right] \right\} < 0$$

Comparing with Definition B.13, we see that $X = W^{-1}$ and that $K = SW^{-1}$. $\quad\square$

### B.5.2.3 NB-Uncertainty

A sufficient condition for norm-bounded case can be obtained.

**Theorem B.14** *System (B.9) with NB-uncertainty is quadratically $\mathcal{R}$-stabilizable if there exist matrices $W \in \Re^{n \times n}$, $P \in \Re^{k \times k}$ and a matrix $S \in \Re^{m \times n}$ such that*

$$\begin{bmatrix} L \otimes W + M \otimes Y(W, S) + M' \otimes Y(W, S)' & I_k \otimes D & M \otimes Z(W, S)' \\ I_k \otimes D' & -I_{k.f_1} & 0 \\ M' \otimes Z(W, S) & 0 & -I_{k.f_2} \end{bmatrix} < 0 \tag{B.18}$$

*with $Y(W, S) = A_0 W + B_0 S$ and $Z(W, S) = E_1 W + E_2 S$. A quadratic $\mathcal{R}$-stabilizing gain is given by*

$$K = SW^{-1}$$

*Proof* System (B.9) is quadratically $\mathcal{R}$-stabilizable if

$$L \otimes X + M \otimes \left[ X(A_0 + B_0 K + DF(E_1 + E_2 K)) \right]$$
$$+ M' \otimes \left[ X(A_0 + B_0 K + DF(E_1 + E_2 K)) \right]' < 0 \qquad (B.19)$$

Using the properties of the Kronecker product, we can write

$$M \otimes \left[ XDF(E_1 + E_2 K) \right] = (I_k \otimes X)(M \otimes DF(E_1 + E_2 K))$$
$$= (I_k \otimes X)(I_k \otimes D)(M \otimes F(E_1 + E_2 K)X)$$
$$= (I_k \otimes X)(I_k \otimes D)(I_k \otimes F)(M \otimes (E_1 + E_2 K))$$

Then we can deduce by denoting the left-hand term of (B.19) as $\mathcal{L}$:

$$\mathcal{L} \leq L \otimes X + M \otimes \left[ X(A_0 + B_0 K) \right] + M' \otimes \left[ X(A_0 + B_0 K) \right]'$$
$$+ (I_k \otimes X)(I_p \otimes D)(I_p \otimes D)'(I_k \otimes X)$$
$$+ \left( M \otimes (E_1 + E_2 K) \right)' \left( M \otimes (E_1 + E_2 K) \right)$$

Multiplying (B.18) by

$$\begin{bmatrix} I_k \otimes X & 0 & 0 \\ 0 & I_{k.f_1} & 0 \\ 0 & 0 & I_{k.f_1} \end{bmatrix}$$

with $X = W^{-1}$, and denoting $K = SX$ and using Schur's complement, one obtains

$$L \otimes X + M \otimes \left[ X(A_0 + B_0 K) \right] + M' \otimes \left[ X(A_0 + B_0 K) \right]'$$
$$+ (I_k \otimes X)(I_p \otimes D)(I_p \otimes D)'(I_k \otimes X)$$
$$+ \left( M \otimes (E_1 + E_2 K) \right)' \left( M \otimes (E_1 + E_2 K) \right) < 0$$

and then (B.19) is satisfied. $\qquad \square$

# Appendix C
# Linear Matrix Inequalities (LMI) and Riccati Equations

## C.1 Introduction

In this appendix, several tools used extensively in the chapters of the manuscript are introduced. Some of them are connected with notions introduced in the previous appendices. It is well known that solving stability analysis or control problems for systems described by linear state space models lead to conditions formulated in terms of solutions of Lyapunov equations, Riccati equations or Linear Matrix Inequalities [45, 230]. An immense literature devoted to the study of the basic properties of these equations and inequalities or the associated numerical methods for solving them exists, see for example [45, 414] and bibliographies therein. It is quite impossible to give a complete overview. Here, only the most important properties related to control design or stability analysis problems involved in the methods developed in the book are presented. For more details, the reader may consult the cited works.

This appendix ends with the presentation of two important theorems known as the "S-procedure" and the "Elimination lemma". The elimination lemma is used to eliminate variables in order to obtain more tractable conditions expressed for example in an LMI form while the S-procedure allows to replace a set of inequalities by one inequality [45].

All the results are presented without proofs. All the proofs can be found in the cited references.

## C.2 Algebraic Lyapunov Equation (ALE)

It is well known that stability, observability or controllability are fundamental properties in linear system analysis and control design [64]. Some well-known direct tests exist, for example the Kalman tests for observability or controllability. Sometimes, these tests have to be performed indirectly [64]. In those cases, Lyapunov and Riccati equations play a central role.

S. Tarbouriech et al., *Stability and Stabilization of Linear Systems with Saturating Actuators*, DOI 10.1007/978-0-85729-941-3, © Springer-Verlag London Limited 2011

Norms and their computation (in particular the $H_2$ norm) are used for specifying performances. In all these problems, a key tool involved is "the Lyapunov equation", connected to Lyapunov theory (see Appendix A).

## C.2.1 Continuous Algebraic Lyapunov Equation (CALE) [414]

The continuous-time algebraic Lyapunov equation is defined by

$$A'P + PA + Q = 0 \tag{C.1}$$

where $A$ and symmetric $Q$ are given real $(n \times n)$ matrices. The unknown is $P = P'$ and the CALE has a unique solution if and only if

$$\lambda_i(A) + \lambda_j^*(A) \neq 0, \quad \forall i, j$$

There exist some relationships between the stability of $A$, $Q$ and the solution $P$, summarized in the following lemma.

**Lemma C.1** *Assume that $A$ is Hurwitz. Then, for symmetric matrices $P$ and $Q$, we have the following facts*:

(i) $P = \int_0^\infty e^{A't} Q e^{At} \, dt$.
(ii) $P > 0$ if $Q > 0$ and $P \geq 0$ if $Q \geq 0$.
(iii) *If $Q \geq 0$ then $(Q, A)$ is observable if and only if $P > 0$*.

Part (iii) allows to conclude that a pair $(C, A)$, with $C \in \Re^{p \times n}$, is observable if and only if the solution to the following Lyapunov equation

$$A'P + PA + C'C = 0 \tag{C.2}$$

is positive definite. In this case $P$ is the observability gramian. By duality, a pair $(A, B)$ is controllable if and only if the solution to

$$AS + SA' + BB' = 0 \tag{C.3}$$

is positive definite. $S$ is the controllability gramian. $P$ and $S$ are involved in the computation of the $H_2$ norm. Let us define the system

$$\dot{x} = Ax + Bw$$
$$y = Cx$$

Define the transfer function $T(s) = C(sI - A)^{-1}B$. The $H_2$ norm of $T(s)$ denoted $\|T(s)\|_2$ is defined by

$$\|T(s)\|_2^2 = \frac{1}{2\pi} \int_{-\infty}^{\infty} \text{trace}\left[T^*(j\omega)T(j\omega)\right] d\omega$$

C.2 Algebraic Lyapunov Equation (ALE)

**Lemma C.2** *Suppose that A is Hurwitz. Then*

$$\|T(s)\|_2^2 = \text{trace}(B'PB) = \text{trace}(CSC')$$

*where P and S are, respectively, the observability and controllability gramians.*

In Lemma C.1, it is assumed that $A$ is Hurwitz. In some cases, $P$ is obtained as a solution to a Lyapunov equation and it is possible to conclude about the stability of $A$ as the following lemma shows.

**Lemma C.3** *Suppose that P is the solution of the Lyapunov equation (C.1), then*

(i) $\Re e(\lambda_i(A)) \leq 0$ *if* $P = P' > 0$ *and* $Q = Q' \geq 0$.
(ii) $A$ *is Hurwitz if* $P = P' > 0$ *and* $Q = Q' > 0$.
(iii) $A$ *is Hurwitz if* $P = P' \geq 0$, $Q = Q' \geq 0$ *and* $(Q, A)$ *is detectable.*

## C.2.2 Discrete Algebraic Lyapunov Equation (DALE) [414]

For discrete-time linear systems, the Lyapunov equation reads

$$A'PA - P + Q = 0 \tag{C.4}$$

where $A$ and $Q = Q'$ are given real $(n \times n)$ matrices. $P = P'$ is the unknown and this equation has a unique solution if and only if

$$\lambda_i(A)\lambda_j(A) \neq 1, \quad \forall i, j$$

As for the continuous Lyapunov equation, there exist relationships between the stability of $A$, $Q$ and the solution $P$, summarized in the following lemma.

**Lemma C.4** *Assume that A is Schur–Cohn stable. Then, for symmetric matrices P and Q, we have the following facts:*

(i) $P = \sum_{i=1}^{\infty} (A')^i Q A^i$.
(ii) $P > 0$ *if* $Q > 0$ *and* $P \geq 0$ *if* $Q \geq 0$.
(iii) *If* $Q \geq 0$ *then* $(Q, A)$ *is observable if and only if* $P > 0$.

From part (iii), the pair $(C, A)$ is observable if and only if the solution to the following Lyapunov equation

$$A'PA - P + C'C = 0 \tag{C.5}$$

is positive definite, $P$ being the observability gramian. By duality, the pair $(A, B)$ is controllable if and only if the solution to

$$ASA' - P + BB' = 0 \tag{C.6}$$

is positive definite, $S$ being the controllability gramian. $P$ and $S$ allows the computation of the $H_2$ norm of the system

$$x_{k+1} = Ax_k + Bw_k$$
$$y_k = Cx_k$$

Define the transfer function $T(z) = C(zI - A)^{-1}B$. The $H_2$ norm of $T(z)$ denoted $\|T(z)\|_2$ is defined by

$$\|T(z)\|_2^2 = \frac{1}{2\pi} \int_0^{2\pi} \text{trace}\big[T^*\big(e^{j\theta}\big)T\big(e^{j\theta}\big)\big]\, d\theta$$

**Lemma C.5** *Suppose that $A$ is stable. Then*

$$\|T(s)\|_2^2 = \text{trace}\big(B'PB\big) = \text{trace}\big(CSC'\big)$$

*where $P$ and $S$ are, respectively, the observability and controllability gramians.*

The discrete-time counterpart of Lemma C.3 can be stated as follows.

**Lemma C.6** *Suppose that $P$ is the solution of the Lyapunov equation (C.1), then*

(i)  $|\lambda_i(A)| \leq 1$ *if* $P = P' > 0$ *and* $Q = Q' \geq 0$.
(ii)  $A$ *is Schur if* $P = P' > 0$ *and* $Q = Q' > 0$.
(iii)  $A$ *is Schur if* $P = P' \geq 0$, $Q = Q' \geq 0$ *and* $(Q, A)$ *is detectable.*

## C.3 Algebraic Riccati Equation (ARE)

In control theory, algebraic Riccati equations have a special role. Numerous control design problems have a solution which can be expressed through the solution of an algebraic Riccati equation. This is the case, for example, in the $H_2$ and $H_\infty$ control synthesis problems, but also in linear optimal filtering (Kalman filtering) or LQG designs.

### C.3.1 Continuous Algebraic Riccati Equation (CARE) [414]

A continuous algebraic Riccati equation can be written as

$$A'X + XA + XRX + Q = 0 \tag{C.7}$$

where $A$, $Q$ and $R$ are given real $(n \times n)$ matrices with $Q$ and $R$ symmetric and $X$ is the unknown. Associated with the Riccati equation is a $(2n \times 2n)$ matrix defined

C.3 Algebraic Riccati Equation (ARE) 393

by

$$H = \begin{bmatrix} A & R \\ -Q & -A' \end{bmatrix}$$

Considering the $(2n \times 2n)$ matrix

$$J = \begin{bmatrix} 0 & -I \\ I & 0 \end{bmatrix}$$

It follows that

$$J^2 = -I$$

and

$$J^{-1}HJ = -JHJ = -H'$$

which means that $H$ and $-H'$ are similar. Thus if $\lambda$ is an eigenvalue of $H$, $-\lambda^*$ is also an eigenvalue of $H$. The matrix $H$ is said to be Hamiltonian.

### C.3.1.1 Characterization of Solutions of CARE

Two important theorems relate the solutions of a CARE to the invariant subspaces of $H$.

**Theorem C.1** *Let $\mathcal{X} \subset \mathcal{C}^{2n}$ be an n-dimensional invariant subspace of $H$ and let $X_1$, $X_2 \in \mathcal{C}^{n \times n}$ be two complex matrices such that*

$$\mathcal{X} = \text{Im} \begin{bmatrix} X_1 \\ X_2 \end{bmatrix}$$

*If $X_1$ is invertible, then $X = X_2 X_1^{-1}$ is a solution to the Riccati equation (C.7) and the spectrum of $A + RX$ is the same as the spectrum of $H|\mathcal{X}$. The solution $X$ is independent of the bases of $\mathcal{X}$.*

The following theorem is the converse of the previous one.

**Theorem C.2** *If $X \in \mathcal{C}^{n \times n}$ is a solution of the CARE (C.7), then there exist matrices $X_1$, $X_2 \in \mathcal{C}^{n \times n}$ with $X_1$ invertible such that $X = X_2 X_1^{-1}$ and the columns of $\begin{bmatrix} X_1 \\ X_2 \end{bmatrix}$ form a basis of an n-dimensional invariant subspace of $H$.*

These two theorems give a characterization of the solutions of a CARE but nothing is said about the structure of the solutions. There exists a sufficient condition for a solution to be hermitian and a necessary and sufficient one for a solution to be real. Among the possible solutions a CARE has, the solutions called "stabilizing solutions" are of a major interest in a lot of problems. In the following paragraph, we discuss the properties of such solutions.

## C.3.1.2 Stabilizing Solution—Riccati Operator

We say that a solution of the CARE is stabilizing if $\lambda(A + RX) \subset C_-$. If we suppose that $H$ has no eigenvalues on the imaginary axis, it must have $n$ eigenvalues in $\Re e(s) < 0$ and $n$ in $\Re e(s) > 0$. We consider the two $n$-dimensional spectral subspaces $\mathcal{X}_-(H)$ and $\mathcal{X}_+(H)$ corresponding, respectively, to the invariant subspaces associated to eigenvalues in $\Re e(s) < 0$ and in $\Re e(s) > 0$. A basis for $\mathcal{X}_-(H)$ can be found and, partitioning accordingly the matrix formed by the eigenvectors which can be selected real, we can write

$$\mathcal{X}_-(H) = \mathrm{Im}\begin{bmatrix} X_1 \\ X_2 \end{bmatrix}$$

where $X_1, X_2 \in C^{n \times n}$. Moreover, if $X_1$ is nonsingular or equivalently if the subspaces

$$\mathcal{X}_-(H) \quad \text{and} \quad \mathrm{Im}\begin{bmatrix} X_1 \\ X_2 \end{bmatrix}$$

are complementary, we can set $X = X_2 X_1^{-1}$ and $X$ is uniquely defined by $H$. If $H$ is such that:

- it has no eigenvalues on the imaginary axis,
- the subspaces $\mathcal{X}_-(H)$ and $\mathrm{Im}\begin{bmatrix} X_1 \\ X_2 \end{bmatrix}$ are complementary,

we say that $H$ belongs to $\mathrm{Dom}(\mathrm{Ric}) \subset \Re^{2n \times 2n}$. As in this case, $X$ is uniquely defined by $H$, we can also define the function

$$\mathrm{Ric} : \mathrm{Dom}(\mathrm{Ric}) \longmapsto \Re^{n \times n}$$

$$H \longmapsto X = \mathrm{Ric}(H)$$

The following fundamental theorem gives a characterization of the stabilizing solutions.

**Theorem C.3** *Suppose that $H \in \mathrm{Dom}(\mathrm{Ric})$ and $X = \mathrm{Ric}(H)$. Then*

(i) *$X$ is real symmetric,*
(ii) *$X$ satisfies the following CARE:*

$$A'X + XA + XRX + Q = 0$$

(iii) *$A + RX$ is stable.*

We can remark that in this theorem no assumption is done on the positive definiteness of the matrix $R$. Under certain restrictions on the sign of $R$, a necessary and sufficient condition for the existence of a unique stabilizing solution can be stated.

C.3 Algebraic Riccati Equation (ARE) 395

**Theorem C.4** *Suppose that $H$ has no eigenvalues on the imaginary axis and that $R$ is either positive semi-definite or negative semi-definite. Then $H \in \text{Dom(Ric)}$ if and only if the pair $(A, R)$ is stabilizable.*

Important particular cases are presented in the following theorem and its corollary.

**Theorem C.5** *If*

$$H = \begin{bmatrix} A & -BB' \\ -C'C & -A' \end{bmatrix}$$

*then $H \in \text{Dom(Ric)}$ if and only if $(A, B)$ is stabilizable and $(C, A)$ has no unobservable imaginary modes. Furthermore*

(i) *$X = \text{Ric}(H) \geq 0$ if $H \in \text{Dom(Ric)}$,*
(ii) *$\text{Ker}(X) = 0$ if and only if $(C, A)$ has no stable unobservable modes.*

**Corollary C.1** *Suppose that $(A, B)$ is stabilizable and $(C, A)$ is detectable. Then the CARE*

$$A'X + XA - XBB'X + C'C = 0$$

*has a unique positive semi-definite solution. Moreover, the solution is stabilizing.*

## C.3.2 Discrete Algebraic Riccati Equation (DARE) [414]

All the results presented in this paragraph are similar to the ones corresponding to the continuous algebraic Riccati equation. A discrete algebraic Riccati equation can be written

$$A'XA - X - A'XR(I + XR)^{-1}XA + Q = 0 \tag{C.8}$$

where $A$, $Q$ and $R$ are given real $(n \times n)$ matrices with $Q$ and $R$ symmetric and $X$ is the unknown. Although the results of this section are valid for the case where $A$ is singular, for the sake of simplicity, we consider $A$ invertible in the sequel. Associated with the Riccati equation is a $(2n \times 2n)$ matrix defined by

$$S = \begin{bmatrix} A + R(A')^{-1}Q & -R(A')^{-1} \\ -(A')^{-1}Q & (A')^{-1} \end{bmatrix}$$

By introducing the $(2n \times 2n)$ matrix

$$J = \begin{bmatrix} 0 & -I \\ I & 0 \end{bmatrix}$$

it follows that

$$J^{-1}S'J = S^{-1}$$

We say that the matrix $S$ is symplectic. A symplectic matrix has no eigenvalues at the origin and if $\lambda$ is an eigenvalue of a symplectic matrix $S$, $\lambda^*$, $1/\lambda$, $1/\lambda^*$ are also eigenvalues of $S$. It is possible to characterize the solutions in terms of the invariant subspaces of $S$ like in the continuous case. In the following, we concentrate the presentation on the stabilizing solutions.

### C.3.2.1 Stabilizing Solution—Riccati Operator

Assume that $S$ has no eigenvalues on the unit circle, it must have $n$ eigenvalues in $|z| < 1$ and $n$ in $|z| > 1$. We consider the two $n$-dimensional spectral subspaces $\mathcal{X}_-(H)$ and $\mathcal{X}_+(H)$ corresponding, respectively, to the invariant subspace associated to eigenvalues in $|z| < 1$ and in $|z| > 1$. A basis for $\mathcal{X}_-(H)$ can be found and, partitioning accordingly the matrix formed by the eigenvectors which can be selected real, we can write

$$\mathcal{X}_-(H) = \text{Im} \begin{bmatrix} Y_1 \\ Y_2 \end{bmatrix}$$

where $Y_1$, $Y_2 \in C^{n \times n}$. Moreover if $Y_1$ is nonsingular or equivalently if the subspaces

$$\mathcal{X}_-(H) \quad \text{and} \quad \text{Im} \begin{bmatrix} Y_1 \\ Y_2 \end{bmatrix}$$

are complementary, we can set $X = Y_2 Y_1^{-1}$ and $X$ is uniquely defined by $S$. If $S$ is such that:

- it has no eigenvalues on the unit disk,
- the subspaces $\mathcal{X}_-(H)$ and $\text{Im}\begin{bmatrix} X_1 \\ X_2 \end{bmatrix}$ are complementary,

we said that $S$ belongs to $\text{Dom}(\text{Ric}) \subset \mathfrak{R}^{2n \times 2n}$. As in this case, $X$ is uniquely defined by $S$, we can also define the function

$$\text{Ric} : \text{Dom}(\text{Ric}) \longmapsto \mathfrak{R}^{n \times n}$$

$$S \longmapsto X = \text{Ric}(S)$$

**Theorem C.6** *Suppose that $S \in \text{Dom}(\text{Ric})$ and $X = \text{Ric}(S)$. Then*

(i) *$X$ is unique and symmetric.*
(ii) *$I + RX$ is invertible and $X$ satisfies the DARE*

$$A'XA - X - A'XR(I + XR)^{-1}XA + Q = 0$$

(iii) *$A - R(I + XR)^{-1}XA = (I + RX)^{-1}A$ is stable.*

### C.4 Linear Matrix Inequalities (LMI) 397

*Remark C.3*

– We may remark that invoking the inversion matrix lemma the DARE can be written as

$$A'(I + XR)^{-1}XA - X + Q = 0$$

– When $A$ is singular, the eigenvalue problem of $S$ is replaced by the generalized eigenvalue problem

$$\lambda \begin{bmatrix} I & R \\ 0 & A' \end{bmatrix} - \begin{bmatrix} A & 0 \\ -Q & I \end{bmatrix}$$

Now if we restrict our attention to positive semi-definite $R$ and $Q$ matrices, we obtain the following theorems.

**Theorem C.7** *Suppose that $R$ and $Q$ are positive semi-definite. Then $S \in \mathrm{Dom}(\mathrm{Ric})$ if and only if $(A, R)$ is stabilizable and $S$ has no eigenvalues on the unit disk.*

**Theorem C.8** *Let*

$$S = \begin{bmatrix} A + R(A')^{-1}Q & -R(A')^{-1} \\ -(A')^{-1}Q & (A')^{-1} \end{bmatrix}$$

*where $R$ and $Q$ are positive semi-definite. Then $S$ has no eigenvalues on the unit disk if and only if $(A, R)$ has no unreachable modes and $(Q, A)$ has no unobservable modes on the unit disk.*

We end by the following important theorem.

**Theorem C.9** *Let*

$$S = \begin{bmatrix} A + R(A')^{-1}Q & -R(A')^{-1} \\ -(A')^{-1}Q & (A')^{-1} \end{bmatrix}$$

*where $R$ and $Q$ are positive semi-definite. Then $S \in \mathrm{Dom}(\mathrm{Ric})$ if and only if $(A, R)$ is stabilizable and $(Q, A)$ has no unobservable modes on the unit disk. Furthermore*

(i) $X = \mathrm{Ric}(S) \geq 0$ *if $S \in \mathrm{Dom}(\mathrm{Ric})$,*
(ii) $X > 0$ *if and only if $(C, A)$ has no stable unobservable modes.*

## C.4 Linear Matrix Inequalities (LMI)

Numerous problems in control design lead to conditions which can be formulated in terms of linear matrix inequalities. Recently, some powerful techniques to solve

LMIs were developed (interior point algorithms) and this is one of the reasons why a lot of work has been done since the end of the twentieth century in order to formulate analysis or control design problems into LMIs. An LMI can be written in a standard way as [45]

$$F(x) = F_0 + \sum_{i=1}^{m} x_i F_i > 0 \tag{C.9}$$

where $x \in \Re^m$ is the unknown and $F_i \in \Re^{n \times n}$ are symmetric matrices, $i = 0, \ldots, m$. In some cases, the inequality is non strict, that is,

$$F(x) \geq 0 \tag{C.10}$$

There exist connections between (C.9) and (C.10). These relations are discussed in details in [414]. In the sequel, we consider strict LMIs. First we can remark that if we have multiple LMIs, say, $F^{(1)}(x) > 0, \ldots, F^{(p)}(x) > 0$, we can replace them by a single LMI, say $\text{diag}(F^{(1)}(x), \ldots, F^{(p)}(x)) > 0$. Strictly speaking, (C.9) is equivalent to a set of $n$ polynomial inequalities.

*Example C.1* Consider the Lyapunov equation

$$A'P + PA + Q = 0$$

where $A$ and $Q$ are given real $(n \times n)$ matrices. If $Q > 0$, the Lyapunov equation can be transformed to a strict LMI

$$A'P + PA < 0 \tag{C.11}$$

The previous inequality can be converted to the standard form. Introduce the matrices $P_1, \ldots, P_m$ with $m = \frac{n(n+1)}{2}$ which is a basis for symmetric $n \times n$ matrices. Then take $F_0 = 0$ and $F_i = -A'P_i - P_i A$, $i = 1, \ldots, m$. We can write

$$P = x_1 P_1 + \cdots + x_m P_m$$

Replacing it in (C.11), we obtain

$$F(x) = F_0 + \sum_{i=1}^{m} x_i F_i > 0$$

## C.5 Schur's Complement

It is possible to convert nonlinear convex inequalities into LMIs using Schur's complements. Suppose that we have the following LMI:

$$\begin{bmatrix} Q(x) & S(x) \\ S'(x) & R(x) \end{bmatrix} > 0$$

## C.5 Schur's Complement

where $Q(x) = Q'(x)$, $R(x) = R'(x)$, and $S(x)$ are function matrices affine in $x$. If $Q(x)$ is nonsingular for all $x$, we have

$$\begin{bmatrix} Q(x) & S(x) \\ S'(x) & R(x) \end{bmatrix} = \begin{bmatrix} I & 0 \\ S'(x)Q^{-1}(x) & I \end{bmatrix} \begin{bmatrix} Q(x) & 0 \\ 0 & \Delta(x) \end{bmatrix} \begin{bmatrix} I & Q^{-1}(x)S(x) \\ 0 & I \end{bmatrix} > 0$$

with $\Delta(x) = R(x) - S'(x)Q^{-1}(x)S(x)$. If $R(x)$ is nonsingular for all $x$, we have

$$\begin{bmatrix} Q(x) & S(x) \\ S'(x) & R(x) \end{bmatrix} = \begin{bmatrix} I & S(x)R^{-1}(x) \\ 0 & I \end{bmatrix} \begin{bmatrix} \nabla(x) & 0 \\ 0 & R(x) \end{bmatrix} \begin{bmatrix} I & 0 \\ R^{-1}(x)S'(x) & I \end{bmatrix} > 0$$

with $\nabla(x) = Q(x) - S(x)R^{-1}(x)S'(x)$. From these equations, with an appropriate invertibility assumption, we can deduce that

$$\begin{bmatrix} Q(x) & S(x) \\ S'(x) & R(x) \end{bmatrix} > 0 \quad \Leftrightarrow \quad Q(x) > 0, \quad R(x) - S'(x)Q^{-1}(x)S(x) > 0$$

$$\begin{bmatrix} Q(x) & S(x) \\ S'(x) & R(x) \end{bmatrix} > 0 \quad \Leftrightarrow \quad R(x) > 0, \quad Q(x) - S'(x)R^{-1}(x)S(x) > 0$$

These equivalences are often used to convert nonlinear inequalities to LMIs.

*Example C.2* Consider the quadratic matrix inequality

$$A'P + PA + PBB'P + Q < 0$$

where $A$ is a $(n \times n)$ matrix, $B$ is a $(n \times m)$ matrix and $Q = Q'$. $P = P'$ is the unknown. Using the Schur's complement with

$$Q(x) = -A'P - PA - Q, \qquad S(x) = PB, \qquad R(x) = I$$

we obtain

$$\begin{bmatrix} -A'P - PA - Q & PB \\ B'P & I \end{bmatrix} > 0$$

Introducing

$$F_0 = \begin{bmatrix} -Q & 0 \\ 0 & I \end{bmatrix}, \qquad F_i = \begin{bmatrix} -A'P_i - P_iA & P_iB \\ B'P_i & 0 \end{bmatrix}$$

and writing

$$P = x_1 P_1 + \cdots + x_m P_m$$

we obtain

$$F(x) = F_0 + \sum_{i=1}^{m} x_i F_i > 0$$

## C.6 Bilinear Matrix Inequalities (BMI)

Some control design problems cannot be easily formulated through an LMI formulation but lead to a more complex formulation known as bilinear matrix inequality (BMI). Some of the problems treated in the book belong to this class. A bilinear matrix inequality can be expressed as

$$F(x, y) = F_{00} + \sum_{i=1}^{n} x_i F_{i0} + \sum_{j=1}^{m} y_j F_{0j} + \sum_{i=1}^{n} \sum_{j=1}^{m} x_i y_j F_{ij} > 0 \qquad (C.12)$$

where $F_{ij} = F'_{ij} \in \Re^{p \times p}$, $x$ and $y$ are the unknowns. A good overview including methodological, structural and computational aspects can be found in [171] where a wide bibliography is also provided. We can remark that fixing $x$ or $y$ converts a BMI into an LMI. This suggests that some relaxation algorithms to solve BMIs can be implemented in the spirit of $D - K$ iterations, consisting in alternately fixing $x$ and then determine $y$ and vice versa, fixing $y$ and determine $x$. Even when this scheme converges, there is no guarantee to achieve a globally or even a locally optimal solution. When it converges, the only certainty is that the solution is feasible. For more details on computational aspects see [171].

*Example C.3* A system described by

$$\dot{x} = Ax + Bu$$
$$y = Cx$$

where $A \in \Re^{s \times s}$, $B \in \Re^{s \times q}$ and $C \in \Re^{p \times s}$, is stabilizable by a static output feedback if there exist a control $u = Ly = LCx$, $L \in \Re^{q \times p}$ and a positive definite symmetric matrix $P$ such that

$$(A + BLC)'P + P(A + BLC) < 0 \qquad (C.13)$$

Up to now an LMI formulation of this problem does not exist. We show that in fact a bilinear matrix inequality is obtained. For that, introduce the matrices $P_1, \dots, P_m$ with $m = \frac{s(s+1)}{2}$ which is a basis for symmetric $s \times s$ matrices and $L_1, \dots, L_n$ with $n = q \times p$ a basis for the $q \times p$ matrices. We have

$$P = y_1 P_1 + \cdots + y_m P_m$$
$$L = x_1 L_1 + \cdots + x_n L_n$$

Replacing in (C.13), we have

$$-\left(A + B \sum_{i=1}^{n} x_i L_i C\right)' \sum_{j=1}^{m} y_j P_j - \sum_{j=1}^{m} y_j P_j \left(A + B \sum_{i=1}^{n} x_i L_i C\right)$$

## C.6 Bilinear Matrix Inequalities (BMI)

$$= -\sum_{j=1}^{m} y_j \left(A' P_j + P_j A\right) - \sum_{i=1}^{n}\sum_{j=1}^{m} x_i y_j \left(C' L_i B' P_j + P_j B L_i C\right) > 0$$

Denoting

$$F_{00} = 0, \qquad F_{i0} = 0, \quad i = 1, \ldots, n$$
$$F_{0j} = -\left(A' P_j + P_j A\right), \quad j = 1, \ldots, m$$
$$F_{ij} = -\left(C' L_i B' P_j + P_j B L_i C\right), \quad i = 1, \ldots, n, \ j = 1, \ldots, m$$

we obtain

$$F(x, y) = F_{00} + \sum_{i=1}^{n} x_i F_{i0} + \sum_{j=1}^{m} y_j F_{0j} + \sum_{i=1}^{n}\sum_{j=1}^{m} x_i y_j F_{ij} > 0$$

### C.6.1 Eigenvalues and Generalized Eigenvalues Problems

We list here some interesting problems used in the sequel. These problems can be solved in polynomial time. For a meaning of the term "polynomial time", see [45]; to be concise, they can be solved in practice very efficiently.

#### C.6.1.1 EigenValue Problems (EVP)

This problem consists of minimizing the maximum eigenvalue of a matrix that depends affinely on an unknown. Its general form is

$$\min \ \lambda$$
$$\text{subject to} \quad \begin{cases} \lambda I - F_1(x) > 0 \\ F_2(x) > 0 \end{cases}$$

where $F_1(x)$ and $F_2(x)$ are symmetric matrices affine in $x$.

*Example C.4* An example of such a problem is described by

$$\min \ \gamma$$
$$\text{subject to} \quad \begin{cases} -A'X - XA - C'C - \gamma^{-1} X B B' X > 0 \\ X > 0 \end{cases}$$

From the Schur complement, this optimization problem can be written as

$$\min \ \gamma$$
$$\text{subject to} \quad \begin{cases} \begin{bmatrix} -A'X - XA - C'C & X B \\ B' X & \gamma I \end{bmatrix} > 0 \\ X > 0 \end{cases}$$

## C.6.1.2 Generalized EigenValue Problems (GEVP)

A generalization of the previous problems is presented here and can be stated as

$$\min \ \lambda$$

$$\text{subject to} \quad \begin{cases} \lambda F_1(x) - F_2(x) > 0 \\ F_1(x) > 0 \\ F_3(x) > 0 \end{cases}$$

where $F_1(x)$, $F_2(x)$ and $F_3(x)$ are symmetric matrices affine in $x$. If $\lambda_{\max}(X, Y)$ denotes the largest generalized eigenvalue of the pencil $\lambda Y - X$ with $Y > 0$ which is also the largest eigenvalue of $Y^{-1/2} X Y^{1/2}$, the previous optimization problem can be written as

$$\min \ \lambda_{\max}\big(F_2(x), F_1(x)\big)$$

$$\text{subject to} \quad \begin{cases} F_1(x) > 0 \\ F_3(x) > 0 \end{cases}$$

In fact the GEPV is a quasiconvex optimization problem since the constraints are convex, but the criterion is quasiconvex. This means that for a feasible $x$, $\hat{x}$ and $0 \le \mu \le 1$

$$\lambda_{\max}\big(F_2\big(\mu x + (1 - \mu)\hat{x}\big), F_1\big(\mu x + (1 - \mu)\hat{x}\big)\big)$$

$$\le \max\big\{\lambda_{\max}\big(F_2(x), F_1(x)\big), \ \lambda_{\max}\big(F_2(\hat{x}), F_1(\hat{x})\big)\big\}$$

An equivalent alternate form for a GEVP is

$$\min \ \lambda$$

$$\text{subject to} \quad F_2(x, \lambda) > 0$$

where $F_2(x, \lambda)$ is affine in $x$ when $\lambda$ is fixed, and affine in $\lambda$ when $x$ is fixed. Moreover $F_2(x, \lambda)$ satisfies the monotonicity condition

$$\lambda > \beta \quad \Rightarrow \quad F_2(x, \lambda) \ge F_2(x, \beta)$$

*Example C.5* An example of a GEPV is given below

$$\max \ \lambda$$

$$\text{subject to} \quad \begin{cases} -A'P - PA - 2\lambda P > 0 \\ P > 0 \end{cases}$$

this problem is involved in the largest bound of the decay rate problem, see [45].

## C.7 The Elimination Lemma and the S-Procedure

In this section, we present two important results extensively used in the robust control literature and used in the book.

### C.7.1 The Elimination Lemma [45]

This lemma allows one to eliminate some variables appearing in a matrix inequality leading to inequalities without the eliminated variable.

**Lemma C.7** *Consider the matrix inequality*

$$T(x) + U(x)KV'(x) + V(x)K'U'(x) > 0 \tag{C.14}$$

$T(x) \in \Re^{n \times n}$ *and* $U(x)$, $V(x)$ *and* $K$ *are matrices of appropriate dimensions. We suppose that the functions matrices* $T$, $U$ *and* $V$ *do not depend of* $K$. *Let* $\bar{U}(x)$ *and* $\bar{V}(x)$, *respectively, for every* $x$, *the orthogonal complements of* $U(x)$ *and* $V(x)$ *(i.e. maximal rank matrices such that* $U'(x)\bar{U}(x) = 0$ *and* $V'(x)\bar{V}(x) = 0$). *Then* (C.14) *holds for some* $x = x_0$ *and* $K$ *if and only if the inequalities*

$$\bar{U}'(x)T(x)\bar{U}(x) > 0$$
$$\bar{V}'(x)T(x)\bar{V}(x) > 0$$

*hold with* $x = x_0$. *Moreover, using Finsler's lemma* [45], *we have for some* $\lambda \in \Re$

$$T(x) - \lambda U(x)U'(x) > 0$$
$$T(x) - \lambda V(x)V'(x) > 0$$

A classical example of the use of this lemma is now presented.

*Example C.6* Consider the problem of stabilization by state feedback of the following system

$$\dot{x} = Ax + Bu$$

This system is stabilizable by state feedback if and only if there exist $W = W' > 0$ and a matrix $K$ of appropriate dimensions such that

$$(A + BK)W + W(A + BK)' < 0$$

Applying the elimination lemma, this inequality is equivalent to

$$\bar{B}'(AW + WA')\bar{B} < 0, \quad W > 0$$

which is an LMI independent of $K$. We have eliminated the unknown $K$. Moreover there exists $\lambda \in \Re$ such that

$$AW + WA' < \lambda BB', \quad W > 0$$

which is also an LMI. Now multiplying on the right and on the left by $W^{-1}$, we obtain

$$A'W^{-1} + W^{-1}A - \lambda W^{-1}BB'W^{-1} < 0$$

which can be translated into a Lyapunov inequality as

$$\left(A - \frac{1}{2}\lambda BB'W^{-1}\right)'W^{-1} + W^{-1}\left(A - \frac{1}{2}\lambda BB'W^{-1}\right) < 0$$

and

$$K = -\frac{1}{2}\lambda B'W^{-1}$$

This example illustrates the potentialities of the elimination lemma.

## C.7.2 The S-Procedure [45]

The S-procedure is also a tool extensively used in robust control literature. In some problems, we find that some quadratic function must be negative whenever some other quadratic functions are all negative. With the S-procedure, we can replace this problem by one inequality to be satisfied by introducing some positive scalars to be determined. There exists different versions of the S-procedure. In what follows, we present two cases. For further details and bibliography comments, the reader is referred to [45].

### C.7.2.1 The S-Procedure for Quadratic Functions and Nonstrict Inequalities

**Lemma C.8** *Let $F_0, \ldots, F_p$ be quadratic functions of the variable $x \in \Re^n$ defined by*

$$F_i(x) = x'T_i x + 2a_i'x + b_i, \quad i = 0, \ldots, p \text{ and } T_i = T_i'$$

*If there exists $\lambda_1 \geq 0, \ldots, \lambda_p \geq 0$, such that $\forall x$*

$$F_0(x) - \sum_{i=1}^{p} \lambda_i F_i(x) \geq 0$$

*then*

$$F_0(x) \geq 0 \quad \forall x \text{ such that } F_i(x) \geq 0, \ i = 1, \ldots, p$$

*When $p = 1$, the converse holds if there exists $x_0$ such that $F_1(x_0) > 0$.*

C.8 Ellipsoids and Polyhedral Sets 405

### C.7.2.2 The S-Procedure for Quadratic Forms and Strict Inequalities

**Lemma C.9** *Let $T_0, \ldots, T_p \in \mathfrak{R}^{n \times n}$ be symmetric matrices. If there exist $\lambda_1 \geq 0, \ldots, \lambda_p \geq 0$, such that*

$$T_0 - \sum_{i=1}^{p} \lambda_i T_i > 0$$

*then*

$$x' T_0 x > 0 \quad \forall x \neq 0 \text{ such that } x' T_i x \geq 0, \ i = 1, \ldots, p$$

*When $p = 1$, the converse holds if there exists $x_0$ such that $x_0' T_1 x_0 > 0$.*

## C.8 Ellipsoids and Polyhedral Sets

To take into account the limitations of the actuators on a control law, in almost all cases, it is necessary to constraint the states (state of the system or state of the controller) to belong to polyhedral sets. These constraints can be directly considered, as for example, in a saturation regions model, or inner approximated with the help of a quadratic Lyapunov function which defines an ellipsoid as level set. In the last case, an ellipsoidal state domain is considered and the resulting ellipsoid has to be included in a polyhedral set. Hence in the sequel, some properties concerning ellipsoids and polyhedral sets are invoked as well as outer or inner approximations of a polytope.

### C.8.1 Ellipsoid and Invariant Ellipsoid [291, 346]

Consider the linear system

$$\dot{x} = Ax \tag{C.15}$$

Let $Q \in \mathfrak{R}^{n \times n}$ be a positive definite matrix. The resulting ellipsoid centered at the origin can be defined as

$$\mathcal{E}(Q, \eta) = \left\{ x \in \mathfrak{R}^n; \ x' Q x \leq \eta^{-1} \right\}$$

There is no loss of generality to consider $\eta = 1$, because rescaling matrix $Q$ (i.e. $Q\eta$) we have

$$\mathcal{E}(Q, \eta) = \left\{ x \in \mathfrak{R}^n; \ x' \underbrace{(Q\eta)}_{P} x \leq 1 \right\} = \mathcal{E}(P, 1)$$

**Theorem C.10** *The ellipsoid $\mathcal{E}(P, 1)$ is positively invariant (or invariant) for the system (C.15) if and only if $P = P'$ satisfies*

$$A'P + PA \leq 0$$

*or equivalently*

$$AS + SA' \leq 0, \quad S = P^{-1}$$

Note that these conditions are LMIs. It is clear that there are connections between the concept of invariance and Lyapunov stability. In fact, for a stable system, a quadratic Lyapunov function can be used to define invariant ellipsoids.

## C.8.2 Convex Polyhedron and Invariant Polyhedron [135, 136, 344]

A convex polyhedron is a non empty subset of $\mathfrak{R}^n$ which can be characterized by a matrix $Q \in \mathfrak{R}^{r \times n}$ and a vector $\rho \in \mathfrak{R}^r$, $r$ and $n$ belonging to $\mathcal{N} - \{0\}$ in the following way

$$S(Q, \rho) = \left\{ x \in \mathfrak{R}^n; \ Qx \leq \rho \right\}$$

The inequality between vector is componentwise. We suppose that the set of inequalities defining $S(Q, \rho)$ is nonredundant. A polytope is a particular case of a polyhedron defined by

$$S(Q, \rho, \beta) = \left\{ x \in \mathfrak{R}^n; \ -\beta \leq Qx \leq \rho \right\}$$

we may note that

$$S(Q, \rho, \beta) = S\left( \begin{bmatrix} Q \\ -Q \end{bmatrix}, \begin{bmatrix} \rho \\ \beta \end{bmatrix} \right)$$

When $\rho = \beta$, we said that the polytope is symmetric and we denote $S(Q, \rho, \beta) = S(|Q|, \rho)$. Concerning positive invariance of a polyhedron, we have the following result.

**Theorem C.11** *The polyhedral set $S(|Q|, \rho)$ is a positively invariant set for the system (C.15) if and only if there exists a matrix $H \in \mathfrak{R}^{r \times r}$ satisfying $H_{ij} \geq 0$, $\forall i \neq j, 1 \leq i, j \leq r$ (essentially non-negative matrix) and such that*

$$QA - HQ = 0$$

$$H\rho \leq 0$$

This theorem is also related to Lyapunov stability theory. In this case, polyhedral Lyapunov functions are involved [31].

## C.8 Ellipsoids and Polyhedral Sets

### C.8.3 Maximum Volume Ellipsoid Contained in a Symmetric Polytope

Consider the following ellipsoid:

$$\mathcal{E}(S, 1) = \left\{ x \in \mathfrak{R}^n; \ x'Sx \leq 1 \right\}$$

and the symmetric polytope

$$S(|Q|, \rho) = \left\{ x \in \mathfrak{R}^n; \ |Qx| \leq \rho \right\}$$

We denote by $Q_{(i)}$ the $i$th row of $Q$, and by $\rho_{(i)}$ the $i$th component of vector $\rho$. We can write

$$S(|Q|, \rho) = \left\{ x \in \mathfrak{R}^n; \ |Q_{(i)}x| \leq \rho_{(i)}, \ \forall i \right\}$$

We have

$$x \in \mathcal{E}(S, 1) \quad \text{if } x'Sx \leq 1$$

$$x \in S(|Q|, \rho) \quad \text{if } x' \frac{Q'_{(i)} Q_{(i)}}{\rho_{(i)}^2} x \leq 1; \ \forall i$$

then

$$\mathcal{E}(S, 1) \subset S(|Q|, \rho) \quad \text{if } 1 \geq x'Sx \geq x' \frac{Q'_{(i)} Q_{(i)}}{\rho_{(i)}^2} x, \ \forall i$$

Suppose now that $Q$ and $\rho$ are given. We want to characterize the ellipsoids $\mathcal{E}(S, 1)$ that are contained in $S(|Q|, \rho)$. These ellipsoids are characterized by matrices $S$ which satisfy

$$\frac{Q'_{(i)} Q_{(i)}}{\rho_{(i)}^2} - S \leq 0, \quad \forall i, \ S > 0$$

Using Schur's complement, this equivalent to

$$\left\{ \begin{array}{l} \begin{bmatrix} S & Q'_{(i)} \\ Q_{(i)} & \rho_{(i)}^2 \end{bmatrix} \geq 0, \quad \forall i \\ S > 0 \end{array} \right.$$

which in turn is equivalent to

$$Q_{(i)} S^{-1} Q'_{(i)} \leq \rho_{(i)}^2, \quad \forall i$$

The volume of ellipsoid $\mathcal{E}(S, 1)$ is proportional to $\sqrt{\det(S^{-1})}$. The maximum volume ellipsoid is obtained solving the convex optimization problem [45]

$$\min \ \log\big(\det(S)\big)$$

$$\text{subject to} \quad \begin{cases} Q_{(i)} S^{-1} Q'_{(i)} \le \rho_{(i)}^2, & \forall i \\ S > 0 \end{cases}$$

or

$$\min \ \log(\det(S))$$

$$\text{subject to} \quad \begin{cases} \begin{bmatrix} S & Q'_{(i)} \\ Q_{(i)} & \rho_{(i)}^2 \end{bmatrix} \ge 0, & \forall i \\ S > 0 \end{cases}$$

## C.8.4 Smallest Volume Ellipsoid Containing a Symmetric Polytope

Under the same assumptions as the previous paragraph, we have

$$S\big(|Q|, \rho\big) \subset \mathcal{E}(S, 1) \quad \text{if } x' S x \le x' \frac{Q'_{(i)} Q_{(i)}}{\rho_{(i)}^2} x \le 1, \ \forall i$$

and then

$$\frac{Q'_{(i)} Q_{(i)}}{\rho_{(i)}^2} - S \ge 0, \quad \forall i, \ S > 0$$

which leads to

$$\begin{cases} \begin{bmatrix} \rho_{(i)}^2 & Q_{(i)} \\ Q'_{(i)} & S^{-1} \end{bmatrix} \ge 0, & \forall i \\ S > 0 \end{cases}$$

which is equivalent to

$$Q_{(i)} S Q'_{(i)} \le \rho_{(i)}^2, \quad \forall i$$

The minimum volume ellipsoid is obtained solving the convex optimization problem [45]

$$\min \ \log\big(\det(S^{-1})\big)$$

$$\text{subject to} \quad \begin{cases} Q_{(i)} S Q'_{(i)} \le \rho_{(i)}^2, & \forall i \\ S > 0 \end{cases}$$

or

$$\min \ \log\left(\det(S^{-1})\right)$$

$$\text{subject to} \quad \begin{cases} \begin{bmatrix} \rho_{(i)}^2 & Q_{(i)} \\ Q'_{(i)} & S^{-1} \end{bmatrix} \geq 0, \quad \forall i \\ S > 0 \end{cases}$$

# References

1. Alamo, T., Cepeda, A., Limon, D.: Improved computation of ellipsoidal invariant sets for saturated control systems. In: Conference on Decision and Control, Sevilla, Spain, December, pp. 6216–6221 (2005)
2. Alamo, T., Cepeda, A., Limon, D., Camacho, E.F.: Estimation of the domain of attraction for saturated discrete-time systems. Int. J. Syst. Sci. **37**(8), 575–583 (2006)
3. Alamo, T., Cepeda, A., Fiacchini, M., Camacho, E.F.: Convex invariant sets for discrete-time Lur'e systems. Automatica **45**, 1066–1071 (2009)
4. Alvarez-Ramirez, J., Suárez, R.: Global stabilization of discrete-time linear systems with bounded linear state feedback. Int. J. Adapt. Control Signal Process. **10**(4–5), 409–416 (1996)
5. Alvarez-Ramirez, J., Suárez, R., Alvarez, J.: Semiglobal stabilization of multi-input linear system with saturated linear state feedback. Syst. Control Lett. **23**, 247–254 (1994)
6. Amato, F., Consentino, C., Merola, A.: On the region of attraction of nonlinear quadratic systems. Automatica **43**, 2119–2123 (2007)
7. Angeli, D., Mosca, E.: Command governors for constrained nonlinear systems. IEEE Trans. Autom. Control **44**(4), 816–820 (1999)
8. Anon: Why the grippen crashed. Aerospace Am. 11 (1994)
9. Arcak, M., Larsen, M., Kokotovic, P.: Circle and Popov criterion as tools for nonlinear feedback design. Automatica **39**(4), 643–650 (2003)
10. Åström, K.J., Rundqwist, L.: Integrator windup and how to avoid it. In: American Control Conference, Pittsburgh, PA, pp. 1693–1698 (1989)
11. Athans, M., Falb, P.L.: Optimal Control: An Introduction to the Theory and Its Applications. McGraw-Hill, New York (1966)
12. Aubin, J.P.: A survey of viability theory. SIAM J. Control Optim. **28**(4), 749–788 (1990)
13. Aubin, J.P., Cellina, A.: Differential Inclusions. Comprehensive Studies in Mathematics, vol. 264. Springer, Berlin (1984)
14. Barbu, C., Reginatto, R., Teel, A.R., Zaccarian, L.: Anti-windup for exponentially unstable linear systems with inputs limited in magnitude and rate. In: American Control Conference, Chicago, IL, June 2000
15. Barmish, B.R.: Necessary and sufficient conditions for quadratic stabilizability of an uncertain system. J. Optim. Theory Appl. **46**(4), 399–408 (1985)
16. Bateman, A., Lin, Z.: An analysis and design method for linear systems under nested saturation. Syst. Control Lett. **48**(1), 41–52 (2003)
17. Beker, O., Hollot, C.V., Chait, Y.: Plant with an integrator: an example of reset control overcoming limitations of linear feedback. IEEE Trans. Autom. Control **46**, 1797–1799 (2001)
18. Bemporad, A., Teel, A.R., Zaccarian, L.: Anti-windup synthesis via sampled-data piecewise affine optimal control. Automatica **40**(4), 549–562 (2004)

19. Bender, F.A., Gomes da Silva Jr., J.M.: Acceleration-bounded control design for actuator fault prevention. In: American Control Conference, New York, NY, pp. 5218–5223 (2007)
20. Bender, F.A., Gomes da Silva Jr., J.M.: Output feedback controller design for systems with amplitude and rate control constraints. Asian J. Control (2012, to appear)
21. Bender, F.A., Gomes da Silva Jr., J.M., Tarbouriech, S.: A convex framework for the design of dynamic anti-windup for state-delayed systems. In: American Control Conference, Baltimore, USA, pp. 6763–6768 (2010)
22. Benzaouia, A.: The resolution of equation $XA + XBX = HX$ and the pole assignment problem. IEEE Trans. Autom. Control **39**(10), 2091–2095 (1994)
23. Benzaouia, A., Burgat, C.: Regulator problem for linear discrete-time systems with nonsymmetrical constrained control. Int. J. Control **48**(6), 2442–2451 (1988)
24. Benzaouia, A., Akhrif, O., Saydy, L.: Stability and control synthesis of switched systems subject to actuator saturation. In: American Control Conference, Boston, USA, pp. 5818–5823 (2004)
25. Bernstein, D.S., Michel, A.N.: A chronological bibliography on saturating actuators. Int. J. Robust Nonlinear Control **5**, 375–380 (1995)
26. Bernussou, J., Peres, P.L.D., Geromel, J.C.: A linear programming oriented procedure for quadratic stabilization of uncertain systems. Syst. Control Lett. **13**, 65–72 (1989)
27. Biannic, J.-M., Tarbouriech, S.: Optimization and implementation of dynamic anti-windup compensators with multiple saturations in flight control systems. Control Eng. Pract. **17**, 703–713 (2009)
28. Biannic, J.-M., Tarbouriech, S., Farret, D.: A practical approach to performance analysis of saturated systems with application to fighter aircraft flight controllers. In: IFAC Symposium on Robust Control Design (ROCOND), Toulouse, France, July 2006
29. Biannic, J.-M., Roos, C., Tarbouriech, S.: A practical method for fixed-order anti-windup design. In: IFAC Symposium on Nonlinear Control Systems (NOLCOS), Pretoria, South Africa (2007)
30. Bitsoris, G.: On the positive invariance of polyhedral sets for discrete-time systems. Syst. Control Lett. **11**(4), 243–248 (1988)
31. Bitsoris, G.: Existence of positively invariant polyhedral sets for continuous-time linear systems. Control Theory Adv. Technol. **7**(3), 407–427 (1991)
32. Bitsoris, G., Vassilaki, M.: Constrained regulation of linear systems. Automatica **31**(2), 223–227 (1995)
33. Blanchini, F.: Feedback control for linear time-invariant systems with state and control bounds in the presence of disturbances. IEEE Trans. Autom. Control **35**(11), 1231–1234 (1990)
34. Blanchini, F.: Ultimate boundedness control for uncertain discrete-time systems via set-induced Lyapunov functions. IEEE Trans. Autom. Control **39**(2), 428–433 (1994)
35. Blanchini, F.: Nonquadratic Lyapunov functions for robust control. Automatica **31**(3), 451–461 (1995)
36. Blanchini, F.: Set invariance in control. Automatica **35**(11), 1747–1767 (1999)
37. Blanchini, F., Miani, S.: Best transient estimate for linear discrete-time uncertain systems. In: European Control Conference, Rome, Italy, pp. 1010–1015 (1995)
38. Blanchini, F., Miani, S.: Constrained stabilization of continuous-time linear systems. Syst. Control Lett. **28**, 95–102 (1996)
39. Blanchini, F., Miani, S.: Set-Theoretic Methods in Control. Birkhäuser, Basel (2008)
40. Bliman, P.A., Oliveira, R.C.L.F., Montagner, V.F., Peres, P.L.D.: Existence of homogeneous polynomial solutions for parameter-dependent linear matrix inequalities with parameters in the simplex. In: Conference on Decision and Control, San Diego, USA, pp. 1486–1491 (2006)
41. Boada, J., Prieur, C., Tarbouriech, S., Pittet, C., Charbonnel, C.: Anti-windup design for satellite control with microthrusters. In: AIAA Guidance, Navigation, and Control Conference, Chicago, IL, USA (2009)
42. Boada, J., Prieur, C., Tarbouriech, S., Pittet, C., Charbonnel, C.: Multi-saturation anti-windup structure for satellite control. In: American Control Conference, Baltimore, USA, June 2010

References 413

43. Bobrow, J.E., Jabbari, F., Thai, K.: An active truss element and control law for vibration suppression. Smart Mater. Struct. **4**(4), 264–269 (1995)
44. Boulingand, G.: Introduction À la Géométrie Infinitésimale. Gauthiers-Villars, Paris (1932)
45. Boyd, S., El Ghaoui, L., Féron, E., Balakrishnan, V.: Linear Matrix Inequalities in System and Control Theory. SIAM Studies in Applied Mathematics. SIAM, Philadelphia (1994)
46. Brieger, O., Kerr, M., Leissling, D., Postlethwaite, I., Sofrony, J., Turner, M.C.: Anti-windup compensation of rate saturation in an experimental aircraft. In: American Control Conference, New York (2007)
47. Bryson, A.E., Ho, Y.C.: Applied Optimal Control. Wiley, New York (1969)
48. Bupp, R.T., Bernstein, D.S., Chellaboina, V.S., Haddad, W.M.: Resetting virtual absorbers for vibration control. In: American Control Conference, New Mexico, USA, vol. 5, pp. 2647–2651 (1997)
49. Burgat, C., Tarbouriech, S.: Global stability of a class of linear systems with saturated controls. Int. J. Syst. Sci. **23**(1), 37–56 (1992)
50. Burgat, C., Tarbouriech, S.: Stability and control of saturated linear systems. In: Fossard, A.J., Normand-Cyrot, D. (eds.) Non-linear Systems, vol. 2. Chapman & Hall, London (1996). Chap. 4
51. Burgat, C., Tarbouriech, S.: Intelligent anti-windup for systems with input magnitude saturation. Int. J. Robust Nonlinear Control **8**, 1085–1100 (1998)
52. Burgat, C., Tarbouriech, S., Klaï, M.: Continuous-time saturated state feedback regulators: theory and design. Int. J. Syst. Sci. **25**(2), 315–336 (1994)
53. Campo, P.J., Morari, M., Nett, C.N.: Multivariable anti-windup and bumpless transfer: a general theory. In: American Control Conference, Pittsburgh, pp. 1706–1711 (1989)
54. Cao, Y.Y., Lin, Z.: Min-max MPC algorithm for LPV systems subject to input saturation. IEE Proc. Control Theory Appl., 152–266 (2005)
55. Cao, Y.Y., Lin, Z., Hu, T.: Stability analysis of linear time-delay systems subject to input saturation. IEEE Trans. Circuits Syst. I, Fundam. Theory Appl. **49**, 233–240 (2002)
56. Cao, Y.Y., Lin, Z., Ward, D.G.: An antiwindup approach to enlarging domain of attraction for linear systems subject to actuator saturation. IEEE Trans. Autom. Control **47**(1), 140–145 (2002)
57. Cao, Y.Y., Lin, Z., Chen, B.M.: An output feedback $h_\infty$ controller for linear systems subject to sensor nonlinearities. IEEE Trans. Circuits Syst. I **50**(7), 914–921 (2003)
58. Castelan, E.B., Hennet, J.C.: Eigenstructure assignment for state constrained linear continuous time systems. Automatica **28**(3), 605–611 (1992)
59. Castelan, E.B., Hennet, J.C.: On invariant polyhedra of continuous-time linear systems. IEEE Trans. Autom. Control **38**(11), 1680–1685 (1993)
60. Castelan, E.B., Tarbouriech, S.: On positive invariance and output feedback stabilization of input constrained linear systems. In: American Control Conference, Baltimore, USA, vol. 3, pp. 2740–2744 (1994)
61. Castelan, E.B., Gomes da Silva Jr., J.M., Cury, J.E.R.: A reduced order framework applied to linear systems with constrained controls. IEEE Trans. Autom. Control **41**(2), 249–255 (1996)
62. Castelan, E.B., Tarbouriech, S., Gomes da Silva Jr., J.M., Queinnec, I.: $\mathcal{L}_2$-stabilization of continuous-time systems with saturating actuators. Int. J. Robust Nonlinear Control **16**, 935–944 (2006)
63. Castelan, E.B., Tarbouriech, S., Queinnec, I.: Control design for a class of nonlinear continuous-time systems. Automatica **44**(8), 2034–2039 (2008)
64. Chen, C.T.: Linear System Theory and Design. Holt, Rinehart & Winston, New York (1984)
65. Chesi, G.: Computing output feedback controllers to enlarge the domain of attraction in polynomial systems. IEEE Trans. Autom. Control **49**(10), 1846–1850 (2004)
66. Chiang, H.-D., Hirsch, M.W., Wu, F.F.: Stability regions of nonlinear autonomous dynamical systems. IEEE Trans. Autom. Control **33**(1), 16–27 (1988)
67. Chiasson, J., Loiseau, J.-J. (eds.): Applications of Time-Delay Systems. LNCIS, vol. 352. Springer, Berlin (2007)

68. Chilali, M., Gahinet, P.: $H_\infty$ design with pole placement constraints: an LMI approach. IEEE Trans. Autom. Control **41**(3) (1996)
69. Chilali, M., Gahinet, P., Apkarian, P.: Robust pole placement in LMI regions. IEEE Trans. Autom. Control **44**, 2257–2270 (1999)
70. Chitour, Y., Lin, Z.: Finite gain $l_p$ stabilization of discrete-time linear systems subject to actuator saturation: the case of $p = 1$. IEEE Trans. Autom. Control **48**(12), 2196–2198 (2003)
71. Clarke, F.H.: Optimization and Non Smooth Analysis. Wiley, New York (1983)
72. Clegg, J.C.: A non-linear integrator for servomechanisms. Trans. Am. Inst. Electr. Eng. **77**, 41–42 (1958)
73. Colaneri, P., Kučera, V., Longhi, S.: Polynomial approach to the control of SISO periodic systems subject to input constraint. Automatica **39**, 1417–1424 (2003)
74. Colenari, P., Geromel, J.C., Locatelli, A.: Control Theory and Design: An $RH_2$ and $RH_\infty$ Viewpoint. Academic Press, San Diego (1997)
75. Coutinho, D.F., Gomes da Silva Jr., J.M.: Computing estimates of the region of attraction for rational control systems with saturating actuators. IET Control Theory Appl. **4**(3), 315–325 (2010)
76. Crawshaw, S.: Global and local analyses of coprime-factor based anti-windup for stable and unstable plants. In: European Control Conference, Cambridge (2003)
77. Daafouz, J.: Robustesse en performance des systèmes linéaires incertains: placement de pôles et coût garanti. PhD thesis, Institut National des Sciences Appliquées de Toulouse, Toulouse, France (November 1997). Rapport LAAS No. 97472
78. Dai, D., Hu, T., Teel, A.R., Zaccarian, L.: Piecewise-quadratic Lyapunov functions for systems with dead zones or saturations. Syst. Control Lett. **58**, 365–371 (2009)
79. D'Amato, F.J., Rotea, M.A., Megretski, A.V., Jönsson, U.T.: New results for analysis of systems with repeated nonlinearities. Automatica **37**(5), 739–747 (2001)
80. de Klerk, E.: Aspects of Semidefinite Programming. Interior Point Algorithms and Selected Applications. Kluwer Academic, Dordrecht (2002)
81. de Oliveira, M.C., Skelton, R.E.: Stability tests for constrained linear systems. In: Reza-Moheimani, S.O. (ed.) Lecture Notes in Control and Information Sciences, vol. 268, pp. 241–257. Springer, New York (2001)
82. de Oliveira, M.C., Geromel, J.C., Hsu, L.: A new absolute stability test for systems with state dependent perturbations. Int. J. Robust Nonlinear Control **12**(4), 1209–1226 (2002)
83. de Souza, C.E., Trofino, A., de Oliveira, J.: Parametric Lyapunov function approach to $\mathcal{H}_2$ analysis and control of linear parameter-dependent systems. IEE Proc., Control Theory Appl. **150**(5), 501–508 (2003)
84. Dolphus, R.M., Schmitendorf, W.E.: Stability analysis for a class of linear controllers under control constraints. In: Conference on Decision and Control, Brighton, UK, pp. 77–80 (1991)
85. Dórea, C.E.T., Hennet, J.C.: On (A,B)-invariance of polyhedral domains for discrete-time systems. In: Conference on Decision and Control, Kobe, Japan, pp. 4319–4324 (1996)
86. Duda, H.: Prediction of pilot-in-the-loop oscillations due to rate saturation. J. Guid. Control Dyn. **20**(3), 581–587 (1997)
87. Edwards, C., Postlethwaite, I.: Anti-windup and bumpless-transfer schemes. Automatica **34**(2), 199–210 (1998)
88. El Ghaoui, L., Oustry, F., Rami, M.A.: A cone complementarity linearization algorithm for static output-feedback and related problems. IEEE Trans. Autom. Control **42**(8), 1171–1176 (1997)
89. Fang, H., Lin, Z., Hu, T.: Analysis of linear systems in the presence of actuator saturation and $\mathcal{L}_2$-disturbances. Automatica **40**(7), 1229–1238 (2004)
90. Fertik, H.A., Ross, C.W.: Direct digital control algorithm with anti-windup feature. ISA Trans. **6**, 317–328 (1967)
91. Feuer, A., Goodwin, G.C., Salgado, M.: Potential benefits of hybrid control for linear time invariant plants. In: American Control Conference, New Mexico, USA, vol. 5, pp. 2790–2794 (1997)

References 415

92. Fiacchini, M.: Convex difference inclusions for systems analysis and control. PhD thesis, Universidad de Sevilla, Spain, January 2010
93. Fiacchini, M., Tarbouriech, S., Prieur, C.: Ellipsoidal invariant sets for saturated hybrid systems. In: American Control Conference, San Francisco, USA, July 2011
94. Fischman, A., Gomes da Silva Jr., J.M., Dugard, L., Dion, J.M., Tarbouriech, S.: Dynamic output feedback under state and control constraints. In: European Control Conference, Brussels, Belgium (1997)
95. Fliegner, T., Logemann, H., Ryan, E.P.: Time-varying and adaptive integral control of linear systems with actuator and sensor nonlinearities. In: International Symposium on Mathematical Theory of Networks and Systems (MTNS), Perpignan, France (2000)
96. Fong, I.-K., Hsu, C.-C.: State feedback stabilization of single input systems through actuators with saturation and dead zone characteristics. In: Conference on Decision and Control, Sydney, Australia (2000)
97. Forni, F., Galeani, S., Zaccarian, L.: A family of global stabilizers for quasi-optimal control of planar linear saturated systems. IEEE Trans. Autom. Control $55(5)$, 1175–1180 (2010)
98. Fridman, E., Pila, A., Shaked, U.: Regional stabilization and $H_\infty$ control of time-delay systems with saturating actuators. Int. J. Robust Nonlinear Control $13$, 885–907 (2003)
99. Fuller, A.T.: In the large stability of relay and saturated control systems with actuator saturation. Int. J. Control $10(4)$, 457–480 (1969)
100. Gahinet, P., Apkarian, P.: A linear matrix inequality approach to $\mathcal{H}_\infty$ control. Int. J. Robust Nonlinear Control $4$, 421–448 (1994)
101. Gahinet, P., Nemirovski, A., Laub, A., Chilali, M.: LMI control toolbox user's guide. The Math Works (1995)
102. Galeani, S., Massimetti, M., Teel, A.R., Zaccarian, L.: Reduced order linear anti-windup augmentation for stable linear systems. Int. J. Syst. Sci. $37(2)$, 115–127 (2006)
103. Galeani, S., Onori, S., Teel, A.R., Zaccarian, L.: Further results on static linear anti-windup design for control systems subject to magnitude and rate saturation. In: Conference on Decision and Control, San Diego, CA, USA, December 2006
104. Galeani, S., Onori, S., Teel, A.R., Zaccarian, L.: Regional, semiglobal, global nonlinear anti-windup via switching design. In: European Control Conference, Kos, Greece, pp. 5403–5410 (2007)
105. Galeani, S., Onori, S., Zaccarian, L.: Nonlinear scheduled control for linear systems subject to saturation with application to anti-windup control. In: Conference on Decision and Control, New Orleans, LA, USA, pp. 1168–1173 (2007)
106. Galeani, S., Teel, A.R., Zaccarian, L.: Constructive nonlinear anti-windup design for exponentially unstable linear plants. Syst. Control Lett. $56(5)$, 357–365 (2007)
107. Galeani, S., Onori, S., Teel, A.R., Zaccarian, L.: A magnitude and rate saturation model and its use in the solution of a static anti-windup problem. Syst. Control Lett. $57(1)$, 1–9 (2008)
108. Galeani, S., Tarbouriech, S., Turner, M.C., Zaccarian, L.: A tutorial on modern anti-windup design. Eur. J. Control $15(3–4)$, 418–440 (2009)
109. Garcia, G.: Quadratic guaranteed cost and disk pole location control for discrete time uncertain systems. IEE Proc., Control Theory Appl. $144$, 545–548 (1997)
110. Garcia, G., Bernussou, J.: Pole assignment for uncertain systems in a specified disk by state feedback. IEEE Trans. Autom. Control $40$, 184–190 (1995)
111. Garcia, G., Bernussou, J.: Pole assignment for uncertain systems in a specified disk by output feedback. Math. Control Signals Syst. $9$, 152–161 (1996)
112. Garcia, G., Tarbouriech, S.: Stabilization with eigenvalues placement of a norm bounded uncertain system by bounded inputs. Int. J. Robust Nonlinear Control $9$, 599–615 (1999)
113. Garcia, G., Tarbouriech, S.: Nonlinear bounded control for time-delay systems. Kybernetika $37$, 381–396 (2001)
114. Garcia, G., Bernussou, J., Arzelier, D.: Robust stabilization of discrete-time linear systems with norm bounded time varying uncertainty. Syst. Control Lett. $22$, 327–339 (1994)
115. Garcia, G., Bernussou, J., Arzelier, D.: A LMI solution for disk pole location with $H_2$ guaranteed cost. Eur. J. Control $1$, 54–61 (1995)

116. Garcia, G., Bernussou, J., Arzelier, D.: Stabilization of an uncertain linear dynamic system by state and output feedback: a quadratic stabilizability approach. Int. J. Control **64**, 839–858 (1996)
117. Garcia, G., Daafouz, J., Bernussou, J.: Output feedback disk pole assignment for systems with positive real uncertainty. IEEE Trans. Autom. Control **41**, 1385–1391 (1996)
118. Garcia, G., Tarbouriech, S., Suárez, R., Alvarez-Ramirez, J.: Non linear bounded control for norm bounded uncertain system. IEEE Trans. Autom. Control **44**(6), 1254–1258 (1999)
119. Garcia, G., Tarbouriech, S., Gomes da Silva Jr., J.M.: Dynamic output controller design for linear systems with actuator and sensor saturation. In: American Control Conference, New York, USA, July 2007
120. Garcia, G., Tarbouriech, S., Gomes da Silva Jr., J.M., Eckhard, D.: Finite $L_2$ gain and internal stabilisation of linear systems subject to actuator and sensor saturations. IET Control Theory Appl. **3**(7), 799–812 (2009)
121. Gatley, S.L., Turner, M.C., Postlethwaite, I., Kumar, A.: A comparison of rate-limit compensation schemes for pilot-induced-oscillation avoidance. Aerosp. Sci. Technol. **10**(1), 37–47 (2006)
122. Gavrilyako, V.M., Korobov, V.I., Sklyar, G.M.: Designing a bounded control of dynamic systems in entire space with the aid of a controllability function. Autom. Remote Control **11**, 1484–1490 (1986)
123. Geromel, J.C., Garcia, G., Bernussou, J.: $H_2$ robust control with pole placement. In: IFAC World Congress, Sidney, Australia, pp. 283–288 (1993)
124. Ghiggi, I., Bender, A., Gomes da Silva Jr., J.M.: Dynamic non-rational anti-windup for time-delay systems with saturating inputs. In: IFAC World Congress, Seoul, Korea (2008)
125. Gilbert, E.G., Kolmanovsky, I.: Fast reference governors for systems with state and control constraints and disturbance inputs. Int. J. Robust Nonlinear Control **9**(15), 1117–1141 (1999)
126. Gilbert, E.G., Tan, K.T.: Linear systems with state and control constraints: the theory and application of maximal output admissible sets. IEEE Trans. Autom. Control **36**(9), 1008–1020 (1991)
127. Glattfelder, A.H., Schaufelberger, W.: Control Systems with Input and Output Constraints. Springer, London (2003)
128. Goebel, R., Prieur, C., Teel, A.R.: Smooth patchy control Lyapunov functions. Automatica **45**(3), 675–683 (2009)
129. Goh, K.C., Safonov, M.G., Ly, J.H.: Robust synthesis via bilinear matrix inequalities. Int. J. Robust Nonlinear Control **6**(9/10), 1079–1095 (1996)
130. Gomes da Silva Jr., J.M.: Sur la stabilité locale de systèmes linéaires avec saturation des commandes. PhD thesis, Université Paul Sabatier, Toulouse, France, October 1997. Rapport LAAS No. 97383
131. Gomes da Silva Jr., J.M., Tarbouriech, S.: Analysis of local stability of linear systems with saturating controls: a polyhedral approach. In: IFAC Conference on Systems, Structure and Control, Bucharest, Romania, pp. 168–173 (1997)
132. Gomes da Silva Jr., J.M., Tarbouriech, S.: Polyhedral regions of local asymptotic stability for discrete-time linear systems with saturating controls. In: Conference on Decision and Control, San Diego, USA, pp. 925–930 (1997)
133. Gomes da Silva Jr., J.M., Tarbouriech, S.: Local stabilization of discrete-time linear systems with saturating controls: an LMI-based approach. In: American Control Conference, Philadelphia, USA, pp. 92–96 (1998)
134. Gomes da Silva Jr., J.M., Tarbouriech, S.: Stability regions for linear systems with saturating controls. In: European Control Conference, Karlsruhe, Germany (1999). F924
135. Gomes da Silva Jr., J.M., Tarbouriech, S.: Contractive polyhedra for linear continuous-time systems with saturating controls. In: American Control Conference, San Diego, USA, pp. 2007–2010 (1999)
136. Gomes da Silva Jr., J.M., Tarbouriech, S.: Polyhedral regions of local stability for linear discrete-time systems with saturating controls. IEEE Trans. Autom. Control **44**(11), 2081–2085 (1999)

# References

137. Gomes da Silva Jr., J.M., Tarbouriech, S.: Local stabilization of discrete-time linear systems with saturating controls: an LMI-based approach. IEEE Trans. Autom. Control **46**(1), 119–124 (2001)
138. Gomes da Silva Jr., J.M., Tarbouriech, S.: Anti-windup design with guaranteed regions of stability: an LMI-based approach. In: Conference on Decision and Control, Hawaii, USA (2003)
139. Gomes da Silva Jr., J.M., Tarbouriech, S.: Anti-windup design with guaranteed region of stability: an LMI-based approach. IEEE Trans. Autom. Control **50**(1), 106–111 (2005)
140. Gomes da Silva Jr., J.M., Tarbouriech, S.: Anti-windup design with guaranteed regions of stability for discrete-time linear systems. Syst. Control Lett. **55**(3), 184–192 (2006)
141. Gomes da Silva Jr., J.M., Fischman, A., Tarbouriech, S., Dion, J.M., Dugard, L.: Synthesis of state feedback for linear systems subject to control saturation by an LMI-based approach. In: IFAC Symposium on Robust Control Design (ROCOND), Budapest, Hungary, pp. 229–234 (1997)
142. Gomes da Silva Jr., J.M., Paim, C.C., Castelan, E.B.: Stability and stabilization of linear discrete-time systems subject to control saturation. In: IFAC Symposium on System Structure and Control, Prague, Czech Republic (2001)
143. Gomes da Silva Jr., J.M., Tarbouriech, S., Reginatto, R.: Analysis of regions of stability for linear systems with saturating inputs through an anti-windup scheme. In: IEEE Conference on Control Applications (CCA), Glasgow, UK (2002)
144. Gomes da Silva Jr., J.M., Tarbouriech, S., Garcia, G.: Local stabilization of linear systems under amplitude and rate saturating actuators. IEEE Trans. Autom. Control **48**(5), 842–847 (2003)
145. Gomes da Silva Jr., J.M., Tarbouriech, S., Garcia, G.: Anti-windup design for time-delay systems subject to input saturation: an LMI-based approach. Eur. J. Control **6**, 1–13 (2006)
146. Gomes da Silva Jr., J.M., Lescher, F., Eckhard, D.: Design of time-varying controllers for discrete-time linear systems with input saturation. IET Control Theory Appl. **1**, 155–162 (2007)
147. Gomes da Silva Jr., J.M., Limon, D., Alamo, T., Camacho, E.F.: Dynamic output feedback for discrete-time systems under amplitude and rate actuator constraints. IEEE Trans. Autom. Control **53**(10), 2367–2372 (2008)
148. Gomes da Silva Jr., J.M., Bender, F.A., Tarbouriech, S., Biannic, J.-M.: Dynamic anti-windup synthesis for state delayed systems: an LMI approach. In: Conference on Decision and Control, Shanghai, P.R. China (2009)
149. Graham, A.: Kronecker Product and Matrix Calculus with Applications. Chichester, Ellis Horwood (1981)
150. Grimm, G., Postlethwaite, I., Teel, A.R., Turner, M.C., Zaccarian, L.: Linear matrix inequalities for full and reduced order anti-windup synthesis. In: American Control Conference, Arlington, VA, pp. 4134–4139 (2001)
151. Grimm, G., Hatfield, J., Postlethwaite, I., Teel, A.R., Turner, M.C., Zaccarian, L.: Anti-windup for stable linear systems with input saturation: an LMI based synthesis. IEEE Trans. Autom. Control **48**(9), 1509–1525 (2003)
152. Grimm, G., Teel, A.R., Zaccarian, L.: The $\mathcal{L}_2$ anti-windup problem for discrete-time linear systems: definition and solutions. In: American Control Conference, Denver, CO, pp. 5329–5334 (2003)
153. Grimm, G., Teel, A.R., Zaccarian, L.: Robust linear anti-windup synthesis for recovery of unconstrained performance. Int. J. Robust Nonlinear Control **14**(13–14), 1133–1168 (2004)
154. Grimm, G., Teel, A.R., Zaccarian, L.: The $\mathcal{L}_2$ anti-windup problem for discrete-time linear systems: definition and solutions. Syst. Control Lett. **57**(4), 356–364 (2008)
155. Gu, K., Kharitonov, V.L., Chen, J.: Stability of Time-Delay Systems. Birkhäuser, Boston (2003)
156. Gußner, T., Adamy, J.: Controller design for polynomial systems with input constraints. In: Conference on Decision and Control, Shanghai, China, December 2009

157. Gußner, T., Adamy, J.: Designing nonpolynomial controllers for polynomial systems with input constraints using convex optimization and passivity. In: IEEE Multiconference on Systems and Control, Yokohama, Japan (2010)
158. Gutman, P.O., Hagander, P.: A new design of constrained controllers for linear systems. IEEE Trans. Autom. Control **30**, 22–33 (1985)
159. Haddad, W.M., Bernstein, D.S.: Explicit construction of quadratic Lyapunov functions for the small gain, positively, circle and Popov theorems and their application to robust stability. Part I: continuous-time theory. Int. J. Robust Nonlinear Control **3**(4), 313–339 (1993)
160. Haddad, W.M., Chellaboina, V.S., Kablar, N.A.: Active control of combustion instabilities via hybrid resetting controllers. In: American Control Conference, pp. 2378–2382 (2000)
161. Hahn, W.: Stability of Motion. Springer, Berlin (1967)
162. Hanus, R.: Anti-windup and bumpless transfer: a survey. In: 12th IMACS World Congress, Paris, France, vol. 2, pp. 59–65 (1988)
163. Hanus, R., Kinnaert, M.: Control of constrained multivariable systems using conditioning technique. In: American Control Conference, Pittsburgh, pp. 1712–1718 (1989)
164. Hanus, R., Kinnaert, M., Henrotte, J.L.: Conditioning technique, a general anti-windup and bumpless transfer method. Automatica **23**, 729–739 (1987)
165. Heath, W.P., Guang, L.: Lyapunov functions for the multivariable Popov criterion with indefinite multipliers. Automatica **45**(12), 2977–2981 (2009)
166. Hennet, J.C.: Une extension du lemme de Farkas et son application au problème de régulation linéaire sous contraintes. C. R. Acad. Sci., Ser. I **308**, 415–419 (1989)
167. Hennet, J.C., Béziat, J.P.: A class of invariant regulators for the discrete-time linear constrained regulation problem. Automatica **27**(3), 549–554 (1991)
168. Hennet, J.C., Castelan, E.B.: Robust invariant controllers for constrained linear systems. In: American Control Conference, Chicago, USA, vol. 2, pp. 993–997 (1992)
169. Hennet, J.C., Castelan, E.B.: Constrained control of unstable multivariable linear systems. In: European Control Conference, Groningen,The Netherlands, pp. 2039–2043 (1993)
170. Henrion, D.: Stabilité des systèmes linéaires incertains à commande contrainte. PhD thesis, Institut National des Sciences Appliquées de Toulouse, Toulouse, France (October 1999). Rapport LAAS No. 99449
171. Henrion, D., Lasserre, J.B.: Convergent relaxations of polynomial matrix inequalities and static output feedback. IEEE Trans. Autom. Control **41**(2), 192–202 (2006)
172. Henrion, D., Tarbouriech, S.: LMI relaxations for robust stability of linear systems with saturating controls. Automatica **35**, 1599–1604 (1999)
173. Henrion, D., Garcia, G., Tarbouriech, S.: Piecewise linear robust control of systems with bounded inputs. In: American Control Conference, Philadelphia, USA (1998)
174. Henrion, D., Garcia, G., Tarbouriech, S.: Piecewise-linear robust control of systems with input constraints. Eur. J. Control **5**(1), 157–166 (1999)
175. Henrion, D., Tarbouriech, S., Garcia, G.: Output feedback robust stabilization of uncertain linear systems with saturating controls: an LMI approach. IEEE Trans. Autom. Control **44**(11), 2230–2237 (1999)
176. Henrion, D., Tarbouriech, S., Kučera, V.: Control of linear systems subject to input constraints: a polynomial approach. Automatica **37**, 597–604 (2001)
177. Henrion, D., Tarbouriech, S., Kučera, V.: Control of linear systems subject to time-domain constraints with polynomial pole placement and LMIs. IEEE Trans. Autom. Control **50**(9), 1360–1364 (2005)
178. Herrmann, G., Turner, M.C., Postlethwaite, I.: Practical implementation of a novel anti-windup scheme in a HDD-dual-stage servo-system. IEEE/ASME Trans. Mechatron. **9**(3), 580–592 (2004)
179. Herrmann, G., Turner, M.C., Postlethwaite, I.: Some new results on anti-windup conditioning using the Weston–Postlethwaite approach. In: Conference on Decision and Control (2004)
180. Herrmann, G., Turner, M.C., Menon, P.P., Bates, D.G., Postlethwaite, I.: Anti-windup synthesis for nonlinear dynamic inversion controllers. In: IFAC Symposium on Robust Control Design (ROCOND), Toulouse (2006)

# References

181. Herrmann, G., Turner, M.C., Postlethwaite, I.: Discrete-time and sampled-data anti-windup synthesis: stability and performance. Int. J. Syst. Sci. **37**(2), 91–113 (2006)
182. Herrmann, G., Turner, M.C., Postlethwaite, I.: Performance oriented anti-windup for a class of neural network controlled systems. IEEE Trans. Neural Netw. **18**(2), 449–465 (2007)
183. Hindi, H., Boyd, S.: Analysis of linear systems with saturation using convex optimization. In: Conference on Decision and Control, Tampa or, USA, pp. 903–908 (1998)
184. Hippe, P.: Windup in Control. Its Effects and Their Prevention. AIC/Springer, Berlin (2006)
185. Horrowitz, I.: A synthesis theory for a class of saturating systems. Int. J. Control **38**(1), 169–187 (1983)
186. Howitt, J.: Application of non-linear dynamic inversion to rotorcraft flight control. In: Proceedings of the American Helicopter Society Conference (2005)
187. Hu, T., Lin, Z.: On enlarging the basin of attraction for linear systems under saturated linear feedback. Syst. Control Lett. **40**(1), 59–69 (2000)
188. Hu, T., Lin, Z.: Control Systems with Actuator Saturation: Analysis and Design. Birkhäuser, Boston (2001)
189. Hu, T., Lin, Z.: Composite quadratic Lyapunov functions for constrained control systems. IEEE Trans. Autom. Control **48**(3), 440–450 (2003)
190. Hu, T., Lin, Z.: Controlled invariance of ellipsoids: linear vs. nonlinear feedback. Syst. Control Lett. **53**(3–4), 203–210 (2004)
191. Hu, Q., Rangaiah, G.P.: Anti-windup schemes for uncertain nonlinear systems. IEE Proc., Control Theory Appl. **147**(3), 321–329 (2000)
192. Hu, T., Lin, Z., Chen, B.M.: Analysis and design for discrete-time linear systems subject to actuator saturation. Syst. Control Lett. **45**(2), 97–112 (2002)
193. Hu, T., Lin, Z., Chen, B.M.: An analysis and design method for linear systems subject to actuator saturation and disturbance. Automatica **38**, 351–359 (2002)
194. Hu, T., Lin, Z., Qiu, L.: An explicit description of null controllable regions of linear systems with saturating actuators. Syst. Control Lett. **47**, 65–78 (2002)
195. Hu, T., Teel, A.R., Zaccarian, L.: Regional anti-windup compensation for linear systems with input saturation. In: American Control Conference, Portland, OR, pp. 3397–3402 (2005)
196. Hu, T., Teel, A.R., Zaccarian, L.: Stability and performance for saturated systems via quadratic and nonquadratic Lyapunov functions. IEEE Trans. Autom. Control **51**(11), 1770–1786 (2006)
197. Hu, T., Teel, A.R., Zaccarian, L.: Anti-windup synthesis for linear control systems with input saturation: achieving regional, nonlinear performance. Automatica **64**(2), 512–519 (2008)
198. Hu, T., Thibodeau, T., Teel, A.R.: Analysis of oscillation and stability for systems with piecewise linear components via saturation functions. In: American Control Conference, St. Louis, MO, USA, pp. 1911–1916 (2009)
199. Huang, H., Li, D., Lin, Z., Xi, Y.: An improved robust model predictive control design in the presence of actuator saturation. Automatica **47**, 861–864 (2011)
200. Ichihara, H.: State feedback synthesis for polynomial systems with input saturation using convex optimization. In: American Control Conference, New York, USA (2007)
201. Johansson, M.: Piecewise linear control systems. PhD thesis, Lund Institute of Technology, Depto. of Automatic Control (Lund University), Lund, Sweden (1999)
202. Johansson, M., Rantzer, A.: Computation of piecewise quadratic Lyapunov functions for hybrid systems. IEEE Trans. Autom. Control **43**(4), 555–559 (1998)
203. Johnson, E.N., Calise, A.J.: Limited authority flight control for reusable launch vehicles. AIAA J. Guid. Control Dyn. **26**(6), 906–913 (2003)
204. Kahveci, N.E., Ioanou, P.A., Mirmirani, M.D.: A robust adaptive control design for gliders subject to actuator saturation nonlinearities. In: American Control Conference, New York (2007)
205. Kailath, T.: Linear Systems. Prentice Hall, Englewood Cliffs (1980)
206. Kaliora, G., Astolfi, A.: Nonlinear control of feedforward systems with bounded signals. IEEE Trans. Autom. Control **49**(11), 1975–1990 (2004)

207. Kapasouris, P., Athans, M., Stein, G.: Design of feedback control systems for stable plants with saturating actuators. In: Conference on Decision and Control, Austin, USA, pp. 469–479 (1988)
208. Kapila, V., Grigoriadis, K. (eds.): Actuator Saturation Control. Dekker, New York (2002)
209. Kapila, V., Haddad, W.M.: Fixed-structure controller design for systems with actuator amplitude and rate nonlinearities. Int. J. Control **73**(6), 520–530 (2000)
210. Kapoor, N., Daoutidis, P.: An observer based anti-windup scheme for nonlinear systems with input constraints. Int. J. Control **72**(1), 18–29 (1999)
211. Kapoor, N., Teel, A.R., Daoutidis, P.: An anti-windup design for linear systems with input saturation. Automatica **34**(5), 559–574 (1998)
212. Kerr, M., Villota, E., Jayasuriya, S.: Robust anti-windup design for input constrained SISO systems. In: 8th International Symposium on QFT and Robust Frequency Domain Methods (2007)
213. Kerr, M.L., Turner, M.C., Postlethwaite, I.: Practical approaches to low-order anti-windup compensator design: a flight control comparison. In: IFAC World Congress, Seoul, Korea (2008)
214. Kerrigan, E.C.: Robust constraint satisfaction: Invariant sets and predictive control. PhD thesis, University of Cambridge, Cambridge (2000)
215. Khalil, H.K.: Nonlinear Systems. MacMillan, London (1992)
216. Khargonekar, P., Petersen, I.R., Zhou, K.: Robust stabilization of uncertain linear system: quadratic stabilizability and $H_\infty$ control theory. IEEE Trans. Autom. Control **35**(3), 356–361 (1990)
217. Kiendl, H., Adamy, J., Stelzner, P.: Vector norms as Lyapunov functions for linear systems. IEEE Trans. Autom. Control **37**(6), 839–842 (1992)
218. Kim, J., Bien, Z.: Robust stability of uncertain linear systems with saturating actuators. IEEE Trans. Autom. Control **39**(1), 202–207 (1994)
219. Kiyama, T., Iwasaki, T.: On the use of multi-loop circle for saturating control synthesis. Syst. Control Lett. **41**, 105–114 (2000)
220. Klaï, M.: Stabilisation des systèmes linéaires continus contraints sur la commande par retour d'état et sortie saturés. PhD thesis, Université Paul Sabatier, Toulouse, France (September 1994). Rapport LAAS No. 94323
221. Kolmanovsky, I., Gilbert, E.G.: Maximal output admissible sets for discrete-time systems with disturbances inputs. In: American Control Conference, Seattle, USA, pp. 1995–1999 (1995)
222. Koplon, R.B., Hautus, M.L.J., Sontag, E.D.: Observability of linear systems with saturated outputs. Linear Algebra Appl. **205–206**, 909–936 (1994)
223. Köse, I.E., Jabbari, F.: Scheduled controllers for linear systems with bounded actuators. Automatica **39**, 1377–1387 (2003)
224. Kothare, M.V., Morari, M.: Stability analysis of anti-windup control scheme: a review and some generalizations. In: European Control Conference, Brussels, Belgium (1997)
225. Kothare, M.V., Morari, M.: Multiplier theory for stability analisys of anti-windup control systems. Automatica **35**, 917–928 (1999)
226. Kothare, M.V., Campo, P.J., Morari, M., Nett, C.N.: A unified framework for the study of anti-windup designs. Automatica **30**(12), 1869–1883 (1994)
227. Kreisselmeier, G.: Stabilisation of linear systems in the presence of output measurement saturation. Syst. Control Lett. **29**, 27–30 (1996)
228. Krikelis, N.J.: State feedback integral control with inteligent integrator. Int. J. Control **32**, 465–473 (1980)
229. Kulkarni, V.V., Safonov, M.G.: All multipliers for repeated monotone nonlinearities. In: American Control Conference, Arlington, USA, June 2001
230. Kwakernak, H., Sivan, R.: Linear Optimal Control Systems. Wiley-Interscience, New York (1972)
231. Labit, Y., Peaucelle, D., Henrion, D.: Sedumi interface 1.02: a tool for solving LMI problems with sedumi. In: Proceedings of the CACSD Conference, Glasgow, Scotland (2002)

# References

232. Laloy, M., Rouche, N., Habets, P.: Stability Theory by Lyapunov's Direct Method. Springer, New York (1977)
233. Langouët, P.: Sur la stabilité locale de systèmes linéaires soumis à des actionneurs limités en amplitude et en dynamique. PhD thesis, University of Toulouse, France (November 2003). Rapport LAAS No. 03576
234. Langouët, P., Tarbouriech, S., Garcia, G.: Pilots evaluation by taking into account both limited actuator and incidence on stability analysis. Technical report, Number Grant F/20062/SA (2002) (in French, limited diffusion)
235. LaSalle, J.P.: The Stability of Dynamical Systems. Regional Conference Series in Applied Mathematics, vol. 25. SIAM, Philadelphia (1976)
236. Lasserre, J.B.: Reachable, controllable sets and stabilizing control of constrained linear systems. Automatica $29(2)$, 531–536 (1992)
237. Leith, D.J., Leithead, W.E.: Survey of gain-scheduling analysis and design. Int. J. Control $73(11)$, 1001–1025 (2000)
238. Limon, D., Gomes da Silva Jr., J.M., Alamo, T., Camacho, E.F.: Improved MPC design based on saturating control laws. Eur. J. Control $11$, 112–122 (2005)
239. Lin, Z.: Semi-global stabilisation of linear systems with position and rate-limited actuators. Syst. Control Lett. $30$, 1–11 (1997)
240. Lin, Z.: Low gain and low-and-high gain feedback: a review and some recent results. In: Conference on Decision and Control, Shanghai, China, December 2009
241. Lin, Z., Hu, T.: Semi-global stabilisation of linear systems subject to output saturation. Syst. Control Lett. $43$, 211–217 (2001)
242. Lin, Z., Saberi, A.: Semi-global exponential stabilization of linear systems subject to input saturation via linear feedback. Syst. Control Lett. $21(3)$, 225–239 (1993)
243. Lin, Z., Saberi, A.: Semi-global exponential stabilization of linear discrete-time systems subject to input saturation via linear feedback. Syst. Control Lett. $24(2)$, 125–132 (1995)
244. Lin, Z., Saberi, A.: Low-and-high gain design technique for linear systems subject to input saturation—a direct method. Int. J. Robust Nonlinear Control $7$, 1091–1101 (1997)
245. Lin, Z., Mantri, R., Saberi, A.: Semi-global output regulation for linear systems subject to input saturation—a low and high gain design. Control Theory Adv. Technol. $10(4)$, 2209–2232 (1995)
246. Lin, Z., Saberi, A., Teel, A.R.: Simultaneous $\mathcal{L}_p$ stabilization and internal stabilization of linear systems subject to input saturation: state feedback case. Syst. Control Lett. $25$, 219–226 (1995)
247. Lin, Z., Saberi, A., Stoorvogel, A.A.: Semiglobal stabilization of linear discrete-time systems subject to input saturation via linear feedback—an ARE-based approach. IEEE Trans. Autom. Control $41(8)$, 1203–1207 (1996)
248. Lin, Z., Saberi, A., Teel, A.R.: The almost disturbance decoupling problem with internal stability for linear systems subject to input saturation—state feedback case. Automatica $32$, 619–624 (1996)
249. Loquen, T., Tarbouriech, S., Prieur, C.: Stability analysis for reset systems with input saturation. In: Conference on Decision and Control, New Orleans, LA, USA, pp. 3272–3277 (2007)
250. Lozier, J.C.: A steady-state approach to the theory of saturable servo systems. IRE Trans. Autom. Control, 19–39 (1956)
251. Lu, B., Wu, F., Kim, S.: Linear parameter varying anti-windup compensation for enhanced flight control performance. AIAA J. Guid. Control Dyn. $28(3)$, 494–504 (2005)
252. Luenberger, D.G.: Linear and Nonlinear Programming. Addison-Wesley, Reading (1984)
253. Lurie, B.J.: The absolutely stable Nyquist stable nonlinear feedback system design. Int. J. Control $40(6)$, 1119–1130 (1984)
254. Ma, C.C.H.: Instability of linear unstable system with inputs limits. ASME J. Dyn. Syst. Meas. Control $113$, 742–744 (1991)
255. Maciejowski, J.M.: Multivariable Feedback Design. Addison Wesley, Reading (1989)

256. Marcopoli, V.R., Phillips, S.M.: Analysis and synthesis tools for a class of actuator-limited multivariable control systems: a linear matrix inequality approach. Int. J. Robust Nonlinear Control **6**(9–10), 1045–1063 (1996)

257. Massimetti, M., Zaccarian, L., Hu, T., Teel, A.R.: LMI-based linear anti-windup for discrete time linear control systems. In: Conference on Decision and Control, San Diego, CA, USA, December, pp. 6173–6178 (2006)

258. Milani, B.E.A.: Piecewise-affine Lyapunov functions for discrete-time linear systems with saturating controls. Automatica **38**, 2177–2184 (2002)

259. Milani, B.E.A., Carvalho, A.N.: Robust optimal linear regulator design for discrete-time systems under state and control constraints. In: IFAC Symposium on Robust Control Design (ROCOND), Rio de Janeiro, Brazil, pp. 273–278 (1994)

260. Miller, R., Pachter, M.: Manual flight control with saturating actuators. IEEE Control Syst. **18**(1), 10–19 (1998)

261. Miyamoto, S., Vinnicombe, G.: Robust control of plants with saturation nonlinearity based on coprime factor representation. In: Conference on Decision and Control, Kobe, Japan, pp. 2838–2840 (1996)

262. Molchanov, A.P., Pyatnitskii, E.S.: Criteria of asymptotic stability of differential and difference inclusions encountered in control theory. Syst. Control Lett. **13**, 59–64 (1989)

263. Montagner, V.F., Oliveira, R.C.L.F., Leite, V.J.S., Peres, P.L.D.: LMI approach for $\mathcal{H}_\infty$ linear parameter-varying state feedback control. IEE Proc., Control Theory Appl. **152**(2), 195–201 (2005)

264. Montagner, V.F., Peres, P.L.D., Tarbouriech, S., Queinnec, I.: Improved estimation of stability regions for uncertain linear systems with saturating actuators: an LMI-based approach. In: Conference on Decision and Control, San Diego, USA, pp. 5429–5434 (2006)

265. Montagner, V.F., Oliveira, R.C.L.F., Peres, P.L.D., Tarbouriech, S., Queinnec, I.: Gain-scheduled controllers for linear parameter-varying systems with saturating actuators: an LMI-based design. In: American Control Conference, New York, USA, pp. 6067–6072 (2007)

266. Morabito, F., Teel, A.R., Zaccarian, L.: Nonlinear anti-windup applied to Euler-Lagrange systems. IEEE Trans. Robot. Autom. **20**(3), 526–537 (2004)

267. Morari, M., Zafiriou, E.: Robust Process Control. Prentice Hall, Englewood Cliffs (1989)

268. Mulder, E.F., Kothare, M.V.: Synthesis of stabilizing anti-windup controllers using piecewise quadratic Lyapunov functions. In: American Control Conference, pp. 3239–3243 (2000)

269. Mulder, E.F., Kothare, M.V., Morari, M.: Multivariable anti-windup controller synthesis using linear matrix inequalities. Automatica **37**(9), 1407–1416 (2001)

270. Nagumo, M.: Uber die lage der integralkurven gewöhnlicher differential-gleichungen. Proc. Phys. Math. Soc. Jpn. **24**(3), 272–559 (1942)

271. Nguyen, T., Jabbari, F.: Disturbance attenuation for systems with input saturation: An LMI approach. IEEE Trans. Autom. Control **44**, 852–857 (1999)

272. Nguyen, T., Jabbari, F.: Output feedback controllers for disturbance attenuation with actuator amplitude and rate saturation. In: American Control Conference, San Diego, USA, pp. 1997–2001 (1999)

273. Nguyen, T., Jabbari, F.: Output feedback controllers for disturbance attenuation with actuator amplitude and rate saturation. Automatica **36**(9), 1339–1346 (2000)

274. Niculescu, S.-I.: Delay Effects on Stability. A Robust Control Approach. Springer, Berlin (2001)

275. Oliveira, R.C.L.F., Peres, P.L.D.: Parameter-dependent LMIs in robust analysis: characterization of homogeneous polynomially parameter-dependent solutions via LMI relaxations. IEEE Trans. Autom. Control **52**(7), 1334–1340 (2007)

276. Oliveira, M.Z., Gomes Da Silva Jr., J.M., Coutinho, D., Tarbouriech, S.: Anti-windup Design for a Class of Nonlinear Control Systems. In: IFAC World Congress, Milan, Italy (2011)

277. O'Reilly, J.: Observers for Linear Systems. Mathematics in Science and Engineering, vol. 170. Academic Press, San Diego (1983)

# References

278. Pagnotta, L., Zaccarian, L., Constantinescu, A., Galeani, S.: Anti-windup applied to adaptive rejection of unknown narrow band disturbances. In: European Control Conference, Kos, Greece, pp. 150–157 (2007)
279. Paim, C.: Analyse et commande de systèmes linéaires soumis à des saturations. PhD thesis, University of Toulouse, France (March 2003). Rapport LAAS No. 03241
280. Paim, C., Tarbouriech, S., Gomes da Silva Jr., J.M., Castelan, E.B.: Control design for linear systems with saturating actuators and $L_2$-bounded disturbances. In: Conference on Decision and Control, Las Vegas, USA (2002)
281. Paim, C., Gomes da Silva Jr., J.M., Castelan, E.B., Tarbouriech, S.: Sintese de observadores de estado para sistemas lineares com entradas saturantes. In: XV Congresso Brasileiro de Automatica (CBA'2004), September 2004
282. Pan, H., Kapila, V.: LMI-based control of discrete-time systems with actuator amplitude and rate nonlinearities. In: Kapila, V., Grigoriadis, K. (eds.) Actuator Saturation Control. Control Engineering Series, pp. 135–159. Dekker, New York (2002). Chap. 6
283. Park, P.: A revisited Popov criterion for nonlinear Lure systems with sector restrictions. Int. J. Control **68**(3), 461–469 (1997)
284. Park, J.-K., Youn, H.Y.: Dynamic anti-windup based control method for state constrained systems. Automatica, 1915–1922 (2003)
285. Park, J.-K., Choi, C.-H., Choo, H.: Dynamic anti-windup method for a class of time-delay control systems with input saturation. Int. J. Robust Nonlinear Control **10**, 457–488 (2000)
286. Park, J.-K., Lim, H., Basar, T., Choi, C.-H.: Anti-windup compensator for active queue management in TCP networks. Control Eng. Pract. **11**(10), 1127–1142 (2003)
287. Peaucelle, D.: Formulation générique de problèmes d'analyse et commande robuste par les fonctions de Lyapunov dépendant des paramètres. PhD thesis, Université Paul Sabatier, Toulouse, France (Juillet 2000). Rapport LAAS No. 00304
288. Petersen, I.R.: A stabilization algorithm for a class of uncertain linear systems. Syst. Control Lett. **8**, 351–356 (1987)
289. Petersen, I.R.: Guaranteed cost LQG control of uncertain linear systems. IEE Proc., Control Theory Appl. **142**, 95–102 (1995)
290. Pittet, C.: Stabilisation des systèmes à commande contrainte. application à une table d'excitation microdynamique. PhD thesis, Université Paul Sabatier, Toulouse, France (October 1998). Rapport LAAS No. 98434
291. Pittet, C., Tarbouriech, S., Burgat, C.: Stability regions for linear systems with saturating controls via circle and Popov criteria. In: Conference on Decision and Control, San Diego, USA, pp. 4518–4523 (1997)
292. Pittet, C., Burgat, C., Tarbouriech, S.: Globally stabilizing controllers synthesis for linear systems with saturating inputs. In: IFAC Conference on Systems, Structure and Control, Nantes, France, vol. 2, pp. 391–396 (1998)
293. Prieur, C., Teel, A.R.: Uniting local and global output feedback controllers. IEEE Trans. Autom. Control **56**(7), 1636–1649 (2011)
294. Prieur, C., Goebel, R., Teel, A.R.: Hybrid feedback control and robust stabilization of nonlinear systems. IEEE Trans. Autom. Control **52**(11), 2103–2117 (2007)
295. Queinnec, I., Tarbouriech, S., Garcia, G.: Anti-windup design for aircraft control. In: IEEE Conference on Control Applications (CCA), Munich, Germany (2006)
296. Ran, A.C.M., Vreugdenhil, R.: Existence and comparison theorems for algebraic Riccati equations for continuous- and discrete-time systems. Linear Algebra Appl. **99**, 63–83 (1988)
297. Rao, V.G., Bernstein, D.S.: Naive control of the double integrator. IEEE Control Syst. Mag. **21**(5), 86–97 (2001)
298. Richard, J.P.: Time-delay systems: an overview of some recent advances and open problems. Automatica **39**, 1667–1604 (2003)
299. Rocha, T.C.T.: Domínios positivamente invariantes de sistemas lineares com restrições na variáveis de controle. Master's thesis, UFSC, Florianópolis, Brazil (July 1994)
300. Romanchuk, B.G.: Computing regions of attraction with polytopes: planar case. Automatica **32**(12), 1727–1732 (1996)

301. Romanchuk, B.G.: Some comments on anti-windup synthesis using LMI's. Int. J. Robust Nonlinear Control **9**(10), 717–734 (1999)
302. Roos, C.: Contribution à la commande des systèmes saturés en présence d'incertitudes et de variations paramétriques. Application au pilotage de l'avion au sol. PhD thesis, University of Toulouse, ISAE, France, December 2007
303. Roos, C., Biannic, J.-M.: A convex characterization of dynamically-constrained anti-windup controllers. Automatica, 2449–2452 (2008)
304. Roos, C., Biannic, J.-M., Tarbouriech, S., Prieur, C.: On-Ground Aircraft Control Design Using an LPV Anti-windup Approach. LNCIS, vol. 365. Springer, Berlin (2007). Chap. 7
305. Roos, C., Biannic, J.-M., Tarbouriech, S., Prieur, C., Jeanneau, M.: On-ground aircraft control design using a parameter-varying anti-windup approach. Aerosp. Sci. Technol. **14**, 459–471 (2010)
306. Rugh, W.J., Shamma, S.J.: Research on gain scheduling. Automatica **36**(10), 1401–1425 (2000)
307. Rundquist, L., Stahl-Gunnarsson, K.: Phase compensation of rate-limiters in unstable aircraft. In: IEEE Conference on Control Applications, Dearlorn, MI, pp. 19–24 (1996)
308. Saberi, A., Stoorvogel, A.: Output regulation of linear plants with actuators subject to amplitude and rate constraints. Int. J. Robust Nonlinear Control **9**(10), 631–657 (1999)
309. Saberi, A., Lin, Z., Teel, A.: Control of linear systems with saturating actuators. IEEE Trans. Autom. Control **41**(3), 368–377 (1996)
310. Saberi, A., Stoorvogel, A.A., Sannuti, P.: Control of Linear Systems with Regulation and Input Constraints. Springer, London (2000)
311. Saberi, A., Han, J., Stoorvogel, A.A.: Constrained stabilization problems for linear plants. Automatica **38**, 639–654 (2002)
312. Saeki, M., Wada, N.: Synthesis of a static anti-windup compensator via linear matrix inequalities. Int. J. Robust Nonlinear Control **12**, 927–953 (2002)
313. Scherer, C.W.: LPV control and full block multipliers. Automatica **37**(3), 361–375 (2001)
314. Scherer, C., Gahinet, P., Chilali, M.: Multiobjective output feedback control via LMI optimization. IEEE Trans. Autom. Control **42**(7), 896–911 (1997)
315. Schmitendorf, W.E.: Designing stabilizing controllers for uncertain systems using Riccati approach. In: American Control Conference, Minneapolis, USA, pp. 502–505 (1987)
316. Schmitendorf, W.E., Barmish, B.R.: Null controllability of linear systems with constrained controls. SIAM J. Control Optim. **18**, 327–345 (1980)
317. Schuster, E., Walker, M.L., Humphreys, D.A., Krstic, M.: Plasma vertical stabilization with actuation constraints in the DIII-D tokamak. Automatica **41**(7), 1173–1179 (2005)
318. Scorletti, G., Folcher, J.P., El Ghaoui, L.: Output feedback control with input saturations: LMI design approaches. Eur. J. Control **7**(6), 567–579 (2001)
319. Sepulchre, R., Jankovic, M., Kokotovic, P.: Constructive Nonlinear Control. Springer, London (1997)
320. Shamma, J.S.: Anti-windup via constrained regulation with observers. Syst. Control Lett. **40**(4), 261–268 (2000)
321. Shewchun, J.M., Féron, E.: High performance bounded control. In: American Control Conference, Albuquerque, New Mexico, USA, vol. 5, pp. 3250–3254 (1997)
322. Shewchun, J.M., Féron, E.: High performance control with position and rate limited actuators. Int. J. Robust Nonlinear Control **9**, 617–630 (1999)
323. Skogestad, S., Postlethwaite, I.: Multivariable Feedback Control: Analysis and Design, 2nd edn. Wiley, New York (2005)
324. Slotine, J.J.E., Li, W.: Applied Nonlinear Control. Prentice Hall, New York (1991)
325. Sofrony, J., Turner, M.C., Postlethwaite, I.: Anti-windup synthesis for systems with ratelimits: a Riccati equation approach. In: SICE Annual Conference, Okayama, Japan (2005)
326. Sontag, E.D.: An algebraic approach to bounded controllability of linear systems. Int. J. Control **39**(1), 181–188 (1984)
327. Sontag, E.D.: Mathematical Control Theory—Deterministic Finite Dimensional Systems. Springer, New York (1990)

# References

328. Sontag, E.D.: Input to state stability: basic concepts and results. In: Nistri, P., Stefani, G. (eds.) Nonlinear and Optimal Control Theory, pp. 163–220. Springer, Berlin (2007)
329. Sontag, E.D., Sussmann, H.J.: Nonlinear output feedback design for linear systems with saturating control. In: Conference on Decision and Control, Honolulu, USA, pp. 3414–3416 (1990)
330. Stein, G.: Bode lecture: respect the unstable. In: Conference on Decision and Control, Tampa, USA (1989)
331. Stoorvogel, A.A., Saberi, A. (eds.): Special issue: control problems with constraints. Int. J. Robust Nonlinear Control **9**(10) (1999)
332. Stoorvogel, A.A., Wang, X., Saberi, A., Sannuti, P.: Stabilization of sandwich non-linear systems with low-and-high gain feedback design. In: American Control Conference, Baltimore, USA, pp. 4217–4222 (2010)
333. Su, Y., Müller, P.C., Zheng, C.: Global asymptotic saturated PID control for robot manipulators. IEEE Trans. Control Syst. Technol. **18**(6), 1280–1288 (2010)
334. Suárez, R., Alvarez-Ramirez, J., Alvarez, J.: Linear systems with single saturated input: stability analysis. In: Conference on Decision and Control, Brighton, UK, pp. 223–228 (1991)
335. Suárez, R., Alvarez-Ramirez, J., Solis-Daun, J.: Linear systems with bounded inputs: global stabilization with eigenvalue placement. Int. J. Robust Nonlinear Control **7**(9), 835–845 (1997)
336. Sun, W., Lin, Z., Wang, Y.: Global asymptotic and finite-gain $\mathcal{L}_2$ stabilization of port-controlled Hamiltonian systems subject to actuator saturation. In: American Control Conference, St. Louis, USA, pp. 1894–1898 (2009)
337. Sussmann, H.J., Yang, Y.: On the stability of multiple integrators by means of bounded controls. In: Conference on Decision and Control, Brighton, UK, pp. 70–72 (1991)
338. Sussmann, H.J., Sontag, E.D., Yang, Y.: A general result on the stabilization of linear systems using bounded controls. IEEE Trans. Autom. Control **39**(12), 2411–2425 (1994)
339. Syaichu-Rohman, A., Middleton, R.H.: Anti-windup schemes for discrete time systems: an LMI-based design. In: Asian Control Conference, Melbourne, VIC, Australia, pp. 554–561 (2004)
340. Sznaier, M.: A set induced norm approach to the robust control of constrained systems. SIAM J. Control Optim. **31**(3), 733–746 (1993)
341. Tarbouriech, S.: Sur la stabilité des régulateurs à retour d'état saturé. PhD thesis, Université Paul Sabatier, Toulouse, France (February 1991). Rapport LAAS No. 91047
342. Tarbouriech, S.: Local stabilization of continuous-time delay systems with bounded inputs. In: Dugard, L., Verriest, E.J. (eds.) LNCIS, vol. 228, pp. 302–317 Springer, Berlin (1997)
343. Tarbouriech, S., Burgat, C.: $(A, B)$-stabilizability conditions with respect to certain Lyapunov functions. In: IEEE Conference on Systems, Man and Cybernetics, Le Touquet, France, pp. 311–318 (1993)
344. Tarbouriech, S., Burgat, C.: Positively invariant sets for constrained continuous-time systems with cone properties. IEEE Trans. Autom. Control **39**(2), 401–405 (1994)
345. Tarbouriech, S., Burgat, C.: Invariance property for linear discrete-time systems with bounded inputs via observer. In: American Control Conference, Seattle, USA, vol. 5, pp. 3914–3915 (1995)
346. Tarbouriech, S., Garcia, G. (eds.): Control of Uncertain Systems with Bounded Inputs. LNCIS, vol. 227. Springer, Berlin (1997)
347. Tarbouriech, S., Garcia, G.: Preliminary results about anti-windup strategy for systems subject to actuator and sensor saturations. In: IFAC World Congress, Prague (2005)
348. Tarbouriech, S., Gomes da Silva Jr., J.M.: Admissible polyhedra for discrete-time linear systems with saturating controls. In: American Control Conference, Albuquerque, USA, vol. 6, pp. 3915–3919 (1997)
349. Tarbouriech, S., Gomes da Silva Jr., J.M.: Synthesis of controllers for continuous-time delay systems with saturating controls via LMIs. IEEE Trans. Autom. Control **45**(1), 105–111 (2000)
350. Tarbouriech, S., Gouaisbaut, F.: Stabilization of quantized linear systems with saturations. In: Conference on Decision and Control, Atlanta, USA, December 2010

351. Tarbouriech, S., Gouaisbaut, F.: $\mathcal{L}_2$ stability for quantized linear systems with saturations. In: IFAC World Congress, Milan, Italy, September 2011
352. Tarbouriech, S., Turner, M.C.: Anti-windup design: an overview of some recent advances and open problems. IET Control Theory Appl. **3**(1), 1–19 (2009)
353. Tarbouriech, S., Gomes da Silva Jr., J.M., Garcia, G.: Stability and disturbance tolerance for linear systems with bounded controls. In: European Control Conference, Porto, Portugal, pp. 3219–3224 (2001)
354. Tarbouriech, S., Garcia, G., Gomes da Silva Jr., J.M.: Robust stability of uncertain polytopic linear time-delay systems with saturating inputs: an LMI approach. Comput. Electr. Eng. **28**, 157–169 (2002)
355. Tarbouriech, S., Gomes da Silva Jr., J.M., Garcia, G.: Delay-dependent anti-windup loops for enlarging the stability region of time-delay systems with saturating inputs. ASME J. Dyn. Syst. Meas. Control **125**(1), 265–267 (2003)
356. Tarbouriech, S., Gomes da Silva Jr., J.M., Garcia, G.: Delay-dependent anti-windup strategy for linear systems with saturating inputs and delayed outputs. Int. J. Robust Nonlinear Control **14**, 665–682 (2004)
357. Tarbouriech, S., Prieur, C., Gomes da Silva Jr., J.M.: An anti-windup strategy for a flexible cantilever beam. In: IFAC World Congress, Prague, CZ (2005)
358. Tarbouriech, S., Gomes da Silva Jr., J.M., Bender, F.A.: Dynamic anti-windup synthesis for discrete-time linear systems subject to input saturation and $\mathcal{L}_2$ disturbances. In: IFAC Symposium on Robust Control Design (ROCOND), Toulouse, France (2006)
359. Tarbouriech, S., Prieur, C., Gomes da Silva Jr., J.M.: Stability analysis and stabilization of systems presenting nested saturations. IEEE Trans. Autom. Control **51**(8), 1364–1371 (2006)
360. Tarbouriech, S., Queinnec, I., Garcia, G.: Stability region enlargement through anti-windup strategy for linear systems with dynamics restricted actuator. Int. J. Syst. Sci. **37**(2), 79–90 (2006)
361. Tarbouriech, S., Garcia, G., Glattfelder, A.H. (eds.): Advanced Strategies in Control Systems with Input and Output Constraints. LNCIS, vol. 346. Springer, Berlin (2007)
362. Tarbouriech, S., Queinnec, I., Garcia, G.: Anti-windup Strategy for Systems Subject to Actuator and Sensor Saturations. LNCIS, vol. 346. Springer, Berlin (2007). Chap. 6
363. Tarbouriech, S., Queinnec, I., Turner, M.C.: Anti-windup design with rate and magnitude actuator and sensor saturations. In: European Control Conference, Budapest, Hungary (2009)
364. Tarbouriech, S., Prieur, C., Queinnec, I., Simões dos Santos, T.: Global stability for systems with nested backlash and saturation operators. In: American Control Conference, Baltimore, USA, pp. 2665–2670 (2010)
365. Teel, A.R.: Global stabilization and restricted tracking for multiple integrators with bounded controls. Syst. Control Lett. **18**, 165–171 (1992)
366. Teel, A.R.: Semi-global stabilizability of linear null controllable systems with input nonlinearities. IEEE Trans. Autom. Control **40**(1), 96–100 (1995)
367. Teel, A.R.: Anti-windup for exponentially unstable linear systems. Int. J. Robust Nonlinear Control **9**, 701–716 (1999)
368. Teel, A.R., Buffington, J.B.: Anti-windup for an F-16's daisy chain control allocator. In: AIAA GNC Conference, New Orleans, LA, USA, pp. 748–754 (1997)
369. Teel, A.R., Kapoor, N.: The $\mathcal{L}_2$ anti-windup problem: Its definition and solution. In: European Control Conference, Brussels, Belgium, July 1997
370. Teel, A.R., Kapoor, N.: Uniting local and global controllers. In: European Control Conference, Brussels, Belgium, July 1997
371. Teel, A.R., Zaccarian, L., Marcinkowski, J.J.: An anti-windup strategy for active vibration isolation systems. Control Eng. Pract. **14**(1), 17–27 (2006)
372. Turner, M.C., Postlethwaite, I.: A new perspective on static and low order anti-windup synthesis. Int. J. Control **77**(1), 27–44 (2004)
373. Turner, M.C., Tarbouriech, S.: Anti-windup for linear systems with sensor saturation: sufficient conditions for global stability and $\mathcal{L}_2$ gain. In: Conference on Decision and Control, San Diego, USA (2006)

# References

374. Turner, M.C., Tarbouriech, S.: Anti-windup compensation for systems with sensor saturation: a study of architecture and structure. Int. J. Control **82**(7), 1253–1266 (2009)
375. Turner, M.C., Zaccarian, L. (eds.): Special issue: anti-windup. Int. J. Syst. Sci. **37**(2), 65–139 (2006)
376. Turner, M.C., Herrmann, G., Postlethwaite, I.: Accounting for uncertainty in anti-windup synthesis. In: American Control Conference, Bostron, MA, pp. 5292–5297 (2004)
377. Turner, M.C., Herrmann, G., Postlethwaite, I.: Incorporating robustness requirements into anti-windup design. IEEE Trans. Autom. Control **52**(10), 1842–1855 (2007)
378. Tyan, F., Bernstein, D.S.: Anti-windup compensator synthesis for systems with saturating actuators. Int. J. Robust Nonlinear Control **5**(5), 521–537 (1995)
379. Tyan, F., Bernstein, D.S.: Dynamic output feedback compensation for linear systems with independent amplitude and rate saturation. Int. J. Control **67**(1), 89–116 (1997)
380. Valmorbida, G.: Analyse en stabilité et synthèse de lois de commande pour des systèmes polynomiaux saturants. PhD thesis, University of Toulouse, Toulouse, France (2010). Rapport LAAS No. 10464
381. Valmorbida, G., Tarbouriech, S., Garcia, G.: State feedback design for input-saturating quadratic systems. Automatica **46**(7), 1196–1202 (2010)
382. Valmorbida, G., Tarbouriech, S., Turner, M.C., Garcia, G.: Anti-windup for NDI quadratic systems. In: IFAC Symposium on Nonlinear Control Systems (NOLCOS), Bologna, Italy, pp. 1175–1180 (2010)
383. Vassilaki, M., Bitsoris, G.: Constrained regulation of linear continuous-time dynamical systems. Syst. Control Lett. **13**, 247–252 (1989)
384. Vassilaki, M., Hennet, J.C., Bitsoris, G.: Feedback control of linear discrete-time systems under state and control constraints. Int. J. Control **47**(6), 1727–1735 (1988)
385. Vidyasagar, M.: Nonlinear Systems Analysis, 2nd edn. Prentice Hall, New York (1993)
386. Villota, E., Kerr, M., Jayasuriya, S.: A study of configurations for anti-windup control of uncertain systems. In: Conference on Decision and Control, San Diego, California (2006)
387. Vinograd, R.E.: The inadequacy of the method of characteristic exponents for the study of nonlinear differential equations. Mat. Sb. **41**(83), 431–438 (1957)
388. Wada, N., Saeki, M.: Synthesis of a static anti-windup compensator for systems with magnitude and rate limited actuators. In: IFAC Symposium on Robust Control Design (ROCOND), Prague, Czech Rep., June 2000
389. Wada, N., Saeki, M.: An LMI based scheduling algorithm for constrained stabilization problems. Syst. Control Lett. **57**, 255–261 (2008)
390. Walgama, K.S., Ronnback, S., Sternby, J.: Generalization of conditioning technique for anti-windup compensators. IEE Proc. D **139**(2), 109–118 (1992)
391. Walsh, M.J., Hayes, M.J.: A robust throughput rate control mechanism for an 802.15.4 wireless sensor network—an anti-windup approach. In: American Control Conference, New York (2007)
392. Wang, X., Stoorvogel, A.A., Saberi, A., Grip, H.F., Roy, S., Sannuti, P.: Stabilization of a class of sandwich systems via state feedback. IEEE Trans. Autom. Control **55**(9), 2156–2160 (2010)
393. Weston, P.F., Postlethwaite, I.: Analysis and design of linear conditioning schemes for systems containing saturating actuators. In: IFAC Nonlinear Control System Design Symposium (1998)
394. Weston, P.F., Postlethwaite, I.: Linear conditioning for systems containing saturating actuators. Automatica **36**(9), 1347–1354 (2000)
395. Wredenhagen, G.F., Bélanger, P.R.: Piecewise-linear LQ control for systems with input constraints. Automatica **30**(3), 403–416 (1994)
396. Wu, W., Jayasuriya, S.: A new QFT design methodology for feedback systems under input saturation. ASME J. Dyn. Syst. Meas. Control **123**(2), 225–232 (2001)
397. Wu, F., Lu, B.: Anti-windup control design for exponentially unstable LTI systems with actuator saturation. Syst. Control Lett. **52**, 305–322 (2004)
398. Wu, F., Soto, M.: Extended anti-windup control schemes for LTI and LFT systems with actuator saturations. Int. J. Robust Nonlinear Control **14**, 1255–1281 (2004)

399. Wu, F., Grigoriadis, K.M., Packard, A.: Anti-windup controller design using linear parameter-varying control methods. Int. J. Control **73**(12), 1104–1114 (2000)
400. Wu, F., Lin, Z., Zheng, Q.: Output feedback stabilization of linear systems with actuator saturation. IEEE Trans. Autom. Control **52**(1), 123–128 (2007)
401. Wurmthaler, C., Hippe, P.: Closed-loop design for stable and unstable plants with input saturation. In: European Control Conference, Gröningen, The Netherlands, pp. 1084–1088 (1993)
402. Yang, Y.: Global stabilization of linear systems with bounded controls. PhD thesis, Rutgers University (February 1993)
403. Yang, Y., Sussmann, H.J., Sontag, E.: Stabilization of linear systems with bounded controls. In: IFAC Nonlinear Control Systems Design Symposium (NOLCOS), Bordeaux, France, pp. 15–20 (1992)
404. Zaccarian, L., Teel, A.R.: A common framework for anti-windup, bumpless transfer and reliable designs. Automatica **38**(10), 1735–1744 (2002)
405. Zaccarian, L., Teel, A.R.: Nonlinear scheduled anti-windup design for linear system. IEEE Trans. Autom. Control **49**(11), 2055–2061 (2004)
406. Zaccarian, L., Teel, A.R.: Modern Anti-windup Synthesis. Princeton University Press, Princeton (2011)
407. Zaccarian, L., Nešič, D., Teel, A.R.: $\mathcal{L}_2$ anti-windup for linear dead-time systems. Syst. Control Lett. **54**(12), 1205–1217 (2005)
408. Zaccarian, L., Nešič, D., Teel, A.R.: First order reset elements and the Clegg integrator revisited. In: American Control Conference, Portland, OR, USA, pp. 563–568 (2005)
409. Zaccarian, L., Li, Y., Weyer, E., Cantoni, M., Teel, A.R.: Anti-windup for marginally stable plants and its application to open water channel control systems. Control Eng. Pract. **15**(2), 261–272 (2007)
410. Zheng, Q., Wu, F.: Output feedback control of saturated discrete-time linear systems using parameter-dependent Lyapunov functions. Syst. Control Lett. **57**, 896–903 (2008)
411. Zheng, A., Kothare, M.V., Morari, M.: Anti-windup design for internal model control. Int. J. Control **60**(5), 1015–1024 (1994)
412. Zheng, Y., Chait, Y., Hollot, C.V., Steinbuch, M., Norg, M.: Experimental demonstration of reset control design. Control Eng. Pract. **8**(2), 113–120 (2000)
413. Zhou, B., Duan, G.-R.: Global stabilisation of linear systems with bounded controls by nonlinear feedback. IET Control Theory Appl. **2**(5), 409–419 (2008)
414. Zhou, K., Doyle, J.C., Glover, K.: Robust and Optimal Control. Prentice Hall, Upper Saddle River (1996)
415. Zhou, B., Duan, G., Lin, Z.: Global stabilization of the double integrator system with saturation and delay in the input. IEEE Trans. Circuits Syst. I **57**(6), 1371–1383 (2010)
416. Zhou, B., Lin, Z., Duan, G.-R.: Global and semi-global stabilization of linear systems with multiple delays and saturations in the inputs. SIAM J. Control Optim. **48**(8), 5294–5332 (2010)
417. Zhou, B., Zheng, W.Y., Duan, G.-R.: An improved treatment of saturation nonlinearity with its application to control of systems subject to nested saturation. Automatica **47**, 306–315 (2011)

# Index

**A**

Anti-windup
 compensator. 31, 267, 283
 direct linear (DLAW), 285, 312, 325, 339
 dynamic, 340
  full-order, 311, 317
  reduced-order, 311, 317, 323, 337, 341
 model recovery (MRAW), 352
 static, 311, 312, 314, 316, 331, 337, 340

**B**

Basin of attraction, 13, 52, 78, 146, 194, 276, 298, 360

**C**

Circle and Popov criteria, 20, 279, 364, 365
Computational burden, 132, 199, 321
Contractive ellipsoid, 17, 126, 196
Contractive polyhedra, 116, 189, 203
Control laws
 dynamic output feedback, 7, 95, 161
 gain scheduled, 174
 guaranteed cost, 210, 230, 377
 observer-based feedback, 91, 155
 piecewise linear state feedback, 85, 220, 251, 263
 pole placement, 230, 241, 382
 robust, 367
 state feedback, 6, 76, 146, 211, 216, 373
 static output feedback, 7, 228

**D**

Differential inclusions, 33, 51
Disturbance
 rejection, 30, 89, 149, 169, 298
 tolerance, 30, 73, 86, 148, 169, 243, 298

**E**

Ellipsoid maximization
 in given directions, 61
 minor axis, 60
 trace, 61
 volume, 60
Equilibrium points, 46, 65, 182, 189, 315, 354
Exogenous signals
 amplitude bounded, 66, 86, 136, 148, 162, 243
 energy bounded, 71, 87, 141, 149, 166, 243, 245, 277

**F**

Finsler's Lemma, 151, 157, 346
Function
 positive definite, 357
 positive real, 363
 positive semi-definite, 357

**I**

Invariance principle, 359

**L**

$\mathcal{L}_2$-gain, 87, 89, 151, 169, 276, 290, 294, 298
$\mathcal{L}_2$-norm, 29, 72, 141
Limit cycle, 85, 189, 317, 333
LMI, 397
 parameterized, 26, 224, 260
Lyapunov equation, 389
Lyapunov function, 358
 Lure, 129, 139, 143
 piecewise affine, 203
 polyhedral, 17, 190
 quadratic, 17, 53, 125, 277, 291, 370

S. Tarbouriech et al., *Stability and Stabilization of Linear Systems with Saturating Actuators*, DOI 10.1007/978-0-85729-941-3, © Springer-Verlag London Limited 2011

# Index

**M**
Maximal invariant set, 115
Model predictive control (MPC), 121, 207

**N**
Nonlinearity
    dead-zone, 40, 123, 276, 288, 311, 326,
        331, 335, 339
    nested, 181
    quadratic, 182
Null controllability, 5, 101

**P**
Performance, 80, 99, 382
Positive invariance, 24, 115, 186, 202, 359

**Q**
Quadratic d-stabilizability, 382
Quadratic $\mathcal{R}$-stabilizability, 386

**R**
Region
    of asymptotic stability, 361
    of asymptotic stability (RAS), 14, 52, 124,
        276, 311
        ellipsoidal, 52, 110, 196
        polyhedral, 112, 186, 193
    of attraction, 13, 52, 78, 146, 194, 276,
        298, 360
    of linearity, 12, 211
    of null controllability, 18, 304
Riccati equation, 230, 245, 257, 392
    parameterized, 23, 26, 85, 210

**S**
Saturation
    model
        polytopic, 33, 51
        regions, 44, 185
        sector nonlinearity, 40, 123, 288
    nested, 180, 307, 311

    nonsymmetric, 275
    rate, 121, 306, 325
    sensor, 181, 306, 343
    symmetric, 275
Saturation avoidance problem, 24, 82, 210
Sector condition
    classical, 41
    generalized (modified), 43, 288
Set of initial conditions (shape set), 68, 76,
        146, 227, 299
Stability
    absolute, 362
    asymptotic, 15, 52, 124
        global, 18, 127, 305
        regional (local), 53, 124, 193, 196, 271
    exponential
        global, 18, 305
        regional (local), 305
    external, 28, 65, 136, 322
    input-to-state, 11, 29, 151, 157
    Lyapunov's, 357
    quadratic, 368
Stabilization
    external, 30, 86, 148
    global, 19, 101
    regional (local), 24, 76, 91, 95
    semiglobal, 21
System
    continuous-time, 123
    discrete-time, 109, 178, 202, 230
    hybrid, 182
    nonlinear, 182, 309
    polynomial, 182
    quadratic, 182
    time-delay, 121, 255, 308

**U**
Uncertainty
    norm-bounded, 103, 172, 210, 368
    polytopic, 103, 172, 370